Principles of Biomedical Engineering

Artech House Series
Engineering in Medicine & Biology

Series Editors
Martin L. Yarmush, Harvard Medical School
Christopher J. James, University of Southampton

Principles of Biomedical Engineering

Sundararajan V. Madihally

ARTECH
HOUSE

BOSTON | LONDON
artechhouse.com

Library of Congress Cataloging-in-Publication Data
A catalog record for this book is available from the U.S. Library of Congress.

British Library Cataloguing in Publication Data
A catalogue record for this book is available from the British Library.

ISBN-13: 978-1-60807-055-8

Cover design by Vicki Kane

© 2010 Artech House
685 Canton Street
Norwood, MA 02062

10 9 8 7 6 5 4 3 2 1

In memory of my father,
who guided me through many aspects
of life, including teaching

Contents

CHAPTER 7

CHAPTER 8

CHAPTER 9

CHAPTER 10

CHAPTER 11

Case studies accompanying this book are available at www.ArtechHouse.com; click on "Links."

Acknowledgments

First and foremost, my sincere thanks go to my wife and my two kids for providing insightful distractions and reminding me the importance of preparing this book for the future generation. I would like to acknowledge my mother, my brothers, my sisters, my father in-law, and my mother in-law for their unconditional support and encouragement. I would like to acknowledge the contributions of my students at Oklahoma State University through the years for pointing many errors in the drafts and offering helpful comments and suggestions to improve the description. My thanks go to Dr. R. Russell Rhinehart, the former department head of chemical engineering, for reading the initial drafts of a few chapters. In addition, I would like to thank Ms. Eileen Nelson for painfully reading every chapter, draft after draft. I would like to thank many faculty members in various schools for anonymously reading chapters and providing critical comments. No doubt those comments have enriched each chapter. I would also like to acknowledge the contributions of many faculty members in sharing their exercise problems.

CHAPTER 1

Introduction

1.1 Overview

Bioengineering applies engineering principles and design concepts to medicine and biology with the intention of improving the overall healthcare of society—particularly the lives of those with medical impairments. It is rooted in the life sciences, chemistry, mathematics, and physics. Bringing together knowledge of problem solving from many engineering disciplines, bioengineers design medical instruments, devices, computational tools; and perform population studies or carry out research to acquire the knowledge needed to solve new problems. Bioengineers play a critical role in developing new products, advancing research, and solving clinical problems. In cardiology alone, devices such as diagnostic monitors to measure the electrical activity of the heart (electrocardiographs), as well as systems to analyze gases in the blood, have revolutionized healthcare. Further, pacemakers and defibrillators help the heart to correct its beating pattern, while lasers and other instruments are used for surgery.

In general, biomedical scientists observe natural phenomenon with the intention of enhancing the basic understanding of laws by formulating hypotheses and testing them. However, bioengineers use the established principles and laws to develop new technologies useful to society. For example, bioscientists explore the phenomena of how cardiac cells function whereas bioengineers use these understandings to develop devices that can monitor the activity of cardiac cells and provide information on their healthiness. While these devices are useful in clinical care, they also offer new tools to get more insight into biological processes and new methods of data collection and analysis.

Bioengineering brings together expertise from other engineering disciplines, such as chemical, electrical, and mechanical engineering, to solve biology and healthcare problems. Compared with other engineering disciplines, biomedical engineering is a relatively new field within educational institutions. Many of the leaders have entered the field of bioengineering with degrees in other engineering disciplines. Whether bioengineers are involved in designing, testing, or managing applications, it is collaborative, requiring them to work on teams with physicians, nurses, technicians and others involved in the practice of medicine. Bioengineers need to know the needs of the basic performance measures in any field. Performance affects different roles in different ways. As a device developer, one has to know where to start, how to proceed, and when a device has been optimized enough. As a device tester, one has to validate whether the device supports expected workloads. The objective of this book is to provide a basic understanding of those features based on engineering fundamentals.

In this chapter, some of the roles of bioengineers are introduced first. Then three examples of device development are given with the intention of introducing the interdisciplinary nature of bioengineering. Further, the bioengineering community, which is broad and multidisciplinary, and some of the organizations are described with the intention of knowing where to look for networking. Finally, locations where one could search to get updated with recent developments are given.

1.2   Roles of Bioengineers

Although the fundamental concepts of bioengineering have a significant history, it is a relatively new discipline compared to other engineering disciplines such as chemical, civil, electrical, and mechanical engineering. Bioengineering requires the ability to use engineering approaches to examine complex biological systems spanning the length scale from molecules to organs. Some of the specialized areas in bioengineering are as follows:

- *Bioinstrumentation* (discussed in Chapters 3, 8, and 9) is the application of electrical measurement (discussed in Chapter 3) and electronics engineering principles to develop devices used in diagnosis and treatment of disease. Medical imaging and signal processing are part of bioinstrumentation—for example, the designing and development of medical devices such as cardiac pacemakers, defibrillators, cochlear implants, artificial kidneys, blood oxygenators, joints, arms, and legs. Examples of instrumentation that have benefited patients include: functional magnetic resonance imaging that employs computer algorithms to produce detailed cardiac images without the need for invasive diagnostic procedures; computerized electrocardiograms that allow the diagnosis and treatment of cardiac arrhythmias without open-heart surgery; the application of materials, microprocessors, and computer and battery technologies, which were used to develop cardiac pacemakers; and biosensors for detecting and monitoring metabolites and identifying specific genetic materials.

- *Biomaterials* (discussed in Chapter 6) are synthetic and natural materials used in the design of implantable and extracorporeal devices. Understanding the properties and behavior of living material is vital in the design of implant materials. The selection of an appropriate material to place in the human body may be one of the most difficult tasks faced by the biomedical engineer. Certain metal alloys, ceramics, polymers, and composites that are nontoxic, noncarcinogenic, chemically inert, stable, and mechanically strong have been used as implantable materials. Specialists in biomaterials could be involved in designing and developing therapeutic devices, such as a laser system for eye surgery or a device for automated delivery of insulin; designing and synthesizing biomaterials; and/or determination of the mechanical, transport (discussed in Chapter 2), and biocompatibility properties of implantable artificial materials.

- *Biomechanics* (discussed in Chapter 5) deals with the application of traditional mechanics to biological or medical problems. Better understanding of the forces and their effects on the human body is essential to get more insight into the functioning of various body parts, the effect of load and overload on specific structures, and the mechanics of biomaterial that could be utilized in prosthetic development. One example is the biomechanical analyses during the impact of automobiles, which can be utilized to develop safety devices such as airbags, seatbelts, and helmets. Other applications of biomechanics include designing prosthetic components, improving athletic performance and rehabilitation engineering, and developing tools for individuals with physical disabilities.

- *Cellular engineering* (discussed in Chapter 7) involves solving biomedical problems at the cellular and subcellular levels utilizing the molecular nature of building blocks of the body such as cells, proteins, carbohydrates, fatty acids, and nucleic acids. Cells can be cultured in large quantities outside the body, which can be utilized to relate function and disease occurrence, regenerate an artificial tissue, detect a biological signal, test the toxicity of a chemical, or collect cell-derived products for use as therapeutic molecules. One example of successful cellular engineering is development of devices that help reduce risk in bone marrow transplantation and other cellular therapies to treat various diseases. Cellular engineering can play a significant role in the production of new vaccines from cloned cells and the development of other therapeutic proteins from monoclonal antibodies.

- *Clinical engineering* is the application of technology to healthcare in hospitals or in the industry. The clinical engineer is a member of the healthcare team in a hospital along with physicians, nurses, and other hospital staff. Clinical engineers are responsible for testing, repairing, and maintaining proper and safety operating conditions with the hospital's diagnostic and therapeutic equipment. Sometimes clinical engineers provide prepurchase evaluations of new technology and equipment, research equipment issues for physicians and administrative staff, and assist clinical departments with service contract analysis, negotiations, and management. Clinical engineers also work in the medical product development industry, contributing to product design or sales and support. Primarily their job is to ensure that new products meet the demands of the medical practice.

- *Medical imaging* (Chapter 8) involves the development of various imaging techniques that can be utilized in a clinical setting, and basic physiology and biology research. Noninvasive imaging modalities provide information regarding functional processes or anatomy of some of the internal organs. Imaging techniques such as magnetic resonance imaging (MRI), X-ray computed tomography (CT), and positron emission tomography (PET) have become important tools for the early detection of disease, the understanding of basic molecular aspects of living organisms, and the evaluation of medical treatment. Often, these images can be obtained with minimal or completely noninvasive procedures, making them less painful and more readily repeatable than invasive techniques. With the incorporation of the digital scan converters in real-time instruments, the computer has assumed an expanding role in diagnostic imaging. Imaging for medical and biological purposes has expanded tremendously due to advances in instrumentation and computational mechanisms. Bioengineers play a critical role in designing, constructing, and/or analyzing medical imaging systems.

- *Rehabilitation engineering,* as defined in the Rehabilitation Act of 1973, means the systematic application of engineering sciences to design, develop, adapt, test, evaluate, apply, and distribute technological solutions to problems confronted by individuals with disabilities in functional areas, such as mobility, communications, hearing, vision, and cognition, and in activities associated with employment, independent living, education, and integration into the community. Rehabilitation engineers are involved in prosthetics; the development of home, workplace, and transportation modifications; and the design of assistive technology that enhance seating and positioning, mobility, and communication. Rehabilitation engineers also develop hardware and software computer adaptations and cognitive aids to assist people with cognitive difficulties. These careers require additional training beyond the bachelor's degree in bioengineering.

- *Physiology models* (Chapter 10) deal with developing strategies, techniques, and models (mathematical and physical) to understand the function of living organisms, from bacteria to humans. Computational modeling is used in the analysis of experimental data and in formulating mathematical descriptions of physiological events. Bioengineering creates knowledge from the molecular to the organ systems levels. Physiological models play a significant role in the development and implementation of computational models of physiological systems; development of new diagnostic procedures, particularly those requiring engineering analyses to determine parameters that are not directly accessible to measurements; development of strategies for clinical decision making based on expert systems, such as a computer-based system for managing the care of patients or for diagnosing diseases; and the design of computer systems to monitor patients during surgery or in intensive care.

Frequently, these specialized areas are interdependent. For example, the design of an artificial hip is greatly aided by studies on anatomy, bone biomechanics, gait analysis, and biomaterial compatibility. The forces that are applied to the hip can be considered in the design and material selection for the prosthesis.

1.3 History of Bioengineering

Bioengineering has evolved into a field of its own as a result of contributions from a number of disciplines. From the historical perspective, the revolutionary changes that have occurred in biology and medicine have depended on new methods that are the result of fundamental discoveries in many different fields. A number of individuals with expertise in different disciplines would collaborate in solving a problem related to bioengineering. Major contributions of bioengineering including hip joint replacement, artificial articulated joints, magnetic resonance imaging, heart pacemakers, arthroscopy, heart-lung machines, angioplasty, bioengineered skin, time-release drug capsules, and kidney dialysis have evolved through successful collaborations between medical practitioners and many disciplines including physics, mathematics, chemistry, computer sciences, and engineering. Significant improvements in computational tools and information technologies have already played a greater role with the emergence of telemedicine. The number of medical device establishments registered at the U.S. Food and Drug Administration (FDA) was nearly 11,000 in 2004. To understand the interdisciplinary nature of bioengineering, the history behind three significant contributions are described in the following subsections.

1.3.1   Development of Biomedical Imaging

Various medical imaging modalities provide complementary windows through which most of the organs and their functions can be visualized. It began many centuries ago with the discovery of various fundamental concepts in physics, biology, and chemistry. In 1668, Antoni van Leeuwenhoek, a Dutch scientist and builder of microscopes, confirmed the discovery by Italian anatomist Marcello Malpighi of capillary systems using a simple magnifying lens. He demonstrated how the red corpuscles circulated through the capillaries of a rabbit's ear and the web of a frog's foot. In 1674, Leeuwenhoek gave the first description of red blood corpuscles (RBC). He then discovered the motion of bacteria (he referred to them as animalcules) and sperm cells. In 1729, British pensioner Stephen Gray distinguished conductors of electricity from nonconductors. In 1750, Benjamin Franklin defined positive and negative electricity. In 1800, Italian physicist Alessandro Volta constructed the first electrical battery. In 1827, German physicist Georg S. Ohm formulated Ohm's law, stating the relation between electric current, electromotive force, and resistance. British physicist and chemist Michael Faraday and American scientist Joseph Henry discovered electromagnetic induction (i.e., refined the science of electromagnetism to make it possible to design practical electromagnetic devices). In 1873, Scottish mathematician and theoretical physicist James C. Maxwell published his famous four equations describing electrical and magnetic phenomena in the book *Treatise on Electricity and Magnetism*. In 1893, Joseph J. Thomson, a British physicist widely recognized for discovering electrons, published a supplementary volume to Maxwell's treatise, describing in detail the passage of electricity through gases. Further, he developed the cathode-ray tubes with the help of some graduate students.

Significant change occurred in 1895 when German physicist Wilhelm C. Roentgen accidentally discovered that a cathode-ray tube could make a sheet of paper coated with barium platinocyanide glow, even when the tube and the paper were in separate rooms. Roentgen decided that the tube must be emitting some sort of penetrating rays. He named them X for unknown. Shortly afterward, Roentgen found that if he aimed the X-rays through Mrs. Roentgen's hand at a chemically coated screen, he could see the bones in the hand clearly on the screen. The first radiograph of human anatomy was reported using his wife's left hand. In 1896, French scientist Henry Becquerel discovered that the X-rays were emitted from uranium ore. Further, he found out that another new type of radiation was being spontaneously emanated without the salts of uranium having to be illuminated— a radiation (referred to as radioactivity) that could pass through metal foils and darken a photographic plate. Two of his students, Pierre and Marie Curie, traced the radiation to the element radium.

In 1905, Robert Kienböck, a German radiologist known for identifying Kienböck's disease, used strips of silver bromide photographic paper to estimate the amount of radiation to which patients were exposed in radiation therapy. Over the next few decades, X-rays grew into a widely used diagnostic tool. Within the first year after the announcement of the discovery of X-rays, many papers and books on X-rays were published. By the 1930s, X-rays were being used to visualize most organ systems using radio-opaque materials. With the developments in photography, various films were developed. In 1932, DuPont produced cellulose nitrate and cellulose acetate films that could be used for X-rays. However, with the development of electron imaging and processing in the television industry, processing biomedical images was also developed. Much of the early work on improving image quality occurred because of research in the application of electron imaging to television camera tubes. The concept of image intensification by cascading stages was suggested independently by a number of workers in the field during the same period. In 1951, Russell H. Morgan and Ralph E. Sturm, radiologists from Johns Hopkins University, constructed a brightness intensifier for X-rays with an image intensification tube similar to the tube used in television cameras.

In the meantime, in 1942, Austrian neurologist Karl T. Dussik reported the first use of diagnostic ultrasound (acoustic energy with a frequency above human hearing—20,000 hertz), although high-frequency sound waves were discovered in late 1780. In the late 1940s, American scientist George D. Ludwig systematically started evaluating ultrasound for diagnostic purposes. Ludwig reported that echo patterns could sometimes be confusing, and multiple reflections from soft tissues could make test results difficult to interpret. Later, Ludwig collaborated with physicist Richard H. Bolt, the director of Acoustics Laboratory at the Massachusetts Institute of Technology, neurosurgeon H. Thomas Ballantine, Jr., and physicist Theodor F. Hueter (from Siemens Electromedical Laboratories in Erlangen, Germany) to work on ultrasound propagation characteristics in mammalian tissues. This research concluded that ultrasonograms suffered from a high level of noise problems. However, along with the advancements of sound navigation and ranging (SONAR) in the metal industry, research into using ultrasounds for diagnostic purposes continued. In 1968, British obstetrician and gynecologist Stuart Campbell published an improved method of ultrasonic imaging, which became standard practice in the examination of the fetus during pregnancy.

Based on the nuclear magnetic resonance phenomenon pioneered in the 1930s by American physicist Isidor I. Rabi and American scientists Felix Bloch and Edward M. Purcell, magnetic resonance imaging (MRI) was developed as a new way of taking pictures of the body's interior. MRI relies on a strong magnetic field and radio signals, which together trigger atoms in the body to send out radio signals of different frequencies (discussed in Chapter 8). A major improvement in the quality of the signal was achieved in 1964 by the conception of the Fourier transform developed by French mathematician Jean Baptiste Joseph Fourier in 1822. In 1971, American researcher Raymond V. Damadian showed that nuclear magnetic relaxation time of tissues and tumors differed, motivating scientists to use MRIs to study disease. Later, Damadian also built the MRI machine that was approved by the FDA.

The science of medical imaging grew slowly with incremental improvements in X-ray techniques. All the images lacked spatial information. In the early 1970s, discovery of computerized tomography (CT) or computerized axial tomography (CAT) by many researchers including British engineer Godfrey N. Hounsfield, physicist Allan M. Cormack, American chemist Paul C. Lauterbur, and British physicist Peter Mansfield revolutionized medical diagnosis with the opportunity of a new imaging option. Hounsfield built an instrument that combined an X-ray machine and a computer. A series of X-ray images from various angles were combined with the computer using certain principles of algebraic reconstruction to build a three-dimensional image of that part of the body. Small detectors inside the scanner measured the amount of X-rays that make it through the part of the body being studied. The computer received this information to create several individual images called slices. These images were stored, viewed on a monitor, or printed on film. Physicians could then instruct the computer to display two-dimensional slices from any angle and at any depth.

Lauterbur applied magnetic field gradients in all three dimensions to create magnetic resonance images via CAT-scan projection-reconstruction technique, borrowed from CT scanning. Three-dimensional models of organs were created by stacking the individual slices together. Mansfield showed how the radio signals could be mathematically analyzed and images could be generated extremely fast, which made it possible to develop a useful imaging technique. Swiss physical chemist Richard R. Ernst realized that one could use switched magnetic field gradients with changing time, which led to the basic reconstruction method for MRI. In the early 1980s, MRI was developed as a new way of taking pictures of the body's interior. Many scientists over the next two decades developed MRI into a robust technology of obtaining sophisticated images without using surgery. Developments in magnet technology led to designs that increased both patient acceptance and the range of potential applications. Development of superconducting electromagnets that produce high magnetic field systems has made it possible to obtain better quality images [Figure 1.1(a)]. Another major improvement in MRI is the development of Open MRI [Figure 1.1(b)] making it possible for people with claustrophobia to get the scans without fear. However, Open MRI utilizes low field resistive electromagnets or permanent magnets and the quality of the image obtained is lower than enclosed systems. Another development in 1990 was that Seiji Ogawa, a Japanese biophysicist working with AT&T's Bell Laboratories, reported that deoxygenated hemoglobin, when placed in a magnetic field, would increase the strength of the field in its vicinity, while oxygenated hemoglobin would not. This lead to the development of functional MRI (fMRI), which allows capturing images of an organ in action or studying functions of that organ.

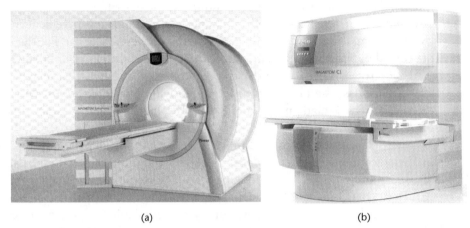

(a)   (b)

Figure 1.1   Magnetic resonance imaging. (a) MAGNETOM Symphony whole body imaging system. (b) MAGNETOM C!, an Open MRI. (Courtesy of Siemens Medical Solutions, Malvern, Pennsylvania.)

Other techniques such as positron emission tomography (PET) and single photon emission computed tomography (SPECT) have also been developed for use in biomedical imaging. Further, substantial improvements in optics hardware and the development of increasingly sensitive electronic imaging sensors have paved the way for the use of light microscopy-based techniques in biomedical imaging. Combined with time-lapse imaging, novel developments have provided powerful tools in unraveling the complex processes underlying the basic building blocks of life. Although there is no perfect imaging method through which everything in the body can be visualized, various medical imaging techniques provide complementary windows through which many tissues and organs can be visualized. The new imaging techniques can produce information about anatomical structures that is linked to functional data.

There has also been significant developments in sophisticated algorithms for extracting structural and functional volumetric information from raw measurements (computed imaging), and for processing, visualizing, and analyzing the image data. Novel imaging methods can provide comprehensive views of the human body in greater depth and detail, while becoming less expensive, faster, and less invasive. For example, innovations in the performance of the computer system used to control data acquisition and image processing are increasing the speed of MRI data acquisition and opening up new possibilities in real-time imaging. Most of the images are constructed using computers from a collection of digital data; image quality depends on the power of the computer and the caliber of its software, and there is significant opportunity for improvement in both. A bioengineer may spend his or her day designing electrical circuits and computer software for medical instrumentation ranging.

1.3.2   Development of Dialysis

The kidneys are vital organs in maintaining the balance of several key electrolytes in the body (including common salt, sodium chloride), eliminating liquid wastes from the body, and filtering over 400 gallons of blood daily. When the kidneys are not functioning correctly, pathological conditions are caused by the accumulation of waste products. The body swells, and if there is no intervention, death results. Acute and chronic kidney failure is an illness that is as old as humanity itself and can lead to death if untreated for several days or weeks. In early Rome and later in the Middle Ages, treatments for uremia (Greek for urine poisoning, or "urine in the blood") included the use of hot baths and sweating therapies. Since an individual can only last a short time with total kidney failure, dialysis was developed as a method of keeping a person alive until a suitable organ donor could be located. Current procedures for the treatment of kidney failure include hemodialysis and peritoneal dialysis, developed on the principles of osmosis, diffusion, and ultrafiltration (separation based on the size of the particles).

- *Hemodialysis* is a more frequently prescribed type than peritoneal dialysis. The treatment involves circulating the patient's blood outside of the body through a dialysis circuit [Figure 1.2(a)]. Two needles are inserted into the patient's vein, or access site, and are attached to the dialysis circuit, which consists of plastic blood tubing, a filter known as a dialyzer, and a dialysis machine that monitors and maintains blood flow and administers dialysate.

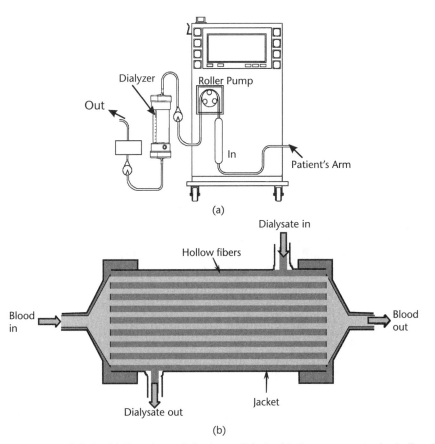

(a)

(b)

Figure 1.2   Hemodialysis. (a) Flow loop of the hemodialysis. (b) Components in the hollow-fiber dialyzer.

Dialysate is a chemical bath that is used to draw waste products out of the blood.

- In *peritoneal dialysis*, the patient's abdominal lining (or peritoneum) acts as a blood filter. A catheter is surgically inserted into the patient's abdomen. During treatment, the catheter is used to fill the abdominal cavity with dialysate. Waste products and excess fluids move from the bloodstream into the dialysate solution. After waiting, depending on the treatment method, the waste-filled dialysate is drained from the abdomen, and replaced with clean dialysate.

Development of dialysers began nearly two centuries ago in 1831 when Scottish chemist Thomas Graham described a procedure he called dialysis to remove solute from a solvent (or purify the blood in case of kidney failure). Using parchment as a membrane, Graham demonstrated that urea can be removed from urine. Osmosis (described in Chapter 2) and dialysis became popular as methods used in chemical laboratories that allowed the separation of dissolved substances or the removal of water from solutions through semipermeable membranes. The blood was spread across the membrane (a thin layer formed between objects and organs) that allowed waste to pass into a balanced fluid, while replenishing substances would pass from the fluid into the blood. In 1855, German physiologist Adolf Fick published a quantitative description of the diffusion process. However, it was Albert Einstein, 50 years later, who derived those empirically defined diffusion laws using the thermodynamics of theory of Brownian molecular motion.

In 1912, John J. Abel (professor of pharmacology at Johns Hopkins University School of Medicine) was investigating byproducts in the blood. He needed a device to extract these materials from the blood. With his colleagues Benjamin Turner and Leonard Rowntree, Abel built the first functioning dialysis machine. This machine circulated blood through 8-mm colloidin (nitrocellulose) tubing, immersed in a salt (saline)-dextrose (a type of naturally occurring sugar) solution, coiled around a rotating drum encased in a glass jacket. Urea (a compound found in urine) and other toxins passed out into the solution and oxygen passed into the blood. Abel called the process vividiffusion. In 1913, they "dialyzed" anesthetized animals by directing the blood outside the body and through tubes with semipermeable membranes. Before Abel could route the animals' blood through the "dialyzer," the blood's ability to clot or coagulate had to be at least temporarily restricted using hirudin, the anticoagulant element in the saliva of leeches.

German physician Georg Haas performed the first dialysis treatment involving humans in 1924. Haas also used a collodion membrane and hirudin as the anticoagulant similar to Abel, but built dialyzers in a variety of models and sizes. When and how much Haas knew about Abel's research group is not known because of the confusion surrounding World War I. By 1928, Haas had dialyzed an additional six patients, none of which survived, likely because of the critical condition of the patients and the insufficient effectiveness of the dialysis treatment. Nevertheless, use of hirudin often led to massive complications arising from allergic reactions since the substance was insufficiently purified and originated in a species distant from humans. In the end, Haas used heparin, which was first isolated in dog livers by American researcher Jay McLean in 1916. Heparin caused substantially fewer complications than hirudin, even when it was insufficiently purified, and it could be produced in larger amounts. Heparin became and remains the anticoagulant of choice, with the development of better separation technologies in the 1930s. In 1923, usage of intact living peritoneum (abdominal lining) was a good dialyzing surface, and German clinical investigator Georg Ganter reported the first use of peritoneal dialysis in clinical therapy. Subsequently, German physician Heinrich Necheles modified his extracorporeal (meaning apparatus carrying the blood outside the body) device by compressing tubes of peritoneal membrane between screens of metal mesh, which reduced the enclosed blood volume and increased the effective surface area. However, due to issues with sterility, peritoneal membranes suffered from increased incidence of infection. Another development in the 1930s was commercial production of cellophane (regenerated cellulose acetate) in large quantities as tubing for sausage casing, which replaced colloidin membranes due to its better performance and mechanical stability. Later, cuprophan (cuprammonium cellulose) membranes replaced cellophane due to its better permeability properties.

Dutch physician Willem J. Kolff secured a success in 1945 when he treated a 67-year-old patient with acute kidney failure for a week, which allowed the patient to be released with normal kidney function. The patient died at the age of 73 from an illness unrelated to the kidney failure. Although Kolff had unsuccessfully treated 16 previous patients in a series of experiments, this success was the first major breakthrough in the treatment of patients with kidney disease. The success is attributed to the technical improvements in the actual equipment used for the treatment. Kolff's device consisted of a rotating wooden drum around which a new membrane made of cellophane was wrapped. Blood was taken from the artery propelled through the tubing when the drum rotated (using the principle of Archimedes' screw). The drum containing blood-filled tubes were partially immersed in an electrolyte solution known as dialysate. As the membranous tubes passed through the bath, the uremic toxins would pass into this rinsing liquid due to diffusion (described in Chapter 2) and osmosis. Blood was returned to the vein, largely cleared of urea and unknown toxic substances.

In 1947, Swedish engineer Nils Alwall published a scientific work describing a modified dialyzer that could better combine the necessary processes of dialysis and ultrafiltration than the traditional Kolff kidney. The cellophane membranes used in the dialyzer could withstand higher pressure because of their positioning between two protective metal grates. All of the membranes were in a tightly closed cylinder so that the necessary pressure did not need to derive from the blood flow but could rather be achieved using lower pressure in the dialysate. However, due to its alleged lack of compatibility with blood (described in Chapter 6), membranes made from unmodified cellulose lost their market share. They have been replaced by modified cellulosic and synthetic dialysis membranes, which show a better compatibility with blood than unmodified cellulose membranes. As an alternative to rotating drum dialysers, American biochemists Leonard T. Skeggs, Jr., and Jack R. Leonards developed parallel plate dialyzers in 1948. In this design, several plates with ridges and grooves were stacked parallel. A semipermeable membrane rested between the grooves and the blood flow. Rather than pumping the blood through the membranous tubes, groves and ridges directed the flow of the dialyzer and alternating layers of membranes material. Later, Skeggs adapted this technology to autoanalyzers for assessing various components in the blood. In 1960, Norwegian surgeon Fredrick Kiil developed an advanced form of parallel-plate dialyzers. These Kiil dialysers were in clinical use until the late 1990s.

By the 1950s, Kolff's invention of the artificial kidney had solved the problem of acute renal failure, but it was not seen as the solution for patients with chronic renal disease. In the mid-twentieth century, American physicians believed it was impossible for patients to have dialysis indefinitely for two reasons: they thought no manmade device could replace the function of kidneys over the long term; and a patient undergoing dialysis suffered from damaged veins and arteries, so that after several treatments, it became difficult to find a vessel to access the patient's blood. However, the scientific development of new materials started in 1900 with the development of steel in 1900; vanadium alloy in 1912; stainless steel in 1926; Bakelite and Pyrex in 1939; and nylon in 1940. New materials lead to new treatment options. Physicians, engineers, and manufacturers joined in various modifications and innovations.

Belding Scribner, a professor of medicine at the University of Washington, came up with the idea of connecting the patient to the dialyzer using plastic tubes, one inserted into an artery and one into a vein [Figure 1.2(a)]. After treatment, the circulatory access would be kept open by connecting the two tubes outside the body using a small U-shaped device (popularly called a Scribner shunt), which would shunt the blood from the tube in the artery back to the tube in the vein. The Scribner shunt was developed using Teflon material. With the shunt, it was no longer necessary to make new incisions each time a patient underwent dialysis. In 1962, Scribner started the world's first outpatient dialysis facility called the Seattle Artificial Kidney Center. Scribner faced an immediate ethical problem of who should be given dialysis, since demand far exceeded the capacity of the six dialysis machines at the center. Scribner decided that the choices should be made by an anonymous admission committee composed of local residents from various walks of life in addition to two physicians who practiced outside of the field of nephrology. It was the creation of the first bioethics committee (discussed in Chapter 11), which changed the approach to accessibility of healthcare in the United States. In collaboration with Albert Babb, a professor of nuclear engineering at the University of Washington, Scribner went on to develop a portable dialysis machine that could be operated by family members in their own homes. The portable unit also included many safety features that would monitor if the system failed. The portable unit made it easier for patients to maintain their daily routines and freed up machines in dialysis centers. Hemodialysis established itself as the treatment of choice worldwide for chronic and acute kidney failure after the early successes in Seattle. However, most of these dialysers had very low efficiency, and the time of dialysis remained relatively long, up to 8 hours. Such a schedule of 8 hours, three times weekly continued to be practiced in some centers. Hence, there was a need for improving efficiency of dialysis machines and industrial production in large numbers.

In 1964, American surgeon Richard D. Stewart built the hollow-fiber dialyzer containing nearly 1 m^2 of surface area of cellulosic material, large enough to fulfill the one end of the dialyzer into thousands of tiny hollow fibers (~225-μm internal diameter) and dialysate is pumped into the cylinder and across the tiny hollow fibers. This technology replaced the traditional membranous tubes and flat membranes with a number of capillary-sized hollow membranes. The development of the related industrial manufacturing technology by Dow Chemical allowed the production of large numbers of dialyzers at a reasonable price. The typical hollow-fiber dialyzers of today, which are equipped with a more effective and better-tolerated membrane made primarily from synthetic polymers, are still based on early concepts developed by Abel. As the clinical use of hemodialysis became increasingly widespread, scientists were better able to investigate the unique attributes of patients with chronic kidney disease. Scientific discoveries in the transport of substances across membranes accompanied the technological refinement of dialyzers and began to include dialysis-specific research. Mathematical models were developed to describe the hemodialysis process and performance characteristics. In contrast to the early years of dialysis, the lack of adequate treatment methods or technologies is no longer a challenge in the treatment of renal patients. Development of physiological models of the dialysis process that resemble the functionality of the body has also played a significant role in improving the performance of the dialyzers in addition to helping the physicians match dialysis therapy to the individual needs of the patient. These models (some discussed in Chapter 8) make it possible to analyze the course of treatment and to predict the effect of dialysis procedures. These efforts made a quantitative description of dialysis possible and allowed the development of dialyzers with clearly defined characteristics. The composition of the dialysate is also adjusted to retain desired components in the blood and cause undesired components to diffuse with the dialysate. Many new membrane materials are also available in high-flux modifications.

As minimally invasive medical procedures evolve, the design requirements for the medical devices used in them become increasingly complex. In 1964, Silastic peritoneal catheters suitable for long-term placement were developed. Further, machines to manufacture and deliver sterile dialysis fluid and to control its inflow and outflow from the peritoneal cavity were developed. Several such devices have become commercially available since that time. Peritoneal dialysis is frequently a better neonatal option to aid babies with compromised renal function than hemodialysis because smaller infants cannot afford to lose the blood volume necessary to prime the hemodialysis blood circuit. Dialy-Nate (Utah Medical Products) is a pre-assembled and closed peritoneal dialysis kit (Figure 1.3) specifically designed for the neonatal and pediatric use. Because it is preassembled, Dialy-Nate is safer and more effective than assembling separate tubing and components. The total priming volume is 63 mL, which includes 58 mL from the bag spike to the catheter and an additional 5 mL from the catheter to the finely graduated dialysate meter. Dialysis can begin much sooner since nursing and technician time is not spent researching hospital dialysis assembly procedures. However, the present challenges are multifold and come from the large number of patients requiring dialysis treatment, the complications resulting from years of dialysis treatment, and a growing population of patients that presents demographic as well as medical challenges. Efforts to relate the patient's outcome to the dialyzer's performance have been difficult and ongoing since 1971.

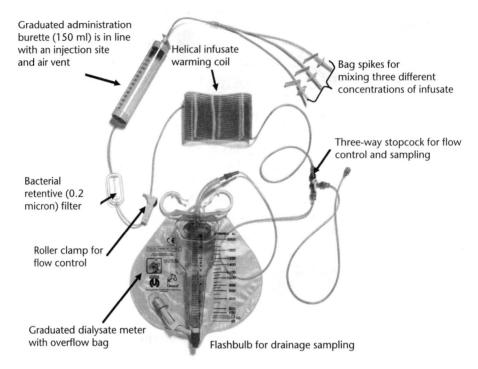

Graduated administration
burette (150 ml) is in line
with an injection site
and air vent

Helical infusate
warming coil

Bag spikes for
mixing three different
concentrations of infusate

Three-way stopcock for flow
control and sampling

Bacterial
retentive (0.2
micron) filter

Roller clamp for
flow control

Graduated dialysate meter
with overflow bag

Flashbulb for drainage sampling

Figure 1.3   Dialy-Nate peritoneal dialysis set with a warming coil. (Courtesy of Utah Medical Products, Inc., Midvale, Utah.)

1.3.3   Development of the Heart-Lung Machine

One of the major advances in biomedical science is the invention and refinement of artificial circulation, also known as the heart-lung machine (also referred as pump oxygenator). The heart-lung machine is the result of the combined efforts of many medical practitioners and engineers. The heart-lung machine maintains the circulation of the blood and the oxygen content of the body while a surgeon operates on the heart. The machine is used in every surgery involving various components of the heart. The heart-lung machine consists of several components (Figure 1.4) including a roller pump, oxygenator, heat exchanger, and a reservoir, which are connected together by plastic tubing. The circuit diverts blood away from the heart and lungs and returns oxygenated blood to the body. The reservoir chamber receives the blood from the body, which is normally the responsibility of the upper right chamber of the heart (called the right atrium). A large drainage tube is placed in the right atrium to drain the deoxygenated blood from the patient into the heart-lung machine (by gravity siphon). This blood is then pumped by the machine through an oxygenator (normally carried out by the lower right chamber of the heart). As the blood passes through the oxygenator, the blood comes into intimate contact with the fine surfaces of the device itself. Oxygen gas is delivered to the interface between the blood and the device, permitting the red blood cells to absorb oxygen molecules directly and release carbon dioxide. Then a roller pump helps deliver this newly oxygenated blood back to the body, which is normally the work of the left heart. The contractions of the heart are halted by running a potassium citrate solution

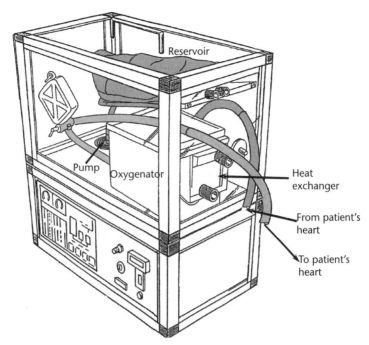

Figure 1.4   Mobile heart-lung machine.

through the coronary vessels. The surgeon is thus enabled to open the heart and make the necessary repairs while the heart is not obstructed by blood. During the surgical procedure, the heart-lung machine is operated by a perfusionist whose role is to maintain the circuit, adjust the flow as necessary, prevent air from entering the circulation, and maintain the various components of the blood within physiologic parameters.

Ideally the heart-lung machine has to meet many requirements including:

1.   The transfer of adequate oxygen;
2.   Blood flow rates that approach normal without danger of air;
3.   The maintenance of a stable blood volume and body metabolism without significant change in pH, blood pressure, and body temperature;
4.   A lack of serious damage to blood elements.

The efforts to bring this concept into reality began many decades ago. At the end of the nineteenth century, many experiments by physiologists such as von Schroder, the German scientist who developed the oxygenator, Max von Frey, Max Gruber, who described the artificial blood oxygenation flow system, and C. Jacobj, who developed the apparatus that relied on donor lungs for gas exchange, laid the foundation for different artificial oxygenation devices.

The pump component of the heart-lung machine circuitry serves a major function of providing blood flow through the circuit and maintaining an artificial circulation throughout the body. The pump found in the heart-lung machine uses a roller that progresses along a blood-containing resilient tube propelling the blood column in front of the roller and propelling out of the pump. A roller pump (also referred as peristaltic pump) could be used to gently provide continuous flow. The first roller pump was patented in 1855 by Porter and Bradley and was hand operated. E. E. Allen modified the first roller pump in 1877 and named it the "surgical pump." In 1891, C. H. Truax, who also distributed and promoted the Allen pump, developed the first double roller pump. However, these suffered from the problem of blood leakage along the roller circuit. Subsequently, several researchers including American surgeon Michael E. DeBakey refined the roller pump design. DeBakey recalled in an interview that the information available in the medical library in Tulane University was not helping in redesigning the pump and he had to go to an engineering library to get more information on pump design. These interactions helped in improving the design of hand-cranked roller pumps for better usage in blood transfusion and other applications including dialysis.

American folk hero Charles A. Lindbergh is partly credited for the development of the heart-lung machine. Years after his pioneering transatlantic flight, Lindbergh teamed up with a vascular surgeon, Alexis Carrel, to find a way to put oxygen into the blood stream. A primitive oxygen exchange device was made, confirming the possibility of using machinery for the function of the lung. John H. Gibbon, Jr. developed the first clinically successful heart-lung machine in 1937 with the help of research technician Mary Hopkinson, who later became his wife. Gibbons found that a pulsing flow, like that produced by the heart, was not necessary. Using a spinning hollow cylinder into which the blood was trickled, they employed centrifugal force to spread the blood in a layer thin enough to absorb the required amounts of oxygen, which were fed in under pressure. Partially replacing the circulation between the heart and the lungs, he managed to keep a cat alive for four hours. This experimental machine used two roller pumps with the capacity to replace the heart and lung action of a cat. He discovered that simply passing the blood through the external artificial circuit caused death in less than 12 hours from multiple small blood clots. To remove these blood clots, a fine metal mesh filter was incorporated in the circuit. After World War II in 1946, Gibbon persuaded the chairman of International Business Machines (IBM) and engineer Thomas J. Watson to give financial and technical support to improve the heart-lung machine. Along with five IBM engineers, Gibbon and Watson invented an improved machine that minimized red blood cell death and prevented air bubbles from entering the circulation using continuous suction. When this device was tested on dogs, it resulted in a 12% mortality rate.

In 1945, Clarence Dennis, an American physiologist at the University of Minnesota, built a modified Gibbon pump that consisted of a nest of vertically revolving stainless steel cylinders mounted over a revolving funnel in which the blood, oxygenated on the walls of the cylinders, was collected. During this time, Dennis also attempted to use Kolff's dialysis machine. Nevertheless, the new pump permitted a complete bypass of the heart and lungs during surgical operations of the heart. However, that machine was hard to clean, caused infections, and never reached human testing. Viking O. Bjork, a Swedish physician who later became popular for Bjork-Shiley mechanical heart valves, invented an oxygenator with multiple screen discs that rotated slowly in a shaft, over which a film of blood was injected. Oxygen was passed over the rotating discs and provided sufficient oxygenation for an adult human. Bjork, with the help of a few chemical engineers, one of which was his wife, prepared a blood filter and an artificial inner layer of silicon under the trade name UHB 300. This was applied to all parts of the perfusion machine, particularly the rough red rubber tubes, to delay clotting and save platelets. Bjork took the technology to the human testing phase.

Gibbon also performed the first successful open heart operations using heart-lung machines on human patients in 1953. A 45-minute operation was performed to correct a defect; the patient made a complete recovery and was alive and well at least five years later. Initially, Gibbon's machine was massive, complicated, and difficult to manage. The blood elements were damaged by the machines actions, causing bleeding problems and severe consumption of red blood cells. However, in view of its ability to permit corrective operations to be performed inside of the human heart for the first time, these side effects of the heart-lung bypass were acceptable. In 1955, British physician Dennis G. Melrose and his colleagues developed a technique to safely stop the heart and reduce the leakage of blood using potassium citrate and then potassium chloride. Subsequent experiments by Melrose and British biochemist David J. Hearse, using the isolated hearts of rabbits and rats, established optimum concentrations of potassium chloride to stop the heart, and ways of preserving the heart while starved of blood. It is now commonplace for surgeons to stop the heartbeat even for several hours while the circulation is maintained by modern, commercially available heart-lung support equipment.

The safety and ease-of-use of heart-lung equipment has improved since its first usage. With better understanding of the flow properties of blood (discussed in Chapter 4) and impact of stress on the components of blood, the major cause of blood damage in the oxygenator was determined to be the direct exposure of blood to gases. Interposing a gas exchange membrane of plastic or cellulose between the flowing blood and the gas solved much of the blood-damage problems. However, the devices required very large surface areas due to limited gas permeability. In the 1960s, availability of gas permeable thin sheets of silicone rubber (dimethyl polysiloxane polymer) changed the design of artificial lungs with potential clinical application. By eliminating the gas interface it was possible to use the heart-lung machine for days at a time. In the early 1980s, an integrated reservoir was introduced to accomplish oxygenation by more sophisticated methods. The modern heart-lung machine is actually more sophisticated and versatile, and can perform a number of other tasks necessary for safe completion of an open-heart operation. First, blood loss is minimized by scavenging all the blood spilled around the heart using suction pressure and returning the blood to the pump. Returning shed blood into the heart-lung machine greatly preserves the patient's own blood throughout the operation. Nevertheless, risks associated with blood clotting within some internal organs still persist. Second, the patient's body temperature can be controlled by selectively cooling or heating the blood as it moves through the heart-lung machine. Thus, the surgeon can use low body temperatures as a tool to preserve the function of the heart and other vital organs during the period of artificial circulation. However, heart-lung machines need a large priming volume (i.e., the amount of blood or blood substitute required to fill the device prior to use). With the advances in computational fluid dynamics (CFD), a full map of the velocity and pressure changes in a conduit of complex geometry such as a blood heat exchanger and oxygenator can be generated. Heat and gas transfer could then be calculated from the CFD velocity field by using mass transfer correlations. These will allow future design of heart-lung machines.

1.3.4   Other Devices

There are a number of other life-saving devices that have a similar history. For example, the development of ventricular defibrillators dates back to the discovery of ventricular fibrillation by Scottish physician John A. MacWilliam. American surgeon Claude S. Beck first used the defibrillator in 1947. In the late 1960s and early 1970s, Michael Mirowskis' team developed the implantable cardioverter-defibrillator in collaboration with an industry. Similarly, the developments of other devices such as stents, defibrillators [Figure 1.5(a)], pacemakers [Figure 1.5(b)], total heart, prosthetic components [Figure 1.5(c)], and cochlear implants have similar interdisciplinary interactions where biomedical engineering principles have made a significant impact.

1.4   Literature Sources

The Food and Drug Administration (FDA) is a U.S. government agency under the Department of Health and Human Services (DHHS) that plays a complex and important role in the lives of all Americans, as well as in the lives of consumers of many products grown, processed, and/or manufactured in the United States. As mandated by federal law, the FDA helps ensure that the food we eat is safe and wholesome, the cosmetics we use are pure and safe, and medicines and medical products are safe and effective for uses indicated in the product labels. The FDA is also responsible for overseeing veterinary medical and food products. From the Food and Drug Administration (www.fda.gov) home page, one can locate consumer education materials, press releases, industry guidance, bulletins for health professionals, and other useful documents and data from FDA's centers and offices. The FDA Web site provides authoritative information on food, cosmetics, human and animal drugs, biologics, and medical devices.

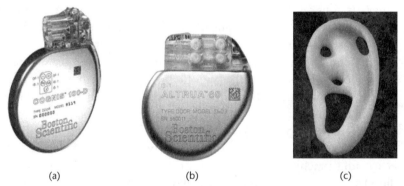

(a)   (b)   (c)

Figure 1.5   Other devices. (a) COGNIS 100-D Cardiac Resynchronization Therapy High Energy Defibrallator. (b) ALTRUE 60 DR Multiprogrammable Pacemaker (Model S603). (Used with permission from Boston Scientific Corporation, St. Paul, Minnesota.) (c) MEDPOR S-Ear. (Used with permission from Porex Surgical, Inc., Newnan, Georgia.)

One has to become familiar with some of the literature resources available on the Internet at MEDLINE and Science Citation Index. MEDLINE and Science Citation Index are search engines, which allow researchers to conduct multidisciplinary broad-based, comprehensive searches that uncover relevant information they need with respect to a particular area of interest.

MEDLINE (Medical Literature Analysis and Retrieval System Online) is an online database of indexed citations and abstracts to medical, nursing, dental, veterinary, healthcare, and preclinical sciences journal articles. MEDLINE can be accessed freely through PubMed (http://www.ncbi.nlm.nih.gov/PubMed/). MEDLINE is developed by the U.S. National Library of Medicine (NLM) at the U.S. National Institutes of Health (NIH). PubMed compiles the individual databases, accesses additional (non-MEDLINE) journals, and is accessible from any computer that can access the World Wide Web. Please read the help screen or take the tutorial found at http://www.nlm.nih.gov/bsd/pubmed_tutorial/m1001.html.

The Science Citation Index (SCI) was developed by the Institute for Scientific Information and covers topics in science, medicine, agriculture, technology, and the behavioral sciences. SCI is owned by Thompson Scientific, and can be accessed online through Web of Science (http://www.lib.sfu.ca/researchtools/databases/dbofdb.htm?DatabaseID=328) after registration or through a registered portal. SCI interprets the impact that a specific article has on its field, based on the number of times it has been cited by other articles, and the names of those articles. It also provides a list of all references cited by a given article, and therefore a means to locate key related articles.

Problems

1.1   Name five specialized areas of biomedical engineering.

1.2   Name four organizations related to biomedical engineering.

1.3   Perform a literature search on Medline for new concepts in blood oxygenators this year. List five articles from different research groups.

1.4   Perform a literature search on Science Citation Index for new publications from this year on magnetic resonance imaging. Write five sentences about the problems they are addressing.

1.5   Visit the FDA Web site. What is the most recent FDA-approved biomedical device? Who is the manufacturer? What does it do?

1.6   Visit five society Web sites and write when their annual meetings are. Are there special meetings for students? List them.

Selected Bibliography

Boettcher, W., F. Merkle, and H. H. Weitkemper, "History of Extracorporeal Circulation: The Invention and Modification of Blood Pumps," *Journal of Extra Corporeal Technology*, Vol. 35, No. 3, 2003, pp. 184–191.

Cooley, D. A., "Development of the Roller Pump for Use in the Cardiopulmonary Bypass Circuit," *Texas Heart Inst. Journal,* Vol. 14, No. 2, 1987, pp. 112–118.

Hogness, J. R., and M. Van Antwerp, (eds.), *The Artificial Heart: Prototypes, Policies, and Patients,* Washington, D.C.: National Academy Press, 1991.

Maher, J. F., *Replacement of Renal Function by Dialysis: A Textbook of Dialysis,* 3rd ed., New York: Springer, 1989.

Pastuszko, P., and R. N. Edie, "John H. Gibbon, Jr., the Inventor of the First Successful Heart-Lung Machine," *Journal of Cardiac Surgery,* Vol. 19, No. 1, 2004, pp. 65–73.

Radegran, K., "The Early History of Cardiac Surgery in Stockholm," *Journal of Cardiac Surgery,* Vol. 18, 2003, p. 564.

Vienken, J., et al., "Artificial Dialysis Membranes: From Concept to Large Scale Production," *American Journal of Nephrology,* Vol. 19, 1999, pp. 335–362.

Wizemann, V., "Hemodialysis: 70 Years," *Journal of Molecular Medicine,* Vol. 72, No. 9, 1994, pp. 1432–1440.

CHAPTER 2

Biotransport

2.1 Overview

The human body is made of complex and delicate structures. These structures are continuously subjected to various disturbances. For example, all cells require a constant supply of nutrients such as oxygen, glucose, and water to support various functions. There is unceasing continuous molecule and ion traffic in and out of the cells and the body. Various biological membranes lining cells and tissues act as barriers for transport of molecules while regulating biological functions at each level. Thus, the transport of molecules can be visualized to occur at five levels (Figure 2.1):

1. Across different tissues;
2. Between different cell types;
3. Between the cells of the same type;
4. From the outside to the inside of the cell;
5. Within a cell.

For example, the flow of molecules and ions between the cell and its environment is precisely regulated by specific transport systems. Regulated movement of various ions across the cell membranes leads to the presence of ionic gradients, which are vital in stimulating nerves and muscles. Furthermore, vital functional molecules such as insulin secreted by pancreatic β-cells needs to be transported to the entire body.

A fundamental understanding of how homeostasis (a Greek word meaning "the same" and "standing," a state of relative constancy of its internal environment) is maintained in various compartments of the body is critical to the design and development of synthetic substitutes, which could effectively interfere and/or replace the diseased or dysfunctional tissue. For example, external support devices such as dialysis machines, heart-lung machines, mechanical ventilators, and contrast agents used during biomedical imaging have to assist in maintaining homeostasis. Dialysate solution used in dialysis should only remove the toxins and excess water—not important electrolytes. In addition, modeling various physiological processes such as alterations in calcium fluxes during a cardiac cycle (and subsequent events) helps to understand the process of tissue regeneration. Although molecular transfer

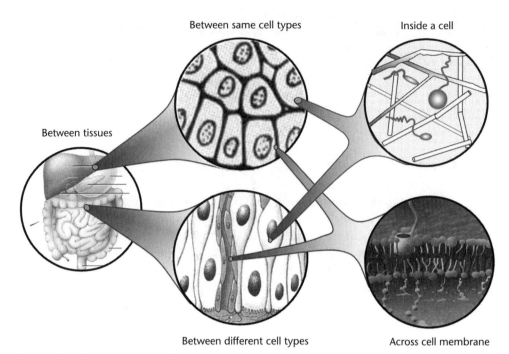

Figure 2.1 Biotransport at multiple levels.

over large distances occurs in conjunction with fluid flow (termed as convective flow), we will discuss the basic transport mechanisms across various biological barriers in this chapter. First, the basic terminologies such as different compartments, pH, pressure, and concentration are defined. Next, diffusion, osmosis, and the transport of molecules across different tissues are described. Finally, movement of macromolecules across single cells is discussed.

2.2 Fundamental Factors

2.2.1 Fluid Compartments

Water is the main fluid in the body through which biomolecules move from one region to the other. Fluid in the body is either intracellular or extracellular. Intracellular fluid is present inside the cells. The extracellular fluid is present outside the cell body. Extracellular fluid is subdivided into the following compartments:

1. The plasma, which occupies the extracellular space within blood vessels and the heart.
2. The interstitial fluid, which occupies the spaces between tissues excluding blood vessels. A large fraction of blood in the capillaries surrounding a tissue space is filtered into the tissue space as interstitial fluid. Thus, interstitial fluid bathes the cells in the tissue space. Thus, all the requirements of cells such as nutrients and signals and their synthesized products are transported through interstitial space.

3. The transcellular fluid, which includes the fluids formed during transport activities (i.e., not inside cells but separated from plasma and interstitial fluid by cellular barriers). Examples include fluid in the bladder, fluid in gastrointestinal tract, cerebrospinal fluid, and sweat.

A general distribution of fluid compartments for a 70-kg male is shown in Figure 2.2(a). The total amount of intracellular fluid is the largest fraction of the fluid volume in the body, contributing to nearly 40% of the body weight. In relation to the body weight, water content decreases from infancy (~75% of body weight) to old age, with the greatest decrease occurring within the first 10 years of life. In adult males, water accounts for nearly 60% of body weight, whereas adult females have less water content (nearly 50% body weight) in their body. This is due to more subcutaneous (under the skin) adipose (fat) tissue, which contains less water.

The human body temperature is maintained at 37.4°C and alterations in body temperature are regulated by many mechanisms including sweating during exercise. Proper redistribution and replenishment of lost water due to various physiological processes such as the metabolism is essential to normal health. An imbalance in day-to-day water distribution can cause abnormalities in body function. Accumulation of water in various parts of the body causes a number of life threatening diseases including lung edema and cerebral edema.

As the water in sweat evaporates from the skin, evaporative cooling takes with it some of the body heat, helping to cool the body. Along with water, the sweat

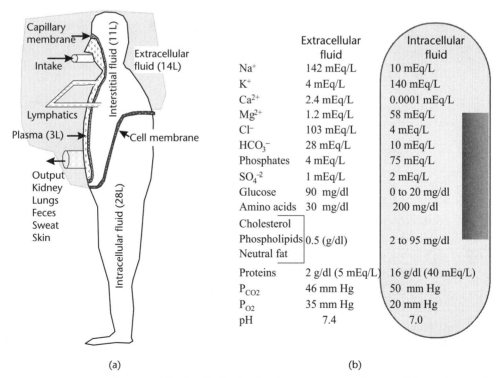

	Extracellular fluid	Intracellular fluid
Na^+	142 mEq/L	10 mEq/L
K^+	4 mEq/L	140 mEq/L
Ca^{2+}	2.4 mEq/L	0.0001 mEq/L
Mg^{2+}	1.2 mEq/L	58 mEq/L
Cl^-	103 mEq/L	4 mEq/L
HCO_3^-	28 mEq/L	10 mEq/L
Phosphates	4 mEq/L	75 mEq/L
SO_4^{-2}	1 mEq/L	2 mEq/L
Glucose	90 mg/dl	0 to 20 mg/dl
Amino acids	30 mg/dl	200 mg/dl
Cholesterol Phospholipids Neutral fat	0.5 (g/dl)	2 to 95 mg/dl
Proteins	2 g/dl (5 mEq/L)	16 g/dl (40 mEq/L)
P_{CO2}	46 mm Hg	50 mm Hg
P_{O2}	35 mm Hg	20 mm Hg
pH	7.4	7.0

(a)   (b)

Figure 2.2   (a) Distribution of fluid in the body. (b) The composition of extracellular and intracellular fluids.

contains electrolytes including sodium, potassium, calcium, magnesium, chloride, bicarbonate, phosphate, and sulfate. Typically, 1 liter of sweat contains 0.02g calcium, 0.05g magnesium, 1.15g sodium, 0.23g potassium, and 1.48g chloride, but the electrolyte composition of sweat is variable from individual to individual. Excessive loss of water could lead to dehydration and exhaustion. The loss of essential electrolytes, which can result in cramping, could lead to reduced nutrient transfer and water in and out of the cell (known as cellular respiration) because cells require a precise balance for maximum cellular efficiency. Hence, replacing the lost water is essential for healthy living.

2.2.2   Concentration

Concentration is the term used to relate the mass of a component in a mixture to the volume of the mixture. In a flow system, concentration relates the mass flow rate of the component to the volumetric flow rate. If the concentration is defined by the mass per unit volume of the mixture (kg/m^3), then it is called mass concentration. Concentration can also be defined in moles per unit volume where one mole is the molecular weight of that component ($kgmol/m^3$). This molar concentration definition is more useful in evaluating systems where reaction is taking place. For charged molecules, concentration is also defined in equivalents per liter to represent the alkalinity or acidity of those components. An equivalent is the amount of a substance that will combine with or replace one mole of H+ ions in a chemical reaction. Thus, Eq/L can be converted in terms of g/L or mol/L using the valency of the component.

EXAMPLE 2.1

A patient has a serum magnesium concentration of 2.7 mEq/L. What is the concentration in mg/dL?

Solution: From the definition of mEq, mEq = mmol × valence

mmol = 2.7 (mEq/L)/2 = 1.35 (mmol/L)

= 1.35 × 24.3 mg/mmol = 32.8 mg/L or 3.28 mg/dL

The cell membrane is a semipermeable membrane (i.e., it allows some substances to pass through but not all). Hence, the intracellular fluid has a distinct ionic composition, different from extracellular fluid [Figure 2.2(b)]. The intracellular fluid has similar composition in all cell types and contains high levels of K^+ ions and low levels of Na^+ ions. On the contrary, extracellular fluid contains high levels of Na^+ ions and low levels of K^+ ions. These differences are important for various biological functions, which will be described later in this chapter and in Chapter 3. The composition of interstitial fluid is similar to that of blood plasma, except it contains a smaller concentration of proteins than plasma and thus a somewhat greater concentration of water. Lymph has almost the same composition as that of original interstitial fluid.

Differences in concentrations across a barrier are one of the driving forces for the transport of molecules in biological systems. The second law of thermodynamics

states that natural processes of the universe proceed from states that are more ordered to states that are more disordered. The movement of any molecule up (to a
higher concentration) or down (to a lower concentration) a concentration gradient
involves a change in free energy (ΔG—down releases energy so ΔG is negative; up
consumes energy so ΔG is positive). Ideally, the amount of free energy released or
consumed is calculated using

$$\Delta G[J/mol] = R[J/mol.K]T[K]\ln\frac{C_{in}}{C_{out}} + zF[C/mol]\Delta V[V] \tag{2.1}$$

where ΔG is the change in Gibbs free energy per mole; R is the universal gas constant (1.987 cal.mol^{-1}.K^{-1} or 8.314 J.mol^{-1}.K^{-1}); T is the absolute temperature,
which must be in Kelvin ($K = 273 + °C$) or Rankine ($°R = 459.67 + °F$); C_{in} is the
concentration inside the cell (mM); C_{out} is the concentration outside the cell (mM);
z is the electrical charge of the transported species; F is Faraday's constant and ΔV is
the voltage across the membrane (discussed in Chapter 3). Faraday's constant is the
number of units of static charge per mol of charge, which is 96,484.5 Coulomb/mol
($= 23,062$ calories/V.mol) released as one mole of ions moves down a voltage gradient of 1 volt (1,000 mV). Equation (2.1) presumes that concentrations represent
the activity of that component in each compartment and there are no interactions
with other species.

The first term on the right side of (2.1) is due to the chemical gradient across
the membrane and is called chemical potential. It is the amount by which the free
energy of the system would change if an additional particle is introduced without
altering the distributions and volume of the system. The second term on the right
side of (2.1) provides energy due to the electrical gradient, normally referred to as
drift, and is discussed in Chapter 3. For the transport of uncharged molecules, the
contribution from the second term is neglected and the total Gibbs free energy is
calculated by knowing the total change in the number of moles in the system.

EXAMPLE 2.2

The concentration of glucose inside the cell is 0.5 mM, and the concentration of glucose
outside the cell is 5 mM. Assuming a body temperature of 37°C, calculate ΔG.
  Solution: Using (2.1) without the electrical gradient

$$\Delta G = RT\ln\frac{C_{in}}{C_{out}} = 8.314*310*\ln\frac{0.5}{5} = -5{,}934.545 \text{ J/mol } = 5.9 \text{ kJ/mol}$$

As the process proceeds with the release of free energy, it can proceed spontaneously
if the phospolipid bilayer allows glucose to pass through it. However, the phospholipid
bilayer is not permeable to glucose and needs other means of transport (discussed in Section 2.3.2).

2.2.3 Pressure

Pressure is defined as the force per unit area. The SI unit of pressure is N/m^2 and is called Pascal. Other units of pressure are atmosphere, bar, torr, millimeters of mercury (1 atm = 760 mmHg) and dynes per square-centimeter. Due to contraction of the heart, the blood pressure in the aorta normally rises to 120 mmHg during systole, whereas it falls to 80 mmHg during diastole. The pressure of arterial blood is largely dissipated when the blood enters the capillaries. Although the diameter of a single capillary is small (7–10 μm), the number of capillaries supplied by a single arteriole is so large that the total cross-sectional area available for the flow of blood is significantly more. Hence, the pressure of the blood as it enters the capillaries decreases. When blood leaves the capillaries and enters the venules and veins, little pressure remains to force it along. Pressure in the veins is nearly one-tenth of that in the arteries. Thus, blood in the veins below the heart is helped back up to the heart by the squeezing effect of contracting muscles present on the veins. Veins also have one-way valves called venous valves to prevent backflow within the veins. One of the functions of the kidney is to monitor blood pressure and take corrective action if it should drop by secreting the enzyme renin.

Pressure also arises due to Earth's gravity. The pressure at the bottom of a stagnant liquid of height (Δh) is expressed in terms of the density (ρ) of the fluid (for blood = 1,050 kg/m^3) as

$$\Delta P = \rho g_c \Delta h \tag{2.2}$$

where g_c is the gravitational constant and is 9.81 m/s^2 in SI units. This pressure is also called hydrostatic pressure. If the person is lying down (called a supine position), then the hydrostatic pressure is ignored. If blood pressure is measured at the head height, the systolic/diastolic pressure readings are nearly 35 mmHg less compared to readings taken at heart level, whereas at ground height the pressure readings will be 100 mmHg greater. The pressure at any point in a fluid acts equally in all directions. This is called Pascal's law of fluid pressure.

Blood pressure is monitored using a device called the sphygmomanometer, which does not require surgical intervention. It consists of an inflatable cuff, a rubber bulb to inflate the cuff, a valve to control pressure, and a pressure gauge such as a monometer. The cuff is wrapped around the upper arm at roughly the same vertical height as the location of the heart. Then the cuff is inflated by squeezing the rubber bulb to a pressure higher (typically 180 mmHg) than the systolic pressure. The pressure difference between the inside (P_{inside}) and the outside ($P_{outside}$) of a vessel is called the transmural pressure, which regulates the shape of a vessel. When the applied external pressure is higher than the systolic pressure, the artery collapses and blood flow is temporarily halted. Then, the valve is opened to release the pressure of the cuff very slowly. When the cuff pressure reaches systolic pressure, the artery reopens allowing blood to pulse in the artery, which makes a whooshing sound due to vibrations against the artery walls. This sound is heard for the first time with the help of a stethoscope. With the help of a monometer, the pressure corresponding to that sound is recorded. The blood flow through the artery increases steadily until the pressure of the cuff drops below the diastolic pressure.

The pressure at which sound is heard the second time is the diastolic pressure. With digital meters, these pressures are recorded and displayed directly. The difference in pressure between two different sites within a vessel is called perfusion pressure. This pressure difference is necessary for fluid to flow.

In the body there is also the exchange of gases. Blood transports CO_2 away from cells and delivers O_2 to cells. When discussing the movement of gases, using concentration to describe the driving force is not the appropriate measure because much of the dissolved gases in body fluids are bound to another component (e.g., oxygen to hemoglobin) or transformed into another chemical form (e.g., CO_2 to bicarbonate). Using the total concentration gives an erroneously high driving force, since it includes bound molecules which are not free to move alone. When discussing movement of gases, an accurate measure of driving force is the partial pressure. In a mixture of gases, the partial pressure of a component is the pressure exerted by that component if it was present alone in the same volume and temperature. If P is the total pressure of a system that contains component A to a mole fraction of y_A, then partial pressure is given by

$$p_A = y_A P$$

The total pressure due to all components is the sum of the individual partial pressures. This is called Dalton's law. If there are four components, then total pressure is obtained using

$$P = p_1 + p_2 + p_3 + p_4$$

Dalton's law assumes that the molecules do not interact with each other, which is valid for respiratory gases (O_2, CO_2, N_2, and H_2O). On a molar basis, dry air at sea level contains 21% O_2, 79% N_2, and nearly 0% CO_2. Since total pressure is about 760 mmHg (or 1 atm), the principles above, say, the partial pressure of oxygen (P_{O_2}) is 160 mmHg ($= 0.21 \times 760$ mmHg). Similarly, the partial pressure of nitrogen (P_{N_2}) is 600 mmHg at sea level. For gases, the chemical potential between adjacent gas phase systems in (2.1) can be calculated using

$$\Delta G_{chem} = RT \ln \frac{p_{in}}{p_{out}} \tag{2.3}$$

where p_{in} and p_{out} are the partial pressures of a component. To convert the pressure to a concentration at a known temperature, the most widely used relationship is the ideal (or perfect) gas law. According to ideal gas law, a system occupying a volume V at an absolute pressure P, and absolute temperature T are related by

$$PV = nRT \text{ or } C[\text{mol/m}^3] = \frac{P[Pa]}{R[J/\text{mol.K}]T[K]} \tag{2.4}$$

where n is the number of moles and C is the concentration. Absolute pressure refers to the pressure measured relative to a perfect vacuum. Using ideal gas, one mole of air at a standard temperature and pressure (STP), which are 273°C and 1 atm, occupies a volume of 22.415L (or 359.05 ft^3). The ideal gas approximation is adequate for most of the biomedical applications involving gases, which occurs above 0°C and near 1 atmosphere.

EXAMPLE 2.3

Mary Joy is inhaling 500 mL of air with each breath at a rate of 12 breaths per minute. The molar compositions of the inspired and expired gases are as given in the table.

The inspired gas is at 25°C and 1 atm, and the expired gas is at body temperature and pressure (i.e., 37°C and 1 atm). Nitrogen is not transported into or out of the blood in the lungs.

(a) Calculate the masses of O_2, CO_2, and H_2O transferred from the pulmonary gases to the blood or vice versa (specify which) per minute.

(b) Calculate the volume of air exhaled per milliliter inhaled.

Species	O_2	CO_2	N_2	H_2O
Inspired gas (%)	21	0	77	2
Expired gas (%)	15	4	75	6

(c) At what rate (g/min) is Mary Joy losing weight by breathing?

Solution: The volume of gases changes with temperature and pressure. However, number of moles does not change if there is no reaction. Hence, the first step is to convert volume to moles.

At STP, 1 mol of gas occupies 22.414L.

Hence,

$$n_{in} = 6,000 \text{(mL/min)} * 10^{-3} \text{(L/mL)} \frac{273(K)}{298(K)} * \frac{1 \text{ mol}}{22.4L} = 0.245 \text{ mol/min}$$

Nitrogen is not absorbed by body fluids. Hence, the same amount of nitrogen that enters the system leaves the system. Using this relation, total moles exhaled can be calculated.

That is,

0.77*0.2246 = 0.75*n_{out}

Rearranging, n_{out} = 0.252 mol exhaled/min

Knowing the total moles exhaled,

O_2 transferred to blood: (0.245*0.21 – 0.252*0.15 (mol O_2/min) = 0.439 g/min

CO_2 transferred from blood: (0.252*0.04 (molCO_2/min))*44 g/mol = 0.443 g/min

H_2O transferred from blood: (0.252*0.06 – 0.245*0.02 (mol H_2O/min)) = 0.184 g/ min.

By breathing, this individual is losing 0.184g in a minute.

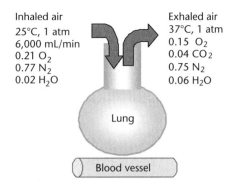

Inhaled air
25°C, 1 atm
6,000 mL/min
0.21 O$_2$
0.77 N$_2$
0.02 H$_2$O

Exhaled air
37°C, 1 atm
0.15 O$_2$
0.04 CO$_2$
0.75 N$_2$
0.06 H$_2$O

Lung

Blood vessel

Gases dissolve in liquids when in contact. The solubility of a gas in a liquid depends on temperature, the partial pressure of the gas over the liquid, properties of the liquid, and the gas. Understanding the solubility of gases in water is important in many applications. For example, assessing the solubility of an inhalation anesthetic agent in blood is important to assess the time required to establish the unconsciousness necessary for a surgical procedure. Alternatively, measuring the blood alcohol level is also possible. When molecules such as oxygen and CO_2, which do not condense into a liquid at STP are in contact with a solution, the amount of that molecule dissolved in liquids is estimated using Henry's law, particularly for molecules that have a very low solubility (near zero concentration). According to Henry's law, the mole fraction of a gas to stay in the solution is dependent on its partial pressure, p_A (Figure 2.3). The mole fraction of a component, x_A, in the solution is given by

$$x_A = \frac{p_A[\text{atm}]}{H_A[\text{atm}]} = \frac{Py_A}{H_A} \qquad (2.5)$$

where H_A is the Henry's law constant for solute A in a specific solvent, P is the total pressure, and y_A is the mole fraction of A in the gas phase. The Henry's law constant for a few of the common gases encountered in biomedical applications is given in Table 2.1. For other elements, the constant can be obtained from various

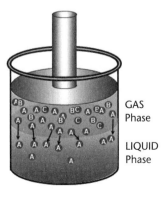

GAS
Phase

LIQUID
Phase

Figure 2.3   Henry's law.

Table 2.1 Henry's Law Constant (atm) for Common Gases

Temp	O_2	N_2	CO_2	CO	H_2	H_2S	C_2H_4
25°C	43,016	84,382	1,701	56,783	69,050	560	40,717
37°C	51,118	99,252	2,209	66,695	73,579	708	49,313

Source: [1].

handbooks. However, one has to check the units as Henry's law can be expressed in different forms. Equation (2.5) is expressed in mole fractions. When Henry's law is expressed in terms of molality (moles solute/kg of solution), the constant will have additional moles of solute per kilogram of solvent units. When Henry's law is expressed in concentration, the constant will have additional moles per liter of solution units. Nevertheless, Henry's law constant is a function of temperature and generally increases with increasing temperature. Henry's law constant does not include the solubility products of ions.

EXAMPLE 2.4

Calculate the concentration of oxygen dissolved in air-saturated water under normal atmospheric conditions at 25°C and at 37°C.

Solution: Normal atmospheric conditions are 21% mole oxygen, which makes the partial pressure of oxygen 0.20948 atm or 20.67 kPa. From (2.5),

$$x_{O_2} = \frac{p_{O_2}}{H_{O_2}} = \frac{0.209\,[\text{atm}]}{43,016\,[\text{atm}]} = 4.86 * 10^{-6}$$

Consider 1 liter of water, which is equivalent to 998g, with the density of 0.998 g/mL at 25°C.

Then, the number of moles is

$$\text{moles of water in } 1\text{L} = \frac{998\,[\text{g}]}{18.02\,[\text{g/mol}]} = 55.383\,\text{mol}$$

Using the definition of the mole fraction,

$$x_{O_2} = \frac{n_{O_2}}{n_{O_2} + n_{H_2O}} \rightarrow 4.86 * 10^{-6} = \frac{n_{O_2}}{n_{O_2} + 55.383\,[\text{mol}]} \rightarrow n_{O_2} = 2.7 * 10^{-4}\,[\text{mol}]$$

Since 1.0L of solvent was considered, the O_2 concentration at 25°C is 2.7×10^{-4} mol/L = 0.27 mmol/L

Similarly at 37°C, O_2 concentration is 2.27×10^{-4} mol/L = 0.227 mmol/L. Note that the solubility decreases with increase in temperature.

2.2.4 Body pH

pH is a scale that measures how acidic or alkaline a solution is. The logarithmic of a concentration of H^+ ions in a fluid is expressed in pH units. The scale ranges from 1 to 14 with 1 being very acid, 7 neutral, and 14 very alkaline. Higher pH numbers indicate a greater potential for absorbing more H^+ ions whereas lower numbers indicate a greater potential for donating H^+ ions. Different fluids in the body have different pH values. The fluid in the stomach is between one and three, urine is slightly acidic, bile has a pH of 7.8, and human blood has a pH of about 7.4. Although the pH of stomach fluids is highly acidic, the surface of the stomach is stable at a pH of 8. In the intestine, pH gradients exist along the epithelial barrier.

Body pH is very important because it controls the speed of body's biochemical reactions by controlling enzyme activity and electrical signals. Hence, the pH of the body fluids must be regulated. The body's pH level is a direct result of what a person eats or drinks. Activities such as drinking black coffee (pH = 5) or alcohol, severe exercise, which produces lactic acid, alter the local pH. Typically, a blood pH below 7.0 or above 7.8 is dangerous. Acidosis is the word used when the body is exposed to acidic conditions and alkalosis is the word used when the body is exposed to alkaline conditions.

The body has two mechanisms to adjust the altered pH: excretion and buffering. Excretion occurs in the lungs as CO_2 or by the kidneys. There are buffering agents that limit pH changes by accepting the produced H^+ ions inside the cell as well as outside the cell. An acid A can dissociate into an anion A^- and H^+ ions, which is written in the reaction form as

$$A \rightleftharpoons H^+ + A^-$$

At equilibrium the rates of the reaction in each direction are equal. We can describe this relationship with an equilibrium constant K_A,

$$K_A = \frac{[H^+][A^-]}{[A]} \text{ or } [H^+] = \frac{K_A[A]}{[A^-]}$$

Since the reaction is written for dissociation of the acid, the equilibrium constant is referred to as the acid dissociation constant, K_A. Taking the negative log on both sides gives

$$-\log[H^+] = -\log K_A - \log \frac{[A]}{[A^-]} \text{ or } pH = pK_A + \log \frac{[A^-]}{[A]} \qquad (2.6)$$

Equation (2.6) is called the Henderson-Hasselbach equation, which is used as a guideline to prepare buffer solutions and to estimate charges on ionizable species in solution, such as amino acid side chains in proteins. Caution must be exercised in using this equation because pH is sensitive to changes in temperature and the salt concentration in the solution being prepared.

Buffering action is more efficient if pK values are closer to pH values. One pH control mechanisms in the body (both in intracellular and extracellular fluids) is buffering by bicarbonates. CO_2 released from the tissues enters the red blood cells via plasma, and combines with water to form carbonic acid. This reaction is catalyzed by an enzyme called carbonic anhydrase, which also dissociates carbonic acid to CO_2 and water. Carbonic acid dissociates into HCO_3^- and H^+ ions.

$$CO_2 + H_2O \xrightleftharpoons{\text{Carbonic anhydrase}} H_2CO_3 \rightleftharpoons H^+ + HCO_3^-$$

Further, CO_2 can react with the bases to form bicarbonate ions, which help buffer in the plasma.

$$CO_2 + OH^- \rightleftharpoons HCO_3^-$$

CO_2 exchange provides pH control in the blood and tissues. For these reactions, the Henderson-Hasselbach equation is written as

$$pH = pK_A + \log \frac{\left[HCO_3^- \right]}{\left[CO_2 \right]} \tag{2.7}$$

From (2.7), an increase in the CO_2 concentration lowers the pH (i.e., acidosis) and an increase in HCO_3^- concentration increases the pH (i.e., alkalosis). While using heart-lung machines, (2.7) is used to predict bicarbonate levels from pH and CO_2 by directly measuring the pH using electrodes. However, a more immediate application is remembering the consequences of rising or lowering one of the components of the system. Although H_2CO_3 and HCO_3^- are involved in pH control, bicarbonate buffering is not an important buffer system because its pK_A is 6.1, which is far from the blood pH.

EXAMPLE 2.5

Your thesis advisor has no equipment money, and the only incubator for culturing cells at 37°C in the laboratory is broken. After ascertaining that it is not repairable, you decide to construct one yourself. This low-tech incubator consists of a box in which you placed a temperature controller, which maintains a uniform 37°C temperature. A pan with water is placed at the bottom of the box to humidify the atmosphere. Since there is no CO_2-level controller available, you will gas the chamber with a 5% CO_2 balance air mixture. How much $NaHCO_3$ would you add in order to obtain a physiological pH? The following data are available (all values measured at 37°C):

$$H_2O + CO_2 \xrightleftharpoons{K_1} H_2CO_3 \xrightleftharpoons{K_2} H^+ + HCO_3^-$$

where K_1 is 4.7×10^{-3} and K_2 is 1.7×10^{-4}. Solubility of CO_2 in water is 0.03 mM/mmHg and the vapor pressure of water at 37°C is 47.4 mmHg. (Courtesy of Dr. Berthiaume, Harvard Medical School.)

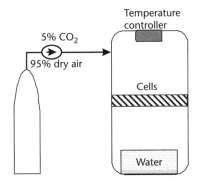

Solution: Equilibrium constants for the two reactions can be written as

$$K_1 = \frac{[H_2CO_3]}{[CO_2]}, \; K_2 = \frac{[H^+][HCO_3^-]}{[H_2CO_3]}$$

Solving for

$$\left[HCO_3^-\right] = \frac{K_1 K_2 [CO_2]}{\left[H^+\right]} \qquad\qquad (E2.1)$$

To determine $[CO_2]$
   $[CO_2]$ = (Solubility of CO_2 in H_2O)* P_{CO_2}

   From Dalton's law, the total pressure inside the incubator, $p_{O_2} + p_{N_2} + p_{CO_2} + p_{H_2O}$

Therefore,
   Composition of the given gas is 5% p_{CO_2} and 95% dry air

   Hence p_{CO_2} = 0.05* 712.6 = 35.63 mmHg

   Thus $[CO_2]$ = CO_2 solubility in H_2O * P_{CO_2} = 0.03 [mM/mmHg] 35.63 [mmHg] = 1.07 $\times 10^{-3}$M
   To determine $[H^+]$: pH = $-$log $[H^+]$ = 7.4 or $[H^+]$ = $10^{-7.4}$ = 4 $\times 10^{-8}$M
   Substituting in (E.2.1), the amount of HCO_3^- required to maintain the physiological pH is
   $[HCO_3^-]$ = (4.7 $\times 10^{-3}$)(1.7 $\times 10^{-4}$)(1.07 $\times 10^{-3}$)/(4 $\times 10^{-8}$) = 0.021M
   Then, amount of $NaHCO_3$ = 84.01 (g/mol) $\times$ 0.021(mol/L) = 1.76 g/L

There are a few intracellular proteins (e.g., hemoglobin) and phosphate containing molecules (e.g., phosphocreatine), which help in buffering. Proteins are the most important buffers in the body. In particular, hemoglobin plays a critical role in pH buffering, and an even higher capacity when deoxygenated. The plasma proteins are also buffers but their quantity is small compared to intracellular proteins. Protein molecules possess basic and acidic groups, which act as H^+ ion acceptors or donors, respectively, if the H^+ ion is added or removed. Phosphate buffers ($H_2PO_4^-$ and $H_2PO_4^{2-}$) are moderately efficient when the pK is 6.8 and closer to plasma pH. The phosphate concentration is low in the extracellular fluid but the phosphate buffer system is an important urinary buffer. These mechanisms are complementary to the long term (days) changes regulated by the kidneys. Essentially, kidneys act like size-separation units for blood, making use of the different sizes of the particles and filtering them. Kidneys change the extracellular fluid electrolytes to maintain the pH at 7.4. Unlike the pH changes due to CO_2, which the respiratory system can produce (minutes), the renal compensation is slow. Further, there are limitations to the acidity of urine. The kidneys cannot produce a urine pH of much less than 4.4. Strong acids can be removed from the blood and excreted in the urine by reacting with the basic salt of phosphoric acid in the urine without producing a reduction in the pH of the urine, or by the addition of NH_3 (a base) to the urine. Buffering by kidney occurs only when the components that change the pH are in excess. If the pH change is due to a deficiency of a component other than carbonic acid, pH regulation by kidneys is not possible unless that deficiency is restored by supplementation.

2.3 Diffusion

Diffusion is a random movement of molecules, which results in molecules moving from an area of higher concentration to an area of lower concentration. Once the concentration becomes uniform, there is no net transfer of molecules from one region to the other although random movement continues to occur. Consider Figure 2.4 where two regions, L and R, containing a solvent (e.g., water) are separated by a membrane, which is permeable to a solute A (e.g., common salt). If a high amount of A is added into region R, then molecules start randomly moving, which results in molecules crossing to region L. Since region L did not have any A, diffusion results in the net gain of A.

The transfer of molecules by diffusion is one of the main processes through which molecules move in many physiological processes. For example, the lipid bilayer and basement membranes are permeable to water molecules and a few uncharged small molecules like O_2, CO_2, N_2, NH_3, fatty acids, and some alcohols. These diffuse freely in and out of the cell. However, the lipid bilayer is not permeable to ions such as K^+, Na^+, Ca^{2+} (called cations as they migrate toward the cathode), Cl^-, HCO_3^- (called anions because they migrate toward the anode), or small hydrophilic molecules like glucose, proteins, and RNA. Removal of blood

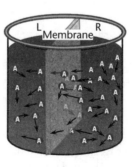

Figure 2.4 The diffusion process.

toxins by dialysis also works by diffusion between the blood and the dialysate, the external fluid used in the process.

2.3.1 Free Diffusion

Diffusion that occurs spontaneously due to the presence of favorable concentration or pressure gradients is free diffusion. Adolf Fick (German physiologist) observed that the laws used for conduction of heat and electricity also apply to free diffusion in a bulk homogeneous solution. According to Fick, the amount of a substance A crossing a given area in time dt is proportional to the concentration gradient (dc_A/dx in kgmoles/m^4) of A in the x direction. This is commonly called *Fick's first law* and written mathematically as

$$J_D[\text{kgmol/m}^2\text{s}] = -D_{AB}[\text{m}^2/\text{s}]\frac{dC_A}{dx}\frac{[\text{kgmol/m}^3]}{[\text{m}]} \tag{2.8}$$

where J_D is the diffusional flux (i.e., the number of molecules moving from a specified region across a unit area of the boundary per unit time) and D_{AB} is the diffusivity constant of solute A through solvent B. From (2.8), the net movement of molecules doubles if the concentration gradient doubles. Also, the time required for a molecule to diffuse between two points is proportional to the square of the distance separating the two points. Fick's first law applies to all fluids, although molecular diffusion is much more rapid in a gas than in a liquid. Hormones such as progesterone, testosterone, and estrogen cross the cell wall by free diffusion.

One example of free diffusion is respiration where air enters the lungs (inspiration) and outflows from the lungs (expiration). During each breath, the exchange of gases occurs between the alveolar spaces of the lungs and the blood in pulmonary capillaries, called external (or pulmonary) respiration (Figure 2.5). The blood gains O_2 and loses CO_2 via diffusion. Partial pressure of oxygen (p_{O_2} in alveolar air is 105 mmHg and p_{O_2} in deoxygenated blood entering pulmonary capillaries is 40 mmHg. Since O_2 is at a higher concentration in the alveolus, it diffuses from alveolus into blood until p_{O_2} reaches 105 mmHg. On the other hand, p_{CO_2} in pulmonary deoxygenated blood is 45 mmHg and p_{CO_2} in alveoli reaches 40 mmHg. Hence, CO_2 diffuses from blood into alveolus until p_{CO_2} reaches 40 mmHg. Diffusion during external respiration is aided by a reduced thickness of the respiratory membrane, large surface area, large numbers of capillaries, and maximum exposure of blood cells. The exchange of gases between blood in small capillaries and tissue cells, called internal respiration, is also by free diffusion. p_{O_2} in arterial capillaries is 105 mmHg but 40 mmHg in surround tissue. Thus, oxygen diffuses from blood into tissues. On the other hand, p_{CO_2} in surrounding tissue is 45 mmHg and 40 mmHg in blood. Within the cells, called cellular respiration, biochemical processes consume O_2 and give off CO_2 during the production of high-energy phosphates such as adenosine triphosphate (ATP). Continuous supply and removal of O_2 and CO_2 is ensured by differences in the partial pressures between the inside and outside. The diffusivity constant (D_{AB}) of a solute varies with the solvent, temperature, and viscosity. To quantify diffusion in a complex biological medium, it is necessary

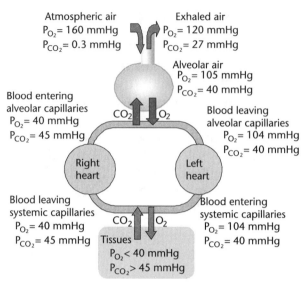

Figure 2.5 Free diffusion in the exchange of oxygen and carbon dioxide in various parts of the body.

to apply the diffusion equation through the use of some macroscopically effective properties, such as the local average concentration, and free solution diffusivity. Using the fundamentals of the equation of motion (described in Chapter 4) for the diffusivity of a rigid spherical solute of radius r_{se} in a dilute solution, Einstein (also known as Stokes-Einstein relation) arrived at the expression

$$D_\infty = \frac{RT}{6\pi\mu r_{se} N_A}$$ (2.9)

where D_∞ is called free solution diffusivity [m²/s], N_A is Avogadro's number (6.023 × 10²³ molecules/mol), and μ is the solvent viscosity (kg/m·s, Pa.s or N.s/m²). Since the gas constant is the product of Boltzmann's constant and Avogadro's number, (2.9) is also written in terms of Boltzmann's constant. Stokes radius rse is also referred to as the hydrodynamic radius of the diffusing molecule whose shape is assumed to be rigid and spherical. Equation (2.9) has been shown to provide good diffusivity predictions for large spherical molecules or particles in solvents, which show less slippage. Equation (2.9) is also used to estimate the effective Stokes radius of the solute if its diffusivity is known. Values for a few of the molecules are given in Table 2.2.

To determine the diffusivity of a solute, its effective Stokes radius is estimated assuming that it is equal to the molecular radius of the solid spherical solute, which gives the expression

$$r_{se} = \left(\frac{3MW}{4\pi\rho N_A} \right)^{1/3}$$ (2.10)

Table 2.2   Physical Parameters for Common Biological Molecules

	MW (Daltons)	Diameter (Å)	D_∞ at 37°C $[m^2/s]$
Urea	60.1	5.69	1.81×10^{-9}
Creatinine	113.1	7.19	1.29×10^{-9}
Glucose	180.16	8.52	9.55×10^{-10}
Vitamin B_{12}	1355.42	16.63	3.79×10^{-10}
Epidermal growth factor (EGF)	6,600	37 (at 34°C)	1.67×10^{-10} (at 34°C)
Nerve growth factor (NGF)	26,500	49 (at 34°C)	1.27×10^{-10} (at 34°C)
Bovine serum albumin	69,000	65	6.7×10^{-11}

Source: [2–4].

where ρ is the density of the solute and typically varies between 1.265–1.539 g/cm^3 for globular proteins. Equations (2.9) and (2.10) estimate free diffusivity of solutes in a homogeneous medium. However, the presence of other molecules affects the diffusivity constant. To account for some of the alterations, free solution diffusivity is related to the diffusivity constant in Fick's first law by

$$D_{AB} = k_D D_\infty$$

where k_D is called a hindrance coefficient for diffusion. In the absence of specific interactions such as adsorption and electrical effects, k_D for spherical solutes in membranes with cylindrical pores is a function of the ratio of the solute radius to the membrane pore radius. However, the temperature dependence of the diffusion coefficient of a spherical molecule is not linear. The temperature dependence of diffusivity constant also arises from the viscosity changes.

EXAMPLE 2.6

A main protein component in serum is 4.5g per 100 mL of albumin (MW = 75,000; diameter = 70Å). Calculate the diffusion coefficient of component at 37°C in water.
Solution: To solve for the diffusion coefficients:
m = viscosity of water at 37°C = 0.76 cP = 0.76 * 10^{-3} Ns/m^2
N_A = 6*1,023 K/J
r = 70Å = 70*10^{-10}m

$$D_\infty = \frac{RT}{6\pi r \mu N_A} = \frac{8.3143 \, J/mol.K * 310K}{6\pi * 70 * 10^{-10} m * 0.76 * 10^{-3} Ns/m^2 * 6 * 10^{23} molecules/mol}$$

= 8.54*10^{-11} m^2/s = 8.54*10^{-7} cm^2/s
The experimentally reported D_∞ value (Table 2.2) is slightly less than this value.

An alternative approach is the development of empirical equations, which relate the diffusivity constant to the molecular weight (MW) of the solute molecule. One such empirical correlation for solutes in a dilute aqueous solution at 37°C is

$$D_{AB}[\text{cm}^2/\text{s}] = 1.013 \times 10^{-4}(MW)^{-0.46} \qquad (2.11)$$

Nevertheless, combining (2.10) and (2.8) gives

$$J_D = \frac{-RT}{6\pi\mu r N_A}\frac{dC}{dx} \text{ or } J_D = -\frac{RT}{f}\frac{dC}{dx} \qquad (2.12)$$

where f is the frictional coefficient. When the concentration gradient across a porous membrane is linear in a location, Fick's first law (2.8) can be written in a simple form as

$$J_{local}[\text{kgmol/m}^2\text{s}] = -P_{mem}[\text{m/s}](C_1 - C_2)[\text{kgmol/m}^3] \qquad (2.13)$$

where P_{mem} is the membrane permeability to the molecule (also known as the local mass transfer coefficient). Permeability takes into account the chemical nature of the permissive porous media. Permeability depends upon the diffusivity of the substrate and the effective thickness, d, through which the substrate must diffuse.

$$P_{mem}[\text{m/s}] = \frac{D_{AB}[\text{m}^2/\text{s}]}{\delta[\text{m}]}$$

Typically, effective thickness is assumed to be the thickness of the barrier. When a number of barriers are present in a series then the effective permeability can be obtained using

$$\frac{1}{P_{mem}} = \sum_{i=1 \text{ to } n} \frac{1}{P_{mem,i}} \qquad (2.14)$$

where $P_{mem,i}$ is the permeability for individual barrier.

EXAMPLE 2.7

A bioengineering company is evaluating a film (150 mm thick) made of polyethylene and nylon for two applications: (1) packaging a drug, which needs to avoid transfer of oxygen at 30°C, and (2) as membrane oxygenator in a heart-lung machine. Partial pressure of O_2 in air is 0.21 atm and the allowable inside pressure is 0.001 atm. Calculate the diffusion of O_2 at steady state. Which one is more suitable in each application? The permeability of oxygen at 30°C in polyethylene film is 4.17 × 10^{-12} m/s.atm and in nylon is 0.029 × 10^{-12} m/s.atm).

Solution: Assume that the resistances to diffusion outside the film and inside are negligible compared to the resistance of the film. Equation (2.14) is defined for concentrations, and to use it for this problem, pressure needs to be expressed in concentration. Assuming an ideal gas law, at a standard temperature (273K) and pressure (1 atm), 1 mol of gas occupies 22.414L.

$$J_{local} = \frac{P_{mem}(p_{A1}-p_{A2})}{22.414*d} = \frac{4.17*10^{-12}(0.21-0.001)}{22.414*0.00015} = 2.480 \times 10^{-10} kgmol / s.m^2$$

A nylon film has a much smaller permeability value for oxygen and would make a more suitable barrier. However, for a membrane oxygenator application, polyethylene would be more suitable.

Equation (2.14) assumes that the solute is easily soluble in the medium in which permeability is defined. However, the molecular transfer reduces in scenarios where solubility is limited. For example, concentration gradient exists for glucose between the extracellular fluid and intracellular fluid but the membranes do not permit the transfer of glucose from one compartment to the other. To cross the lipid bilayer of the cell membrane, a substance must enter (or partition) into the chemical matrix of the membrane. Lipids do not interact effectively with polar molecules, such as water or substances such as glucose that readily dissolve in water. Therefore, polar molecules experience significant resistance when entering the lipid bilayer. Consequently, even in the presence of a favorable chemical concentration gradient, the diffusion of glucose across a cell membrane is very slow. A general approach to include the effect of solubility in the permeability calculation is by incorporating a factor called the partition coefficient, φ, which is defined as the ratio of the solubility of the compound in lipids to that in water. Then

$$P_{mem}[m/s] = \frac{D_{AB}[m^2/s]}{\delta[m]} \varphi \tag{2.15}$$

2.3.2 Facilitated Diffusion

The lipid bilayer is a semipermeable barrier. A number of molecules cannot diffuse freely, despite the presence of electrochemical gradients, as the lipid bilayer is impermeable. One way through which some molecules are transported in and out of cells is by using specialized structures called transporters. This process is called facilitated diffusion and does not require an expenditure of energy.

Transporters are typically proteins, or assemblies of proteins, embedded in the lipid bilayer. Transporters are broadly categorized into two groups: carriers and channels. Carriers are highly specific proteins, which interact with the molecule to be transported at a specific sites called receptor sites. Those sites have a three-dimensional shape, permitting interaction with molecules that have a matching three-dimensional configuration. This arrangement provides specificity, similar to a key specific to a lock. Molecules with a shape that does not match a binding site cannot bind to the transporter and hence are not transported. Unlike free diffusion where transport of a molecule is linearly proportional to its concentration [Figure 2.6(a)], carrier mediated transport of a molecule is proportional to its concentration only when carriers are in excess. When all the carriers are bound to the molecule (or saturated), an increase in the molecule concentration does not alter the rate of transport. Consider glucose transport, which needs transporters to pass through the lip. Facilitated diffusion of glucose can be considered to occur in three steps [Figure 2.6(b)]:

1. Glucose in the extracellular environment binds to a carrier.
2. The binding causes the carrier to undergo a transformational change resulting in opening at the other end.
3. The glucose is released on the other side.

If all the carriers are engaged in transporting glucose, the presence of excess glucose does not alter the rate of glucose transport. Then the rate at which the transporter engages and disengages from a glucose molecule determines the rate of transport. Thus characteristics of the carrier influence the transfer rate, and understanding transporter characteristics is important in many therapeutic applications.

Facilitated diffusion of molecules also occurs through transmembrane proteins called channels, which form a hydrophilic pore through the membrane. These channels open or close (hence they are called gated) by external stimuli such as mechanical stress, changes in membrane potential, or chemicals. Channels are less selective than carriers and transport, and are not limited by the number of channels, unlike carriers. Many channels open or close in response to the binding of a small molecule (Figure 2.7), which is not the substance transported when the channel opens. These are called ligands and they can originate in extracellular environments or intracellularly based on the need of the cell. An example is the neuronal nicotinic Acetylcholine receptor (nAChR), which is widely distributed in the vertebrate brain. The nAChR binds to the ligand acetylcholine as well as the ligand nicotine (n). Binding of ligands causes a conformational change and opens the ion channel. These channels transport sodium, calcium, and potassium ions and propagate synaptic signals. Thus, the nAChRs mediate the rewarding properties of nicotine. In

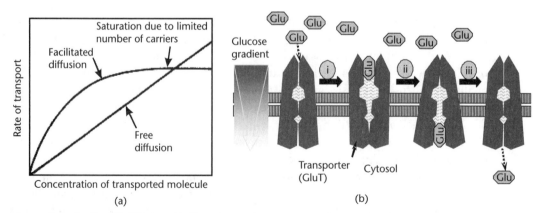

Figure 2.6 Facilitated diffusion: (a) alteration in the transport kidneys and (b) steps in glucose transport.

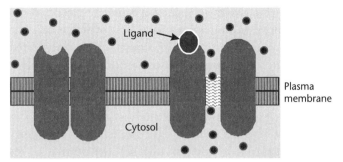

Figure 2.7   Transport through a ligand-gated channel.

addition, ligand-gated channels regulate several central functions such as memory and attention, sleep and wakefulness, anxiety, and pain. They are fast responders.

In excitable cells like neurons and muscle cells, some channels open or close in response to changes in membrane potential across the plasma membrane. These channels are referred as voltage-gated ion channels. Examples include the sodium and potassium voltage-gated channels of nerve and muscle, and the voltage-gated calcium channels. There are mechanically gated channels (example by sound waves) and light-gated channels like channelrhodopsin, which opens by the action of light. Voltage-gated channels are fast messengers.

2.3.3   Active Transport

The transport of molecules against their concentration gradient or electrical potential with the expenditure of energy is called active transport. The energy source can be ATP, electron transport, or light, and many molecules in the body are transported in the body by this mechanism of active transport. The use of energy to drive molecules implies that the transport system itself must somehow be able to harness energy in a particular form and use it to transport a solute against a concentration or electrical gradient. Based on how energy is harnessed, active transport is grouped into two main types, primary and secondary.

In primary transport, energy is directly coupled to movement of the desired substance across a membrane, independent of any other species. The free energy required for the transport process may be provided by high-energy phosphate bonds. The Na^+ ion gradient across the two sides of the plasma membrane is created by the active transport of the Na^+ ion out of the cell by the Na^+/K^+ ATPase [Figure 2.8(a)], the single most user of energy in the body. First, the binding of a phosphate group causes the transporter (also called pumps as they use energy) to undergo a transformational change, and the release of three Na^+ ions to the extracellular fluid. Further, two K^+ ions are transported into the cytosol when the phosphate group decouples from the transporter. An ATP-driven active transporter includes Na^+ ions, K^+ ions, Ca^{2+} ions, Vacuolor H^+ ATPases, and multidrug-resistant glycoproteins. Examples for light and electron flow are H^+, Na^+, or Cl^- transporters, and cytochrome oxidases of mitochondria.

In secondary active transport, the transport process is coupled to another transport process with a downhill concentration gradient. The transport of solutes across a membrane could occur in the opposite direction, which is called counter-transport (antiport) or in the same direction, which is called cotransport (symport). In countertransport [Figure 2.8(b)], one solute is transferred from a high concentration region to a low concentration region, which yields the energy to drive the transport of the other solute from a low concentration region to a high concentration region. An example for countertransport is the sodium-calcium exchanger or antiporter, which allows three sodium ions into the cell to transport one Ca^{2+} out. Many cells also posses a calcium ATPase, enzymes that can operate at low intracellular concentrations of calcium, and sets the normal or resting concentration of this important second messenger. But the ATPase exports calcium ions more slowly: only 30 per second versus 2,000 per second by the exchanger. The exchanger comes into service when the calcium concentration rises steeply and enables rapid recovery.

Cotransport [Figure 2.8(c)] uses the flow of one solute species from a high to low concentration to move another molecule. An example is the glucose symporter, which cotransports two Na^+ ions for every molecule of glucose it imports into the cell. Also, the active transport of small organic molecules such as amino acids is mediated by a Na^+ ion cotransport. Active transport can also transport in a group, which is called group translocation.

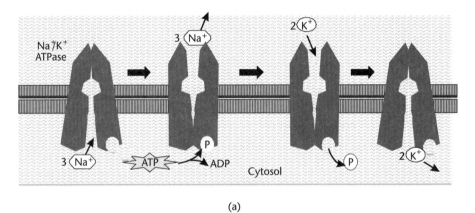

(a)

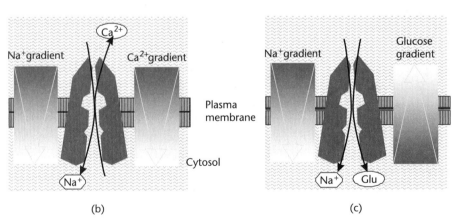

(b)   (c)

Figure 2.8 Active transport of molecules: (a) primary transport, (b) secondary countertransport, and (c) secondary cotransport.

EXAMPLE 2.8

A cell is in an environment where the outside Na^+ ion concentration is 140 mM and glucose concentraton is 0.005 mM. The inside of the cell has 10 mM of Na^+ ions and 5 mM of glucose. The potential difference across the membrane is −70 mV. The cellular transport system facilitates cotransport of Na^+ ions and glucose. For every glucose molecule intake, two Na^+ ions are imported. If both molecules need to be transported into the cell, will the energy released by the transfer of Na^+ ions be sufficient for the transport of glucose up its gradient?

From (2.1), $\Delta G = RT \ln \dfrac{C_{in}}{C_{out}} + zF\Delta V$

$$\Delta G = 8.314 * 310 * \ln \frac{10}{140} + (+1)(96,484.5)(-0.07) = -13,555.7 \text{ J/mole} = 13.6 \text{ kJ/mol}$$

For two molecules moved, energy released is 27.2 kJ.
For glucose, there is no electrical gradient. Hence,

$$\Delta G = RT \ln \frac{C_{in}}{C_{out}} = 8.314 * 310 * \ln \frac{5}{0.005} = 17,803.63 \text{ J/mole} = 17.8 \text{ kJ/mol}$$

Since one molecule of glucose is moved, the energy released by two Na^+ ions is sufficient to import glucose into the cell.

2.4 Osmosis

Osmosis is a special type of diffusion that involves the movement of water into an area of higher solute concentration. Consider a U tube (Figure 2.9) where two aqueous salt solutions, 10 mM (10 millimoles per liter) in limb B and 100 mM in limb A, are separated by a membrane selectively permeable (or semipermeable) to water only. Since salt molecules cannot cross the membrane, they stay in limb A but water molecules diffuse from the region of lower solute concentration to the region of higher solute concentration. Thus, the water would move from limb B to limb A in the diagram. From the perspective of water, its concentration is higher in the 10-mM salt solution than in the 100-mM salt solution. Therefore, water molecules would move down their concentration gradient. The pressure exerted by one solution upon another across the membrane is called osmotic pressure, π. From Figure 2.9, it can be seen that solution in limb A exerts a pressure across the semipermeable membrane. The number of particles formed by a given solute determines osmotic pressure. Each nondiffusing particle in the solution contributes the same amount to the osmotic pressure regardless of the size of the particle.

Unlike diffusion, osmosis is a reversible process (i.e., the direction of water flow through the membrane can be reversed by controlling the external pressure on the solution). In Figure 2.9, the osmotic pressure (P) can also be viewed as the

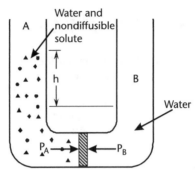

Figure 2.9 Process of osmosis.

hydrostatic pressure required to stop the osmotic flow. If the pressure is increased to more than the osmotic pressure, then the flow reverses. This is the principle used in reverse osmosis (hyperfiltration) filtration units.

The two solutions in Figure 2.9 are analogous to the intracellular and extracellular fluids. Water transfers across cell membranes within the body either from red cells to the blood plasma, or from within cells of the various tissues in the body (like muscles) to interstitial fluid (the fluid in between cells). The osmotic pressure difference between the interstitial and plasma fluids is due to the plasma proteins since the proteins do not readily pass through the capillary wall. The osmotic pressure created by the proteins is given the name of colloid osmotic pressure or oncotic pressure. For human plasma, the oncotic pressure is 28 mmHg, 19 mmHg caused by the plasma proteins and 9 mmHg caused by the cations within the plasma that are retained through electrostatic interaction with the negative surface charges of the proteins. Albumin is the most plentiful of the plasma proteins and globulins account for the majority of the remainder.

2.4.1 Osmolarity

Osmole is defined as one mole of a nondiffusing and nondissociating substance. One mole of a dissociating substance such as NaCl is equivalent to two osmoles. The number of osmoles per liter of solution is called osmolarity. (Osmolality is a measure of the osmoles of solute per kilogram of solvent.) A solution that is isosmotic (iso = "same") to another solution has the same concentration of solute particles. A solution that is hyperosmotic (hyper = "over") to another solution has a greater solute concentration, and a solution that is hyposmotic (hypo = "under") to another solution has a lower solute concentration. For physiological solutions, it is convenient to work in terms of milliosmoles (mOsm) or milliosmolar (mOsM).

Table 2.3 lists the compositions of the body fluids in milliosmolar. About 80% of the total osmolarity of the interstitial fluid and plasma is produced by sodium and chloride ions. The composition of these two fluids is very similar since the interstitial fluid arises from the filtration of plasma. The extracellular and intracellular fluids have the same osmotic pressure.

Changes in water volume in body fluids alter the osmolarity of body fluids, blood pressure, and interstitial fluid pressure. The sensation of thirst results from

Table 2.3   Composition of the Extracellular and Intracellular Fluids

Solute	Plasma (mOsm/L)	Interstitial (mOsm/L)	Intracellular (mOsm/L)
Na^+	142	139	14
K^+	4.2	4.0	140
Ca^{2+}	1.3	1.2	0
Mg^{2+}	0.8	0.7	20
Cl^-	108	108	10
HCO_3^-	24	28.3	10
$HPO_4^-, H_2PO_4^-$	2	2	11
SO_4^-	0.5	0.5	1
Phosphocreatine			45
Carnosine			14
Amino acids	2	2	8
Creatine	0.2	0.2	9
Lactate	1.2	1.2	1.5
Adenosine triphosphate			5
Hexose monophosphate			3.7
Glucose	5.6	5.6	
Protein	1.2	0.2	4
Urea	4	4	4
Others	4.8	3.9	10
Total (mOsmole/L)	301.8	300.8	301.2
Corrected osmolar activity (mOsmole/L)	282.5	281.3	281.3
Total π (mmHg)	5,443	5,443	5,443

an increase in osmolarity of the extracellular fluids and from a reduction in plasma volume. Cells within the hypothalamus can detect an increased extracellular fluid osmolarity and initiate activity in neural circuits that results in a conscious sensation of thirst. Baroreceptors, specially adapted groups of nerve fibers within the walls of a few blood vessels and the heart, can also influence the sensation of thirst. When they detect a decrease in blood pressure, signals are sent to the central nervous system along sensory neurons to influence the sensation of thirst. Low blood pressure associated with hemorrhagic shock, for example, is correlated with an intense sensation of thirst.

EXAMPLE 2.9

0.9% NaCl, commonly called saline solution, is used clinically. What is the osmolarity of this solution?

Solution: NaCl dissociates into Na^+ and Cl^- ions. 0.9% NaCl solution contains 0.9 grams NaCl per 100 mL of water or 9 grams per liter. MW of NaCl is 58.5. Thus, there is 9/58.5 or 0.154 mol or 154 mmol/L of NaCl of solution. Hence, osmolarity is $154 \times 2 = 308$ mOsM.

2.4.2 Tonicity

It refers to the effect of a solution on the shape of a cell. A solution that does not change the shape of a cell is said to be isotonic. In an isotonic solution, the cell is in equilibrium (i.e., the movement of water and solutes into the cell are equal to the movement of water and solutes out of the cell). Thus the cell remains the same size. A cell placed in a hypotonic solution will swell due to water moving into the cell. Alternatively, if a cell is placed in a hypertonic solution, the cell will shrink due to water osmotically moving out.

To understand the difference between tonicity and osmolarity, consider the intracellular fluid of an animal cell, which has an osmolarity of 290 mOsm. Animal cells typically contain no urea although their cell membranes are permeable to urea. When an animal cell is placed in a 290 mOsm urea solution, urea moves into the cell. Presence of urea inside causes the osmotic pressure to increase, which in turn helps more water to get into the cell by osmosis. Increased water and urea content within the cell causes the cell to swell (and burst). Thus an isosmotic urea solution is hypotonic to animal cells. Inside cells, lysosomes contain osmotically active material and rapidly swell and break if suspended in hypotonic solutions. Lysosomes suspended in an isoosmotic solution of a nonpermeant solute are in osmotic balance and are stable.

Large volumes of a solution of 5% human albumin are injected into patients undergoing a procedure called plasmapheresis. The albumin is dissolved in physiological saline (0.9% NaCl) and is therefore isotonic to human plasma. If 5% solutions are unavailable, pharmacists may substitute a proper dilution of a 25% albumin solution. Mixing 1 part of the 25% solution with 4 parts of a diluent results in the correct 5% solution of albumin. If the diluent used is sterile water, not physiological saline, the resulting solution is strongly hypotonic to human plasma. The result is a massive, life-threatening hemolysis in the patients.

Sports drinks can also be grouped based on the tonicity. Isotonic drinks (e.g., Gatorade) quickly replace fluids lost by sweating and supply a boost of carbohydrates. This drink is typically made for the average athlete. Hypotonic drinks quickly replace fluid lost, and are best for low-perspiration athletes such as jockeys and gymnasts. Hypertonic drinks (e.g., Powerade) supplement daily carbohydrate intake normally after exercise to increase glycogen stores. This drink is used primarily for athletes such as long distance runners who need the extra carbohydrates and electrolytes.

2.4.3 Osmotic Pressure

The pressure difference between regions A and B at equilibrium is the osmotic pressure (π) of region A. The expression for calculating π is derived using thermodynamic principles. Consider the equilibrium state of liquid mixtures in the two regions in Figure 2.9. Water diffuses from region B into region A until the free energy per mole of water on each side of the membrane is the same. For an ideal solution with very small solute concentration, it can be shown that

$$\pi = RT\Delta C \qquad\qquad (2.16)$$

where ΔC is the concentration difference of the solute across the semipermeable membrane (moles/liter). Equation (2.16) is called Van't Hoff's equation and is for a single solute. A common method of dealing with departures from ideal dilute solutions is to use the osmotic coefficient, ϕ. With the osmotic coefficient, the Van't Hoff equation is written as:

$$\pi = \phi R T \Delta C$$

where ϕ is the molal osmotic coefficient. For nonelectrolytes such as glucose, $\phi > 1$ at physiologi cal concentrations. For electrolytes, $\phi < 1$ due to the electrical interactions between the ions (e.g., NaCl at physiological concentrations has an osmotic coefficient of 0.93). For macromolecules, the deviation becomes much more dramatic; hemoglobin has an osmotic coefficient of 2.57, and that of the parvalbumin present in frog myoplasm is nearly 3.7.

EXAMPLE 2.10

Using the values given in Example 2.5, calculate the osmotic pressure due to each component. (These results are inaccurate because of electrical effects.)
Solution: For albumin

$$C = \frac{4.5g}{100 \text{ mL}} * \frac{1,000 \text{ mL}}{1L} * \frac{1 \text{ mol}}{75,000g} = 0.6 * 10^{-3} \text{mol/L}$$

Assuming Van't Hoff's law, $\pi = RT\Delta C$

$$= 0.082 (\text{L.atm/mol.}K) * (273 + 37)K * 0.6 * 10^{-3} \text{mol/L} = 0.01525 \text{ atm}$$
$$\rightarrow 0.01525 \text{ atm} * \frac{760 \text{ mmHg}}{1 \text{ atm}} = 11.6 \text{ mmHg}$$

2.4.4 Osmometers

Monitoring osmolality is important for improving patient care by direct measurement of body fluid osmolarity, transfusion and infusion solutions, and kidney function. Further, osmotic pressure measurements are used to determine the molecular weights of proteins and other macromolecules. Van't Hoff's equation establishes a linear relationship between concentration and osmotic pressure, the slope of which is dependent upon temperature. If the dilute ideal solution contains N ideal solutes then

$$\pi = RT \sum_{i=1}^{N} C_{S,i} \tag{2.17}$$

Equation (2.17) can also be written in terms of the mass concentration ρ_S

$$\pi = RTC_S = \frac{RT\rho_S}{MW_S}$$

where ρ_S is the solute density in the solvent (ration of solute mass to solvent volume and MWS is the molecular weight of the solute. Equation (2.17) is rearranged to calculate the molecular weight of the solute. A number of osmometers are available to measure the osmotic pressures. A simple technique is using membrane osmometers shown in Figure 2.10. The pure solvent is separated from the solute-containing solution by a semipermeable membrane, which is tightly held between the two chambers. At equilibrium, hydrostatic pressure, which corresponds to osmotic pressure, is calculated using the rise in the liquid level (h); ρ is the solution density. Sensitivity of the membrane osmometer measurements are affected by the membrane asymmetry observed when both cells are filled with the same solvent but have a pressure difference between the two cells caused by membrane leakage, compression, solute contamination, or temperature gradients. Ballooning caused by pressure differentials is another problem and is detected by measuring the pressure changes as the solvent is added or removed from the solvent cell. Alternatively, vapor pressure osmometers are used either alone or in combination with membrane osmometers. However, a more robust method for determining the osmolarity of biological fluids is by measuring the depression in the freezing point between the sample solution and pure water.

2.5   Transport Across Membranes

Interestitial fluid regulates the shape of the vessel by maintain a constant pressure, while providing a suitable transport medium for nutrients and waste products between cells and capillary blood. Porous tissue membranes selectively allow various molecules to pass through them. Small molecules carried in the plasma pass into the interstitial fluid to supply nutrients to the cells, but blood cells and platelets are too large to pass through capillary walls. Similarly, substances (i.e., waste) pass through

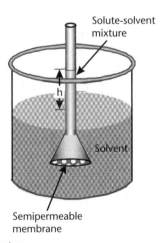

Figure 2.10 Membrane osmometer.

the pores in the capillary walls from the interstitial space. In 1896, British physiologist Ernest H. Starling described the regulation of fluid exchange across the blood capillaries to the tissue spaces. According to Starling, three local factors regulate the direction and rate of fluid transfer between plasma and tissue fluids at a location:

1. The hydrostatic pressures on each side of the capillary membranes;
2. The osmotic pressures of plasma and tissue fluids across the capillary membranes;
3. The physical properties of the capillary membranes.

Consider a blood vessel surrounded by interstial fluid (Figure 2.11). The normal hydrostatic pressure difference (capillary blood pressure, P_{inside}, minus the pressure of the fluid in the interstitial space, $P_{outside}$) across the capillary favors filtration of water out of the capillary into the interstitial space. However, the osmotic pressure difference across the capillary (osmotic pressure due to plasma proteins, π_{inside}, minus the osmotic pressure due to proteins in the interstial space, $\pi_{outside}$) favors the retention. Starling's hypothesis can be used to estimate whether water at any given point along the capillary will have a tendency to enter or leave the capillary. For example, the blood pressure at the arterial end (32 mmHg) is significantly higher than at the venous end (15 mmHg). The net pressure difference at the arterial end (+10 mmHg) favors transport of water out of the capillary where as the net pressure difference (–7 mmHg) at the venous end favors transport of water into the capillary. Understanding these factors is important in determining the water balance throughout the body, particularly in the formation of lung edema and kidney function. Fluid filtered from the vascular space into the interstitium then percolates through the interstitium to enter the initial lymphatics. When the net transcapillary filtration pressure or the permeability of the pulmonary microvascular endothelial barrier increases, such as in acute or chronic pulmonary hypertension or in acute respiratory distress syndrome (ARDS), an imbalance in the rate of fluid filtered into the pulmonary interstitium and the rate of fluid exiting the lymphatic system could occur. This results in the accumulation of fluid within the pulmonary interstitium and leads to a state called pulmonary edema.

The volumetric flow rate (J_v) of water across the membrane can be calculated using

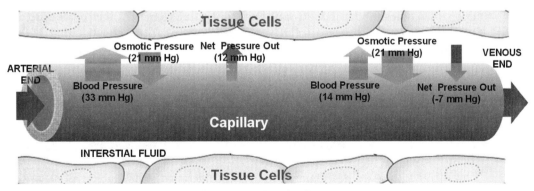

Figure 2.11   Starling forces regulating the transfer of molecules.

$$J_v\left[\mathrm{m}^3/\mathrm{s}\right]=L_p\left[\mathrm{m/s\cdot atm}\right]S\left[\mathrm{m}^2\right]\left(\left(P_{inside}-P_{outside}\right)-\sigma\left(\pi_{inside}-\pi_{outside}\right)\right)\left[\mathrm{atm}\right]\qquad(2.18)$$

where L_p is termed as the hydraulic conductance since water is the typical solvent in biological applications, S is the surface area of the membrane, and σ is the Staverman reflection coefficient, which is a measure of effectiveness of the membrane in keeping plasma proteins without transferring across the vascular bed. The capillary reflection coefficient for albumin is 0.9 and for glucose it is 0.025–0.05. The L_pS is also called the filtration coefficient and J_v in renal transport is referred as the glomarular filtration rate, GFR (see Chapter 10 for more details).

The term $\sigma(\pi_{inside}-\pi_{outside})$ is called the Starling oncotic force. If the membrane is impermeable, then the full oncotic pressure gradient between inside (typically plasma in a capillary or intracellular fluid) and outside (typically interstitial fluid or extracellular fluid) is experienced. This would result in a σ of 1. If, on the other hand, the membrane is completely permeable to all protein, then no gradient exists, and fluid leaks out entirely as expressed by the hydrostatic pressure gradient. This would result in a σ of 0. Equation (2.18) can be rearranged to understand the direction of solvent flow

$$J_v=L_pS\left[\left(P_{inside}-\sigma\pi_{inside}\right)-\left(P_{outside}-\sigma\pi_{outside}\right)\right]$$

When $P_{inside}-\sigma\pi_{inside}>P_{outside}-\sigma\pi_{inside}$, the flow is from the inside to the outside. Normal practice is to express the transport per unit area and estimate the osmotic pressure by the Van't Hoff equation. Thus

$$\frac{J_v}{S}=L_p\left[\left(P_{inside}-P_{outside}\right)-\sigma\left(RT\Delta C_{in-out}\right)\right]\qquad(2.19)$$

The transport of solute across a membrane occurs by two processes: diffusion and accompanied by the flow of the solvent. Using (2.13) for diffusive term, solute transport can be calculated using the equation

$$J_s\left[\mathrm{mol/s}\right]=P_{mem}S\Delta C+\left(1-\sigma_f\right)CJ_v\qquad(2.20)$$

where P_{mem} is the diffusive permeability, σ_f is the solvent drag or ultrafiltration coefficient, which describes the retardation of solutes due to membrane restriction, and C is the mean intramembrane solute concentration. Equations (2.18) and (2.19) are popularly referred to as Kedem–Katchalsky equations in literature. Although these equations are derived on the basis of the linear thermodynamics of irreversible processes, they have been broadly applied in studies on substance permeation across artificial and biological membranes.

EXAMPLE 2.11

A biomedical device development company is interested in developing a cellulose acetate asymmetric membrane for use in peritoneal dialysis. The pure water flux through the membrane at 37°C is determined to be 75 μL/min when the pressure difference is 10 cmH$_2$O. Since the osmotic pressure is important, performing an experiment using a 1% glucose solution was considered with P_{inside} = 200 cmH$_2$O and $P_{outside}$ = 20 cm H$_2$O at 37°C. Find J_v if the reflection coefficient for glucose is 0.1. If 100-μm-thick membranes are used with a surface area of 1 m^2, how much of glucose is lost? Assume that the solvent drag coefficient is zero and the hindrance coefficient is 1 (density of glucose = 1.54 g/cm^3).

Solution: From (2.18), we can determine the product L_pS from pure water flow data when $(\pi_{feed} - \pi_{permeate}) = 0$

$$L_pS = \frac{J_v}{P_{inside} - P_{outside}} = \frac{75[\mu L/min]}{10[cm\ H_2O]} = 7.5\,mL/min.cm\ H_2O$$

When the feed side is 1 wt% glucose solution, the osmotic pressure must be determined. Using the Van't Hoff's equation, the osmotic pressure on the feed side is given by

$$\pi = RT\,\Delta C_S$$

For this purpose, first the concentration of glucose needs to be determined. The density of glucose is 1.54 g/cm^3, and the density of water is 1 g/cm^3.

For 1,000g of 1 wt% glucose solution we have 990g of water and 10g of glucose. Since the molecular weight of glucose is 180.16, C_S is computed as

$$C_S = \frac{10/180.16}{990/1+10/1.54} = 5.57 \times 10^{-5}\,mol/cm^3$$

With the ideal gas constant R = 82.057 cm^3·atm/mol·K, the osmotic pressure for the feed side is

π = (82.057)(298.15)(5.57 × 10^{-5}) = 1.147 atm = 1.147 [atm]*760*1.36 [cmH$_2$O/atm] = 1,465.234 cmH$_2$O

The solvent flow rate is then determined

$$J_v = L_pS[(P_{inside} - P_{outside}) - \sigma(\pi_{inside} - \pi_{outside})]$$
$$J_v = 7.5[(200 - 20) - 0.1(1,465.234 - 0)] = 251\ mL/min = 0.251\ cm^3/min$$

From Table 2.2, the solvent-free diffusivity of glucose is 9.55 × 10^{-10} m^2/s

Assuming effective thickness to be the thickness of the membrane and the partition coefficient to be 1, P_{mem} can be calculated from (2.16)

$$P_{mem}\,[m/s] = \frac{9.55\times10^{-10}\,[m^2/s]}{1\times0^{-4}\,[m]} = 9.55\times10^{-6}\,[m/s] = 5.73\times10^{-2}\ cm/min$$

From (2.21),

$$J_S = 5.73\times10^{-2}\,[cm/min]\times1\times10^4\,[cm^2](5.57\times10^{-5} - 0)[mol/cm^3]$$
$$+(1-0)5.57\times10^{-5}\,[mol/cm^3]0.25[cm^3/min] = 0.032\ mol/cm$$

For transendothelial transport, σ_f is similar to the reflection coefficient σ. The permeability can be calculated by multiplying (2.16) with the number of pores present in the membrane. The transport equation can be applied for the transport of plasma between the capillary with subscript C and the surrounding interstitial fluid with subscript if.

$$J = L_p S\left[\left(P_c - P_{if}\right) - \left(\pi_p - \pi_{if}\right)\right]$$

(2.21)

Hydraulic conductance is a function of the membrane characteristics. Diffusion is also centered on the permeability of the matrix (i.e., membrane porous architecture). Since different tissue membranes in the body have different hydraulic conductances and perform different functions, modeling membrane characteristics is useful in understanding the fundamental differences between different membranes and developing interventional therapies. Although detailed discussion of this topic is beyond the scope of this book, the one-dimensional model for hydraulic conductance for a membrane containing a cylindrical pore can be described by

$$L_p = \frac{N\pi R^4}{8\,\mu L}$$

(2.22)

where N is the number of pores per unit surface area of the vessel wall, R is the pore radius, L is the thickness of the vessel wall or the depth of the cleft measured from the lumen to the tissue, and μ is the viscosity of the fluid. Equation (2.22) is obtained using the Poiseuille flow (described in Chapter 4) assumption.

If the pores are assumed to be rectangular with a total length L_{jt}, width W, and fractional opening f, then the hydraulic conductance can be written as

$$L_p = \frac{L_{jt}fW^3}{12\,\mu L}$$

(2.23)

The fiber matrix theory is used to describe the molecular transfer across membranes, as the membranes are made of fibers. The hydraulic conductivity of a fibrous membrane is written as

$$L_p = \frac{S_{fiber}}{L}\frac{K_p}{\mu}$$

(2.24)

where S_{fiber} is the area of fiber-filled pathway and K_p is the Darcy hydraulic conductance, which relates to the gap spacing between the fibers. However, these models have to be tested for each application and cannot be used in every scenario.

2.6   Transport of Macromolecules

Tissue membranes offer routes between blood and tissue for the diffusion of molecules, which are attracted to lipids (lipophilic) and small-molecular-weight molecules, which are attracted to water (hydrophilic). For example, liver cells secrete albumin, the principal blood plasma protein necessary for homeostasis. Everyday nearly 2% of plasma albumin is replaced. The diffusion of albumin (see Example 2.6) is significantly smaller than oxygen and other smaller molecules. Further, vascular walls are impermeable to macromolecules such as large proteins, polysaccharides, or polynucleotides. This impermeability of microvascular walls is a prerequisite for the establishment and maintenance of a fluid equilibrium between the intravascular and extravascular fluid compartments. On the other hand, a large number of other macromolecular substances must continually have access to the interstitial space to be ultimately returned to the plasma via the lymphatic system. This is of particular importance in inflammatory conditions and for immune response. Transport of protein across the alveolar epithelial barrier is a critical process in recovery from pulmonary edema and is also important in maintaining the alveolar milieu in the normal healthy lung.

One strategy used by cells to selectively transport macromolecules is called transcytosis, also referred to as vesicular transport. Membrane components such as lipids and proteins are continuously transported from the Golgi complex to the cell membrane through vesicles continuously. Transcytosis is of critical importance to a number of physiological processes, including bone formation (endochondral ossification) and wound healing. Fibroblasts in the connective tissue secrete large procollagen, which matures into collagen in the connective tissue, and transport using transcytosis. Albumin is transported in large quantities through transcytosis. The existence of transcytosis was first postulated by Palade in the 1950s. Transcytosis involves the sequential formation and fusion of membrane-bounded vesicles. Cells ingest molecules using a part of the cell membrane to form a vesicle. The substance to be ingested is progressively enclosed by a small portion of the plasma membrane, which first invaginates and then pinches off to form an intracellular vesicle containing the ingested material. The process is called endocytosis. The nature of the internalized constituents, either fluids or particles of different sizes, delineates the various internalization processes. Based on the engulfed particle size, endocytosis is referred as phagocytosis (cell eating), and pinocytosis (cell drinking).

Phagocytosis involves the entry of large particles, typically 1 mm or more, including particles as diverse as inert beads, apoptotic cells, and microbes. The early events leading to particle engulfment during phagocytosis is complex and involves different mechanisms. In general, cells invaginate their cell surface to form a phagosome and ingestion occurs in specialized regions, called caveolae, present within the plasma membrane. A vacuole is formed that contains the engulfed material, which is fused with the lysosome to be degraded or transported to other parts of the cells. Specialized cells (also referred as phagocytes) such as neutrophils, dendritic cells, and macrophages use phagocytosis as a mechanism to remove pathogens. Recognition of specific sites on the particle by cell-surface molecules of the phagocytes initiates the phagocytic process. Phagocytosis and the subsequent killing of microbes in phagosomes form the basis of an organism's innate defense against intracellular pathogens. Pinocytosis refers to engulfing smaller macromolecules from the extracellular environment. As in phagocytosis, a vesicle is formed, which contains the molecules that were brought into the cell. Vacuoles and vesicles produced by phagocytosis and pinocytosis can fuse with lysosomes, and the ingested molecule is degraded. In some cases, the degradation of pathogens in the phagosome produces peptides, which are presented by phagocytic cells to activate B-lymphocytes or T-lymphocytes. Phagocytosis and pinocytosis remove membrane from cell surface to form vacuoles that contain the engulfed material. The formed vacuole is then moved across the cell to eject the molecule through the other end of the cell membrane by the reverse process. Cells release macromolecules (exocytosis) and particles by a similar mechanism, but with the sequence reversed. An important feature of both processes is that the secreted or ingested macromolecules are sequestered in vesicles and do not mix with other macromolecules or organelles in the cytoplasm. Further, they fuse with the plasma membrane without intersecting other membrane pathways in the cell.

Cells have other strategies not involving membrane vesicles to selectively move across smaller cellular barriers using receptors. Further, paracellular transport, the movement between adjacent cells, is accomplished by regulation of tight junction permeability. Further details on these topics can be found in advanced texts or papers (see the Selected Bibliography).

Problems

2.1   The serum lactate concentration is 15 mg/dL. Express this concentration in mEq/L. The molecular weight of lactate is 89 and valency is 1.

2.2   Approximate the concentrations in mg/dL of Na^+, K^+, Cl^-, and HCO_3^- electrolytes in the extracellular fluid shown Figure 2.2.

2.3   Falk et al. [5] tested a new drug delivery device to deliver paracetamol. If the spherical radius is 0.36 nm, what is the diffusivity of the molecule at 37°C in a solution of viscosity similar to water?

2.4   If a dry inspired gas mixture is made up of 2% CO_2, 21% O_2, and balance N_2, what is the partial pressure of O_2 for a barometric pressure of 730 mmHg. The inspired gas temperature is 21°C.

2.5   When air is inspired, it is warmed to 37°C and moisture is added to the point of saturation. Under these conditions, the water vapor exerts a partial pressure of 47 mmHg. What is the total pressure at the alveoli?

2.6   How much oxygen (in g/mL) is taken up into the lungs from the air at a temperature of 25°C, within 1 minute in a resting state? Assume that pO_2 in the air is 21 kPa and the volume of each breath is 0.5L. The frequency of breathing is 15 breaths/minute.

2.7   The CO_2 equivalent to 5 mmHg is released from the blood to the air at 37°C every breath in a resting state through respiration. Assume the volume of each breath is 0.5L and the frequency of breathing is 15 breaths/minute. How many gm/L of CO_2 are released in 1 minute?

2.8   AlcoMate is developing a new handheld breath analyzer for detection of alcohol employing a newly available microchip oxide sensor to determine equivalent blood alcohol concentration (BAC). The detection range of the model Prestige AL 6000 is 0.00% to 0.40% in increments of .01% from a measured breath sample with a sensor accuracy rate of +/− 0.01% at 0.10%. The system works by detecting the alcohol in the breath and correlating that to alcohol in the blood using Henry's law constant. Derive a relationship that relates the amount of alcohol in the breath to the blood alcohol concentration (BAC in mg/100 mL). These are designed assuming the breath temperature to be 34°C. Typically plasma has 94% water and red blood cells 72% water. On average, blood consists of 44% hematocrit (amount of red blood cells). Henry's law constant at 34°C is 25.75 atm.

AlcoMate Protege. (Courtesy of AK Solutions USA, LLC, Palisades Park, New Jersey.)

(a) If a person is said to have 0.08% BAC (g ethanol per 100 mL of blood), what should the amount of alcohol in the breath be?

(b) If the person lives in a higher altitude and the hematocryt content is 54%, what BAC (g ethanol per 100 mL of blood) reading will she get for the same breath concentration?

2.9   Sevoflurane (MW = 200) is claimed to be a more potent inhalation anesthetic with less side effects approved for clinical use relative to isoflurane (MW = 185). To compare the effectiveness of the two drugs, Wissing et al. [6] administered sevoflurane to patients for the first 10 minutes at 2 L/min with 2.5% volume fraction. The rest of the gaseous mixture had nitrous oxide, which does not interact with both the anesthetics. The vapor mixture is at 25°C and 1 atm.

(a) Calculate the partial pressure of the anesthetic in the vapor mixture and the total moles of anesthetic. (Hint: for gases, mole fraction = volume fraction = pressure fraction.)

(b) The Henry's law constant [7] for sevoflurane and isoflurane at 37°C are 4.7 kPa.dm^3/mol and 3.65 kPa.dm^3/mol, respectively. Which anesthetic content will be higher in the blood content after 10 minutes? Show your calculations. You could assume that the hematocrit is 44%, total blood volume is 5L, and no loss of the anaesthetic during that time period.

2.10   When food enters the stomach, it stimulates the production and secretion of hydrochloric acid for digestion, reducing the stomach pH from 4 to 2. What is the concentration of the acid (assuming all the pH is due to HCI) before and after the change in pH?

2.11   Proprionic acid is an antioxidant and is added as a food preservative to prevent contamination with molds. It is a weak organic acid with a pKa of 4.9 (HP $\Leftrightarrow$ H$^+$ + P$^-$). Shortly after adding 0.1 mM HP to a suspension of red blood cells, the compound distributes itself by passive diffusion to an equilibrium distribution across cell membranes. At the steady-state the intracellular pH is 7.0, and the extracellular pH is 7.4. The concentration of total proprionic acid ([HP] and [P$^-$]) in the cell should be: (a) less than that outside the cell; (b) equal to that outside the cell; and (c) more than that outside the cell

2.12   How much oxygen is transported at 37°C from the alveoli (pO$_2$ alveoli = 104 mmHg) into the pulmonary capillaries (pO$_2$ blood = 40 mmHg) in 1 minute? Leave the answer in mol/m^2. Henry's constant for O$_2$ in water is 51,118 L.mmHg/mol, the diffusivity of O$_2$ in water at 37°C is 2.5 × 10^{-9} m^2/s, and the "respiratory membrane" is 0.4 μm thick. (Hint: On the alveoli side, it is gas phase and on the capillary side, it is liquid phase.)

2.13   Each red blood cell contains 280 × 10^6 molecules of hemoglobin and each gram of hemoglobin can bind 1.34 mL of O$_2$ in arterial blood.

(a) What is the oxygen carrying capacity (in mL) of 100 mL of blood (40% of blood is made up of red blood cells)? Assume the volume of one red blood cell is 90 mm^3.

(b) The teaching assistant is choking on a pretzel, limiting his O$_2$ intake to 100 mL/min. Assume the body consumes 250 mL/min of O$_2$ and has 5L of arterial blood. How long before he runs out of oxygen?

2.14   The diffusion coefficient of oxygen in water is approximately 2 × 10^{-5}cm^{2}s^{-1}. If the rate of oxygen exchange across a 1-m^2 area of a 1-micrometer-thick membrane is 1 mol/s when the oxygen concentration on one side of the membrane is 10 mmol/l, then what is the concentration of oxygen on the other side of the membrane?

2.15   Repeat Example 2.4 for 2.0g per 100 mL of globulin (MW = 170,000, diameter = 110Å).

2.16   A group in a new biomedical company is interested in developing therapeutic molecules delivered to the brain. The drug has to pass from the blood through the barriers to the target sites. The measured permeability of sucrose (MW =

342) across endothelial cells and the basement membranes are 7.2×10^{-3} cm/min and 8.2×10^{-3} cm/min, respectively.

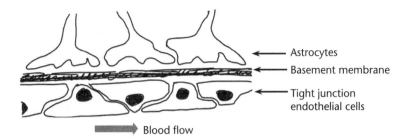

(a) Calculate the effective permeability for the system.
(b) If solvent free diffusions are used, what is the thickness of the basement membrane?

2.17   Administering corticosteroids such as hydrocortisone is an approach taken in few disease states. For delivering hydrocortisone across the epithelial barrier in the intestine, understanding the permeability characteristics of intestine is important. The net permeability of epithelial cells and the basement membrane is found to be 2.7 cm/s. If the permeability of epithelial cells is 4.2 cm/s, what is the permeability of the basement membrane?

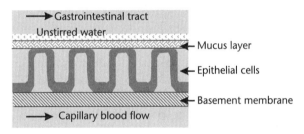

2.18   What is the osmolarity of a 0.45% NaCl in 2.5% dextrose (MW = 180) solution? How many mEq/L of sodium are present? How many mEq/L of chloride are present?

2.19   (a) A report from the endocrinology laboratory indicates that a patient has a serum ionized calcium concentration of 1 mmol/L. Express this concentration in mEq/L and in mg/dL.

(b) The serum calcium concentration is expressed as mg/dL on the biochemical profile. Express a 10-mg/dL serum calcium concentration in mmol/L and in mEq/L.

2.20   How many grams of glucose are required to make a solution that has an osmolarity of 1 Osm/L or 1,000 mOsm/L? What is the osmolarity of a 5% dextrose solution?

2.21   (a) How many mmol/L of sodium are present in a 0.9% NaCl solution? How many mEq/L Na^+ ions are present in this solution? What is the osmolarity of this solution?

(b) How many milligrams of $MgCl_2$ are needed to supply 1 mEq of Mg^{+2}?

(c) How many mg/mL of calcium are in a 10% calcium chloride solution? How many mmol/mL and mEq/mL of calcium and chlorides are in the same solution? What is the osmolarity?

2.22   You have a 25% solution of magnesium sulfate. How many mEq/mL of magnesium are in this solution? How many mEq/mL of sulfate?

You want to administer 0.75 mEq/kg/day of magnesium to a 10-kg magnesium-depleted patient with arrhythmias. The patient is already receiving 0.45% NaCl in 2.5% dextrose at a rate of 40 mL/kg/day. How many mL of the magnesium sulfate solution do you need to add to 500 mL of 0.45% NaCl in 2.5% dextrose to achieve this rate of magnesium infusion?

2.23   Plasmalyte has the following composition: 140-mEq/L sodium, 98-mEq/L chloride, 5-mEq/L potassium, 3.0-mEq/L magnesium, 27-mEq/L acetate, and 23-mEq/L gluconate. What is the osmolarity of plasmalyte? How many mg/dL magnesium does it contain?

2.24   Hypernatremia in elderly or comatose patients results due to loss of water or accumulation of NaCl. Manifestations include thirst, dry tongue, restlessness, and weight changes. A method used to bring this condition to normalcy is to use 3% NaCl solution. How many mmol/L of sodium are present in a 3% NaCl solution? How many mEq/L Na^+ ions are present in this solution? What is the osmolarity of this solution? Why do you think the method works?

2.25   (a) Lactated Ringer's solution has the following composition: 130-mEq/L sodium, 109-mEq/L chloride, 4-mEq/L potassium, 2.7-mEq/L calcium, and 28-mEq/L lactate. What is the osmolarity of lactated Ringer's solution? How many mg/dL calcium does it contain?

(b) You add 20 mEq of KCl to 1L of lactated Ringer's solution and enough 50% dextrose to make a 2.5% dextrose solution. What is the potassium concentration of the resulting solution in mEq/L? The chloride concentration in mEq/L? The osmolarity? Ignore the volume contributed by the addition of the 50% dextrose solution.

(c) How many mL of 50% dextrose must you add to 1L of lactated Ringer's solution to make it a 5% dextrose solution? A 2.5% dextrose solution?

2.26   A treatment for a 60-kg hypokalemic patient (a serum potassium concentration of 1.6 mEq/L) is to receive a potassium infusion of 0.2-mEq potassium/kg/hr. The patient is receiving lactated Ringer's solution at 80 mL/kg/day. How much KCl do you need to add to 1L of lactated Ringer's solution to achieve this rate of potassium infusion? How many mEq/L of potassium are in the final solution? How many mEq/L of chloride are in the final solution? What is the osmolarity of the final solution?

2.27   A 10-kg patient is diagnosed with diabetic ketoacidosis, which is caused by an insufficient amount of insulin resulting in increased blood sugar and accumulation of organic acids and ketons. Since the patient has a reduced phosphate level, she has to receive 0.02 mmol/kg/hr of phosphate. The plan is to use a potassium phosphate solution that contains 236 mg K_2HPO_4 and 224 mg KH_2PO_4 per mL. The patient's rate of fluid infusion is 80 mL/kg/day.

(a) How many mL of the potassium phosphate solution is added to 500 mL of 0.9% NaCl to achieve this rate of phosphate supplementation? Assume that the phosphate remains as HPO_4-2 and $H2PO_4$-1.

(b) What rate of potassium supplementation will the patient be receiving (assuming that you have not added any other potassium to the fluids)?

2.28   If the osmotic pressure in human blood is 7.7 atm at 37°C, what is the solute concentration assuming that $\sigma = 1$? What would be the osmotic pressure at 4°C?

2.29   One of the dialysate buffers used in hemodialysis contains the following components:

Component	Na^+	Cl^-	Ca^{+2}	Acetate	K+	HCO_3	Mg^{+2}	Dextrose
Amount (Eq/L)	105	137	3	4	2	33	0.75	200

To make this buffer NaCl, $CaCl_2$, KCl, $MgCl_2$, acetic acid, dextrose, and $NaHCO_3$ are added. Determine how much salt should be added to obtain the final composition. What is the osmotic pressure of the solution?

2.30   An ideal semipermeable membrane is set up as shown in the figure. The membrane surface area is S and the cross-sectional area of the manometer tube is s. At $t = 0$, the height of fluid in the manometer is 0. The density of the fluid is ρ. Show that the fluid height rises to a final value with an exponential behavior. Find the final value and the time constant.

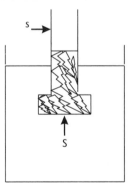

2.31   In some cases of trauma, the brain becomes very swollen and distended with fluid, a condition known as cerebral edema. To reduce swelling, mannitol may be injected into the bloodstream. This reduces the driving force of water in the blood, and fluid flows from the brain into the blood. If 0.01 mol 1^{-1} of mannitol is used, what will be the approximate osmotic pressure? If a 20% mannitol solution is available and the dosage is 0.5 g/kg to a 60-kg patient, how many mL of this solution do you need to administer?

2.32   What is the free energy needed to move glucose back from the tubular fluid to the blood when the concentration in the tubular fluid has dropped to 0.005 mM?

2.33   Calculate J_v (in mL.min^{-1}.100g tissue^{-1}) for the typical values given below. Extrapolate this to the whole body of a 70-kg person. What proportion of plasma volume turns over each day? Calculate lymph flow for the tissue and for the whole body, assuming a steady state (no swelling). Lp is 0.16×10^{-7} cm.sec^{-1}.mmHg^{-1} and the surface area is 100 cm^{2}g^{-1}. Capillary pressure is 20 mmHg, interstitial hydrostatic pressure is zero, plasma colloid osmotic pressure is 25 mmHg, interstitial oncotic pressure is 5 mmHg, oncotic reflection coefficient is 0.95, and hematocrit is 45%.

2.34   Consider the transport of glucose from capillary blood to exercising muscle tissue. The glucose consumption of the tissue is 0.01 mmol.s^{-1}g^{-1}. Blood flow to the region is 0.01 mL.s^{-1}g^{-1}. The arterial glucose concentration is 5 μmol. cm^{-3}.

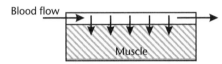

(a) Calculate the glucose concentration in the exiting blood.
(b) If the product of glucose permeability and surface area is 0.0033 cm^3 s^{-1}, then what is the glucose concentration in the tissue space? You can use 1g of tissue as the basis.

2.35   The hydraulic permeability of the human red cell is 1.2×10^{-11} cm^3/dyne/sec. At time = 0, the impermeant concentration inside the cell is suddenly made 1-molar greater than the outside concentration. If the red cell surface area is 150 square microns and the initial cell volume is 65 μm^3, how long (in seconds) will it take for the cell volume to increase to twice its initial size? Assume that the internal concentration does not change as the cell swells.

2.36   Calculate the lymph flow in mL/hr given that the colloid osmotic pressure of the interstitial fluid is 20 mmHg, the colloid osmotic pressure of the capillary is 28 mmHg, the hydrostatic pressure difference is 26 mmHg, K is 2 mL/ hr.mmHg, and sigma is 2. (Answer: 20 mL/hr.) After an endotoxin shock, the flow rate increased to five times and lead to protein-rich pulmonary edema. What do you think happened?

2.37   A 55-kg healthy male has 290 mOsm/L extracellular osmolality, 22.5L intracellular fluid volume, and 11.3L extracellular fluid volume. By accident, he becomes the victim of severe burns and loses 2.5L of water.
(a) Calculate the new extracellular fluid osmolality following the water loss.
(b) Following total restitution of the water compartments the patient undergoes surgery with skin grafts. During the long procedure he receives sufficient water by glucose infusion, but he loses 900 mOsmol NaCl. Calculate the new osmolality.

2.38   You are interested in developing a new membrane oxygenator for a heart-lung machine. The goal is to transfer 200 mL/min of oxygen into the blood flowing at an average flow rate of 5 L/min. Assume that the blood enters the oxygenator at venous pressure of 40 mmHg. An approach you are considering is to stack 10 cm by 10 cm by 5-μm-thick membranes. O$_2$-containing channels are filled with pure O$_2$.

(a) Calculate the oxygen concentration in the arterial blood leaving the oxygenator.

(b) How many membrane units are required [8]?

2.39   The lung capillary pressure varies between 2 to 12 mmHg and the lung interstitial pressure varies from –7 to 1 mmHg. The plasma oncotic pressure varies from 20 to 35 mmHg and interstitial oncotic pressure varies from 5 to 18 mmHg. If two liters per day of water is transferred, what is the coefficient of filtration when (a) the reflection coefficient is 1 and (b) the reflection coefficient is 0.9?

2.40   In Example 2.11, if 1% (by wt) albumin is used instead of glucose, what is the change in the solvent transfer? For albumin, the reflection coefficient is 0.9.

References

[1]   Hines, A. L., and R. N. Maddox, *Mass Transfer Fundamentals and Applications,* Upper Saddle River, NJ: Prentice-Hall, 1985.

[2]   Collins, M. C., and W. F. Ramirez, "Mass Transport Through Polymetric Membranes," *J. Phys. Chem.,* Vol. 83, 1979, pp. 2294–2301.

[3]   Thorne, R. G., S. Hrabetova, and C. Nicholson, "Diffusion of Epidermal Growth Factor in Rat Brain Extracellular Space Measured by Integrated Optical Imaging," *J. Neurophysiol.,* Vol. 92, No. 6, 2004, pp. 3471–3481.

[4]   Stroh, M., et al., "Diffusion of Nerve Growth Factor in Rat Striatum as Determined by Multiphoton Microscopy," *Biophys. J.,* Vol. 85, No. 1, July 2003, pp. 581–588.

[5]   Falk, B., S. Garramone, and S. Shivkumar, "Diffusion Coefficient of Paracetamol in a Chitosin Hydrogel," *Materials Letter,* Vol. 58, No. 26, 2004, pp. 3261–3265.

[6]   Wissing, H., "Pharmacokinetics of Inhaled Anaesthetics in a Clinical Setting: Comparison of Desflurane, Isoflurane and Sevoflurane," *British Journal of Anaesthesia,* Vol. 84, No. 4, 2000, pp. 443–449.

[7]   Floate, S., and C. E. W. Hahn, "Electrochemical Reduction of the Anaesthetic Agent Sevoflurane (Fluromethyl 2,2,2-Trifluoro-1-[Trifluoromethyl] Ethyl Ether) in the Presence of Oxygen and Nitrous Oxide," *Sensors and Actuators B, Chemical,* Vol. 99, No. 2–3, 2004, pp. 236–252.

[8]   Ethier, C. R., and C. A. Simmons, *Introduction to Biomechanics: From Cells to Organisms,* Cambridge, U.K.: Cambridge University Press, 2007.

Selected Bibliography

Amidon, G. L., et al., "Quantitative Approaches to Delineate Passive Transport Mechanisms in Cell Culture Monolayers," *Transport Processes in Pharmaceutical Systems,* 2000, pp. 219–316.

Bird, R. B., E. N. Lightfoot, and W. E. Stewart, *Transport Phenomena,* 2nd ed., New York: John Wiley & Sons, 2006.

Chung, S. H., O. S. Andersen, and V. Krishnamurthy, (eds.), *Biological Membrane Ion Channels: Dynamics, Structure, and Applications,* New York: Springer, 2006.

Crank, J., *The Mathematics of Diffusion,* 2nd ed., Oxford, U.K.: Clarendon Press, 1979.

Deen, W. M., *Analysis of Transport Phenomena,* Oxford, U.K.: Oxford University Press, 1998.

Fournier, R. L., *Basic Transport Phenomena in Biomedical Engineering,* 2nd ed., London, U.K.: Taylor and Francis, 2006.

Rechkemmer, G., "Transport of Weak Electrolytes," in Field, M., and R. A. Frizzell, (eds.), *Handbook of Physiology, Sect. 6: The Gastrointestinal System, Vol. IV, Intestinal Absorption and Secretion*, Bethesda, MD: American Physiology Society, 1991, pp. 371–388.

Sten-Knudsen, O., *Biological Membranes: Theory of Transport, Potentials and Electric Impulses*, New York: Cambridge University Press, 2002.

Truskey, G. A., F. Yuan, and D. F. Katz, *Transport Phenomena in Biological Systems*, 2nd ed., Upper Saddle River, NJ: Prentice-Hall, 2009.

CHAPTER 3

Bioelectrical Phenomena

3.1 Overview

As described in the previous chapter, there is a difference in the composition of electrolytes in the intracellular and extracellular fluids. This separation of charges at the cell membrane and the movement of charges across the membrane are the source of electrical signals in the body. Exchange of ions through ion channels in the membrane is one of the primary signal mechanisms for cell communication and function. For example, normal function in the brain is a matter of the appropriate relay of electrical signals and subsequent neurotransmitter release. Through electrical signals, neurons communicate with each other and with organs in the body. Neurons and muscle cells were the first to be recognized as excitable in response to an electrical stimulus. Later, many other cells were found to be excitable in association with cell motion or extrusion of material from the cells, for example secretion of insulin from the pancreatic beta cells. A few uses of evaluating bioelectrical phenomena are summarized in Table 3.1.

It is important to understand the conductivity of electricity in the body and time-dependent changes in bioelectrical phenomena. Abnormal electrical activity in the tissue leads to many diseases including that in the heart, brain, skeletal muscles, and retina. Substantial progress in research and clinical practice in electrophysiology has occurred through the measurement of electrical events. Understanding that the body acts as a conductor of the electrical currents generated by a source such as the heart and spreads within the whole body independently of an electrical source position led to the possibility of placing electrodes on the body surface noninvasively (i.e., without surgical intervention) to measure electrical potentials. ECG has become a common diagnostic tool to monitor heart activity. Understanding the movement of ions, relating the ion transfer to the function of cells with knowledge about mechanisms of tissues and organ function, and the complex time-dependent changes in electrical signals, and propagation in excitable media is important for development of advanced computational tools and models. This chapter provides an introduction to these concepts.

Table 3.1 Uses of Bioelectrical Flow Analysis

At the overall body level	Body composition and body hydration by measuring electrical characteristics of biological tissue.
	Transcutaneous electrical nerve stimulation (TENS), a noninvasive electroanalgesia used in physiotherapy to control both acute and chronic pain arising from several conditions.
At the tissue level	Electrocardiogram (ECG or EKG), pacemakers, defibrillators.
	Electroencephalogram (EEG): records spontaneous neural activity in the brain and understand brain functions with time and spatial resolution.
	Electroneurogram (ENG): records neural activity at the periphery.
	Electromyogram (EMG): records activity in the muscle tissue.
	Electrogastrogram (EGG): records signals in the muscles of the stomach.
	Electroretinogram (ERG): detects retinal disorders.
At the cell level	Understanding cellular mechanisms in transport of various molecules, developing novel therapeutic molecules.

3.2 Membrane Potential

The bilayered cell membrane functions as a selectively permeable barrier separating the intracellular and extracellular fluid compartments. The difference in the ion concentration between the intracellular and extracellular compartments creates an electrical potential difference across the membrane, which is essential to cell survival and function. The electrical potential difference between the outside and inside of a cell across the plasma membrane is called the membrane potential ($\Delta\Phi_m$). The electric potential difference is expressed in units of volts (named after Alessandro Volta, an Italian physicist), which is sometimes referred to as the voltage. Under resting conditions, all cells have an internal electrical potential (Φ_i), usually negative to outside (Φ_o) (Figure 3.1). This is because of a small excess of negative ions inside the cell and an excess of positive ions outside the cell. Convention is to define the internal electrical potential referenced to the external fluid, which is assumed to be at ground (or zero) potential. When the cell is not generating other electrical signals, the membrane potential is called the resting potential ($\Delta\Phi_{rest}$) and it is determined by two factors:

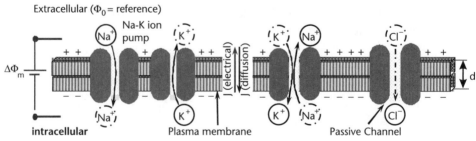

Figure 3.1 A part of the membrane of an excitable cell at rest with part of the surrounding intracellular and extracellular media. The direction of electrical and diffusional fluxes is opposing. In the figure, the assumption is that cell membrane is permeable to only K+ ions.

- Differences in specific ion concentrations between the intracellular and extracellular fluids;
- Differences in the membrane permeabilities for different ions, which is related to the number of open ion channels.

At rest, separation of charges and ionic concentrations across the membrane must be maintained for the resting potential to remain constant. At times, membrane potential changes in response to a change in the permeability of the membrane. This change in permeability occurs due to a stimulus, which may be in the form of a foreign chemical in the environment, a mechanical stimulus such as shear stress, or electrical pulses. For example, when the axon is stimulated by an electrical current, the Na^+ ion channels at that node open and allow the diffusion of Na^+ ions into the cell (Figure 3.1). This free diffusion of Na^+ ions is driven by the greater concentration of Na^+ ion outside the cell than inside [Figure 2.2(b)]. Initially, diffusion is aided by the electric potential difference across the membrane. As the Na^+ ion concentration inside the cell increases, the intracellular potential rises from −70 mV towards zero. A cell is said to be depolarized when the inside becomes more positive and it is said to be hyperpolarized when the inside becomes more negative. At the depolarized state, the Na^+ ion diffusion is driven only by the concentration gradient, and continues despite the opposing electric potential difference. When the potential reaches a threshold level (+35 mV), the K^+ ion channels open and permit the free diffusion of K^+ ions out of the cell, thereby lowering and ultimately reversing the net charge inside the cell. When the K^+ ion diffusion brings the intracellular potential back to its original value of resting potential (−70 mV), the diffusion channels close and the Na^+ and K^+ ion pumps turn on. Na^+/K^+ ion pumps are used to pump back Na^+ and K^+ ions in the opposite direction of their gradients to maintain the balance at the expense of energy. The Na^+ ion pump transports Na^+ ions out of the cell against the concentration gradient and the K pump transports K^+ ions into the cell, also against the concentration gradient. When the original K^+ ion and Na^+ ion imbalances are restored the pumps stop transporting the ions. This entire process is similar in all excitable cells with the variation in the threshold levels.

3.2.1   Nernst Equation

The movement of ions across the cell wall affects the membrane potential. To understand the changes in membrane potential, relating it to the ionic concentration change is necessary. Consider the movement of molecules across the membrane, which is driven by two factors: the concentration gradient and the electrical gradient (Figure 3.1). Analogous to the concentration flux discussed in Chapter 2, (2.8), electrical flux ($J_{electric}$) across a membrane with a potential gradient ($d\Delta\Phi$), in the x-dimension (dx) is written as

$$J_{electric} = -u\frac{z}{|z|}C\frac{d\Delta\Phi}{dx}$$

(3.1)

where z is the valency on each ion (positive for cations and negative for anions), $|z|$ is the magnitude of the valency, C is the concentration of ions [kgmol/m³], and u is the ionic mobility [m²/V.s]. Ionic mobility is the velocity attained by an ion moving through a gas under a unit electric field (discussed in Section 3.4.1), similar to the diffusivity constant described in (2.8). Ionic mobility is related to the diffusivity constant by

$$u = \frac{D_{AB}|z|F}{RT} \tag{3.2}$$

where F is the Faraday's constant, which is equal to the charge on an electron (in Coulombs), times the number of ions per mole given by the Avogadro's number (6.023×10^{23}). Substituting (3.2) into (3.1),

$$J_{electric} = \frac{-zFD_{AB}}{RT}C\frac{d\Delta\Phi}{dx} \tag{3.3}$$

In (3.3), product zFC is the amount of charge per unit volume or charge density [C/m³]. The total flux is given by the sum of electrical fluxes and the diffusional fluxes (3.8).

$$J_{total} = -D_{AB}\frac{dC}{dx} - \frac{zFD_{AB}}{RT}C\frac{d\Delta\Phi}{dx} \tag{3.4}$$

Equation (3.4) is known as the *Nernst-Planck* equation and describes the flux of the ion under the combined influence of the concentration gradient and an electric field. At equilibrium conditions, J_{total} is zero. Hence, (3.4) reduces to

$$D_{AB}\frac{dC}{dx} = -\frac{zFD_{AB}}{RT}C\frac{d\Delta\Phi}{dx} \tag{3.5}$$

Rearranging (3.5), and integrating between the limits $\Delta\Phi = \Delta\Phi_i$ when $C = C_i$ and $\Delta\Phi = \Delta\Phi_o$ when $C = C_o$ gives

$$\Delta\Phi_{rest}[V] = \Phi_i - \Phi_o = -\frac{R[J/mol.K]T[K]}{zF[C/mol]}\ln\frac{C_i[mM]}{C_o[mM]} \tag{3.6}$$

Equation (3.6) is called the *Nernst equation* (named after the German physical chemist Walther H. Nernst) and is used to calculate a potential for every ion in the cell knowing the concentration gradient. The subscript i refers to the inside of the cell and the subscript o refers to the outside. In addition, the ratio RT/zF is the ratio of thermal energy to electrical energy. RT/F is 26.71 mV at 37°C. Converting natural logarithm to logarithm of base 10, the constant is 61.5 mV at 37°C or 58 mV at 25°C.

EXAMPLE 3.1

What is the resting $\Delta\Phi$ for Na$^+$ ions, if the outside concentration is 140 mM and the inside concentration is 50 mM?

Solution: $\Delta\Phi_{Na+} = \Phi_i - \Phi_o = -\dfrac{RT}{zF}\ln\dfrac{C_i}{C_o} = -26.17\ln\dfrac{50}{140} = 26.95mV$

3.2.2   Donnan Equilibrium

Both sides of the cell membrane have two electrolyte solutions and the cell membrane is selectively permeable to some ions but blocks the passage of many ions. Thus, one charged component is physically restricted to one phase. This restriction can also result from the inherently immobile nature of one charged component, such as fixed proteins. In either case, an uneven distribution of the diffusible ions over the two phases develops, as their concentrations adjust to make their electrochemical potentials the same in each phase. In turn, this establishes an osmotic pressure difference and an electric potential difference between the phases. This type of ionic equilibrium is termed a Donnan equilibrium, named after the Irish physical chemist Frederick G. Donnan. According to the Donnan equilibrium principle, the product of the concentrations of diffusible ions on one side of the membrane equals that product of the concentration of the diffusible ions on the other side. Number of cations in any given volume must be equal to number of anions in the same volume (i.e., space charge neutrality—outside of a cell, net charge should be zero). In other words, to maintain equilibrium, an anion must cross in the same direction for each cation crossing the membrane in one direction, or vice versa. Assuming that a membrane is permeable to both K$^+$ and Cl$^-$ but not to a large cation or anion (Figure 3.2), the Nernst potentials for both K$^+$ and Cl$^-$ must be equal at equilibrium, that is,

$$-\frac{RT}{F}\ln\frac{C_{K,i}}{C_{K,0}} = \frac{RT}{F}\ln\frac{C_{Cl,i}}{C_{Cl,0}}$$

or

$$\frac{C_{K,0}}{C_{K,i}} = \frac{C_{Cl,i}}{C_{Cl,0}}$$

(3.7)

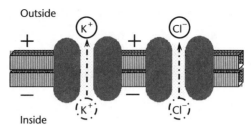

Figure 3.2 Space charge neutrality.

Equation (3.7) is a Donnan equilibrium. The rule is important because it contributes to the origin of resting potential. In general, when the cell membrane is at rest, the active and passive ion flows are balanced and a permanent potential exists across a membrane only if the membrane is impermeable to some ion(s) and an active pump is present. The Donnan equilibrium only applies to situations where ions are passively distributed (i.e., there are no metabolic pumps utilizing energy regulating ion concentration gradients across the membrane).

EXAMPLE 3.2

A cell contains 100 mM of KCl and 500 mM of protein chloride. The outside of the cell medium has 400 mM KCl. Assuming that the cell membrane is permeable to Cl^- and K^+ ions and impermeable to the protein, determine the equilibrium concentrations and the membrane potential. Also assume that these compounds can completely dissociate.

Solution: $C_{K,0} + C_{K,i} = 500, C_{Cl,0} + C_{Cl,i} = 1,000,$ and $C_{K,0} = C_{Cl,0}$

From (3.7) we have $\dfrac{C_{K,0}}{500 - C_{K,0}} = \dfrac{1,000 - C_{Cl,0}}{C_{Cl,0}}$

or

$C_{K,0} = C_{Cl,0} = 333$ mM, $C_{K,i} = 167$ mM, and $C_{Cl,i} = 667$ mM

Also,

$$\Phi_K = \Phi_{Cl} = 26.71 \ln \frac{333}{167} = 18.4 \text{ mV}$$

The Donnan equilibrium is useful in pH control in red blood cells (RBCs). When RBCs pass through capillaries in the tissue, CO_2 released from tissues and water diffuse freely into the red blood cells. They are converted to carbonic acid, which dissociates into hydrogen and bicarbonate ions, as described in Section 3.2.4. H^+ ions do not pass through cell membranes but CO_2 passes readily. This situation cannot be sustained as the osmolarity and cell size will rise with intracellular H^+ and bicarbonate ion concentration, and rupture the cell. To circumvent this situation, the bicarbonate ion diffuses out to the plasma in exchange for Cl^- ions due to the Donnan equilibrium. The discovery of this chloride shift phenomenon is attributed to Hamburger HJ, which is popularly called Hamburger's chloride shift. An ion exchange transporter protein in the cell membrane facilitates the chloride shift. A build up of H^+ ions in the red blood cell would also prevent further conversion and production of the bicarbonate ion. However, H^+ ions bind easily to reduced hemoglobin, which is made available when oxygen is released. Hence, free H^+ ions are removed from the solution. Reduced hemoglobin is less acidic than oxygenated hemoglobin. As a result of the shift of Cl^- ions into the red cell and the buffering of H^+ ions onto reduced hemoglobin, the intercellular osmolarity increases slightly and water enters causing the cell to swell. The reverse process occurs when the red blood cells pass through the lung.

3.2.3 Goldman Equation

Most biological components contain many types of ions including negatively charged proteins. However, the Nernst equation is derived for one ion after all permeable ions are in Donnan equilibrium. The membrane potential experimentally measured is often different than the Nernst potential for any given cell, due to the existence of other ions. An improved model is called the *Goldman equation* (also called the Goldman-Hodgkin-Katz equation), named after American scientist David E. Goldman, which quantitatively describes the relationship between membrane potential and permeable ions. According to the Goldman equation, the membrane potential is a compromise between various equilibrium potentials, each dependent on the membrane permeability and absolute ion concentration. Assuming a planar and infinite membrane of thickness (L) with a constant membrane potential ($\Delta\Phi_m$),

$$\frac{d\Phi}{dx} = \frac{\Delta\Phi_m}{L} \tag{3.8}$$

Substituting (3.8) into (3.4) and rearranging gives

$$dx = \frac{-dC}{\dfrac{J_{total}}{D_{AB}} + \dfrac{z\Delta\Phi_m F}{RTL}C} \tag{3.9}$$

Equation (3.9) is integrated across the membrane with the assumption that total flux is constant and $C = C_i$ when $x = 0$, $C = C_O$ when $x = L$. Furthermore, considering the solubility of the component using (3.14), and rearranging an equation to the total flux is obtained as:

$$J_{total} = \frac{-P_{mem}zF\Delta\Phi_m}{RT}\ln\left[\frac{C_i - C_o e^{\left(-\frac{zF\Delta\Phi_m}{RT}\right)}}{1 - e^{\left(-\frac{zF\Delta\Phi_m}{RT}\right)}}\right] \tag{3.10}$$

For multiple ions, the ion flux through the membrane at the resting state is zero. For example, consider the three major ions involved in the nerve cell stimulation, Na^+, K^+, and Cl^- ions. Then,

$$J_{total} = J_{K^+} + J_{Na^+} + J_{Cl^-} = 0$$

Rewriting (3.10) for each molecule and simplification provides

$$\Delta\Phi_m = \Delta\Phi_i - \Delta\Phi_o = -\frac{RT}{F}\ln\left(\frac{P_K[K^+]_i + P_{Na}[Na^+]_i + P_{Cl}[Cl^-]_o}{P_K[K^+]_o + P_{Na}[Na^+]_o + P_{Cl}[Cl^-]_i}\right) \qquad (3.11)$$

where P_{Na}, P_K, and P_{Cl} are the membrane permeabilities of Na^+, K^+, and Cl^- ions, respectively. If only one ion is involved then (3.11) reduces to (3.7).

EXAMPLE 3.3

A mammalian cell present in an extracellular fluid similar to Figure 3.2. The intracellular fluid composition is also similar to Figure 3.2. If the cell is permeable to Cl^-, K^+, and Na^+ ions with a permeability ratio of 0.45: 1.0: 0.04, what is the resting potential of the membrane?

Solution: From (3.11),

$$\Delta\Phi_m = \Delta\Phi_i - \Delta\Phi_o = -26.71*\ln\left(\frac{140+0.04*10+0.45*103}{4+0.04*142+0.45*4}\right) = -74.5 \text{ mV}$$

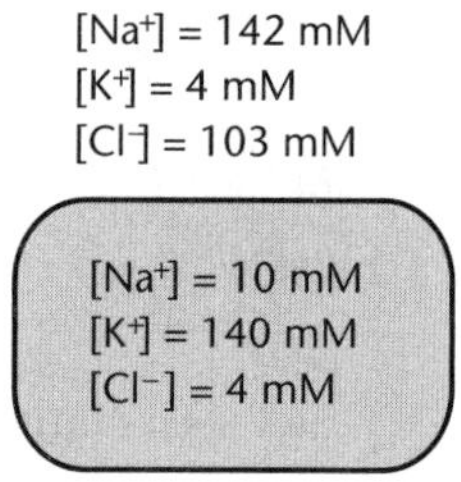

$[Na^+]$ = 142 mM
$[K^+]$ = 4 mM
$[Cl^-]$ = 103 mM

$[Na^+]$ = 10 mM
$[K^+]$ = 140 mM
$[Cl^-]$ = 4 mM

3.3   Electrical Equivalent Circuit

3.3.1   Cell Membrane Conductance

The flow of ions through the cell membrane is controlled by channels and pumps. Hence, the rate of flow of ions depends on the membrane permeability or conductance (G) to a species in addition to the driving force (i.e., membrane potential difference). Rate of flow of ions (or rate of charge of charge) is called the current and Ampere is a unit in SI units. The relationship between current I, potential difference, and conductance is written as

$$I = G\Delta\Phi \qquad (3.12)$$

where G is the conductance (siemens, S = A/V) of the circuit and $\Delta\Phi$ is the membrane potential (i.e., the voltage between inside and outside of the cell membrane). Equation (3.12) is called *Ohm's law*. The reciprocal of conductance is called resistance and a widely used form of Ohm's law in electrical engineering is:

$$I = \frac{\Delta\Phi}{R} \tag{3.13}$$

where R is the resistance. All substances have resistance to the flow of an electric current. For example, metal conductors such as copper have low resistance whereas insulators have very high resistances. Pure electrical resistance is the same when applying direct current and alternating current at any frequency. In the body, highly conductive (or low resistance electrical pathway) lean tissues contain large amounts of water and conducting electrolytes. Fat and bone, on the other hand, are poor conductors with low amounts of fluid and conducting electrolytes. These changes form the basis of measuring the body composition in a person.

When the cell membrane is "at rest," it is in the state of dynamic equilibrium where a current leak from the outside to inside is balanced by the current provided by pumps and ion channels so that the net current is zero. In this dynamic equilibrium state, ionic current flowing through an ion channel is written as

$$I_i = G(\Delta\Phi_m(t) - \Delta\phi_{rest,i}) \tag{3.14}$$

$\Delta\phi_{rest,i}$ is the equilibrium potential for the ith species, and $\Delta\phi_m(t) - \Delta\phi_{rest,i}$ is the net driving force.

EXAMPLE 3.4

If a muscle membrane has a single Na^+ ion channel conductance of 20 pS, find the current flowing through the channel when the membrane potential is −30 mV. The Nernst potential for the membrane is +100 mV.
   Solution:

$$I = g\left(\Delta\Phi_m(t) - \Delta\Phi_{rest}\right)$$
$$= \left(20 \times 10^{-12}\,S\right)\left(\left(-30 \times 10^{-3}\,V\right) - \left(100 \times 10^{-3}\,V\right)\right) = -2.6 \times 10^{-12}\,A = -2.6\ pA$$

3.3.2   Cell Membrane as a Capacitor

Intracellular and extracellular fluids are conducting electrolytes, but the lipid bilayer is electrically nonconductive. Thus the nonconductive layer is sandwiched between two conductive compartments. It can be treated as a *capacitor* consisting of two conducting plates separated by an insulator or nonconductive material known as a dielectric. Hence, the cell membrane is modeled as a capacitor, charged by the imbalance of several different ion species between the inside and outside of a cell. A dielectric material has the effect of increasing the capacitance of a vacuum-filled parallel plate capacitor when it is inserted between its plates. The factor K by which the capacitance is increased is called the dielectric constant of that material. K varies from material to material, as summarized in Table 3.2 for few materials.

Consider the simplest dielectric-filled parallel plate capacitor whose plates are of cross sectional area A and are spaced a distance d apart. The formula for the membrane capacitance, C_m, of a dielectric-filled parallel plate capacitor is

$$C_m = \frac{\varepsilon A}{d} \tag{3.15}$$

where $e \, (= K_{\varepsilon 0})$ is called the *permittivity* of the dielectric material between the plates. The permittivity ε of a dielectric material is always greater than the permittivity of a vacuum e_0. Thus, from (3.15), C_m increases with the dielectric constant surface area (space to store charge) and decreases with an increase in membrane thickness. The cell membrane is modeled as a parallel plate capacitor, with the capacitance proportional to the membrane area. Hence, measuring the membrane capacitance allows a direct detection of changes in the membrane area occurring during cellular processes such as exocytosis and endocytosis (discussed in Chapter 2). The spacing between the capacitor plates is the thickness of the lipid double layer, estimated to be 5 to 10 nm. Capacitors store charge of electrons for a period of time depending on the resistance of the dielectric. The amount of charge, Q_{ch}, a capacitor holds is determined by

$$Q_{ch} = C_m \Delta \Phi \tag{3.16}$$

The unit of capacitance is called a *farad* (F), equivalent to coulombs per volt. The typical value of capacitance for biological membranes is approximately 1 mF/cm^2. For a spherical cell with the diameter of 50 mm, the surface area is 7.854×10^{-5} cm^2. Then the total capacitance is 12.73 μF. The amount of reactants is proportional to the charge and available energy of capacitor (described in Section 3.4.1). The number of moles of ions that move across the membranes is calculated using the relation

$$Q_{ch} = nzF \tag{3.17}$$

where n is the number of moles of charge separated.

Table 3.2 Dielectric Constant of a Few Materials

Material	Vacuum	Air	Water	Paper	Pyrex	Membrane	Teflon
K	1	1.00059	80	3.5	4.5	2	2.1

EXAMPLE 3.5

If a cell membrane has a capacitance of 0.95 mF/cm², how many K⁺ ions need to cross the
membrane to obtain a potential of 100 mV?
    From (3.16),

$$Q_{ch} = C_m \Delta\Phi = 0.95 * 0.1 = 0.095 \, \mu C/cm^2$$

From (3.17),

$$n = \frac{Q_{ch}}{zF} = \frac{0.095 \times 10^{-6} (C/cm^2)}{(+1) * \dfrac{96,484}{6.023 \times 10^{23}} C/ions} = 5,930 \text{ ions per } \mu m^2 \text{ of membrane}$$

Capacitors oppose changes in voltage by drawing or supplying current as they
charge or discharge to the new voltage level. In other words, capacitors conduct
current (rate of change of charge) in proportion to the rate of voltage change. The
current that flows across a capacitor is

$$I_C = \frac{dQ_{ch}}{dt} = C_m \frac{d\Delta\Phi_m}{dt} \text{ or } \Delta\Phi_m = \frac{1}{C_m} \int_0^t i_c dt \qquad (3.18)$$

Ionic current described by (3.12) is different than capacitative current. In a
circuit containing a capacitor with a battery of $\Delta\Phi_{rest}$ [Figure 3.3(a)], the capacitor
charges linearly until the voltage reaches a new steady state voltage, $\Delta\Phi_{SS}$ [Figure
3.3(b)]. After reaching the new voltage, the driving force is lost. Since the voltage does not change, there is no capacitative current flow. The capacitance of the
membrane per unit length determines the amount of charge required to achieve a
certain potential and affects the time needed to reach the threshold. In other words,
slop of the line in Figure 3.3(b) is affected by capacitance. When the conductance
is increased by two, steady state voltage is reduced. However, the slope of the line
is much higher.
    Capacitive reactance (R_C) (generally referred to as reactance and expressed in
ohms, Ω) is the opposition to the instantaneous flow of electric current caused by
capacitance. Reactances pass more current for faster-changing voltages (as they

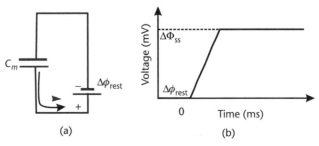

(a)   (b)

Figure 3.3   (a, b) Circuit with a capacitor.

charge and discharge to the same voltage peaks in less time), and less current for slower-changing voltages. In other words, reactance for any capacitor is inversely proportional to the frequency of the alternating current. In alternating current (AC) circuit, reactance is expressed by

$$R_C = \frac{1}{2\pi f C_m} \tag{3.19}$$

where f is the frequency of the alternating current in hertz. Equation (3.19) suggests that reactance is the reciprocal of frequency (i.e., reactance decreases as frequency increases). At very low frequencies reactance is nearly infinite. However, capacitance is independent of frequency and indirectly defines cell membrane volume. Ideally reactance is expressed in capacitance at a given frequency. Large capacitance values mean a slower rate of ion transfer.

3.3.3 Resistance-Capacitance Circuit

Based on the above discussion, the cell membrane has both resistive and capacitative properties. Then, a cell membrane can be represented by an equivalent circuit of a single resistor and capacitor in parallel (Figure 3.4) with a resting potential calculated using the Nernst equation for an individual ionic species. When multiple ionic species are involved in causing changes in electrical activity, resistance of each species is considered separately. It is conceived that cells and their supporting mechanisms are not a series of extracellular and intracellular components but reside in parallel with a small series effect from the nucleolus of the cells. Simplification of circuits with resistors in series or parallel by replacing them with equivalent resistors is possible using the *Kirchoff's laws* in electrical circuits. Two resistors in series are replaced by one whose resistance equals the sum of the original two resistances, since voltage and resistance are proportional. Reciprocals of resistances of two resistors in parallel are added to obtain the equivalent resistance since current and resistance are inversely proportional. Conversely, capacitor values are added normally when connected in parallel, but are added in reciprocal when connected in series, exactly the opposite of resistors.

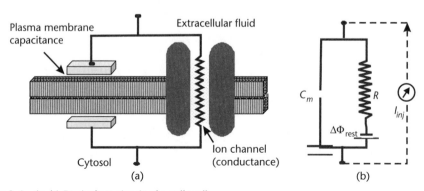

Figure 3.4 (a, b) Equivalent circuit of a cell wall.

To describe the behavior of an RC circuit, consider K^+ ion conductance in series with its Nernst potential ($\Delta\Phi_{rest}$). A positive stimulus current, I_{inj}, is injected from time t_0 to t. Then using Kirchoff's current rule, the equivalent circuit of a cell is written as

$$I_C + I_i - I_{inj} = 0$$

where I_C is current through the capacitor and I_i is the ionic current through the resistor. The injected current is subtracted from the other two. This is based on the convention that the current is positive when positive charge flows out of the cell membrane. Substituting (3.14) and (3.18),

$$C_m \frac{d\Delta\Phi_m}{dt} + \frac{\Delta\Phi_m - \Delta\Phi_{rest}}{R_m} - I_{inj} = 0 \qquad (3.20)$$

Due to injected current, the membrane reaches a new state potential, $\Delta\Phi_{SS}$. At steady state $d\,\Delta\Phi_m/dt = 0$, (3.20) reduces to

$$\Delta\Phi_{SS} = I_{inj}R_m + \Delta\Phi_{rest}$$

Rearranging and integrating with the limits $\Delta\Phi = \Delta\Phi_o$ when $\Delta t = \Delta t_o$ and $\Delta\Phi = \Delta\Phi$ when $\Delta t = \Delta t$

$$\ln\frac{\left(\Delta\Phi_{SS} - \Delta\Phi_0\right)}{\left(\Delta\Phi_{SS} - \Delta\Phi\right)} = \frac{t - t_0}{\tau}$$

where τ is the product of R_m and C_m and has the units of time. In generally, τ is called the time constant. An alternative form is

$$\Delta\Phi = \Delta\Phi_{SS} + \left(\Delta\Phi_0 - \Delta\Phi_{SS}\right)e^{\left(\frac{t_0 - t}{\tau}\right)} \qquad (3.21)$$

Equation (3.21) is used for both charging and discharging of an RC circuit for any ionic species. In general, the majority of experiments are conducted by injecting a current as a step input to understand the property of a cellular membrane with respect to a particular ionic species. A frequently measured property is τ to understand the resistance and capacitance properties of cells. At the instant when the current is first turned on ($t = 0$) the first term in square brackets $\Delta\Phi_{SS}$ is zero and $\Delta\Phi = \Delta\Phi_{rest}$. For $t \gg 0$, the second term is nearly zero and $\Delta\Phi = \Delta\Phi_{SS}$.

EXAMPLE 3.6

In a current step experiment, a 100-ms-long pulse of 1 nA of current is applied into a cell using a microelectrode and thereby drive the membrane potential of the cell from its resting level of −70 mV to a steady state level of +30 mV. The capacitance of the cell membrane is 0.5 nF. What is the membrane time constant? What is the membrane potential 50 ms after the end of the current pulse?

$$\Delta\Phi_0 = -70 \text{ mV}, \ \Delta\Phi_{SS} = +30 \text{ mV}$$

$$C = 0.5 \text{ nF} = 5 \times 10^{-10} \text{ F}$$

$$I = 1 \text{ nA} = 1 \times 10^{-9}[\text{A}]$$

$$R_m = \frac{\Delta\Phi_{SS} - \Delta\Phi_{rest}}{I_{inj}} = \frac{[30-(-70)]10^{-3}}{1 \times 10^{-9}} = 1 \times 10^8 \ \Omega$$

$$\tau = RC = (1 \times 10^8 \Omega) \times (5 \times 10^{-10}) = 0.05 \text{ second} = 50 \text{ ms}$$

When the current is turned off,

$$\Delta\Phi_0 = 30 \text{ mV and } \Delta\Phi_{SS} = -70 \text{ mV}$$

From (3.21),

$$\Delta\Phi = -70 + (30 - (-70)) e^{\left(\frac{-50}{50}\right)} = -33.22 \text{ mV}$$

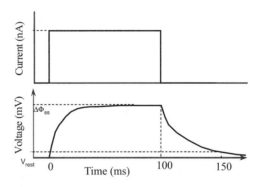

Using the equivalent circuit model, the Goldman equation can be linearized using conductances and individual ion potentials in the resting membrane. The conductance is related to membrane permeability but is not the same. $\Delta\Phi_{rest}$ is constant and there is no net current crossing in or out of the cell, although each individual conductance may be carrying a net current. So, in steady state, the total current must be zero $(I_{K^+} + I_{Na^+} + I_{Cl^-} = 0)$. Hence,

$$g_K \left(\Delta\Phi_m - \Delta\Phi_K (t) \right) + g_{Na} \left(\Delta\Phi_m - \Delta\Phi_{Na} (t) \right) + g_{CL} \left(\Delta\Phi_m - \Delta\Phi_{Cl} (t) \right) = 0$$

Solving for the membrane potential,

$$\Delta\Phi_m = \frac{g_K \Delta\Phi_K + g_{Na}\Delta\Phi_{Na} + g_{Cl}\Delta\Phi_{Cl}}{g_K + g_{Na} + g_{Cl}} \tag{3.22}$$

Since K^+, Na^+, and Cl^- ions are not at their equilibrium potentials, (3.22) gives a steady state potential and not an equilibrium state potential. There is a continuous flux of those ions at the resting membrane potential. Often the equilibrium potential is not computed explicitly but defined as an independent leakage potential $\Delta\Phi_L$ and determined by the resting potential. If the sum of current across the membrane is not zero, then charge builds up on one side of the membrane while changing the potential. Then, measuring the current flowing across a cell helps in determining the capacitative and the resistive properties if the membrane potential is known. Conductance properties of two ionic species can also be obtained in terms of the membrane potentials. Consider a cell with a transfer of K^+ ions and Na^+ ions. Then

$$g_K\left(\Delta\Phi_m - \Delta\Phi_K(t)\right) + g_{Na}\left(\Delta\Phi_m - \Delta\Phi_{Na}(t)\right) = 0$$

Rearranging this gives

$$\frac{g_{Na}}{g_K} = \frac{-\left(\Delta\Phi_m - \Delta\Phi_K\right)}{\left(\Delta\Phi_m - \Delta\Phi_{Na}\right)} \tag{3.23}$$

Thus, in steady-state, the conductance ratio of the ions is equal to the inverse ratio of the driving forces for each ion.

EXAMPLE 3.7

Electrical activity of a neuron in culture was recorded with intracellular electrodes. The experiments were carried out at 37°C. The normal saline bathing the cell contained 145-mM Na^+ ions and 5.6-mM K^+ ions. In this normal saline, the membrane potential was measured to be −46 mV, but if the Na was replaced with an impermeant ion (i.e., the cell was bathed in a Na-free solution) the membrane potential increased to −60 mV. In the normal saline the E_{Na} was +35 mV, so that the membrane is impermeable to Cl^-. To simplify the issue you can ignore Na efflux in the Na-free saline condition. What ratio of g_{Na} to g_K would account for the normal membrane potential?
Solution: From (3.23),

$$\frac{g_{Na}}{g_K} = \frac{(-46 - (-60))}{(-46 - (+35))} = 0.17$$

3.3.4 Action Potential

The transport of ions causes cells to change their polarization state. Many times a small depolarization may not create a stimulus. Typically, cells are characterized by threshold behavior (i.e., if the signal is below the threshold level, no action is observed). The intensity of the stimulus must be above a certain threshold to trigger an action. The smallest current adequate to initiate activation of a cell is called the *rheobasic current* or *rheobase*. The threshold level can be reached either by a very high strength signal for a short duration or a low strength signal applied for a long duration. The membrane potential may reach the threshold level by a short, strong stimulus or a longer, weaker stimulus. Theoretically, the rheobasic current needs an infinite duration to trigger activation. The time needed to excite the cell with twice rheobase current is called *chronaxy*. If the membrane potential is given a sufficiently large constant stimulus, the cell is made to signal a response. This localized change in polarization, which is later reset to the original polarization, is called an *action potential* [Figure 3.5(a)]. This large local change in polarization triggers the same reaction in the neighborhood, which allows the reaction to propagate along the cell.

The transfer of Na^+ and K^+ ions across the membrane are primarily responsible for the functioning of a neuron. When the neuron is in a resting state, the cell membrane permeabilities are very small as many of the Na^+ and K^+ ion channels are closed; the cell interior is negatively charged (at -70 mV) relative to the exterior. When a stimulus is applied that depolarizes a cell, the cell surface becomes

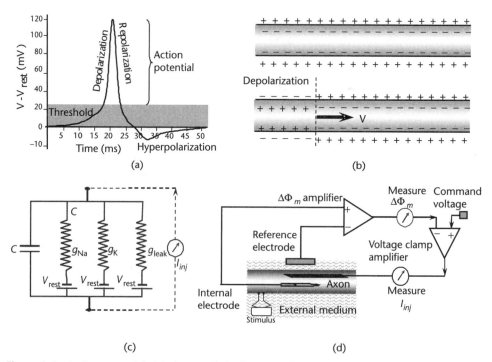

Figure 3.5 Action potential: (a) characteristics features of an action potential curve; (b) changes in charge distribution during depolarization; (c) Hodgkin-Huxley equivalent circuit for a squid axon; and (d) voltage clamp measurement circuit.

permeable to Na$^+$ ions, which rush into the cell. This results in a reversal of polarization at the points where Na$^+$ ions entered the cell [Figure 3.5(b)]. The activation of the ion channels in turn accelerates the depolarization process, producing the rising phase of the action potential. The rise of the membrane potential ultimately triggers the process of ion channel inactivation, which prevents further membrane depolarization. At the same time, the voltage-dependent K$^+$ ion channels are activated, repolarizing the cell and producing the falling phase of the action potential. During depolarization, K$^+$ ion channels become permeable to K$^+$ ions, which rush outside through potassium channels. Because the K$^+$ ions move from the inside to outside of the cell, the polarization at the surface of the cell is again reversed, and is now below the polarization of the resting cell. Since the pumps are slow, the excess Na$^+$ ion concentration inside the cell can diffuse along the length of the axon. This build up of positive charge inside the next node causes the potential at that node to rise. K$^+$ ions also diffuse in the opposite direction, partially canceling the effect of the Na$^+$ ion diffusion. However, as a K atom is nearly twice as heavy as an Na atom, the Na diffusion is faster, producing a net current. When the potential at the next node exceeds the threshold (somewhere between –70 mV and zero), the Na$^+$ ion channel is stimulated to open. At this point, the process described above repeats itself at the next node. Thus the electrochemical nerve impulse jumps from node to node.

An RC circuit model with single ionic species [Figure 3.4(b)] is insufficient to describe all the changes in the membrane potential. To develop a model for the action potential, British biophysicist Alan L. Hodgkin and British physiologist Andrew F. Huxley studied the changes in the membrane potentials of a squid giant axon. They extended the RC equivalent circuit to include gated conductances for Na$^+$ and K$^+$ ion species and all other conductances lumped as leakage [Figure 3.5(c)]. Each of these conductances is associated with an equilibrium potential, represented by a battery in series with the conductance. The net current, which flows into the cell through these channels, has the effect of charging the membrane capacitance, giving the interior of the cell a membrane potential $\Delta\Phi_m$ relative to the exterior. When a current I_{inj} is injected into the system, the equivalent circuit of the cell is written using Kirchoff's law as

$$C_m \frac{d\Delta\Phi_m}{dt} + I_{Na} + I_K + I_{Leak} - I_{inj} = 0$$

where I_{Leak} is the leakage current required to maintain the constant resting potential in the absence of any depolarization. Substituting Ohm's law for currents due to individual ions, we get

$$C_m \frac{d\Delta\Phi_m}{dt} = I_{inj} - (\Delta\Phi_m - \Delta\Phi_{Na})g_{Na} - (\Delta\Phi_m - \Delta\Phi_K)g_K - (\Delta\Phi_m - \Delta\Phi_{leak})g_{leak} \quad (3.24)$$

Hodgkin and Huxley measured potassium and sodium conductance [Figure 3.5(d)] using a voltage clamp and space clamp techniques (described in Section

3.3.4) and empirically fitted curves to the experimental data. They modeled the time dependency on the ionic conductance by a channel variable or activation coefficients, which indicate the probability of a channel being open. In general, the conductance for a time-dependent channel is written in terms of the channel variable (x), ranging from zero to one, and the maximum conductance. They developed a set of four differential equations to demonstrate action potential of the membrane. There are sophisticated programs, which allow simulating various action potentials. Whenever a short (millisecond range) inward current pulse (in the nano-ampere range) is applied to a patch of axonal membrane, the membrane capacitance is charged and the membrane potential depolarizes. As a result n and m are increased. If the current pulse is sufficient in strength then the generated I_{Na} will exceed I_K, resulting in a positive feedback loop between activation m and I_{Na}. Since t_m is very small at these potentials, the sodium current shifts the membrane potential beyond 0 mV. As sodium inactivation h and I_K increase, the membrane potential returns to its resting potential and even undershoots a little due to the persistent potassium current. If $I_{inj} = 0$, it can be proved that the rest state is linearly stable but it is excitable if the perturbation from the steady state is sufficiently large. For $I_{inj} = 0$ there is a range where repetitive firing occurs. Both types of phenomena have been observed experimentally in the giant axon of the squid. Hodgkin and Huxley performed their voltage clamp experiments at 6.3°C. Higher temperatures affect the reversal potentials since the Nernst equation is temperature dependent. Temperature affects the transition states of ion channels dramatically. Higher temperatures lead to lower time constants and decreased amplitudes. A multiplication factor for each time constant has been developed to account for temperature. Although the Hodgkin-Huxley model is too complicated to analyze, it has been extended and applied to a wide variety of excitable cells. There are experimental results in both skeletal muscle and cardiac muscle, which indicate that a considerable fraction of the membrane capacitance is not "pure," but has a significant resistance in series with it. This leads to modifications in the electrical equivalent circuit of the membrane. Understanding the behavior of ion channels has also allowed simplification of the model. One such is the Fitzhugh-Nagumo model, based on the time scales of the Na⁺ and K⁺ ion channels. The Na⁺ channel works on a much faster time scale than the K⁺ channel. This fact led Fitzhugh and Nagumo to assume that the Na⁺ channel is always in equilibrium, which allowed them to reduce the four equations of the Hodgkin and Huxley model to two equations.

3.3.5 Intracellular Recording of Bioelectricity

Measurements of membrane potential, both steady-state and dynamic changes, have become key to understanding the activation and regulation of cellular responses. Developing mathematical models that depict physiological processes at the cellular level depends on the ability to measure required data accurately. One has to measure or record the voltage and/or current across the membrane of a cell, which requires access to inside the cell membrane. Typically, the tip of a sharp microelectrode (discussed in Section 3.5) is inserted inside the cell. Since such measurements involve low-level voltages (in the millivolts range) and have high source resistances, usage of amplifiers is an important part of bioinstrumentation signals

[Figure 3.5(d)]. Amplifiers are required to increase signal strength while maintaining high fidelity. Although there are many types of electronic amplifiers for different applications, *operational amplifiers* (normally referred to as *op-amps*), are the most widely used devices. They are called operational amplifiers because they are used to perform arithmetic operations (addition, subtraction, multiplication) with signals. Op-amps are also used to integrate (calculate the areas under) and differentiate (calculate the slopes of) signals. An op-amp is a DC-coupled high-gain electronic voltage amplifier with differential inputs (usually two inputs) and, usually, a single output. *Gain* is the term used for the amount of increased signal level (i.e., the output level divided by the input level). In its ordinary usage, the output of the op-amp is controlled by negative feedback, which almost completely determines the output voltage for any given input due to amplifier's high gain. Modern designs of op-amps are electronically more rugged and normally implemented as integrated circuits. For complete discussion about op-amps, the reader should refer to textbooks related to electrical circuits.

One technique to measure electrical activity across a cell membrane is the *space clamp technique*, where a long thin electrode is inserted into the axon (just like inserting a wire into a tube). Injected current is uniformly distributed over the investigated space of the cell. After exciting the cell by a stimulus, the whole membrane participates in one membrane action potential, which is different from the propagating action potentials observed physiologically, but describes the phenomenon in sufficient detail. However, cells without a long axon cannot be space-clamped with an internal longitudinal electrode, and some spatial and temporal nonuniformity exists.

Alternatively, the membrane potential is clamped (or set) to a specific value by inserting an electrode into the axon and applying a sufficient current to it. This is called the *voltage clamp technique,* the experimental setup used by Hodgkin and Huxley [Figure 3.5(d)]. Since the voltage is fixed, (3.24) simplifies to

$$I_{inj} - \left(\Delta\Phi_m - V_{N2} \right) g_{N2} - \left(\Delta\Phi_m - V_K \right) g_K - \left(\Delta\Phi_m - V_{leak} \right) g_{leak} = 0$$

American biophysicist Kenneth S. Cole in the 1940s developed the *voltage clamp* technique using two fine wires twisted around an insulating rod. This method measures the membrane potential with a microelectrode (discussed in Chapter 9) placed inside the cell, and electronically compares this voltage to the voltage to be maintained (called the command voltage). The clamp circuitry then passes a current back into the cell though another intracellular electrode. This electronic feedback circuit holds the membrane potential at the desired level, even when the permeability changes that would normally alter the membrane potential (such as those generated during the action potential). There are different versions of voltage clamp experiments. Most importantly, the voltage clamp allows the separation of membrane ionic and capacitive currents. Therefore, the voltage clamp technique indicates how membrane potential influences ionic current flow across the membrane with each individual ionic current. Also, it is much easier to obtain information about channel behavior using currents measured from an area of membrane with a uniform, controlled voltage, than when the voltage is changing freely with time and between different regions of membrane. This is especially so as the opening and closing (gating) of most ion channels is affected by the membrane potential. The voltage clamp method is widely used to study ionic currents in neurons and other cells. However, the errors in the voltage clamp experiment are due to the current flowing across the voltage-clamped membrane, which generates a drop in the resistance potential in series with the membrane, and thus causes an error in control of the potential. This error, which may be significant, can only be partially corrected by electronic compensation. All microelectrodes act as capacitors as well as resistors and are nonlinear in behavior. Subsequent correction by computation is difficult for membrane potentials for which the conductance parameters are voltage dependent.

A recent version of voltage-clamp experiment is the *patch clamp technique*. This method has a resolution high enough to measure electrical currents flowing through single ion channels, which are in the range of pico-amperes (10^{-12} A). Patch-clamp refers to the technique of using a blunt pipette to isolate a patch of membrane. The patch-clamp technique was developed by German biophysicist Erwin Neher and German physiologist Bert Sakmann to be able to study the ion currents through single ion channels. The principle of the patch-clamp technique is that a small patch of a cell membrane is electrically isolated by the thin tip of a glass pipette (0.5–1-μm diameter), which is pressed towards the membrane surface. By application of a small negative pressure, a very tight seal between the membrane and the pipette is obtained. The resistance between the two surfaces is in the range of 10^9 ohms (G ohm, which is then called the giga-seal). Most techniques for monitoring whole-cell membrane capacitance work by applying a voltage stimulus via a patch pipette and measuring the resulting currents. Because of the tightness of this seal, different recording configurations can be made.

Patch-clamp technique measurements reveal that each channel is either fully open or fully closed (i.e., facilitated diffusion through a single channel is all or none). Although the initial methods applied a voltage step (similar to Problem 3.7) and analyzed the exponential current decay in the time domain, the most popular methods use sinusoidal voltage stimulation. Patch-clamp technique is also used to study exocytosis and endocytosis from single cells. Nevertheless, cell-attached and excised-patch methods suffer from uncertainty in the area of the membrane patch, which is also likely to be variable from patch to patch, even with consistent pipette geometry. The number of ion channels per patch is another source of variability, in particular for channels that are present at low density. There is also a concern that patch formation and excision may alter channel properties. Thus, conductance density estimates obtained from patches are not very reliable.

Another way to observe the electrical activity of a cell is through *impedance measurements*, which are measured after applying an alternating current. Impedance (Z) is a quantity relating voltage to current, and is dependent on both the capacitative and resistive qualities of the membrane. The units of impedance are ohms. In the series pathway, two or more resistors and capacitors are equal to impedance as the vector sum of their individual resistance and reactance.

$$Z^2 = R_c^2 + R^2 \qquad (3.25)$$

In the parallel pathway, two or more resistors and capacitors are equal to impedance as the vector sum of their individual reciprocal resistances and reactance.

$$\frac{1}{Z^2} = \frac{1}{R_c^2} + \frac{1}{R^2} \qquad (3.26)$$

Impedance is also dependent on the frequency of the applied current. For instance, if a stimulus is applied, the cell is affected if the impedance measurement changes. A small amplitude AC signal is imposed across a pair of electrodes onto which cells are deposited and the impedance of the system measured. By measuring the current and voltage across a small empty electrode, the impedance, resistance, and capacitance can be calculated. When cells cover the electrode, the measured impedance changes because the cell membranes block the current flow. From the impedance changes, several important properties of the cell layer can be determined such as barrier function, cell membrane capacitance, morphology changes, cell motility, electrode coverage, and how close the cells are to the surface.

Quantitative high-resolution imaging methods show great promise, but present formidable challenges to the experimentalist, and probably cannot be used to extract kinetic data. Immunogold methods can only provide relative densities of specific channel subunits; determining the absolute densities of the channels, and their functional properties require complementary, independent methods.

3.4   Volume Conductors

So far, only time-dependent variations and concentration-dependent variations in bioelectrical activity were discussed with the assumption that the same membrane potential is distributed independent of the location of the source. However, the stimulus propagates in a direction and location becomes an important factor. All biological resistances, capacitances, and batteries are not discrete elements but *distributed* in a conducting medium of electrolytes and tissues, which continuously extend throughout the body in 3D. Hence they are referred to as *volume conductors*. For example, electrical activity of the brain spreads (spatially) over the entire head volume. The head is a volume conductor with the conductive brain inside the skull, with another thin conductor on the outside, the skin. Thus, it is modeled with an equivalent source (e.g., a single current dipole) in a specific volume conductor.

3.4.1   Electric Field

A fundamental principle which must be understood in order to grasp volume conductors is the concept of the electric field. Every charged object is surrounded by a field where its influence can be experienced. An electrical field is similar to a gravitational field that exists around the Earth and exerts gravitational influences upon all masses located in the space surrounding it. Thus, a charged object creates an electric field, an alteration of the space or field in the region that surrounds it. Other charges in the electrical field would feel the unusual alteration of the space.

The electric field is a vector quantity whose direction is defined, by convention, as the direction that a positive test charge would be pushed in when placed in the field. Thus, the electric field direction about a positive source charge is always directed away from the source, and the electric field direction about a negative source charge is always directed toward the source [Figure 3.6(a)].

A charged object can have an attractive effect upon an oppositely charged object or a repulsive effect upon the same charged object even when they are not in contact. This phenomenon is explained by the electric force acting over the distance separating the two objects. Consider the electric field created by a positively charged (Q) sphere [Figure 3.6(b)]. If another positive charge Q_1 enters the electrical field of Q, a repulsive force is experienced by that charge. According to *Coulomb's law*, the amount of electric force is dependent upon the amount of charge and the distance from the source (or the location within the electrical field). The electric force (F_E) acting on a point charge Q_{ch} as a result of the presence of a second point charge Q_1 is given by

$$F_E = \frac{k_C Q_{ch} Q_1}{r^2} \qquad (3.27)$$

where k_C is the Coulomb's law proportionality constant. The units on k_C are such that when substituted into the equation the units on charge (coulombs) and the units on distance (meters) will be canceled, leaving a Newton as the unit of force. The value of k_C is dependent upon the medium that the charged objects are immersed in. For air, k_C is approximately 9×10^9 Nm^2/C^2. For water, k_C reduces by as much as a factor of 80. k_C is also related to permittivity (ε_0) of the medium by the

relation $k_C = \dfrac{1}{4\pi\varepsilon_0}$ and normally Coulomb's law with permittivity is used.

If two objects of different charges, with one being twice the charge of the other, are moved the same distance into the electric field, then the object with twice the charge requires twice the force. The charge that is used to measure the electric field strength is referred to as a test charge since it is used to *test* the field strength. The magnitude of the electric field or electrical field strength E is defined as the force per charge of the test charge.

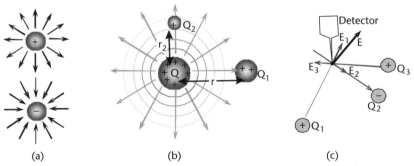

Figure 3.6   Electric field: (a) convention of direction; (b) influence of point source; and (c) multiple point charges.

$$E = \frac{F}{q} = \frac{1}{4\pi\varepsilon_0} \frac{Q_{cb}Q_1}{Q_1 r^2} = \frac{1}{4\pi\varepsilon_0} \frac{Q_{cb}}{r^2} \qquad (3.28)$$

The electric field strength in a given region is also referred by the electrical flux, similar to molar flux. Electrical flux is the number of electrical field lines per unit area. The electric field from any number of point charges is obtained from a vector sum of the individual fields [Figure 3.6(c)]. A positive charge is taken to be an outward field and negative charge to be an inward field.

Work is defined as the component of force in the direction of motion of an object times the magnitude of the displacement of the object

$$W_g = F_\perp d$$

For example, if a box is moving in the horizontal direction across the floor, the work done by the force F is $F\cos\theta$ (Figure 5.2) times the magnitude of the displacement traveled, d, across the floor. Work done by gravitation force is independent of the path and the work done by that force is written as the negative of a change in potential energy, that is,

$$W_g = -\left(PE_f - PE_i\right)$$

where PE_f is the final potential energy and PE_i is the initial potential energy. In the gravitational field, objects naturally move from high potential energy to low potential energy under the influence of the field force. On the other hand, work must be done by an external force to move an object from low potential energy to high potential energy. Similarly, to move a charge in an electric field against its natural direction of motion would require work. The exertion of work by an external force would in turn add potential energy to the object. Work done by electrical force can also be written, similar to gravitational field, as

$$W_E = -\left(PE_f - PE_i\right)$$

The work done by the electric force when an object Q_1 is moved from point a to point b in a constant electric field is given by

$$W_E = Q_1 E \cos\theta d$$

where d is the distance between the final position and the initial position, and cosq is the angle between the electric field and the direction of motion. This work would change the potential energy by an amount which is equal to the amount of work done. Thus, the electric potential energy is dependent upon the amount of charge on the object experiencing the field and upon the location within the field.

EXAMPLE 3.8

A positive point charge 20C creates an electric field with a strength 4 N/C in the air, when measured at a distance of 9 cm away. What is the magnitude of the electric field strength at a distance of 3 cm away? If a second 3C positive charge, located at 9 cm from the initial first, is moved 4 cm at an angle of 30° to the electrical field, what is the work done on that charge?

Solution: From (3.38),

$$E_1 = \frac{k_C Q_{ch}}{d_1^2} = 4[\text{N/C}] = \frac{k_C Q_{ch}}{9^2[\text{cm}]^2} \tag{E3.1}$$

At location 2,

$$E_2 = \frac{k_C Q_{ch}}{3^2[\text{cm}]^2} \tag{E3.2}$$

Dividing (E3.2) by (E3.1),

$$E_2 = \frac{4[\text{N/C}]9^2[\text{cm}]^2}{3^2[\text{cm}]^2} = 36[\text{N/C}]$$

$$W_E = 3[\text{C}] \times 4[\text{N/C}] \times \cos 30 \times 0.04[\text{m}] = 0.415\text{J}$$

An *electric dipole* is when two charges of the same magnitude and opposite sign are placed relatively close to each other. Dipoles are characterized by their dipole moment, a vector quantity with a magnitude equal to the product of the charge of one of the poles and the distance separating the two poles. When placed in an electric field, equal but opposite forces arise on each side of the dipole creating a torque (or rotational force, discussed in Section 5.2.2), T_{ele}:

$$T_{ele} = pE$$

for an electric dipole moment p (in coulomb-meters). Understanding the changes in the direction of the dipole is one of the most important parts of the electrocardiogram (see Section 3.4.5).

3.4.2 Electrical Potential Energy

While electric potential energy has a dependency upon the charge of the object experiencing the electric field, electric potential is purely location dependent. Electric potential is the amount of electric potential energy per unit of charge that would be possessed by a charged object if placed within an electric field at a given location. The electric potential difference ($\Delta\Phi$) is the difference in electric potential between the final and the initial location when work is done upon a charge to change its potential energy. The work to move a small amount of charge dQ_{ch} from the negative side to the positive side is equal to $\Delta\Phi dQ_{ch}$. As a result of this change in potential energy, there is also a difference in electric potential between two locations. In the equation form, the electric potential difference is

$$\Delta\Phi = \frac{dW_E}{dQ_{ch}}$$

$\Delta\Phi$ is determined by the nature of the reactants and electrolytes, not by the size of the cell or amounts of material in it. As the charge builds up in the charging process, each successive element of charge dQ_{ch} requires more work to force it onto the positive plate. In other words, charge and potential difference are interdependent. Summing these continuously changing quantities requires an integral.

$$W_E = \int_0^{Q_{ch}} \Delta\Phi dQ_{ch} \tag{3.29}$$

Substituting (3.16) into the above equation and integrating gives

$$W_E = \frac{1}{2}\frac{Q_{Ch}^2}{C_m} \text{ or } W_E = \frac{C_m}{2}\Delta\Phi^2 \tag{3.30}$$

The only place a parallel plate capacitor could store energy is in the electric field generated between the plates. This insight allows energy (rather the energy density) calculation of an electric field. The electric field between the plates is approximately uniform and of magnitude, E, which is s/e_0, where $\sigma(= Q_{Ch}/A)$ is the surface charge density in C/cm². The electric field elsewhere is approximately zero. The potential difference between the plates is

$$\Delta\Phi = Ed \tag{3.31}$$

where d is the distance between the plates. The energy stored in the capacitor is written as

$$W_E = \frac{C_m}{2}\Delta\Phi^2 = \frac{\varepsilon_0 A E^2 d}{2} \tag{3.32}$$

Ad is the volume of the field-filled region between the plates, so if the energy is stored in the electric field then the energy per unit volume, or *energy density*, of the field is given by

$$w_E = \frac{W}{Ad} = \frac{\varepsilon_0 E^2}{2} \tag{3.33}$$

This equation is used to calculate the energy content of any electric field by dividing space into finite cubes (or elements). Equation (3.43) is applied to find the energy content of each cube, and then summed to obtain the total energy. One can demonstrate that the energy density in a dielectric medium is $w_E = \dfrac{\varepsilon E^2}{2}$ where $\varepsilon (= K\varepsilon_0)$ is the *permittivity* of the medium. This energy density consists of two elements: the energy density $\dfrac{\varepsilon_0 E^2}{2}$ held in the electric field, and the energy density $\dfrac{(K-1)\varepsilon_0 E^2}{2}$ held in the dielectric medium. The density represents the work done on the constituent molecules of the dielectric in order to polarize them.

EXAMPLE 3.9

Defibrillators are devices that deliver electrical shocks to the heart in order to convert rapid irregular rhythms of the upper and lower heart chambers to normal rhythm. In order to rapidly deliver electrical shocks to the heart, a defibrillator charges a capacitor by applying a voltage across it, and then the capacitor discharges through the electrodes attached to the chest of the patient undergoing ventricular fibrillation. In a defibrillator design, a 64-μF capacitor is charged by bringing the potential difference across the terminals of the capacitor to 2,500V.
(a) How much energy is stored in the capacitor?
(b) How much charge is then stored on the capacitor? What is the average electrical current (in amperes) that flows through the patient's body if the entire stored charge is discharged in 10 ms?
Solution:
(a) Using (3.32), the energy stored in the capacitor

$$W_E = \frac{1}{2}\left(64 \times 10^{-6}\right)\left(2.5 \times 10^3\right)^2 = 200J$$

(b) Charge stored is

$$Q_{ch} = C_m \Delta\Phi = \left(64 \times 10^{-6}\right)\left(2.5 \times 10^3\right) = 400 \times 10^{-3}C$$

Alternatively Q_{ch} can also be calculated using (3.27)

Average electrical current $= \dfrac{Q_{ch}}{time} = \dfrac{400 \times 10^{-3}}{10 \times 10^{-3}} \dfrac{C}{s} = 40A$

3.4.3 Conservation of Charge

Differences in potential within the brain, heart, and other tissues reflect the segregation of electrical charges at certain locations within 3D conductors as nerves are excited, causing cell membrane potentials to change. While the potential measured at some distance from an electrical charge generally decreases with increasing distance, the situation is more complex within the body. For example, generators of the EEG are not simple point-like charge accumulations but rather are dipole-like layers. Moreover, these layers are convoluted and enmeshed in a volume conductor with spatially heterogeneous conductivity. The particular geometry and orientation of these layers determines the potential distribution within or at the surface of the 3D body. Nevertheless, charged conductors, which have reached electrostatic equilibrium [Figure 3.7(a)] follow these characteristics:

1. The charge spreads itself out when a conductor is charged.
2. The excess charge lies only at the surface of the conductor.
3. Charge accumulates, and the field is strongest on pointy parts of the conductor.
4. The electric field at the surface of the conductor is perpendicular to the surface.
5. The electric field is zero within the solid part of the conductor.

Gauss' law of electricity (named after Carl F. Gauss, a German physicist) is a form of one of Maxwell's equations, the four fundamental equations for electricity and magnetism. According to Gauss' law of electricity, for any closed surface that surrounds a volume distribution of charge, the electric flux passing through that surface is equivalent to the total enclosed charge. In other words, if E is the electric field strength [Figure 3.7(b)] and Q is the total enclosed charge, then

$$\oint_S E \bullet dA = \frac{Q}{\varepsilon_0} \tag{3.34}$$

where dA is the area of a differential square on the closed surface S with an outward direction facing normal to the surface. ε_0, the permittivity of free space, is 8.8542×10^{-12} C^2/Nm^2. The dot product is used to get the scalar part of E as the direction is known to be perpendicular to the area. To prove the Gauss theorem, imagine a sphere with a charge in the center. Substituting for E from (3.28) and denoting dA in spherical coordinates, for ease of use:

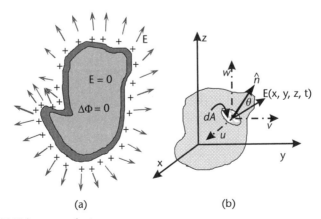

(a)   (b)

Figure 3.7   (a, b) Volume conductors.

$$\oint_S \frac{Q}{4\pi\varepsilon_0 r^2} \cdot r^2 \sin\theta d\theta d\phi = \frac{Q}{\varepsilon_0}$$

In other words, the area integral of the electric field over any closed surface is equal to the net charge enclosed in the surface divided by the permittivity of space. Gauss' law is useful for calculating the electrical fields when they originate from charge distributions of sufficient symmetry to apply it. Gauss' law permits the evaluation of the electric field in many situations by forming a symmetric Gaussian surface surrounding a charge distribution and evaluating the electric flux through that surface. Since charge is enclosed in a 3D space, a normal practice is to express charge either by unit surface area called surface charge density (σ_{ch} with units C/cm^2) or by unit volume basis called charge density (ρ_{ch} with units C/cm^3) and then integrate accordingly. When charge enclosed is electric charge density of ρ_{ch}, then total charge,

$$Q = \oint_V \rho_{ch} \bullet dV \qquad (3.35)$$

Consider an infinitely long plate with surface charge density of 2 C/m^2 in a medium that has a permittivity of 25×10^{-12} C^2/Nm^2. What is the electrical field strength?

Solution: Because the plate is infinite and symmetrical, the only direction the E field can go is perpendicular to the plate. So,

$$|E| \oint_S dA = 2A|E|$$

$$Q_{ch} = \sigma_{ch}\left[C/cm^2\right] * A\left[cm^2\right] = 2A[C]$$

From (3.34),

$$2A|E| = \frac{\sigma_{ch}}{\varepsilon}A$$

Hence

$$|E| = \frac{\sigma_{ch}}{2\varepsilon} = \frac{2\left[C/cm^2\right]}{2 \times 25 \times 10^{-12}\ [C^2/Nm^2]} = 4 \times 10^{10}\ [N/C]$$

and

$$E = \frac{\sigma_{ch}}{2\varepsilon}\hat{n}$$

where $\hat{n}$ is a unit vector perpendicular to the plate.

The integral form of Gauss' law finds application in calculating electric fields around charged objects. However, Gauss's law in the differential form (i.e., divergence of the vector electric field) is more useful. If E is a function of space variables x, y, and z in Cartesian coordinates [Figure 3.7(b)], then

$$\nabla \bullet E = u\frac{\partial E}{\partial x} + v\frac{\partial E}{\partial y} + w\frac{\partial E}{\partial z} \qquad (3.36)$$

where u is the unit vector in the x direction, v is the unit vector in the y direction, and w is the unit vector in the z direction. The divergence theorem is used to convert surface integral into a volume integral and the Gauss law is

$$\oint_S E \bullet dA = \oint_V (\nabla \bullet E)dV = \frac{Q}{\varepsilon_0} \qquad (3.37)$$

Substituting (3.35) for Q_{ch} and simplification results in the differential form of Gauss law

$$\nabla \bullet E = \frac{\rho_{ch}}{\varepsilon_0} \qquad (3.38)$$

Since cross production of the electric field is zero $(\nabla \times E = 0)$, E is represented as a gradient of electrical potential, that is,

$$E = -\nabla\Delta\Phi$$

A negative sign is used based on the convention that electric field direction is away from the positive charge source. Then the divergence of the gradient of the scalar function is

$$\nabla^2\Delta\Phi = \frac{-\rho_{ch}}{\varepsilon_0} \qquad (3.39)$$

or

$$\frac{\partial^2\Delta\Phi}{\partial x^2} + \frac{\partial^2\Delta\Phi}{\partial y^2} + \frac{\partial^2\Delta\Phi}{\partial z^2} = \frac{-\rho_{ch}}{\varepsilon_0} \qquad (3.40)$$

In a region of space where there is no unpaired charge density $(\rho_{ch} = 0)$, (3.39) reduces to $(\nabla^2\Delta\Phi = 0)$, Laplace's equation.

While the area integral of the electric field gives a measure of the net charge enclosed, the divergence of the electric field gives a measure of the density of sources.

It also has implications for the conservation of charge. The charge density is ρ_{ch} on the top plate, and $-\rho_{ch}$ on the bottom plate. Gauss's law is always true, but not always useful as a certain amount of surface symmetry is required to use it effectively. In developing EMG, the volume conductor comprises planar muscle and subcutaneous tissue layers. The muscle tissue is homogeneous and anisotropic while the subcutaneous layer is inhomogeneous and isotropic. However, differential form of Gauss's law is useful to derive conservation of charge.

According to conservation of charge, the total electric charge of an isolated system remains constant regardless of changes within the system. If some charge is moving through the boundary (rate of change of charge is current), then it must be equal to the change in charge in that volume. To obtain a mathematical expression, consider a gross region of interest called control volume (see Chapter 4) with a charge density ρ_{ch}. The rate of change of charge across the control surface is grouped as a change in current density J_s inside a homogeneous finite volume conductor. A charge density is a vector quantity whose magnitude is the ratio of the magnitude of current flowing in a conductor to the cross-sectional area perpendicular to the current flow, and whose direction points in the direction of the current. For a simple case of one inlet and one outlet without generation of charge and consumption of charge, the charge-conservation is written as

$$\frac{dQ_{ch}}{dt} = \dot{Q}_{in} - \dot{Q}_{out} \tag{3.41}$$

where Q_{in} is the rate of flow of net charge into the system and Q_{out} is the charge out of the system. However, for biological applications, obtaining an expression for 3D space is necessary. For this purpose, the law of local conservation of charge is obtained using the divergence theorem.

$$\frac{d\rho_{ch}}{dt} = \nabla \bullet J \tag{3.42}$$

Equation (3.42) is called the *continuity equation of charge*, stating how charge is conserved locally. If the charge density increases in some region locally (yielding a nonzero $\partial\rho/\partial\tau$), then this is caused by current flowing into the local region by the amount $\partial\rho/\partial\tau = -\nabla \bullet J$. If the charge decreases in a local region (a negative $-\partial\rho/\partial\tau$), then this is due to current flowing out of that region by the amount $-\partial\rho/\partial\tau = -\nabla \bullet J$. The divergence is taken as a positive corresponding to J spreading outwards. Current density is related to electric field by $J = \gamma E$, where γ is the electrical conductivity of a material to conduct an electric current. Then

$$\nabla \bullet J = \nabla \bullet \gamma E = \gamma(\nabla \bullet E) = -\gamma(\nabla^2 \Delta\Phi)$$
$$\nabla^2 \Delta\Phi = -\frac{\rho_{ch}}{\varepsilon} = \frac{\nabla \bullet J}{\gamma} \tag{3.43}$$

Equation (3.43) is popularly called the Poisson equation. The Poisson equation describes the electrical field distribution through a volume conductor, and is the equation utilized in analyzing electrical activity of various tissues. It is solved using many computational tools by two approaches: surface-based and volume-based. In both cases, the volume conductor is divided into small homogeneous finite elements. Finite elements are assumed to be isotropic for surface-based analysis (also referred as the boundary-element method), and integral form of the equation is used. For volume-based analysis, finite elements are anisotopic and a typically differential form of equations is used.

Knowing the number, location, orientation, and strength of the electrical charge sources inside the head, one could calculate the reading at an electrode on the surface of the scalp with the Poisson equation. However, the reverse is not true (i.e., the estimation of the electrical charge sources using the scalp potential measurements). This is called the inverse problem and it does not have an unique solution. Hence, it is difficult to determine which part of the brain is active by measuring a number of electrical potential recordings at the scalp.

3.4.4   Measuring the Electrical Activity of Tissues: Example of the Electrocardiogram

The fundamental principles of measuring bioelectrical activity are the same for all tissues or cells: the mapping in time and space of the surface electric potential corresponding to electric activity of the organ. However, for diagnostic monitoring, electrical activity should be measured with less invasiveness, unlike intracellular measurements (described in Section 3.3.5). For this purpose, the concept of volume conduction is used (i.e., bioelectrical phenomena spread within the whole body independent of electrical source position). As the action potential spreads across the body, it is viewed at any instant as an electric dipole (depolarized part being negative while the polarized part is positive). For example, electrical activity of the heart is approximated by a dipole with time varying amplitude and orientation. An electrode placed on the skin measures the change in potential produced by this advancing dipole, assuming that potentials generated by the heart appear throughout the body. This forms the basis of the electrocardiogram (ECG), a simple noninvasive recording of the electrical activity generated by the heart.

In muscle cells, an active transport mechanism maintains an excess of Na^+ and Ca^{2+} ions on the outside, and an excess of Cl^- ions inside. In the heart, resting potential is typically –70 mV for atrial cells and –90 mV for ventricular cells. The positive and negative charge differences across each part of the membrane causes a dipole moment pointing across the membrane from the inside to the outside of the cell. However, each of these individual dipole moments are exactly cancelled by a dipole moment across the membrane on the other side of the cell. Hence, the total dipole moment of the cell is zero. In other words, the resting cell has no dipole moment.

If a potential of 70 mV is applied to the outside of the cell at the left hand side, the membranes' active transport breaks down at that position of the cell, and the ions rush across the membrane to achieve equilibrium values. The sinoatrial (S-A) node, natural pacemaker of the heart) sends an electrical impulse that starts the wave of depolarization [Figure 3.8(a)]. The equilibrium values turn out to be a slight excess of positive ions inside the cell, giving a Nernst potential inside the cell of about +10 mV. This potential difference at the left side of the cell causes the adjacent part of the cell membrane to similarly break down, which in turn causes its adjacent part to break down. Thus, a wave of depolarization sweeps across the cell from left to right. Now the individual dipole moments across the cell membrane do not cancel each other out, and there is a net dipole moment of the cell pointing in the direction of the wave of depolarization. As a wave of depolarization spreads through the cell, it is electrically represented by a time varying electric dipole moment that goes to zero after the resting membrane potential is restored in the process of repolarization.

In the depolarized state, there are individual dipole moments across the membrane, this time pointing from the outside to the inside, unlike the muscle cell at rest. As all the individual cells are aligned in the same direction in the cardiac tissue, when the wave of depolarization reaches the right side of the cell, that potential causes the adjacent cell to go through the wave of depolarization. Thus, a

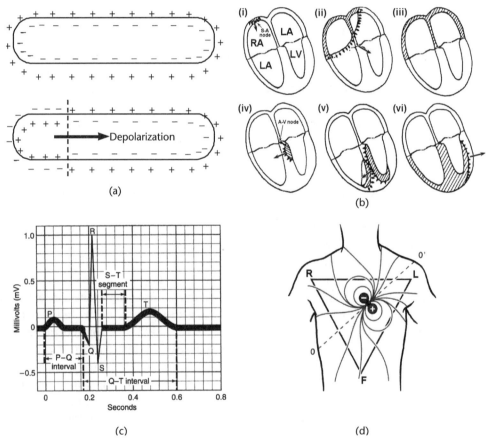

Figure 3.8 Electrical activity in the heart and ECG: (a) muscle cell at rest and changes in charge distribution during membrane depolarization; (b) depolarization and repolarization of the heart during a cardiac cycle (arrows indicate the direction of the resultant dipole); (c) electrical potentials during a cardiac cycle; and (d) dipole moment and distribution of electrical field during initial depolarization.

wave of depolarization sweeps the entire tissue, across the right atria (RA) and left atria (LA) as shown in Figure 3.8(b). Increased concentration of Ca^{2+} ions inside the cell causes the muscle to contract. This depolarization produces the P wave [Figure 3.8(c)] and is usually 80–100 ms in duration. After about 250 ms, the cell membrane begins to pump positive ions outside and negative ions inside the cell. A wave of repolarization follows with the velocity pointing in the opposite direction. By convention, a wave of depolarization moving toward the positive electrode is recorded as a positive voltage (upward deflection in the recording). The wave of repolarization and muscle relaxation of the heart follows behind the polarization wave in the same order. However, there is no distinctly visible wave representing atrial repolarization in the ECG because it occurs during ventricular depolarization. Because the wave of atrial repolarization is relatively small in amplitude, it is masked by the much larger ventricular depolarization (QRS signal).

The atria and the ventricles are not connected except at the A-V node [Figure 3.8(b)] where the depolarized atrial tissue triggers a wave of depolarization across the ventricles. The brief isoelectric (zero voltage) period after the P wave represents the time in which the pulse is traveling within the A-V node where the conduction velocity is greatly decreased. Ventricle depolarization produces a QRS signal, which is normally 60 to 100 ms. The period of time from the onset of the P wave to the beginning of the QRS complex is termed the P-Q interval, which normally ranges from 120 to 200 ms in duration. This relatively short duration indicates that ventricular depolarization normally occurs very rapidly. The constant electric period [ST segment in Figure 3.8(c)] following the QRS is the time at which the entire ventricle is depolarized and roughly corresponds to the plateau phase of the ventricular action potential. The repolarization of ventricular contractile forces produce a T curve and is longer in duration than depolarization (i.e., conduction of the repolarization wave is slower than the wave of depolarization). The *Q-T interval* represents the time for both ventricular depolarization and repolarization to occur, and therefore roughly estimates the duration of an average ventricular action potential. This interval can range from 200 to 400 ms depending upon heart rate. At high heart rates, ventricular action potentials shorten in duration, which decreases the Q-T interval. In practice, the Q-T interval is expressed as a "corrected Q-T (Q-Tc)" by taking the Q-T interval and dividing it by the square root of the R-R interval (interval between ventricular depolarizations). This allows an assessment of the Q-T interval that is independent of heart rate. Normal corrected Q-Tc intervals are less than 440 ms. After the wave of depolarization sweeps across the entire muscle mass, all the cells on the outside are negative, and once again, no potential difference exists between the two electrodes.

The negative charge, which results from the positive charge transferring inside the cell, is detected at the surface. Electrodes placed on the surface of arms and legs sense the negativity due to the positive charge inside heart cells. However, it is not as though the entire heart causes a negative charge to be detected. When the atrial cells are depolarized, there is a negative charge detected there, but still a positive charge around the ventricular cells. One side of the heart is sensed as negative, the other positive [Figure 3.8(d)]. There are two poles or a dipole. The state of the dipole depends on the state of the heart beat. To begin with, the left side of the septum is excited, then the entire septum. Hence, the dipole is pointing left to right, then it is pointing top to bottom. As everything repolarizes, the dipole direction heads back upwards and ends up going bottom to top. The ECG is recorded by measuring the potential difference between any two points on the body. Generally, ECG is recorded at a paper speed of 25 mm/s. The stretch between two points is called a LEAD and it is modeled as a direction. The potential difference measured by the lead depends on the size of the electric dipole (wave of electrical depolarization), the direction of the electrodes and the distance of the electrodes from the dipole.

To reduce variation in distance during clinical evaluations, standard positions for electrode placement have been generated. Dutch physiologist Willem Einthoven in 1903 developed the first practical device for recording cardiac potentials, called the string galvanometer, which became the electrocardiograph. He developed the classical limb lead system, whereby an electrode is placed at each corner of an imaginary equilateral triangle, known as Einthoven's triangle, superimposed on the front of a person with the heart at its center. The three corners represented right arm (RA), left arm (LA), and left leg (LL). Einthoven's lead system is conventionally found based on the assumption that the heart is located in an infinite, homogeneous volume conductor. The stretch between two limb (arm or leg) electrodes constitutes a lead. Each side represents the three standard limb leads of the ECG [Figure 3.9(a)]: Lead I records the cardiac potentials between RA and LA, lead II records between RA and LL and lead III records between LA and LL.

Einthoven studied the relationship between these electrodes, forming a triangle where the heart electrically constitutes the null point. Einthoven's triangle is used when determining the electrical axis of the heart. According to Kirchhoff's law, these lead voltages have the following relationship: lead I + lead III = lead II. Hence, only two are independent at any time. Using this basis, *scalar ECG* is measured using only two electrodes plus a ground (right arm, left arm, and right leg). However, more electrodes are necessary to display the cardiac vector (referred to as *vector ECG*) as a function of time at particular leads (directions). In 1934, American physiologist Frank N. Wilson developed an improved system, which amplified the action potentials, known as the augmented (as the signal is increased) lead system (leads 4, 5, and 6). These are referred to as aVR (right arm), aVL (left arm), and aVF (foot) where V stands for voltage (the voltage seen at the site of the electrode). These leads record a change in electric potential in the frontal plane. Augmented leads [Figure 3.9(b)] are unipolar in that they measure the electric potential at one point with respect to a null point (NP, one which does not register any significant variation in electric potential during the contraction of the heart). This null point is obtained for each lead by adding the potential from the other two leads. For example, in lead aVR, the electric potential of the right arm is compared to a null point, which is obtained by adding together the potential of lead aVL and lead aVF. A wave traveling *towards* the positive (+) lead inscribes an upward deflection of the EKG; conversely, a wave traveling *away* from the positive lead inscribes a downward deflection. For example, a wave traveling from the head to the feet is shown as an upwards deflection in aVF, since it is going towards the aVF+ lead. Waves that are traveling at a 90° angle to a particular lead will create no deflection and are called *isoelectric* leads.

Six electrodes record the electric potential changes in the heart in a cross-sectional plane [Figure 3.9(c)]. These are called precordial leads (V1 through V6),

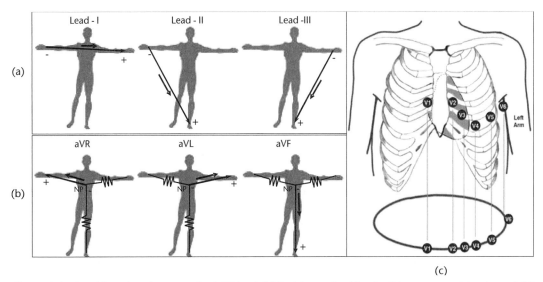

Figure 3.9 Lead locations in a standard 12-lead ECG: (a) standard leads, (b) augmented leads, and (c) precordial leads.

which are unipolar leads and record the electrical variations that occur directly under the electrode look to the six leads. The most commonly used clinical ECG system, the 12-lead ECG system, consists of the following 12 leads: six leads (I, II, III, aVR, aVL, and aVF) record the electrical activity in the vertical plane (head to foot), and the other six leads (V1, V2, V3, V4, V5, V6), record in the horizontal plane (across the chest). Understanding the precordial leads and how they align with the heart is critical to get an accurate lead recording on the ECG. For example, normally the R wave becomes progressively taller moving from V1 to V5. However, misplacement of electrodes could lead to poor R wave progression as electrical potential reading is a function of the inverse of the square of the distance. This perturbation also occurs in some pathological conditions in which the R wave in leads V1–V4 either does not become bigger, or increases very slowly in size. Hence for proper diagnosis, the location of electrodes is the key aspect apart from other noises in the signal (discussed later). The American Heart Association mandates protocols for standard identification techniques for placing the electrodes. Nevertheless, 12-lead EKG may not be sufficient in assessing certain abnormalities and a number of additional leads are used to get a better assessment of electrical variations.

3.4.5 Biopotential Recording Practicalities

To measure biopotentials, a metal electrode (discussed in Chapter 9) is placed in contact with the tissue. Consider the case of EMG where either surface electrodes on the skin or a needle electrode inserted into a muscle record the time variations in electric potential. Needle electrodes probe a single muscle fiber and give a characteristic time record of electric potential with variations of several millivolts. Such recordings of voluntary muscle activity check for normal functioning of nerve stimulation of muscle. More detailed information is obtained with external electrical stimulation of the muscle since an entire group of muscle fibers is simultaneously activated. Measurements at a number of distances along a muscle determine conduction velocities along the stimulating nerve. While not as common as the ECG or EEG, the electromyogram is more directly related to the depolarization of a single cell or small group of cells. In the recorded signal, the stimulation artifact normally strongly dominates in relation to the electrical stimulation response. If the response signal and the artifact overlap, the response signal is lost due to overdrive of the recording amplifier. Every rising edge of the stimulation impulse leads to a charging current and every falling edge leads to a discharging current in the capacitive part of the load impedance. The two edges of a same phase stimulation pulse produce a slowly decreasing discharge current through the tissue after the second (falling) edge.

When measuring biopotentials (say, ECG), everything else (even other biopotentials such as EEG, EMG, EOG) and power line interferences are noise sources. Although some noises have characteristic frequencies and are minimized using band pass filters, one has to understand and develop sophisticated signal processing methodologies. If at a particular measurement site, the resistance of the conductor (tissue) is higher or the conductor size (finger compared to arm) is decreased, then the electrical potential increases. Artifacts also arise due to changes in the amount of intervening tissue (tissue filter function and volume conductor effects) but especially from the dead skin cell layer. Further, movement of the patient causes the skin to stretch, creating a disturbance of the charge distribution at the electrode/electrolyte boundary. This is commonly referred to as the motion artifact and removal of motion artifact is a major task of the monitoring system's filtering algorithms. Complex algorithms try to manipulate incoming signals to produce output signals based solely on the electrical activity of the tissue of interest. However, motion artifact persists as the major component of signal degradation, even with complex algorithmic filtering. Hence, computer algorithms still have a difficult time identifying poor signals due to motion artifact.

Problems

3.1   Determine the Nernst potentials for potassium ions and chloride ions between the inside and the outside of the cell membrane. Use the concentration values shown in Figure 3.2.

3.2   The Nernst potential for magnesium ions (Mg^{2+}) is −50 mV. The extracellular concentration of magnesium C_{out} is 0.6 mM. What is the intracellular concentration of magnesium (in mM)?

3.3   Voltage-dependent L-type Ca^{2+} channels ($I_{Ca,L}$) play vital roles for cardiac functions, including triggering cardiac contraction and pacemaker activity in nodal cells. Using the values given for Ca^{2+}, calculate the resting potential across a cell.

3.4   Assume that a membrane is selectively permeable to SO_4^{-2}. At equilibrium, what is the membrane potential if the concentration of Na_2SO_4 is 120 mM on one side and 10 mM on the other side?

3.5   A patient is given enough Li^+ to raise his plasma concentration to 0.1 mM. If lithium is permeable and at equilibrium across the plasma cell membrane, what is the intracellular Li^+ for a cell that has a membrane potential of –80 mV?

3.6   A cat motor neuron has the following composition

	K^+ ion	Na^+ ion	Cl^- ion
Inside	150	15	9
Outside	5.5	152	124

(a) What is the resting potential if the membrane is permeable to only Cl^- ions?

(b) However, if the relative permeabilities are K^+:Na^+:Cl^- 1:0.04:0.45, then what is the membrane potential?

(c) If you expose the cells to tetrodotoxin (from the puffer fish), which blocks voltage-gated Na^+ ion channels, then what is the membrane potential?

(d) If you expose the cells to 4-aminopyridine, which blocks K channels (and has been tried in MS patients as a means to overcome nerve conduction block), then what is the membrane potential?

(e) What has to vary for the cell's membrane potential to shift to –50 mV?

3.7   A cockroach leg muscle has the following composition

	K^+ ion	Na^+ ion	Cl^- ion
Inside	110	27	44
Outside	27	111	150

(a) What is the resting potential if the membrane is permeable to only Na^+ ions?

(b) However, if the relative permeabilities are K^+:Na^+:Cl^- 1:0.04:0.45, then what is the membrane potential?

(c) If you expose the cells to lidocaine, which block voltage-gated Na ion channels, then what happens to the membrane potential?

(d) If you expose the cells to tetraethylammonium (TEA), applied to the inside of a membrane, it will block voltage-gated K channels. Then what is the membrane potential?

3.8   A cell has two pumps in its plasma membrane, both Na^+ dependent. One is a Na^+/CI symport, the other a Na^+/glucose symport.

(a) By artificial means, you change the membrane potential from –80 mV to 0 mV. What effect will this have on the rate of operation of each pump? Note that one of these pumps is electrogenic.

(b) You have two toxins, one that blocks each pump, but you do not know which is which. How would you determine their identities by applying each to the cell and measuring membrane potential only? Note that one of these pumps is electrogenic.

3.9   Duncan et al. are interested in measuring the changes in the membrane perme-
         ability of eye lenses with age. They incubated in an artificial aqueous humor
         solution (35°C) with the following composition: 130-mM NaCl, 5-mM KC1,
         0.5-mM MgCl$_2$, 1-mM CaCl$_2$, 10-mM NaHCO$_3$, and 5-mM glucose, buff-
         ered with 10 mM HEPES to pH 7.3. They reported the intracellular sodium
         content for a lens from 20-year-old person to be 25 mM where as it is 40 mM
         for a lens from a 60-year-old person. However, there is no significant differ-
         ence in the intracellular content of potassium. The membrane potentials are
         −50 mV and −20 mV for 20-year-old and 60-year-old lenses, respectively. As-
         suming that the lens is permeable to Na$^+$ and K$^+$ ions, calculate the change in
         the permeability of Na$^+$ ions to K$^+$ ions with age.

3.10   Consider a cell membrane with Na$^+$(out) = 150 mM, Na$^+$(in) = 15 mM, K$^+$(out)
         = 5 mM, K$^+$(in) = 150 mM, Cl$^-$(out) = 155 mM, and Cl$^-$(in) = 10 mM. The
         resting potential across this cell membrane is −70 mV. The concentration of
         Na$^+$ and K$^+$ across the cell membrane are maintained by the Na$^+$/K$^+$ pumps,
         each of which pumps 3Na$^+$ out and 2K$^+$ in.

         (a) Calculate the Nernst potentials for Na$^+$, K$^+$, and Cl$^-$, respectively.

         (b) Using the Goldman equation, calculate the membrane potential. The
             concentrations of Na$^+$, K$^+$, and Cl$^-$ inside and outside of the cell are given
             above, and the relative permeabilities of the individual ions are
             $P_{K^+} = 1$, $P_{Na^+} = 0.04$, $P_{Cl^-} = 0.45$.

         (c) What will happen if the flow of K$^+$ is blocked across the leak channels?
             What will be the membrane potential predicted by the Goldman equation
             in this case [personal communication with Dr. Nada Bowtany, Rutgers
             University, 2004]?

3.11   Cartilage consists of extracellular matrix containing collagen and glycosami-
         noglycans, the long sugar chains. Under physiological conditions, collagen is
         electroneutral, while the glycosaminoglycans have a fixed negative charge A$^-$.
         Thus, cartilage can be modeled as a compartment containing fixed negative
         charges that cannot move into the surrounding medium. On the other hand,
         ions in the surrounding medium freely flow in and out of the cartilage matrix.
         Shirin wanted to measure the concentration of the fixed negative charges in-
         side cartilage. She dissected a piece of articular cartilage from a bovine (cow)
         knee and placed it in a 0.15M NaCl solution at pH = 7.4. By placing elec-
         trodes inside the cartilage, and in the surrounding bath, she measured a volt-
         age difference of −25 mV.

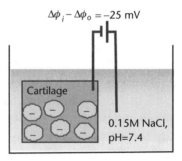

(a) Assume the ions in the experiment consist of Na^+, Cl^-, H^+, and OH^-, and calculate the concentration of each ion *inside* the cartilage sample. The dissociation constant for water is 10^{-14}.

(b) Calculate the concentration of fixed negative charge, G^-, inside the cartilage.

(c) To test the mechanical properties of the cartilage sample, the researchers placed the cartilage sample between two mechanical plates, and compressed the sample such that the final sample volume V_f was half the original volume. Assuming that the volume of the salt bath is significantly larger than the volume of the cartilage sample, the salt concentration in the bath does not change appreciably after the compression. Calculate the concentration of ions inside the cartilage sample after compression. Does the electric potential measured across the cartilage boundary change [personal communication with Dr. Nada Bowtany, Rutgers University, 2004]?

3.12   Calculate the resting membrane potential for squid giant axon for which the following data is available.

Species	Na^+ ion	K^+ ion
Intracellular(mM/L)	14	140
Extracellular(mM/L)	142	4
Nernst potential at 37°C (mV)	+61	−94

3.13   A salmon egg deposited in sea water (Na^+ = 300 mM, K^+ = 2 mM, Ca^{2+} = 1 mM and Cl^- = 310 mM), shows a membrane potential of −43 mV with an intracellular microelectrode. What can you tell about the composition of the salmon eggs cytoplasm, if: (a) the membrane is only permeable to potassium (be quantitative), and (b) the membrane is only permeable to calcium (be quantitative)?

3.14   A novel polymer membrane has been formed with the intention of being used in dialysis. Prior to using, one has to know the membrane permeability. To determine the permeability, an easy technique is to place the membrane between two chambers. Then one chamber is filled with 0.15M KCl, and the other side is filled with 0.03M KCl. The measured potential difference across the membrane is 10 mV with side 2 being more positive than side 1. Then the solutions are replaced by 0.15M NaCl on one side and 0.03M on the other side. Now the measured potential difference is 40.5 mV with side 1 being more positive than side 2. What is the P_{Na}/P_{Cl} and P_{Cl}/P_K for this membrane?

3.15   Mammalian muscle cytoplasm is composed mainly of Na, K, and Cl. These same ions dominate the extracellular fluid as well. The following table lists the composition of cytoplasm and extracellular space of muscle cells:

	Na (mM)	K (mM)	Cl (mM)
Cytoplasm	12	155	4
Extracellular space	145	4	120

(a) What are the Nernst potentials for these ions?

(b) The sum of the cation concentrations does not equal the chloride concentration on either side of the membrane. Why?

3.16 The potential difference across a cell membrane from outside to inside is initially −90 mV (when in its resting phase). When a stimulus is applied, Na^+ ions are allowed to move into the cell such that the potential on the inside changes to +20 mV for a short amount of time (depolarization time).

(a) If the membrane capacitance per unit area is 1 mF/cm^2, how much charge moves through a membrane of area 0.05 cm^2 during the depolarization time?

(b) The charge on a Na^+ ion is positive. How many ions move through the membrane?

(c) How much energy is expended in moving these ions through the membrane to create this change in potential (called the action potential)?

3.17 If the time constant of a typical axon is 10 ms, what would be the resistance of one square centimeter of its membrane? The membrane capacitance of a typical axon is 1 mF/cm^2.

3.18 Dr. Shingo Yashu is studying a neuromusclular interaction using voltage clamp measurements. He is interested in using tetraethylammonium (TEA) to block potassium conductance.

(a) To understand the output, he needs to derive a potassium conductance equation using the Hodgkin-Huxley model. Please help him.

(b) A part of a muscle cell membrane was studied with the voltage clamp measurements with a 30-mV positive voltage step input. He measured the membrane current as 0.5 mA/cm^2 after 5 ms of step input. When the potassium current was blocked with TEA the current was 1 mA/cm^2 after 5 ms of step input. Also, he observed that the flow of K^+ ions could be stopped with a 100-mV increase in resting membrane potential. What is the potassium ion conductance g_K?

3.19 The axons of the myelinated nerve fibers are completely encased in a myelin sheath. Their neural conduction rates are faster than those of the unmyelinated nerve fibers. Myelinated nerve fibers are present in somatic and autonomic nerves. In a patient, one of the myelinated nerves was damaged in their right arm, which reduced the propagation speed to 65% of the original. Damaged and healthy nerves are stimulated with a similar stimulus current. If the smallest current adequate to initiate activation is considered to remain unchanged, how much longer is stimulus needed with the damaged nerve? The propagation speed is inversely proportional to the square root of the membrane capacitance $(\Theta \alpha (C_m)^{-\frac{1}{2}})$.

3.20 A membrane has 75 Na^+ ion channels/mm^2 with a conductance of 4 pS/channel. The capacitance of the membrane is 0.01 pF/mm^2. Calculate the flow and the time of sodium ions across the membrane of surface area of 1 mm^2 with $V_M = -65$ mV, $V_M' = 35$ mV, and $V_{Na} = 62$ mV.

3.21 A defibrillator detects ventricular fibrillation, and shocks the heart to restore the normal rhythm. Phillips is interested in manufacturing a new defibrillator using a 40-mF capacitor charged with an adjustable power supply capable of generating a high voltage between 0 and 2,500V. The capacitor discharges through the patient, who is represented by the equivalent resistance R at the time of defibrillation.

(a) What is the value of the voltage for the energy accumulated in the capacitor to be equal to 100J?

(b) With the high voltage set to the value found in (a), how long will it take for the capacitor voltage to decrease to the value 10V? Assume that the resistance of the patient is 50W.

(c) If the cycle time between two shocks is set for 20 seconds, what is the residual voltage at the end of 20 seconds? Do you see accumulation of voltage for 10 shocks?

3.22 During defibrillation, Marcus measured the supplied current in a patient for the defibrillator voltage. The obtained profiles are shown in the figure. Based on the graphs

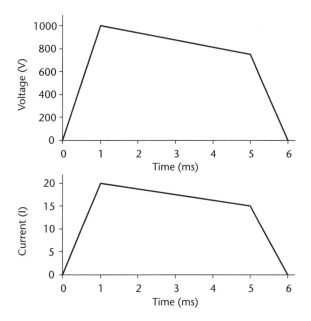

(a) Estimate the patient's resistance. Do you think that the patient's body has a constant resistance? Validate your answer.

(b) Derive the relation between voltage and time for the 0 to 1 ms interval, the 1 and 5 ms interval, and the 5 to 6 ms interval.

(c) Use the results in part (b) to derive the equations for the current in the three intervals.

(d) Draw the graph for the product $I\Delta\phi$ as a function of time t between 0 and 6 ms. (The product $I\Delta\phi$ is the power in watts delivered to the patient at any instant of time.) Estimate the area under the curve of $I\Delta\phi$, which is equal to the total energy (in joules) delivered to the patient.

3.23 The membrane potential can be written as $\Delta\Phi = Q/C$, where C is the membrane capacitance. For the value of C we found, $C = 1.3\ 10^{-11}$ Farads, show that the charge Q needed to generate a -70 mV potential drop across the membrane is very small compared to the bulk concentrations of Na^+ and K^+. The bulk concentrations are Na^+(out) = 145 mM, Na^+(in) = 15 mM, K^+(out) = 5 mM, and K^+(in) = 145 mM. *Hint:* $\rho = z.F.c$ [personal communication with Dr. Nada Bowtany, Rutgers University, 2004].

3.24 In a spherical neuron soma, the values of the conductances to the three main ions were estimated to be: $G_{Na} = 40$ pS, $G_K = 700$ pS, $G_{Cl} = 250$ pS. What is the total membrane resistance?

3.25 A pacemaker (or "artificial pacemaker" not to be confused with the heart's natural pacemaker), is a medical device designed to regulate the beating of the heart. It can be designed as a simple RC device that sends a pulse to a patient's heart every time the capacitor in the pacemaker is charged to a voltage of 0.23V. In a typical design, a pacemaker with a 110-μF capacitance and a battery voltage of 3.0V is used. If 60 pulses per minute are needed, what is the value of the resistance? What is the expected lifetime of such a battery if a typical current of 5 μA is produced and the battery supplies a charge of 0.4 ampere-hour?

3.26 Consider the RC circuit loop shown in the figure and a magnification of the capacitor in the circuit, showing an enclosed volume around one plate of the capacitor. At the capacitor, the principle of conservation of charge states that the rate of change of charge Q, dQ/dt, enclosed within the dotted volume is equal to $I_{in} - I_{out}$, where I_{in} is the current entering the enclosed volume and I_{out} is the current leaving the volume. Thus, $dQ/dt = I_{in} - I_{out}$.

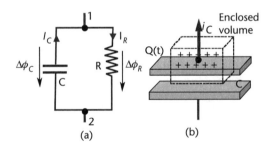

(a)   (b)

(a) Using this conservation of charge principle, derive an expression for i_c in the circuit as a function of the capacitance C, and $\Delta\phi_C$, the voltage drop across the capacitor. Note that $Q = C\Delta\phi_C$ for a capacitor.

(b) Using Ohm's law across the resistor, and Kirchoff's current and voltage laws, write a differential equation for $\Delta\phi_C = \Delta\phi_1 - \Delta\phi_2 = \Delta\phi_R$.

(c) From (b), solve for $\Delta\phi_C(t)$. Plot $\Delta\phi_C$ as a function of time and as a function of $\Delta\phi_o = \Delta\phi(t=0)$ [personal communication with Dr. Nada Bowtany, Rutgers University, 2004].

3.27 The circuit loop in Problem 3.26 is connected to a current source I_s, which outputs a current I into the loop. The forcing current, I, is a step function and is plotted as a function of time in part (b) of the figure. In this new loop $\Delta\phi_C = \Delta\phi_1 - \Delta\phi_2$, however, $I = I_R - I_C$.

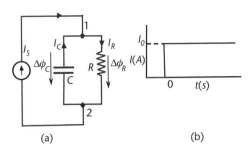

(a)   (b)

(a) Show that: $\dfrac{d\Delta\phi}{dt} + \dfrac{1}{RC}\Delta\phi = \dfrac{I}{C}$ for $t \geq 0$ where $\Delta\phi = \Delta\phi_1 - \Delta\phi_2$.

(b) Assuming $\Delta\phi = 0$ at $t = 0$, solve the differential equation above for $t \geq 0$ to get an expression for $\Delta\phi(t)$. What is $\Delta\phi$ as $t \rightarrow \infty$?

(c) Plot $\Delta\phi$ as a function of time.

(d) Show that at $t = \tau = RC$, V will have reached 63.8% of its final value defined as $\Delta\phi_{final} = \Delta\phi$ $(t \rightarrow \infty)$.

(e) Suppose that the circuit represents the membrane of a neuron receiving an input I from another neuron. Then the voltage $\Delta\phi$ represents the membrane potential of the neuron modeled by the circuit. How will decreasing τ, with $\tau = RC$, affect the speed of the neuron's response to the current input I [personal communication with Dr. Nada Bowtany, Rutgers University, 2004]?

3.28 Hodgkin and Huxley applied a step voltage across the membrane and measured the current in their voltage clamp experiment. To measure conductance as a function of time, they measured current. R is replaced with a time varying resistor $R(t)$ and the current source is replaced with a voltage source consisting of $\Delta\phi$, and a resistor r_m. The voltage clamp is achieved by stepping the voltage up and holding it at a value $\Delta\phi_m$ across the membrane. r_ν is sufficiently small to assume that $\Delta\phi = -\Delta\phi_C = \Delta\phi_1 - \Delta\phi_2 = -\Delta\phi_R$.

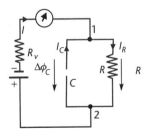

(a) Plot the voltage $\Delta\phi$ ($= \Delta\phi_1 - \Delta\phi_2$) across the membrane as a function of time.

(b) What is the current in the capacitor before the clamping voltage is applied, at $t < 0$?

(c) What is the current in the capacitor after the clamp has been applied, at $t > 0$?

(d) What is the current in the resistor after the clamp has been applied, at $t > 0$?

(e) Assuming that $R(t)$ represents the voltage gated sodium channels, how do you expect the current measured by the current meter to vary as a function of time if the cell generated an action potential?

(f) What should the clamp voltage, V_m, be to elicit an action potential [personal communication with Dr. Nada Bowtany, Rutgers University, 2004]?

3.29   With $\delta = 5$ nm, $\varepsilon_m \sim \varepsilon_o = 8.85 \cdot 10^{-12}$ Farads/m, and the radius of the cell, $R = 25\ \mu$m, calculate the capacitance, C, of the cell in units of Farads. Assume that the cell is a sphere and that the surface area of the cell is equal to the surface of the capacitor plate.

3.30   An air-filled parallel plate capacitor has a capacitance of 5.0 pF. A potential of 100V is applied across the plates, which are 1.0 cm apart, using a storage battery. What is the energy stored in the capacitor? Suppose that the battery is disconnected and the plates are moved until they are 2.0 cm apart. What is the energy stored in the capacitor now? Suppose, instead, that the battery is left connected and the plates are again moved until they are 2.0 cm apart. What is the energy stored in the capacitor in this case?

3.31   Wie wants to modify a defibrillator by replacing the capacitor inside of it with another one. Assume the capacitor is a simple parallel-plate capacitor.

(a) If we wanted a "bigger" capacitor, one that could store more charge per volt of potential difference, would we want to increase or decrease the surface area of the capacitor's plates?

(b) If we wanted a capacitor with *double* the capacitance of the original, 64-mF capacitor, by what factor would we want the area of the plates increased or decreased?

(c) What if a 32-mF capacitor with 5,000V is used?

(d) Suppose we successfully have replaced our original 64-mF capacitor with a 128-mF capacitor. How high would the voltage have to be to apply the same charge to the patient? How much energy would then be discharged through the patient?

3.32   Instead of changing the area (which would be difficult to do without more material), we could try to change the separation of the plates.

(a) In order to increase the capacitance, should we try to separate the plates more, or should we try to squeeze them closer together?

(b) Suppose we squeezed the original capacitor so that the distance between the plates became 1/4 of what it was originally. What would the new capacitance be?

Selected Bibliography

Aidley, D. J., *The Physiology of Excitable Cells*, 4th ed., New York: Cambridge University Press, 1998.

Bronzino, J. D., (ed.), *The Biomedical Engineering Handbook*, 2nd ed., Boca Raton, FL: CRC Press, 1999.

Gillis, K. D., "Techniques for Membrane Capacitance Measurements," in Sakmann, B., and E. Neher, (eds.), *Single-Channel Recording*, 2nd ed., New York: Plenum Press, 1995, pp. 155–198.

Gulrajani, R. M., *Bioelectricity and Biomagnetism*, New York: John Wiley & Sons, 1998.

Hille, B., *Ionic Channels of Excitable Membranes*, 3rd ed., Sunderland, MA: Sinauer Associates, 2001.

Hodgkin, A. L., and A. F. Huxley, "A Quantitative Description of Membrane Current and Its Application to Conduction and Excitation in Nerve," *Journal of Physiology*, Vol. 117, 1952, pp. 500–544.

Koch, C., *Biophysics of Computation: Information Processing in Single Neurons*, New York: Oxford University Press, 2004.

Malmivuo, J., and R. Plonsey, *Bioelectromagnetism*, New York: Oxford University Press, 1995.

Fall, C. P., et al., *Computational Cell Biology*, New York: Springer, 2005.

Plonsey, R., and R. C. Barr, *Bioelectricity*, New York: Plenum Press, 1988.

Webster, J. G., *Medical Instrumentation: Application and Design*, New York: John Wiley & Sons, 2000.

CHAPTER 4

Biofluid Flow

4.1 Overview

In previous chapters, movements of molecules across barriers due to electrochemical gradients were discussed. For molecules to reach larger distances they have to be carried in the blood or other fluids in the body. Molecular transfer over large distances occurs in conjunction with fluid flow. For example, transportation of nutrients and therapeutic components to the tissues, removal of metabolic waste products, and immune surveillance need blood circulation throughout the body. Understanding the dependency of physical variables (which control the flow of body fluids) particularly blood flow dynamics in the body, such as perfusion pressure, flow rate, and the diameter of the elastic vessels carrying the blood, are important to develop surrogate devices. Heat produced by the liver cells during rest or in the muscle cells during exercise is absorbed by the blood and distributed around the body. It is important to understand the fundamentals of energy transfer to the design and development of implanted prosthetic devices such as artificial heart and blood vessels, or external support devices such as dialysis machines and mechanical ventilators. A few applications of fluid flow to biological systems are listed in Table 4.1.

The transport laws governing the nonreactive systems are often useful to make rational predictions of the analogous systems. Cells that are suspended in the flowing medium (e.g., cells in blood) or lining the lumen where fluid flows (e.g., endothelial vessels) are sensitive to the hydrodynamic forces exerted on them. In response to altered hydrodynamic forces, cells can alter their genetic expression profiles, which then can directly or indirectly affect the physical, chemical, and biological properties of the contacting fluid. If nonbiological components are introduced either as drug delivery vehicles or as prosthetic devices, they are exposed to similar hydrodynamic forces. Modeling various physiological processes such as recruitment of blood cells to the site of injury (and subsequent events) helps understand the process of wound healing and tissue regeneration. This chapter will introduce the basic conservation principles of mass, momentum, and energy useful in many biomedical applications.

Table 4.1   Uses of Biofluid Flow Analysis

Analysis of Physiological System	Design of Devices, Instruments, and Diagnostic Tools
Blood flow through the circulatory system	Artificial hearts, prosthetic heart valves, stents, vascular grafts, extracorporeal blood circuits, cardiac bypass, kidney dialysis, cell separator, wound healing and tissue regeneration, nanomedicine, drug delivery systems
Gas flow in lungs	Heart-lung machines, artificial oxygen carriers, mechanical ventilators, mucosal drug delivery, evaluation of pollutant particles, replacement surfactants
Cerebral circulation	Drug delivery systems, forced convection systems
Synovial fluid flow in cartilage	Artificial joints, cartilage regeneration
Eye	Contact lenses

4.2   Fluid Flow Characteristics

4.2.1   Conservation of Mass

In nature, matter exists in three states: solids, liquids, and gases. Liquids and gases are combined as fluids as their shape is determined by the container in which they are present, unlike solids. Further, fluids lack the ability to resist forces in contrast to solids; a deformation force on a solid may result in no effect due to resistance or may cause some defined displacement. Since fluids cannot resist forces, they move or flow under the action of the forces. Their shape changes continuously as long as the force is applied and their characteristics such as density and viscosity could also change. Despite sharing many similar characteristics liquids and gasses possess distinct characteristics. For example, application of pressure on liquids does not change the volume. Also, liquid density does not change with pressure and is often regarded as being incompressible. However, gases compress with increasing pressure (i.e., change volume) and are considered compressible. The density of gas is a function of the pressure and temperature. If the liquid volume is smaller than the container volume, then liquid only occupies its volume, leaving the rest of the container volume unoccupied. However, a gas has no fixed volume and expands to fill the entire volume of the container.

In any system of solid or fluid whether stationary or moving, the total mass of the system does not change and is conserved. This is called the law of conservation of mass. Consider the case of blood flowing through a tubing (Figure 4.1) (or air flowing through ventilator tubing). Based on the conservation of mass principle, the mass of blood entering the tube per unit time ($\dot{m}$) is equal to the mass of blood leaving that tubing assuming there is no accumulation in the tubing. In the mathematical form this can be represented as

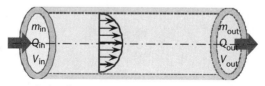

Figure 4.1   Fluid movement in a tube.

$$\dot{m}_{out} = \dot{m}_{in} \tag{4.1}$$

For practical purposes where it is easy to measure volumetric flow, mass is written as the product of density and volumetric flow rate as

$$\left(\rho\dot{Q}\right)_{out} = \left(\rho\dot{Q}\right)_{in} \tag{4.2}$$

Further, the volumetric flow rate is written as a product of the average velocity and cross-sectional area of the tubing, which is of particularly useful to deal with the evaluation of force. Momentum is the product of mass and velocity, which indicates the tendency of an object in motion to keep moving or not to slow down. For a simple case of one inlet and one outlet

$$\dot{m} = \rho_{out} V_{out} A_{out} = \rho_{in} V_{in} A_{in}$$

For incompressible fluids, conservation of mass simplifies to

$$\dot{Q}_{in} = \dot{Q}_{out} \tag{4.3}$$

If there is one inlet but three outlets, (4.1) can be extended to

$$\dot{m}_{in} = \dot{m}_{1,out} + \dot{m}_{2,out} + \dot{m}_{3,out} \tag{4.4}$$

Equation (4.4) is useful when there are multiple branches from one manifold, for example, the airway in the lung dividing into different lobes.

4.2.2   Inertial and Viscous Forces

When a fluid flows, there are several forces acting on it including pressure, gravitational, and surface tension. Similar to solids, there is an inertial force as fluids have a mass, given by Newton's second law. A force applied to a body causes an acceleration (a) of that body of a magnitude proportional to the force, in the direction of the force, and inversely proportional to the body's mass. This is termed the inertial force, which tends to keep the fluid flowing. In the mathematical form, this can be written as

$$F = ma = (Q\rho)a = m\frac{dV}{dt} = \frac{d(Q\rho V)}{dt}$$

In the *English engineering system* (lb_m, lb_f, ft, and s), gravitational constant g_c is used to convert lb_m to slugs (the mass units in the British gravitational system), and g_c is 32.2 (lb_m ft)/(lb s^2). In the SI system, g_c is 1. Hence, g_c is not included in subsequent discussion in this chapter.

Another force depends on the stickiness of the fluids. The velocity of blood in contact with the stationary pipe wall is not the same as in the center of the pipe.

The stationary wall drags the fluid, while the fluid far away from the wall has lesser dragging from the wall. Since fluids are continuous media, changes in velocity within the fluid depends on how fluid stickiness. In any case, the velocity increases continuously from the end of the wall to the center of the pipe and decrease again. A schematic showing various velocities in the fluid medium at different distances from the wall is called a velocity profile. As particles of fluid next to each other are moving with different velocities, one could say that they are sheared. The force acting is called the shear force and depends on the stickiness. Shear stress (t, unit is N/m^2) is the force per unit area of the surface required to produce the shearing action. Due to different velocities in adjacent layers, there is shear stress in every layer. Newton's law of viscosity states that shear stress is proportional to the velocity at which the adjacent layers move with respect to each other, which is called the shear rate $\dot{\gamma}$ (unit is s^{-1}). In the equation form, this is represented as

$$\tau = \mu \frac{dV}{dy} \text{ or } \tau = \mu \dot{\gamma} \tag{4.5}$$

where μ is the Newtonian viscosity (or absolute viscosity) of the fluid. Viscosity is an intrinsic property of a fluid indicating the internal friction of a fluid (i.e., viscosity is a quantitative measure of slowing the motion due to contact with adjacent layers). The SI unit of viscosity is Pa.s, equivalent to N.s/m^2 and kg/m·s. In CGS system, unit of viscosity is poise (P) or dyne·s/cm^2. Since poise is a very large unit, centipoise (cP) is the most commonly used unit. Viscosity of water is around 1 cP (0.76 cP at 37°C) and that of plasma is around 1.2 cP. Viscosity of white blood cells is 3 to 4 cP. Fluids that obey (4.5) are referred to as Newtonian fluids and those that deviate from the law are referred to as non-Newtonian fluids. Blood is considered non-Newtonian although plasma alone is a Newtonian fluid. Mucus, saliva, and synovial fluid are non-Newtonian. Biological fluids used as substitutes in various applications such as buffer solutions are Newtonian fluids. Generally, the viscosity of liquids decreases with increasing temperature. However, viscosity of gases increases with increase in temperature.

Kinematic viscosity is the ratio of the viscosity of a fluid to its density, and indicates the fluid resistance to flow under the influence of gravity. To differentiate from kinematic viscosity, the quantity defined in Newton's law of viscosity is sometimes called dynamic viscosity, or absolute viscosity. The SI unit of kinematic viscosity is m^2/s, which is too large and not commonly used. A commonly used unit is cm^2/s, which is given the name stokes (St) after Irish mathematician and physicist G. G. Stokes.

In the majority of the cases, fluid flow characteristics are controlled by two forces: viscous forces and the inertial forces. When viscous forces dominate at low velocities of fluid flow, the flow is characterized by a smooth motion of the fluid and is called laminar flow. Laminar flow can be thought of as fluid divided into a number of layers that flow parallel to each other without any disturbances or mixing between the layers. On the other hand, when inertial forces dominate, the fluid exhibits a disturbed, random motion called turbulent flow. Turbulent flow can be thought of fluid containing small eddies, like tiny whirlpools that float downstream appearing and disappearing. Laminar flow is well organized and very efficient relative to turbulent flow, which is disturbed and accompanied by high energy losses. Turbulent flow is undesirable in blood circulation due to the excessive workload it would put on the heart in addition to affecting cell viability. Newton's law of viscosity is applicable to laminar flow conditions only. When viscous forces are negligible, flow is said to be inviscid.

A useful index to determine whether the flow is laminar or turbulent is the ratio of the inertial forces to the viscous forces. This dimensionless ratio is known as the Reynolds number (N_{Re}), in honor of British engineer Osborn Reynolds.

$$N_{Re} = \frac{\rho l_c V}{\mu} \tag{4.6}$$

where ρ [kg/m³] is the density of the fluid, l_c is the channel characteristic length [m], V [m/s] is the average velocity of the fluid over the cross-section, and μ [kg/m·s] is the absolute viscosity of the fluid. For flow through circular conduits,

$$N_{Re} = \frac{\rho D V}{\mu} \tag{4.7}$$

where D is the tube diameter. For conduits with noncircular cross-sections, four times the hydraulic radius (R_h) is used as the diameter; R_h is the ratio of the cross-sectional area (A) to the wetted perimeter (Pe). Typically in a smooth-surfaced tube, flow is laminar for N_{Re} less than 2,100. Except through the aorta, the average N_{Re} is sufficiently low for the flow to be laminar throughout the circulatory system. Flow is turbulent for N_{Re} greater than 4,000. Fully developed turbulent flow is observed starting at N_{Re} of 10,000 for a rough-surfaced tube and starting at N_{Re} equal to 10^8 for a smooth tube. Flow where N_{Re} is between 2,100 and 4,000 is called *transitional flow*. The range of N_{Re} for which this type of flow exists varies considerably, depending upon the system configuration, stability, and the roughness of the tube surface.

EXAMPLE 4.1

Calculate the Reynolds number for plasma flowing in a 1.5-mm artery at a velocity of 45 cm/s, and state the flow regime.
    Solution: From (4.3),

$$N_{Re} = \frac{1.026[\text{g/cm}^3]*0.15[\text{cm}]*45[\text{cm/s}]}{0.012[\text{g/cm.s}]} = 577.125$$

The flow is in the laminar region.

4.2.3   Conservation of Momentum

According to the conservation of momentum, the total momentum of a system or an object is a constant (i.e., momentum is neither created nor destroyed). For example, if a moving object collides with a stationary object, the total momentum before the collision, plus any additional impulse from outside the system, will equal the total momentum after the collision. Momentum is changed through the action of forces based on Newton's second law of acceleration. In general, the change in momentum must equal the sum of the forces acting on the mass. Consider an incompressible steady laminar flow in the z-direction in a long rigid tube (Figure 4.2) of radius r. In this case, using cylindrical polar coordinates (r, ϕ, z) is simpler than using Cartesian coordinates. Assume that the flow is a completely developed laminar flow (i.e., velocity profile shows a parabolic shape and the entry effects are stabilized). The forces on the fluid consist of the pressure force, the gravitational force, and the viscous force. If the tube is horizontally placed, the gravitational forces can be neglected. However, if the tube is oriented vertically or at an angle, then gravitation forces ($=\rho gh$) need to be considered and the tube must be balanced by the hydrostatic pressure. For a horizontal tube, the difference in the pressure forces must be balanced by the difference in the viscous forces. Within the flowing fluid, consider a differential element of length Δz in the direction of the flow and radial thickness Δr at r radius in the flow system. The pressure force, given as the product of the area and the pressure acting on it, is

$$\text{Pressure force} = 2\pi r \Delta r(P + \Delta P) - 2\pi r \Delta r P$$

The fluid at $r + \Delta r$ exerts a shear stress on the fluid in the shell in the positive z-direction, which is designated as $\tau_{rz} + \Delta\tau_{rz}$. This notation means "shear stress evaluated at the location $r + \Delta r$." Multiplying the surface area of the differential element at that location yields a force $2\pi (r + \Delta r) \Delta z \tau_{rz}(r + \Delta r)$ in the z-direction. In the same way, the fluid at the location r exerts a stress $t_{rz}(r)$ on the fluid below (i.e., at a smaller r). As a result, the fluid below exerts a reaction on the shell fluid that has the opposite sign. The resulting force is $-2\pi r \Delta z \tau_{rz}$. Therefore, the total force arising from the shear stresses on the two cylindrical surfaces of the shell fluid is written as

$$\text{Viscous forces} = 2\pi(r + \Delta r)(\tau_{rz} + \Delta\tau_{rz})(r + \Delta r)\Delta z - 2\pi r \tau_{rz}\Delta z$$

Using the conservation of momentum principles gives

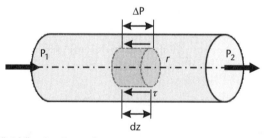

Figure 4.2   Forces in fluid flowing in a tube.

$$\sum F_x = \left\{ 2\pi(r+\Delta r)\left(\tau_{rz} + \Delta\tau_{rz}\right)(r+\Delta r)\Delta z - 2\pi r\tau_{rz}\Delta z \right\} - \left\{ 2\pi r\Delta r(P+dP) - 2\pi r\Delta r.P \right\} = 0$$

Neglecting the product of $\Delta r \cdot \Delta\tau$, and assuming the differential elements approach zero, the above equation is written in the form

$$(r+dr)\tau - r\tau = rdr\frac{dP}{dz}$$

Rearranging the terms using chain rule for τ and r results in

$$\frac{d(r\tau)}{dr} = r\frac{dP}{dz}$$

This equation applies to both laminar and turbulent flow in one-dimension, and to Newtonian and non-Newtonian fluids. For Newtonian fluids and the laminar flow condition, (4.5) is substituted to obtain

$$\frac{1}{r}\frac{d}{dr}\left(r\frac{dV}{dr}\right) = \frac{1}{\mu}\frac{dP}{dz}$$

where V is the velocity in the z-direction. dP/dz can be approximated by $-(P_1-P_2)/L$ where L is the tube length. Since dP/dz is not a function of r, integrating twice with respect to r, and applying the boundary condition that dV/dr at $r = 0$ (due to symmetry) and $V_x = 0$ at $r = R$ (with no slip of the fluid at the wall assumption, which is true for the majority of the laminar flow conditions), we get

$$V = \frac{(P_1 - P_2)(R^2 - r^2)}{4\mu L} \tag{4.8}$$

Equation (4.8) is similar to an equation of a parabola, where V is dependent on r^2. Hence, plotting V at various r values using (4.8) yields a parabolic velocity profile over the tube cross-section. The maximum velocity, V_{max}, occurs at the central axis of the tube where $r = 0$. Integrating (4.8) over the cross-section provides an expression for the volumetric flow rate as

$$Q = \frac{\pi R^4 \Delta P}{8\mu L} \tag{4.9}$$

Equation (4.9) is commonly known as the Poiseuille equation or Hagen-Poiseuille equation due to the work of Hagen in 1839 and Poiseuille in 1840.

Dividing (4.9) by the cross-sectional area, the average velocity, V_{ave}, is obtained. Furthermore, we can show that

$$V_{ave} = V_{max}/2$$

Equation (4.9) is also written in a simple form as $\Delta P = Q*R$ where R is the resistance and $R = \dfrac{8\mu L}{\pi R^4}$. Wall shear stress is calculated using (4.5) in conjunction with (4.9) to obtain

$$\tau_w = \frac{d}{4}\frac{\Delta P}{L} = 2d\frac{\mu Q}{\pi R^4} \tag{4.10}$$

Equation (4.10) is useful to calculate the shear stress experienced by the stent, inserted to support a blood vessel or an intestinal wall. If the adhesion of the stent is weaker than the shear stress, this would lead to movement of the stent in the flow direction.

EXAMPLE 4.2

A Fleisch-type pneumotachometer is a device that measures lung volumes from flow measurements. It consists of a number of parallel capillary tubes that add resistance to the flow. For an athlete, the pressure drop was found to be 95 Pa at a normal level. Using the calibrated resistance value of 0.32 kPa.s/L for air at 25°C and 1 atm, calculate the flow rate. If the breathing rate is 12 per minute, what is the lung volume? If the athlete is practicing in an area where the outside temperature is 3°C and pressure is 0.95 atm, what is the error introduced in the volume measurement?

Solution:

$$Q = \frac{\Delta P}{R} = \frac{95[Pa]}{0.32*10^3[Pa.s/L]} = 0.2969 \text{ L/s}$$

$$\text{Volume} = \int_0^t Q dt$$

Assuming the Q to be independent of t, Volume = Q * Δt
Δt = time taken for each breath = 60/12 = 5 seconds
Hence
V = 0.2969*5 = 1.4843L
Assuming ideal gas law,

$$\frac{P_1 V_1}{T_1} = \frac{P_2 V_2}{T_2} \rightarrow \text{Hence } V_2 = \frac{P_1}{P_2}\frac{T_2}{T_1}V_1 = \frac{1[atm]}{0.95[atm]}\frac{276[K]}{298[K]}*1.4843[L] = 1.447L$$

$$Error = 100 * \left(1 - \frac{V_2}{V_1}\right) = 100 * \left(1 - \frac{1.447}{1.4843}\right) = 2.5\%$$

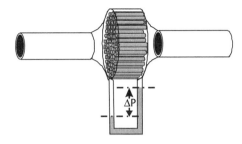

4.2.3.1   Analogy with Electrical Circuits

The Poiseuille law corresponds to Ohm's law for electrical circuits. The pressure difference moves the blood through the system. Similarly, a battery or other power supply develops a *potential difference*, which moves electrons around the circuit. Thus, a fluid pressure drop (energy per unit volume) corresponds to a potential difference or voltage drop (energy per unit charge). The flow rate of a fluid is analogous to the current in a circuit ($I = \delta q / \delta t$). From Ohm's law (discussed in Chapter 3), the quantity in parentheses is the resistance. The electrical analog of this fluid quantity (i.e., the ratio of viscosity to cross sectional area is the resistance) is

$$R = \rho L / r$$

where ρ is the resistivity, L is the length of the conductor, and r is its radius. Ohm's law is more useful than the Poiseuille law as it is not restricted to long straight wires with constant current. However, the Poiseuille law is more useful in knowing the dependency on pipe diameter, pipe length, and liquid viscosity.

   In biological systems, many branches stem from a single supply, either in the circulatory system or in the respiratory system. From the conservation of mass principle (4.15) the flow into a branch equaled the sum of the flows out. In the electrical case, the same is true for currents at a junction of conductors: conservation of charge. To find the total resistance in a branched circuitry in series and parallel configurations, the principles of Kirchoff's law are used. They can also be adapted to flow systems with the following rules:

   When components are *in series* (Figure 4.3),

1.   The total pressure drop is equal to the sum of the pressure drops across each component. That is,

$$\Delta P = \Delta P_1 + \Delta P_2 + \Delta P_3$$

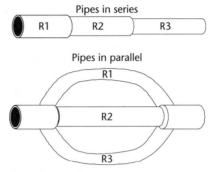

Figure 4.3   Flow in series and parallel circuits.

2. The volumetric flow is the same in each component. If an arteriole and a venule were joined by a single capillary, the fluid flow through them all would be the same.

$$Q = Q_1 = Q_2 = Q_3$$

Substituting the Poiseuille equation, we get

$$R_T = R_1 + R_2 + R_3 \tag{4.11}$$

When components are *in parallel*,

1. The pressure drop across each component is the same (i.e., $\Delta P_1 = \Delta P_2 = \Delta P_3$).
2. The total flow is equal to the sum of the flows through each component, as indicated in (4.13). Substituting the Poiseuille equation (4.9), we get

$$\frac{\pi r^4}{8\,\mu L} = \frac{\pi r_1^4}{8\,\mu L_1} \frac{\pi r_2^4}{8\,\mu L_2} + \frac{\pi r_3^4}{8\,\mu L_3} \tag{4.12a}$$

or

$$\frac{1}{R_T} = \frac{1}{R_1} + \frac{1}{R_2} + \frac{1}{R_3} \tag{4.12b}$$

This implies that the resistances through alternate paths determine the flow rate through each branch. More analogies between electrical systems and fluid flow are discussed in Chapter 10.

EXAMPLE 4.3

Consider cardiac bypass surgery. Blood supply to the heart occurs through coronary arteries as shown in the figure. Consider one of the arteries to be 3 mm in diameter and have a length of 2 cm. The average velocity at which blood flows in that artery is 1.3 cm/s. Assuming the density of blood to be 1.056 g/cm³ and viscosity to be 3 cP.

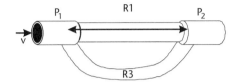

(a)   Calculate the pressure drop over the length of the vessel.
(b)   Calculate the shear stress at the wall.
(c)   Suppose an identical vessel has an atherosclerotic layer on its inner wall, which reduces the diameter to 300 mm. For the same volumetric flow rate, what are the pressure drop and wall shear rate in the diseased vessel?
(d)   To reduce the pressure drop in the diseased vessel back to its value for a healthy vessel, a bypass tube is connected to the start and end of the diseased vessel and the bypass tube. The length of the bypass tube is 4 cm and its diameter is constant. What diameter should the bypass tube have if the pressure drop is constrained to be equal to or lower than the original pressure drop for the healthy vessel?

Solution:

(a)   Using (4.9), $\Delta P = \dfrac{8Q\mu L}{\pi r^4} = \dfrac{8V\mu L}{r^2}$

$$\Delta P = \frac{8*1.3*0.03*2}{0.15^2} = 27.733 \text{ dyne/cm}^2 = 27.733/1{,}333 \text{ mmHg} = 0.020 \text{ mmHg}$$

(b)   From (4.10) $\tau_{rz} = \dfrac{\Delta P}{2L} r$ and shear stress at the wall is

$$\tau_{rz}\big|_{r=R} = \frac{\Delta P}{2L} R = \frac{27.33*0.15}{2*2} = 1.050 \text{ dyne/cm}^2$$

(c)   Let subscript n = normal and d = diseased.
      From (4.1), $Q_n = Q_d$

$$\pi r_n^2 v_n = \pi r_d^2 v_d \text{ or } v_d = v_n \left(\frac{r_n}{r_d}\right)^2 = 100 v_n$$

$$\Delta P = \frac{8*100*1.3*0.03*2}{(0.15*0.1)^2} = 27.333*10^4 \text{dyne/cm}^2 = 208.000 \text{ mmHg}$$

From (4.10),

$$\tau_{wall}\big|_2 = \frac{\Delta P_2 r_2}{2L} = \frac{27.733*10^4*0.15*0.1}{2*2} = 1{,}040.000 \text{ dyne/cm}^2$$

(d) With the bypass in place, the pressure drop is restored to the value for a healthy vessel.
Let subscripts n = normal, d = diseased, and b = bypass.
From (4.26),

$$\frac{\pi r_n^4}{8\mu L_n} = \frac{\pi r_d^4}{8\mu L_d} + \frac{\pi r_b^4}{8\mu L_b} \quad \frac{r_n^4}{L_n} = \frac{r_d^4}{L_d} + \frac{r_b^4}{L_b} \rightarrow \frac{(0.15)^4}{2} = \frac{(0.015)^4}{2} + \frac{r_3^4}{4}$$

Solving for r_3 = 0.1784 cm or 1.784 mm.

4.2.3.2 Flow Past Immersed Objects

Biological fluids contain particles such as cells, large molecules, therapeutic particles, and environmental contaminants. An immersed object moving through a fluid has to overcome resistances which depend on the relative velocity between fluid and solid, the shape of the solid, the density of the gas, and its viscosity. When the fluid flows, these particles would experience a drag force due to the shear effects of flowing fluid. Also, work has to be done to get fluids move around objects and this leads to a pressure difference between the front and back of the object. Understanding the influence of these forces is important to many areas such as recruitment of blood cells to a site of injury, drug delivery, and deposition of toxic substances in the airway. The drag force, F_D (unit is N), caused by both viscous friction and unequal pressure distributions is calculated using an empirical equation

$$F_D = \frac{1}{2} C_D \rho V^2 A \tag{4.13}$$

where C_D is the coefficient of drag, and A is the projected area of the object in the plane perpendicular to the flow. For example, A is πd^2 for a spherical particle. C_D provides a measure of the drag on a solid body, and changes with Reynolds number and shape of the object. For a spherical particle, C_D is 0.44 when N_{Re} is greater than 1,000. At N_{Re} less than 0.1, the flow is called creeping flow and C_D is given by

$$C_D = \frac{24}{N_{Re}} \tag{4.14}$$

Here, N_{Re} is calculated using the fluid density and viscosity values. At these low N_{Re} values, viscous dissipation dominates.

EXAMPLE 4.4

One of the recently developed treatments for asthma is administering immunoglobulins as aerosol drug delivery particles. The goal is to deliver the drug through the nasal cavity.

While studying the drug release, a first task is to understand whether aerosols settle down or not. For this purpose, one could carry out an experiment by suspending the particles (10-mm size) in air at 20°C (density 1.2 kg/m³, 1.8×10^{-4} g/cm.s) inside a 30-cm-tall measuring jar. If the density of the particle is 1.01 g/cm³, how long will it take for the particle to settle from the top of the measuring jar?

Solution: In this case there is no fluid flow. Thus, (4.17) becomes

$$\sum F_y = \sum \dot{m}_{in}V_{in} - \sum \dot{m}_{out}V_{out} = 0$$

For objects in fluids, there are three forces acting on the particle:

(a) Gravity, $F_g = mg = \rho_p \Psi g$ (acting downward)

(b) Buoyant force, $F_b = m_{\text{fluid displaced}}g = \rho_f \Psi g$ (acting upward)

(c) Hydrodynamic drag, $F_D = \dfrac{1}{2}C_D \rho_f V^2 A$ (acting upward opposing settling)

$$\sum F_y = \frac{1}{2}C_D \rho_f V^2 A + \rho_f \Psi g - \rho_p \Psi g = 0$$

Rearranging,

$$V = \sqrt{\frac{2\Psi}{A}\frac{g}{C_D}\frac{(\rho_p - \rho_f)}{\rho_f}} = \sqrt{\frac{4}{3}D_p\frac{g}{C_D}\frac{(\rho_p - \rho_f)}{\rho_f}} \qquad (E4.1)$$

Assuming a $N_{RE} < 0.1$, using (4.14),

$$C_D = \frac{24}{N_{Re}} = \frac{24\mu_p}{\rho_f D_p V}$$

Substituting into (E4.1),

$$V = \frac{1}{18}\frac{(\rho_p - \rho_f)D_p^2 g}{\mu_f} = 0.3 \text{ cm/s}$$

To confirm whether the N_{RE} assumption is true, calculate the N_{RE}.

$$N_{Re} = \frac{\rho_f D_p V}{\mu_f} = \frac{0.001 * 0.001 * 0.3}{0.00018} = 0.0017$$

Hence, the assumption is valid. Furthermore, the time taken for settling is $t = L/V = 30/0.3 = 100$ seconds.

4.3   Nonidealities in Biological Systems

4.3.1   Oscillatory and Pulsating Flows

Majority of the flow systems in the body including blood flow and gas exchange vary with time. This is referred to as *unsteady* flow. If the flow has a periodic behavior and a net directional motion throughout the cycle, it is called *pulsatile* flow. Furthermore, if the fluid flow has a periodic behavior and oscillates back and forth without a net forward output, it is called *oscillatory* flow. Body fluid flow is either oscillatory or pulsatile. This would be in contrast to many manmade piping systems in which fluid flow is continuous. Both oscillatory and pulsatile flows occur naturally in respiratory, vascular, and acoustic flows in the ear. For example, blood is ejected from the heart during systole due to which a pressure pulse propagates along the arterial tree. The pulse propagation is wavelike in character with a variation in velocity (or volumetric flow rate), unlike steady state flow (Figure 4.4). Hence, N_{Re} varies as a function of time and can be higher than 4,000 despite the absence of turbulance. The shape of the pulse wave also changes as it propagates through the arterial system.

Similar to the Reynolds number, British physiologist J. R. Womersley developed a dimensionless number to characterize the dominant features of pulsatile flow. The Womersley number (N_{α}) is the ratio of transient inertial forces associated with the local acceleration to the viscous forces per unit mass. In other words,

$$N_{\alpha} = l_c \sqrt{\omega \rho / \mu} \qquad (4.15)$$

where ω is the angular velocity of the applied pressure gradient. ω is related to the frequency (f) of the pulsatile wave by $\omega = 2\pi f$ and the units are cycles per second or hertz (1 Hz is one cycle per second). While calculating N_{Re}, one needs to use the characteristic length as the radius of the tube rather than the tube diameter if l is the radius of the tube in the N_{α} definition. However, if diameter is used in calculating N_{α}, then N_{Re} is calculated using the diameter.

If α is less than 1, then viscous forces dominate. The velocity profile will be similar to a steady state condition, and flow is said to be in a quasi-steady state. When α is greater than 1, the importance of the transient inertial force relative to the viscous force is greater. Hence N_{α} is used to assess the occurrence of turbulence in a pulsatile flow. Condition in the human aorta can result in an N_{α} of 10 or more;

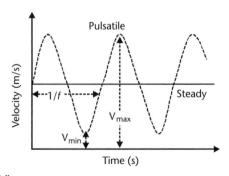

Figure 4.4   Pulsatile fluid flow.

N_α is less than 0.001 in capillaries. The Womersley criterion can also be applied in the respiratory system.

Another dimensionless number used in characterizing the oscillating and pulsatile flow conditions is called the Strouhal number (N_{St}), named after the Czech physicist V. Strouhal. The N_{St} represents a measure of the ratio of inertial forces due to the unsteadiness of the flow or local acceleration to the inertial forces due to changes in velocity from one point to another in the flow field. N_{St} is defined as the

$$N_{St} = fl_c/V \tag{4.16}$$

Flow with a very small N_{St} suggests that the flow is quasi-steady. When the fluid encounters a solid obstacle in the flow path, the fluid flow near that obstacle moves slower than the flow farther away. This forms what is known as the boundary layer. When an adverse pressure gradient is created from the flow at the boundary, the flow separates the boundary. This causes the inception of a vortex, a swirling motion in the fluid. At higher N_{Re}, the vortex on one side of the obstacle grows larger relative to other side and continues to build until it cannot stay attached to the boundary. The vortex separates, which is called vortex shedding. The presence of curvatures and branching in the blood vessels can also favor the formation of vortices. These vortices appear as regular periodic phenomena and do not exhibit a chaotic flow pattern associated with turbulent flow conditions. N_{St} is useful in determining the vortex shedding, which typically occur when N_{St} is approximately 0.2. N_{St} is related to N_{Re} and N_α by

$$N_{St} = \frac{N_\alpha^2}{2\pi N_{Re}}$$

Using a similar derivation for the pulsatile flow, Womersley's modified formula for shear stress at the wall is

$$\tau_w = \frac{N_\alpha}{\sqrt{2}} \frac{\mu Q}{\pi R^3} \tag{4.17}$$

Only if the N_α value is smaller than 1, the pulsatile flow follows the Poiseuille flow.

EXAMPLE 4.5

A Palmaz-Schatz stent was introduced to a patient into a 5 mm in diameter artery. The velocity of blood could be estimated to be 0.1 m/s and the frequency of pulsation to be 0.02 Hz. Calculate the Womersley number and Strouhal number for the scenario assuming the viscosity of blood to be 3 cP and density to be 1.056 g/cm³. What is the stress experienced by the stent? Compare that to the steady-state level wall of shear stress.

Solution: $Q = \pi * 0.25^2 [\text{cm}]^2 * 10 [\text{cm/s}] = 1.963 [\text{cm}^3/\text{s}]$

From (4.4),

$$N_\alpha = 0.5[\text{cm}]\sqrt{\frac{2\pi * 0.02[s^{-1}] * 1.056[g/cm^3]}{0.03[g/cm.s]}} = 0.911$$

$$N_{Re} = N_{Re} = \frac{1.056[g/cm^3] * 0.5[\text{cm}] * 10[cm/s]}{0.03[g/cm.s]} = 176$$

$$N_{St} = N_\alpha^2 / 2\pi N_{Re} = 0.00075$$

Using (4.17)

$$\tau_w = \frac{0.911}{\sqrt{2}} \frac{0.03[g/cm.s] * 1.963[cm^3/s]}{\pi * 0.25^3[\text{cm}]^3} = 0.772 \text{ dyne/cm}^2$$

At steady state using (4.10)

$$\tau_w = 2 * 0.5[\text{cm}] \frac{0.03[g/cm.s] * 1.963[cm^3/s]}{\pi * 0.25^4[\text{cm}]^4} = 4.8 \text{ dyne/cm}^2$$

Thus, the pulsatile flow approximation is nearly five times lower than steady flow calculations.

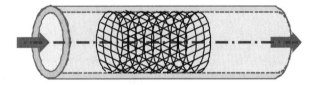

4.3.2 Alterations in Viscosity

Determination of a viscosity change occurring in body fluids, especially blood or other hematologic fluids, is important in many applications. For example, it enables a blood property to be readily determined. Abnormal viscosity values can be used to diagnose and assess the severity of diseases such as sickle cell disease, diabetes, cerebral infarction (i.e., stroke), and myocardial infarction (i.e., heart attack). If shear stress does not vary linearly with shear rate as described in (4.5), then the fluid is described as non-Newtonian. There are several types of non-Newtonian fluids, characterized by the way viscosity changes in response to shear rate [Figure 4.5(a)]. *Bingham plastics* resist a small shear stress but flow easily under larger shear stresses. (e.g., toothpaste, jellies, mashed potatoes, and modeling clay). *Pseudo-plastics* are shear-thinning fluids, which show decreased viscosity with increasing velocity gradient (e.g., polymer solutions and blood). *Dilatant fluids* are shear-thickening fluids where viscosity increases with an increasing velocity gradient. They are uncommon, but suspensions of starch and sand behave in this way.

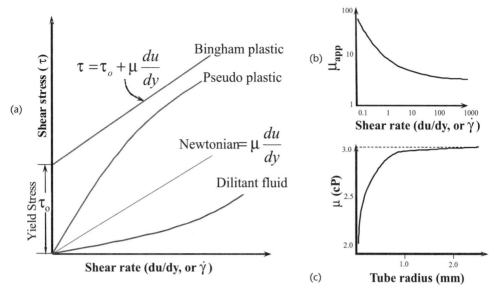

Figure 4.5 Alterations in viscosity: (a) non-Newtonian fluids, (b) apparent viscosity, and (c) the Fahraeus-Lindqvist effect.

Many of the biological fluids are non-Newtonian. Although non-Newtonian behavior has a low effect on flow resistance, its effect on flow separation is more significant near the boundaries. Non-Newtonian fluids can be grouped into two subgroups, one where properties are independent of time under shear and the other dependent on time.

The viscosity of non-Newtonian fluids changes with shear rate. Thus, experimental parameters such as the shear rate affect the measured viscosity while measuring the viscosity of these solutions. To distinguish from Newtonian viscosity, the measured viscosity is called the apparent viscosity (μ_{app}). It is calculated using the local slope of a $\tau - \dot{\gamma}$ curve and accurate when explicit experimental parameters are adhered to. For example, μ_{app} of blood at 37°C is highly sensitive to shear rates less than 100 s^{-1} [Figure 4.5(b)] and decreases with an increase in shear rate. Apparent viscosity of synovial fluids is around 10 kg/m-s at 0.1 s^{-1} shear rate (a rate resembling knee flexes during very slow walking), which decreases to ~ 0.1 kg/m.s at 10 s^{-1} (very fast running). Similarly, μ_{app} of respiratory tract mucus is 1 kg/m-s at 0.1–1 s^{-1} shear rates, which decreases to ~ 0.01 kg/m-s at 100–1,000 s^{-1} shear rates.

Normal blood and some deoxygenated sickle cell blood are classified into shear-thinning (non-Newtonian) fluids, and can be described by Casson's equation [1], an empirical relationship for fluids deviating from the ideal Bingham plastic behavior.

$$\sqrt{\tau} = \sqrt{\tau_0} + \sqrt{\mu_c \dot{\gamma}} \tag{4.18}$$

where τ_0 is the yield stress (0.04 dyne/cm^2 at 37°C) and μ_c is the Casson viscosity, which is the viscosity at high shear rates. Casson's equation suffices for time-dependent, one dimensional simple stream patterns. For complicated flow patterns, it is necessary to acquire experimental data, because there is no general constitutive equation for blood. Also, blood does not comply with Casson's equation at a very low shear rate (< 0.1 s^{-1}).

When blood is allowed to flow through capillaries of decreasing size, a second non-Newtonian characteristic is observed. Below a critical vessel diameter, blood viscosity becomes dependent upon vessel radius [Figure 4.5(c)]. The critical radius is approximately one mm. Viscosity falls sharply down to a vessel diameter of approximately 12–15 mm. This phenomenon is known as the Fahraeus-Lindqvist effect. The dynamic hematocrit, the hematocrit of blood when it is actually moving as opposed to the bulk hematocrit, which is measured after the blood is drawn from a person and spun down for settling, decreases below the bulk hematocrit in tubes down to a diameter about 15 mm.

Dynamic hematocrit determined in arterioles may have a value of only 25%. One reason for this decrease is that the erythrocytes pass through these smaller tubes much faster than the plasma, because they are near the axis (where maximum velocity occurs) while most of the plasma is near the wall. A second reason is that RBCs are excluded from the smaller vessels, in a process called plasma skimming. This is because RBCs, being relatively large compared to the openings in small arterioles, have difficulty gaining entrance. Regardless of the precise mechanism, a relative decrease in an erythrocyte number causes a decrease in the viscosity. Using the marginal zone theory [2], a relationship can be obtained to calculate μ_{app}.

$$\mu_{app} = \mu_p \left[1 - \left(1 - \frac{\delta}{R} \right)^4 \left(1 - \frac{\mu_p}{\mu_b} \right) \right]^{-1} \qquad (4.19)$$

where μ_p is the viscosity of the plasma, μ_b is the central-blood viscosity, δ is the cell-free plasma layer thickness, and R is the radius of the capillary.

Variation in blood viscoelasticity among healthy individuals is very small. Thus, changes due to disease or surgical intervention can be readily identified, making blood viscoelasticity a useful clinical parameter. For example, the viscoelasticity of an individual's blood changes significantly as the result of cardiopulmonary bypass (CPB) surgery. Examination of a group of patients undergoing CPB found that the changes seen are not solely due to changes in hematocrit but also may be a due to the combined effects of:

• Dilution of plasma proteins by the priming solution;
• Changes in plasma viscosity;
• Effects of the priming solution on aggregation and deformability of the RBCs;
• Effect of blood thinning medications.

The Fahraeus-Lindqvist effect provides several important advantages for the body. The fact that blood viscosity declines in smaller vessels is beneficial since less resistance is incurred in the vascular circuit. Therefore lower perfusion pressures, lower transmural (blood) pressures, and a smaller pump (heart) are required to adequately perfuse the tissues. It is fortuitous that the very vessels in which the

Fahraeus-Lindqvist effect occurs are also the vessels with the highest flow resistance (i.e., the arterioles). If a Newtonian fluid of the same viscosity was substituted for non-Newtonian blood, flow resistance would be much higher.

4.3.3   Influence Fluid Flow on Blood

Blood behaves as a Newtonian fluid only in regions of high shear rate (> 100 s^{-1}) and, for flow in large arteries where the shear rate is well above 100 s^{-1}, a value of 3.5 cP is often used as an estimate for the viscosity of blood [Figure 4.5(b)]. In smaller arteries and in the capillaries where the shear rate is very low, blood must be treated as a non-Newtonian fluid. The non-Newtonian behavior is traceable to the elastic red blood cells (RBCs). RBCs take up about half the volume of *whole blood*, and have a significant influence on the flow. The specific gravity of whole blood is 1.056 in which plasma is 1.026, and RBCs are 1.090. The mean hydrostatic pressure falls from a relatively high value of 100 mmHg in the largest arteries to values of the order of 20 mmHg in the capillaries and even lower in the return venous circulation. The parameter used for modeling purposes is the hematocrit level, the volume fraction that RBCs occupy in a given volume of blood. The average hematocrit is 40–52% for men, and 35–47% for women. Average blood volume in humans is about 5 liters, with 3 liters of plasma and 2 liters representing the volume of the blood cells, primarily the RBCs. When the RBCs are at rest, they tend to aggregate and stack together in a space-efficient manner (region 1 in Figure 4.6). In order for blood to flow freely, the size of these aggregates must be reduced, which in turn provides some freedom of internal motion. The forces that disaggregate the cells also produce elastic deformation and orientation of the cells, causing elastic energy to be stored in the cellular microstructure of the blood (region 2 in Figure 4.6). As the flow proceeds, the sliding of the internal cellular structure requires a continuous input of energy, which is dissipated through viscous friction (region 3 in Figure 4.6). These effects make blood a viscoelastic fluid, exhibiting both viscous and elastic properties (discussed in Chapter 5). In certain cases, RBCs stack together in presence of serum proteins (particularly increased fibrinogen and globulins) or other macromolecules. When four or more RBCs are arranged in a linear pattern as a "stack of coins," it is known as *rouleaux formation*. Such long chains of RBCs sediment more readily. This is the mechanism for the sedimentation rate, which increases nonspecifically with inflammation and increased "acute phase" serum proteins.

4.4   Conservation of Energy

4.4.1   Different Energy Forms

Energy is necessary to perform various activities including the functioning of the heart and body movement. According to the conservation of energy principle, energy cannot be created or destroyed, but only transferred from one system to another. In other words, the total energy of interacting bodies or particles in a system remains constant. Some parts are gaining energy while others are losing energy

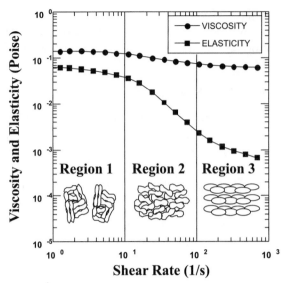

Figure 4.6 Effect of blood flow on its viscosity and elasticity. (Courtesy of Vilastic Scientific, Inc. and [3].)

all through the fluid. Another form of conservation of energy is the first law of thermodynamics.

The total energy of the system encompasses several types of energy including kinetic energy, potential energy, internal energy, and others due to chemical reactions, electrostatic interactions, and magnetic field effects.

Kinetic energy (E_k) is due to the motion of the system. For example, moving fluid has kinetic energy because of its motion, similar to point particle moving with velocity V. If the fluid accelerates, it has more kinetic energy than before. E_K is given by

$$E_k = \frac{1}{2}mV^2 \Rightarrow [\text{kg}][\text{m/s}]^2 = \text{J}$$

$$\dot{E}_k = \frac{1}{2}\dot{m}V^2 \Rightarrow [\text{kg/s}][\text{m/s}]^2 = \text{J/s} = \text{W}$$

Potential energy (E_p) is due to the position of the system in a potential field. For example, flowing fluid has gravitational potential energy or stored energy due to the Earth's gravitational field ($g = 9.8$ m/s^2). Essentially, blood that is at a higher elevation has greater potential energy, and it pushes down on the blood at lower elevations. One could think of the hydrostatic pressure as a gravitational potential energy density.

$$E_p = mgz \Rightarrow \text{J} = [\text{kg}][\text{m/s}^2][\text{m}]$$

$$\dot{E}_p = \dot{m}gz \Rightarrow \text{W} = \text{J/s} = [\text{kg/s}][\text{m/s}^2][\text{m}]$$

Another example of potential energy is by the presence of a particle in an electromagnetic field.

Internal energy (U) is all the energy possessed by a system other than kinetic and potential energy. It includes the energy due to the rotational and vibrational motion of molecules within the system, interactions between molecules within the system, and motion and interactions of electrons and nuclei within molecules. According to convention, a system has no internal energy at absolute zero Kelvin. In general, U is dependent on the composition, physical state (solid, liquid, or gas), and temperature of the system. However, U is independent of pressure for ideal gases, liquids, and solids. Temperature is the depiction of the average vibrational energy of all the molecules within the system. Increase in temperature increases internal energy in solids and liquids. However, in gases and some liquids, increase in temperature at constant pressure is accompanied by a change in volume or increasing temperature at constant volume changes pressures. These changes depend on how molecules or atom rearrange themselves as the space between their neighbors is altered. Since determining U is cumbersome, the term enthalpy (H) is utilized. Enthalpy refers to the summation of both internal energy change and the accompanying changes in volume at constant pressure. Enthalpy is also a function of temperature, chemical composition, and physical state, but a weak function of pressure. U and H are relative quantities, and values are defined with respect to a reference state.

Energy is transferred from one form to another in an open system where molecules move across the boundaries. In a closed system where no mass is transferred across the system boundaries (i.e., batch system), energy may be transferred between the system and the surroundings. Energy is transferred in two ways: heat and work.

1. *Heat* (q) is energy that flows due to a temperature difference between the system and its surroundings. Heat always flows from a high to low temperature, defined to be positive if it flows to a system (i.e., input). Heat is the quantity of energy stored or transferred by thermal vibrations of molecules. Heat is additive. If two fluids with heat energies of 10 joules and 20 joules are added together, the added masses will have a total heat energy of 30 joules. Heat is produced when the cells respire. When a person is not exercising, most of the heat is produced by the liver. During exercise, more heat is produced by the muscles. Blood flowing through these organs absorb the heat and distributes it around the rest of the body. The amount of heat energy lost by the warmer body or system must equal the gain in heat energy by the cooler body or system. The natural process continues until thermal equilibrium is reached and the system settles at a temperature between the original temperatures of the two substances. When no heat transfer occurs across the boundaries, the system is called adiabatic.

2. *Work* (W) is energy flows in response to any driving force (e.g., applied force, torque) other than temperature. Work is a scalar quantity and obtained by multiplying the applied force and the distance traveled due to the applied force. By convention, work is defined as positive if it flows from the system (i.e., output), work may come from a moving piston or moving turbine.

In an open system, material crosses the system boundary as the process occurs (e.g., continuous process at steady state). In an open system, work must be done to push input fluid streams at a pressure P_{in} and flow rate Q_{in} into the system, and work is done by the output fluid streams at pressure P_{out} and flow rate Q_{out} on the surroundings as it leaves the system. Fluid flows due to difference in pressure. To move the fluid, pressure is applied through force. For example, a piston or a pump can be used to apply that force. In order to gain force, energy has to be lost. However, the net energy of the entire fluid is unchanged. To consider these concepts in the conservation of energy principle, a new term "flow work" is defined, which is a measure of energy expended. Flow work is sometimes referred to as the pressure energy although it is not an intrinsic form of energy. One could think of pressure as an energy density (energy per unit volume). This description is from the viewpoint that by virtue of this energy, a mass of fluid with a pressure P at any location is capable of doing work on its neighboring fluids. Consider an artery with a constriction. When blood flows through the artery, the surrounding fluid does work on the part that goes through the constricted region. The accelerating fluid is pushed from behind by the forces that produce pressure. They do work on the accelerating fluid and the accelerating fluid works on the fluid in front of it. When the fluid is accelerating, kinetic energy increases while the pressure decreases. In the equation form,

$$W_f[J] = F[N] \bullet d[m] \tag{4.20}$$

For an incompressible fluid flow in a cross-sectional area A and pressure drop ΔP, force term is replaced with the product of pressure and the cross-sectional area, that is,

$$W_f[J] = \Delta P[N/m^2]A[m^2] \bullet d[m]$$

Since the product of the cross-sectional area and the distance is volume, V,

$$W_f[J] = \Delta P V \tag{4.21}$$

Similarly, work needs to be performed to overcome resistances of the lungs and chest wall to achieve normal breathing. During breathing, the external work of breathing is performed by the ventilator and is computed from the area subtended by the inflation volume and applied pressure from either airway or tracheal. A system does not possess heat or work. Heat or work only refers to energy that is being transferred to the system. The total rate of work ($\dot{W}$) W done by a system on its surroundings is divided into two parts: flow work and shaft work, which is the work done by the process fluid on a moving part within the system (e.g., piston, turbine, and rotor).

$$\dot{W} = \dot{W}_{shaft} + \dot{W}_{flow}$$

4.4.1.1   Open Energy Balance

Consider a liquid flowing through a tank that contains a heating unit and a stirrer to mix the components (Figure 4.7). Conservation of energy equations can be developed similar to conservation of mass and momentum equations to obtain

$$\left(\dot{U}_{out} - \dot{U}_{in}\right) + \left(\dot{E}_{k,out} - \dot{E}_{k,in}\right) + \left(\dot{E}_{p,out} - \dot{E}_{p,in}\right) - \dot{Q} + (\dot{W}_s + P_{out}\dot{V}_{out} - P_{in}\dot{V}_{in}) = \frac{d\left(E_k + E_p + U\right)}{dt}$$

At steady state,

$$\left(\dot{U}_{out} - \dot{U}_{in}\right) + \left(\dot{E}_{k,out} - \dot{E}_{k,in}\right) + \left(\dot{E}_{p,out} - \dot{E}_{p,in}\right) - \dot{Q} + (\dot{W}_s + P_{out}\dot{V}_{out} - P_{in}\dot{V}_{in}) = 0 \quad (4.22)$$

Using the definition of enthalpy, for example, at the outlet,

$$\dot{H}_{out} = \dot{U}_{out} + P_{out}\dot{V}_{out} \qquad (4.23)$$

where $\dot{H}_{out}$ is the rate of enthalpy at the outlet with the units of J/s. In low pressure systems without any reactions taking place between the constituent and without any change in phase (i.e., liquid remains, the change in enthalpy from $T_{initial}$ to T_{final} is calculated using the relation

$$\Delta\dot{H} = \dot{m} \int_{T_{initial}}^{T_{final}} C_p dT \qquad (4.24)$$

where C_p is the specific heat of the fluid at constant pressure and ΔT is the temperature change. Typically, C_p represents the amount energy required to increase the temperature of a unit mass of substance by 1°. The value of C_p depends on the temperature and C_p is typically expressed as a function of temperature, which can be found in many handbooks. For small temperature ranges, accompanying changes in C_p are ignored and an average C_p value (Table 4.2) is used.

Substituting (4.23) into (4.22) and rearranging, the energy balance for a steady state flow system is obtained as

$$\left(\dot{H}_{out} - \dot{H}_{in}\right) + \left(\dot{E}_{k,out} - \dot{E}_{k,in}\right) + \left(\dot{E}_{p,out} - \dot{E}_{p,in}\right) = \dot{Q} - \dot{W}_s \qquad (4.25)$$

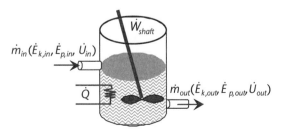

Figure 4.7 Flow of liquid through a tank containing a heater and a mixer.

Table 4.2 Specific Heat Capacities of Few Substances

	Liquid Water	Blood	Body	Air	Protein	Wood
C_p [J/kg°C]	4,187	3,780	3,470	1,006	1,700	1,700

Equation (4.25) is also referred to as the first law of thermodynamics for flow systems. This form is useful in heat transport applications such as heating or cooling physiological solutions. Even in adiabatic processes ($Q = 0$), the increase in the internal energy is due to fluid friction and can be written as head loss.

EXAMPLE 4.6

There is a 10,000-W electrical heater available for heating blood from 4°C to 37°C. Calculate what amount of blood can be heated in one second.

$$\text{Solution: } \dot{Q} = 10,000[W] = 10,000 \text{ [J/s]}$$

From Table 4.2, C_p of blood = 3,780 [J/kg°C]
There are no moving parts and assuming no change in potential energy and kinetic energy, (4.25) reduces to

$$\left(\dot{H}_{out} - \dot{H}_{in} \right) = \dot{Q} = 10,000[W] = 10,000 \text{ [J/s]}$$

From (4.24),

$$\Delta \dot{H} = \dot{m} \int_{4}^{37} 3,780 * dT$$

$$= 3,780(37 - 4)\dot{m} = 10,000$$

$$\dot{m} = 0.080 \text{ kg/s}$$

Let the *outlet* position be denoted as 2 and the *inlet* position be denoted as 1. Rearranging the conservation of energy equation, we get

$$\frac{p_1}{\rho} + \frac{V_1^2}{2} + gz_1 + \frac{\dot{W}_{shaft\,in}}{\dot{m}} = \frac{p_2}{\rho} + \frac{V_2^2}{2} + gz_2 + {}_1loss_2$$

Then dividing by g yields

$$\frac{V_1^2}{2g} + \frac{p_1}{\rho g} + z_1 + h_M = \frac{V_2^2}{2g} + \frac{p_2}{\rho g} + z_2 + h_L \tag{4.26}$$

where z_1 and z_2 are the elevations at two locations, g is the gravitational acceleration, h_M is the rate of work done on the fluid by a pump divided by the mass flow rate, and h_L is the rate of energy dissipation due to viscous losses divided by the mass flow rate.

In ideal flow, static pressure loss is observed as increased kinetic energy (increased velocity) of the fluid. Further, for an ideal inviscid fluid that does not dissipate kinetic energy as thermal energy due to shear within the fluid, the pressure loss is completely recoverable. Bernoulli's equation is the simplified form of (4.26) with assumption of: steady flow, incompressible flow, no heat transfer (either added or removed), and no viscous dissipation of kinetic energy to heat. Bernoulli's equation also states that the pressure drop between two points located along a streamline at similar heights is a function of the velocities. It is valid for flow along a single streamline (i.e., different streamlines may have different h_0).

$$\frac{p}{\rho g} + \frac{V^2}{2g} + z = \text{Constant}$$

The constant is called *total head* (h_0), which has different values on different streamlines. $v^2/2g$ is known as the *velocity head* or *dynamic pressure*, and it is the difference between the locations, which reflects the convective acceleration. z expresses the *hydrostatic* pressure increment resulting from the height differences. Bernoulli's equation applied to two locations, 2 and 1, corresponding to the outlet and inlet, respectively, is written as

$$\frac{p_1}{\rho g} + \frac{V_1^2}{2g} + z_1 = \frac{p_2}{\rho g} + \frac{V_2^2}{2g} + z_2 \qquad (4.27)$$

Although Bernoulli's equation is derived assuming steady inviscid flow, it is used in biomedical engineering with modifications for unsteady-state problems. An alternative form of Bernoulli's equation is derived as

$$\frac{p_1}{\rho g} + \frac{V_1^2}{2g} + z_1 = \frac{p_2}{\rho g} + \frac{V_2^2}{2g} + z_2 + \frac{1}{g}\int_1^2 \frac{\partial V}{\partial t} ds \qquad (4.28)$$

where s is the distance between two locations. The integral term in (4.28) expresses the effect of the temporal or local acceleration between two points, so we need to know its importance in different situations. However, unsteady flows will have unsteady streamlines and the basis for the above integration is invalid in many cases. This form of Bernoulli's equation should be used with caution and a rigorous understanding of unsteady flow is necessary.

EXAMPLE 4.7

Blood is flowing through a 6-mm diameter vessel at an average velocity of 50 cm/s. Mean pressure in the aorta be 100 mmHg. Blood density is 1.056 g/cc.

(a) If the blood were to enter a region of stenosis where the diameter of the blood vessel is only 3 mm, what would be the approximate pressure at the site of narrowing?

(b) If the blood enters an aneurysm region instead with a diameter of 10 mm, determine the pressure in the aneursym. If for conditions of vigorous exercise, the velocity of blood upstream of the aneurysm were four times the normal value (200 cm/s), what pressure would develop at the aneurysm?

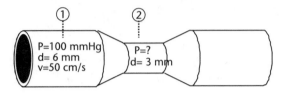

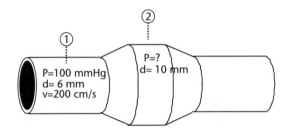

Solution:

(a) For a circular tube, (4.3) can be rewritten as

$$V_2 = \frac{V_1 A_1}{A_2} = V_1 * \left[\frac{d_1}{d_2}\right]^2$$

Since the tube is horizontal $z_1 = z_2$, (4.27) can be rewritten as

$$P_2 = P_1 + \frac{\rho V_1^2}{2}\left[1 - \left[\frac{d_1}{d_2}\right]^4\right]$$

$$P_2 = 100 \text{ mmHg} + \frac{1.056(\text{g/cm}^3)*50^2(\text{cm/s})^2*7.5*10^{-4}(\text{mmHg/(g/cm.s}^2))}{2}\left[1-\left[\frac{6}{3}\right]^4\right] = 85.1 \text{ mmHg}$$

(b) If the blood enters the aneurysm region,

$$P_2 = P_1 + \frac{\rho V_1^2}{2}\left[1 - \left[\frac{d_1}{d_2}\right]^4\right]$$

$$P_2 = 100 \text{ mmHg} + \frac{1.056*50^2*7.5*10^{-4}}{2}\left[1-\left[\frac{6}{10}\right]^4\right]\text{mmHg} = 100.86 \text{ mmHg}$$

During exercise,

$$P_2 = 100 \text{ mmHg} + \frac{1.056*200^2*7.5*10^{-4}}{2}\left[1-\left[\frac{6}{10}\right]^4\right]\text{mmHg} = 113.79 \text{ mmHg}$$

For a real fluid, shear converts kinetic energy to thermal energy in a nonre-coverable way, which is manifested as the net pressure loss. In practice, there are additional losses termed minor losses in addition to the frictional pressure loss. These minor losses are contributed by the entrance and exit effects (or inlets and outlets) of the tubes; expansions and contractions due to different sizes of tubing and catheters needed to adjust the flow rate; bends, elbows, tees, and other fittings used in the flow system; and valves (open or partially closed) used for controlling flow rate. Many times minor losses are the *largest* cause of frictional loss in a flow system. Although there is no adequate theory to predict minor losses, a commonly used empirical relation is

$$h_L = K_L \frac{V^2}{2g}$$

where K_L is the minor loss coefficient. For many standard fittings, K_L is obtained from the manufacturer or from a handbook. If there several fittings in the flow system, all minor losses are summed to obtain the total loss. These minor losses are added to the frictional loss in the tubes to obtain the total head loss as

$$h_L = \sum f \frac{l}{D}\frac{V^2}{2g} + \sum K_L \frac{V^2}{2g} \tag{4.29}$$

where f is Darcy's friction factor and l is the length of the tube. f is typically obtained using either a Moody plot or empirical correlations, which are described in [4].

4.4.2   Energy Balance in the Body

Energy is used continuously in the body and is fundamental to the physiological functions of the body such as body movement, functioning of the heart, respiration, digestion, and functioning of the brain. Metabolism is the sum of many individual chemical reactions by which cells process nutrient molecules and maintain a living state. The metabolism is grouped into *anabolism* and *catabolism*. In anabolism, a cell uses energy to generate complex molecules such as DNA, RNA, and proteins needed by the cells and perform other life functions such as creating cellular structures. In catabolism, a cell breaks down complex molecules such as carbohydrates and fats to simpler molecules to yield energy. Both processes are coupled to each other, linked by energy sources such as adenosine triphosphate (ATP), nicotinamide adenine dinucleotide (NAD), flavin adenine dinucleotide (FAD), and precursor metabolites.

For any particular chemical bond, for example a covalent bond between hydrogen and oxygen, the amount of energy required to break that bond is exactly the same as the energy released when the bond is formed. This energy is called the bond energy. Chemical reactions that release energy are called exergonic reactions, and reactions that require energy in order to occur are called endergonic reactions. The production of energy is primarily derived by the breakdown of chemical bonds from the catabolism of carbohydrates (principally glucose) and fatty acids. For example, glucose in the bloodstream diffuses into the cytoplasm of a cell where it is catabolized into CO_2 and water by the reaction

$$C_6H_{12}O_6 + 6O_2 \rightarrow 6CO_2 + 6H_2O + energy$$

This reaction is also referred as the combustion or burning of glucose. The energy produced by a breakdown of glucose in the presence of oxygen (aerobic respiration) is available for storage and use by the cell with the remaining amount dissipated as heat. This path of glucose breakdown (known as Krebs cycle) produces 30 ATP molecules per glucose molecule and it is the most efficient method of energy production. When respiratory and circulatory systems cannot deliver enough oxygen to sustain muscle contraction, for example during vigorous exercise, breakdown of glucose without oxygen (anaerobic respiration) supplies ATP (6 ATP per glucose molecule). Anaerobic respiration produces lactic acid from the breakdown of glucose, which builds up in cell. This could lead to deleterious effects such as muscle pain and fatigue.

Fatty acids are also concentrated sources of energy. For example, triglycerides are neutral lipids consisting of a glycerol backbone and three long-chain fatty acids. They represent a major form of stored energy in the adipose tissue and skeletal muscle. Unlike glucose metabolism, all ATP production from the metabolism of fatty acids is oxygen-dependent and occurs in the mitochondria. As a result, fatty acid oxidation is not as efficient as glucose as a source of energy and requires more oxygen to produce an equivalent amount of ATP. Normally a balance exists between carbohydrate metabolism to fatty acid metabolism. Typically, a decrease in glucose metabolism is observed with the increase in the metabolism of fatty acids. Obesity is characterized by an imbalance between energy intake and expenditure, resulting in a net increase in the storage of body energy primarily as fat. When there is a limited amount of carbohydrates and fatty acids such as with starvation, proteins are used as energy source.

Within a cell, mitochondria synthesize water using the hydrogen atoms removed from molecules like glucose, and the oxygen taken during cellular respiration. The released catabolic energy utilization occurs in two steps. First, the glucose is metabolized and the energy is stored in the form of ATP. Second, this ATP is used to power many activities including muscle movement. The important role of mitochondria is its ability to release the energy of glucose in small, discrete steps so that some of the energy is trapped in with the formation of ATP. Only about 50% of the energy in glucose is converted to ATP. When the third phosphate group of adenosine triphosphate is removed by hydrolysis, a substantial amount of free energy is released.

$$ATP + H_2O \leftrightarrow ADP + phosphate + energy$$

where ADP is adenosine diphosphate. ATP is used throughout the body to store energy that would otherwise be released as heat. This provides the energy for most of the energy-consuming activities of the cell. The exact amount of energy released depends on the conditions, but a value of 7.3 kcal per mole is an average value. For this reason, this bond is known as a "high-energy" bond. The bonds between the first and second phosphates are weak bonds with low bond energies. Another primary source of energy is NADH. Through the mitochondrial electron transport chain, NADH transfers two electrons and a hydrogen ion to oxygen, releasing energy.

$$NADH + H^+ + 1/2O_2 \leftrightarrow NAD^+ + H_2O + energy$$

Notice that water is necessary to supply hydrogen ions in this reaction and water is also a product. Cells also use the oxidation of flavin adenine dinucleotide (FAD) to release energy

$$FADH_2 + 1/2O_2 \leftrightarrow FAD + H_2O + energy$$

A muscle fiber contains only enough ATP to power a few twitches. A normal metabolism cannot produce energy as quickly as a muscle cell can use it, so extra storage source is needed. The ATP pool is replenished as needed by two other sources of high energy:

- *Creatine phosphate*, a phosphorylated form of creatine, is used by muscle cells to store energy. The phosphate group can be quickly transferred to ADP to regenerate the ATP necessary for muscle contraction.
- *Glycogen*, which is a storage form of glucose. Nearly 70% of glycogen is stored in the muscle, 20% in the liver, and the remaining in blood.

4.4.3   Energy Expenditure Calculations

In many scenarios, energy imbalance occurs either due to a high-energy intake, low-energy expenditure, or a mixture of the two conditions. This could result in obesity, which may be treated effectively by increasing energy expenditure. Requirements of individuals vary in relation to a wide variety of factors, making energy balance an issue of great importance in weight loss purposes or acute as well as chronic illness feeding regimens. Indirect calorimetry is a widely used method for assessing:

- Resting energy expenditure (REE) or the energy required by the body during a nonactive period, primarily used to determine the caloric requirements of patients;

- Modifying nutritional regimens through determination of the respiratory quotient (RQ), the ratio of moles of carbon dioxide produced to moles of oxygen consumed (or V_{CO_2}/V_{O_2}) over a certain time interval.

One of the formulas for predicting energy expenditure is the Harris-Benedict equation, established in 1919 by American physiologists and nutritionists J. A. Harris and F. G. Benedict ; this equation took into account gender, age, height, and weight:

Male (calories/day): (66.473 + 13.752*weight) + (5.003*height) − (6.775*age)
Female (calories/day): (655.095 + 9.563*weight) + (1.850*height) − (4.676*age)

where weight is in kilograms, height is in centimeters, and age is in years. Harris-Benedict is correct 80–90% of the time in healthy, normal volunteers. However, these formulas are skewed towards young and nonobese persons. There is a large variation between individuals, when comparing their measured energy expenditure to the calculated amount. These equations have limited clinical value when tailoring nutrition programs for specific individuals for weight loss purposes or acute as well as chronic illness feeding regimens.

In 1948, British physiologist John B. De V. Weir derived an equation to relate O_2 consumption and CO_2 production to energy released in the body. Experimentally, Weir determined the amounts of calories released from carbohydrates, fats, and proteins via metabolism with one liter of oxygen (Table 4.3). Based on the three fractions, Weir wrote three equations for oxygen consumption, CO_2 produced, and total energy released. Using these equations and neglecting the effect of protein, he arrived at representation for the overall energy (E_E) released per liter O_2 (Kcal/LO$_2$):

$$E_E[\text{Kcal/L}] = 3.941 + 1.106\frac{V_{CO_2}}{V_{O_2}} \tag{4.30}$$

Weir showed that the error from neglecting the effect of protein (i.e., assuming U_N = 0) is negligible and shows little variation between individuals. E_E can therefore be estimated from V_{O_2} and V_{CO_2} and alone. More recent studies have calculated different RQ and energy release equivalent values for each of the three substrates. To account for energy released by proteins, (4.30) was modified as

$$E_E[\text{Kcal/L}] = 3.941 + 1.106\frac{V_{CO_2}}{V_{O_2}} - 2.170\frac{U_N}{V_{O_2}} \tag{4.31}$$

Table 4.3 Weir's Data

	Carbohydrate	Protein	Fat
R.Q.	1	0.802	0.718
Kcal released/L O$_2$	5.047	4.463	4.735

where U_N is the measured urinary nitrogen production. However, the Weir equation is still used to determine the metabolic rate from indirect calorimetry. Clinically used metabolic carts essentially measure the oxygen consumed and the carbon dioxide produced by the patient and then calculate the energy expenditure for the patient using the modified Weir equation.

EXAMPLE 4.8

A person consumes 100 gm of glucose ($C_6H_{12}O_6$, MW is 180) and 100 gm of palmitoyl-stearoyl-oleoyl-glycerol ($C_{55}H_{104}O_6$, MW is 860) fat, which are both completely metabolized into CO_2 and water.

(a) What is the RQ at body temperature?
(b) What is the total energy released according to Weir equation?

Solution:

(a) Metabolic reactions for glucose can be written as

$$C_6H_{12}O_6 + 6O_2 \rightarrow H_2O + 6CO_2$$

100 gm glucose corresponds to 0.556 mol. Hence, the oxygen consumed is 3.333 mol. Since 3.333 moles of CO_2 are released for 3.333 moles of O_2 consumption, the RQ is 1.00. Metabolic reactions for fat can be written as

$$C_{55}H_{104}C_6 + 78O_2 \rightarrow 52H_2O + 55CO_2$$

100 gm of fat corresponds to 0.116 mol. Hence oxygen consumed is 9.070 mol. Number of moles CO_2 released = 0.116*55 = 6.396 mole, RQ = 6.396/9.070 = 0.705

Total O_2 consumed = 3.333 + 9.070 = 12.403 moles
Total CO_2 released = 3.333 + 6.396 = 9.729 moles.
Hence RQ = 0.784

(b) From (4.30)

$$E_E[Kcal/L] = 3.941 + 1.106 * 0.784 = 4.809 \text{ Kcal/L of oxygen}$$

Total oxygen consumed = 12.403 moles
Assuming the ideal gas law at 37°C and 1 atm,

$$V_{O_2} = 12.403[\text{moles}] * 0.082[\text{L.atm/mol.K}] * 310[\text{K}]/1[\text{atm}] = 315.284L$$

Hence, total energy released = 315.284*4.809 = 1,516.061 Kcal

4.5   Fluid Power

Power (P_w) is the rate at which work is done when a force F acts to accelerate a moving body. The units of power are watts (W) or J/s. For a translating process, power is given by the dot product of linear velocity and the force component in the direction of velocity, that is,

$$P_w[J] = F \cdot V$$

Similar to fluid work, power is expressed in the pressure and volume terms as

$$P_w = Q\Delta P \tag{4.32}$$

Thus, power output is calculated as the product of the pressure drop times the volumetric flow rate. The power required to move a fluid through a conduit is a function of the fluid velocity, the diameter of the conduit, and the fluid density and viscosity. Similarly, in the case of electrical flow through a resistor, the power, $P_{Electric}$ (watts) is

$$P_{Electric} = I\Delta\phi$$

where $\Delta\phi$ is the voltage (volts) and I is the current (amps).

4.5.1   Power Calculations in a Cardiac Cycle

Consider the power output of the left ventricle (LV) during which the aortic valves are open and blood is being ejected out of the aorta (Figure 4.8). Cardiac output (CO) is the term used to define the volume of blood pumped by the heart per unit time. The cardiac output of the average an adult male at rest is about 5 L/min. If the normal systolic blood pressure is 120 mmHg (1.60×10^5 dyne/cm²), and cardiac output is 5 L/min, then the work done by the heart is

$$P_w = Ps \times CO$$
$$= 1.60 \times 10^5 \left[\text{dyne/cm}^2\right] * 83.333 \left[\text{cm}^3/\text{s}\right] = 1.333 * 10^7 \text{ dyne/s } = 1.333\text{W}$$

In the right ventricle (RV), the average systolic pressure is 25 mmHg and a similar calculation for the right ventricle will give the power output as 0.278W. These calculations provide the pressure work that the heart must do to overcome the systolic pressure. In addition, the heart must also do kinetic work to accelerate the

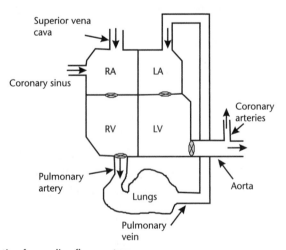

Figure 4.8   Schematic of a cardiac flow system.

blood flow. Assuming a cross-sectional aortic area of 1.5 cm², the velocity of the ejected blood can be calculated as then the velocity is 55.556 cm/s. However, blood is pumped out only in the systole of the cardiac cycle, ejection time corresponds to the systole only. Hence, the velocity is not uniform. For obtaining average velocity, acceleration is calculated and integrated over the time scale. If the length of systole is about 40% of a second, then acceleration is 138.889 cm/s². One could calculate the kinetic power by the rate of change of kinetic energy, that is,

$$P_k = \frac{1}{2}\dot{m}v^2 = \frac{1}{2}(83.33[cm^3/s]*1.06[g/cm])(138.889[cm/s])^2 = 0.125W$$

Since the velocity of blood ejected into the pulmonary artery is nearly equal to that in the aorta, the kinetic power of the right ventricle is also about 0.040W. Thus, the total power output of the heart is the equal to the sum of the pressure and kinetic terms for the right and left ventricles:

Total power = 1.333 + 0.278 + 0.125 + 0.125 = 1.861W

For comparison, a D cell battery is rated at about 4 W.h. Thus, a 100% efficient artificial heart could be run for about 2 hours on one D cell battery for an adult at rest.

EXAMPLE 4.9

The amount of mechanical energy generated per heartbeat for a normal young adult is about 1J. If mean blood pressure is 90 mmHg, calculate the stroke volume.
Solution: During each heartbeat approximately 70 cc of blood is pushed from the heart at an average pressure of 10^5 mm of Hg. Calculate the power output of the heart in watts assuming 60 beats per minute.
Cardiac work equation is $W = P*$stroke volume
$1(J)*10^7$ (ergs/J) = 90 (mmHg)*1,333 (dyne/cm³/mmHg)*SV
Hence, stroke volume = 83.3 mL

$$\Delta V = 70 \times 10^{-6} \ m^3$$

$P = 10^5 \times 10^{-3} \times 13,600 \times 10$ [Pa] = 136×10^5[Pa]
$t = 1$ second
Power = W/t = $P\Delta V/t = 136 \times 10^5 \times 70 \times 10^{-6} = 99,960 \times 10^{-5} = 0.9996W \sim 1W$

4.5.2   Efficiency of a Pump

There is always some inefficiency in converting energy from one form to another or accomplishing work. To account for the loss of energy, efficiency (η) is defined as the ratio of the work done by the system to the energy it requires to accomplish that work.

$$\eta = \frac{\dot{V} \Delta P}{E_{input}} \tag{4.33}$$

Efficiency can also be defined in terms of power as the ratio of total power output to the total power input. The anatomy of the heart allows a relatively direct calculation of the efficiency of the heart muscle. The energy of the heart is primary resourced from the oxidation of glucose, delivered to the heart muscle by the blood flow. Using the Weir equation (4.39), every milliliter (at standard temperature and pressure, STP) of oxygen used releases 20J of energy. Thus, one needs to estimate the rate of oxygen utilization by the heart to calculate the total energy consumed by the heart. Since the heart is supplied with oxygen by coronary arteries, oxygen usage is determined by the product of coronary blood flow and oxygen used per unit of blood. Normal coronary blood flow of an adult resting male is about 0.250 L/min (4.167 cm^3/s), the oxygen concentration in the arterial blood entering the heart is 0.20 cm^3 O_2/cm^3 blood and venous blood leaving the heart muscle has 0.05 cm^3 O_2/cm^3 blood.

Thus, the rate of O_2 usage = (4.167 cm^3/s)*(0.20– 0.05) cm^3 O_2/cm^3 blood = 0.625 cm^3 oxygen/s.

Total energy used = (0.625 cm^3 O_2/s)*(20 J/cm^3 O_2) = 12.5 J/s = 12W

$$\eta = \frac{P_{out}}{P_{in}} = \frac{1.691[W]}{12[W]} = 0.15 \text{ or } 15\%$$

Thus, if the efficiency of cardiac muscle is defined as the efficiency of converting ATP to power, then the efficiency of the cardiac muscle is about 30%. This is significantly higher compared to the efficiency of the best automobiles built on internal combustion engines, which is about 15%.

4.5.3 Pumps in Series and Parallel

In many disease states (e.g., congestive heart failure), where parts of the heart do not function properly, the required flow rate and power cannot be achieved by the heart. Artificial pumps can be used to support functioning of one chamber of the heart: some examples of these are: left ventricular assist devices (LVAD) and right ventricular assist devices (RVAD); left ventricular assist systems (LVAS); or two chamber support [e.g., biventricular assist device (BiVAD)] or the entire heart. While designing these pumps, some design criteria have to be considered in biomedical applications include the material with which they are made of (discussed in Chapter 6), size compactness so that the patient can move around safely, negligible vibration and noise level of the pump so that the patient is at ease, and the continuous power source. Once the pumps are developed, they can be used either in series or parallel. If two identical pumps are placed in series, their flow rate will be equal, and each will provide half of the power.

$$Q_2 = Q_1 = Q$$
$$hp_{total} = hp_1 + hp_2 \tag{4.34}$$

However, when two identical pumps are placed in parallel with their suction and discharges connected, they will produce the same head and equal flow rates.

$$Q_{total} = 2 * Q_{one}$$
$$hp_2 = hp_1 = hp \tag{4.35}$$

EXAMPLE 4.10

An LVAD is an implantable pump that can be used to treat a patient whose heart muscle is too weak to provide adequate blood flow to the circulatory system on its own. An LVAD is connected as a shunt from the pulmonary vein to the aorta. The total blood flow rate is 5.0 L/min and blood has a density of 1.056 g/mL. At rest, the mean velocity and pressure of the blood in the pulmonary vein are 30 cm/s and 6 mmHg, respectively; the average velocity and pressure of the blood in the aorta are 30 cm/s and 90 mmHg, respectively. If the LVAD provides work at the rate of 0.500 mhp at rest, what is the rate of work, in hp, provided by the patient's left atrium and left ventricle at rest?

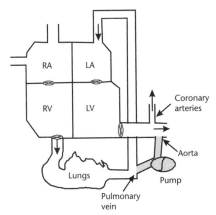

Solution: System = blood in L. atrium, L. ventricle, LVAD for expanding the work term into the heart, LVAD and flow work components

$$\dot{W} = \dot{W}_{heart} + \dot{W}_{LVAD} - P_{in}\dot{V}_{in} + P_{out}\dot{V}_{out} = 0$$
$$\dot{W}_{heart} = -\dot{W}_{LVAD} + \dot{V}(P_{in} - P_{out})$$
$$\dot{W}_{heart} = -(-0.00050 \text{ hp}) + (5.0 \text{ L/min})(6 - 90 \text{ mmHg}) = 0.00050 \text{ hp} - 420 \text{ L.mmHg/min}$$
$$= -0.0007515 \text{ hp} \approx -0.00075 \text{ hp}$$

The magnitude of the heart's work is comparable to that of the LVAD.

4.6   Optimization Principle for Fluid Transport

The total power required to sustain continuous flow of blood, through a segment of circulation, includes the cost of overcoming friction in fluid flowing through a tube and the metabolic cost of producing the fluid flowing through the vessels and the vessels that contain it.

The total power of circulation is

$$P_T = P_f + P_b$$

where P_f is the power required to drive the flow and P_b is the power required to maintain blood supply. In the case of fluid flow Q, driven by a pressure difference, ΔP, from (4.32)

$$P_f = Q\Delta P$$

Using the Poiseuille law (4.17), Q and ΔP are related and power for flow is given by

$$P_f = \frac{8LQ^2\mu}{\pi r^4} \tag{4.36}$$

The cost of maintaining blood includes cost of synthesis of proteins, maintenance of fluid composition, and so forth. The work rate required to maintain the blood supply (P_b) can be assumed to be directly proportional to the blood volume (V_b). Thus,

$$P_b = \alpha_b V_b$$

where α_b is the unit cost of blood. The volume of blood in a circular vessel can be calculated knowing the α_b and hence $P_b = \alpha_b \pi r^2 L$

Thus,

$$P_T = \frac{8LQ^2\mu}{\pi r^4} + \alpha_b \pi r^2 L \tag{4.37}$$

4.6.1   Minimum Work of Circulation

Murray's law [5] asks if there is some design principle that minimizes the cost of maintaining a fluid transport system. The total work of circulation will vary with the vessel radius. As the vessel radius increases, the cost of flow declines, while the cost of blood volume increases. Thus, there will be a minimum work of circulation at the radius where $dP_T/dr = 0$.

Differentiating (4.46) with respect to r,

$$\frac{dP_T}{dr} = \frac{-4 * 8LQ^2\mu}{\pi r^5} + 2\alpha_b \pi rL = 0$$

Rearranging,

$$Q^2 = \frac{\pi^2 r^6 \alpha_b}{16\mu} \tag{4.38}$$

π, α_b, and μ are all constants. Hence

$$Q \propto r^3$$

Obtaining α_b is relatively difficult. However, some of the experimental data suggest that the exponent on the radius could vary from 2.7 to 3.2.

Using (4.16) in conjunction with Murray's law forms the basis for modeling the microvascular branching geometry as a means to providing insight into blood flow distribution, blood pressure distribution, X-ray based indicator for dilution curves, and to get more insight into the impact of changes in the vessel dimensions, capillary recruitment, and functional changes of vessel diameters.

$$Q_0 = Q_1 + Q_2$$
$$r_0^3 = r_1^3 + r_2^3$$

Rearranging $\dfrac{r_1}{r_0} = \left[1 + \left(\dfrac{r_2}{r_1}\right)^3\right]^{-1/3}$ and $\dfrac{r_2}{r_0} = \dfrac{r_2}{r_1}\left[1 + \left(\dfrac{r_2}{r_1}\right)^3\right]^{-1/3}$

EXAMPLE 4.11

Calculate the volumetric flow rate in a 5-mm vascular graft of a length of 5 cm. Assume the viscosity if 3 cP and α_b is 780 erg/mL.s [6]. Then calculate the work required to drive the flow.

Solution: From (4.38),

$$Q^2 = \frac{\pi^2 r^6 \alpha_b}{16\mu} \rightarrow= \frac{\pi^2 (0.5[\text{cm}])^6 * 780[\text{erg.cm}^{-3}\text{s}^{-1}]}{16 * 3 * 10^{-3}[\text{gm.cm}^{-1}\text{s}^{-1}]} = 2,500\left[\text{cm}^6/\text{s}^2\right] = 50\,\text{cm}^3/\text{s}$$

The branching and the predictions of Murray's law agree well with the branching in the lungs. However, it does not predict the angle of bifurcation and the length of each bifurcation. Weibel [7] developed an improved model that is used in approximating network of repeatedly bifurcating tubes in the lungs.

The behavior of fluid flow in the regions where there is branching in the blood vessels and airway, change in fluid properties, change in fluid velocity are intensely investigated areas because of widespread applications. For example, understanding the changes in fluid flow at the athereosclerotic sites are important to decipher the etiology of the disease and developing therapies such as stents to normalize flow patterns in that location. During *rouleaux* formation, the extent of aggregation is determined by opposing forces: the aggregation induced by the presence of macromolecules and the disaggregation induced by the negative surface charge and the flow-induced shear stress. Another example is inhalation of aerosols affecting lung function. The Poiseuille equation is developed with the assumption that the flow is at steady state, occurs in one direction in rigid tubes although many biological conduits are not rigid during flow (but elastic, see Chapter 6), and the fluid is Newtonian. However, at the entry to the tube before the laminar flow has become fully developed, flow is not steady and variations in two or three directions exist. A similar analysis can be performed to describe the fluid flow in 3D. Then for a Newtonian incompressible fluid, one would obtain

$$\text{the } x\text{-component} \quad \rho \frac{Du}{Dt} = \rho g_x - \frac{\partial P}{\partial x} + \mu \left(\frac{\partial^2 u}{\partial x^2} + \frac{\partial^2 u}{\partial y^2} + \frac{\partial^2 u}{\partial z^2} \right) \tag{4.39}$$

$$\text{the } y\text{-component} \quad \rho \frac{Dv}{Dt} = \rho g_y - \frac{\partial P}{\partial y} + \mu \left(\frac{\partial^2 v}{\partial x^2} + \frac{\partial^2 v}{\partial y^2} + \frac{\partial^2 v}{\partial z^2} \right) \tag{4.40}$$

$$\text{the } z\text{-component} \quad \rho \frac{Dw}{Dt} = \rho g_z - \frac{\partial P}{\partial z} + \mu \left(\frac{\partial^2 w}{\partial x^2} + \frac{\partial^2 w}{\partial y^2} + \frac{\partial^2 w}{\partial z^2} \right) \tag{4.41}$$

Equations (4.39) to (4.41) are called the Navier-Stokes equations in the Cartesian coordinate system and if the viscous terms are omitted, one would obtain Euler equations. They can be developed for any other coordinate system that spans the solution space. Alternatively, a transformation can be performed to the developed equations to acquire different coordinate systems from the Cartesian result. These differential equations are difficult to solve analytically, and one could have a number of possible solutions. However, the development of computational fluid dynamics (CFD) has significantly helped understanding the flow fields in biological systems. A number of solutions have been published for various applications. Changes in fluid properties in time and space is obtained by solving incompressible Navier-Stokes equations coupled with mass, and energy balances using finite volume techniques. With the exponential increase in massive computer capabilities, CFD is employed by several researchers to explore the nature of flow stagnation patterns in various biological systems. For further discussion, the reader is referred to [8].

Problems

4.1 A bioengineer is given the task of making a larger scale model of flow through an arteriole, with instructions to duplicate arteriolar N_{Re} in a 0.25-mm diameter tube using a fluid with a 5-cP viscosity and 3.0 g/cm^3 density. For the arteriole fluid, assume viscosity to be 3.25 cP, density to be 1.05 g/cm^3, diameter as 0.005 cm, and mean velocity to be 0.75 cm/s. What U_{mean} should be used for the model?

4.2 Microfluidic devices (see Chapter 10) have attracted a great deal of attention because of their applications in biology and biotechnology. The prevalence of laminar flow in microfluidics enables new technologies. For a cross-sectional rectangular channel that is 100 mm wide and 50 mm deep, and water flowing at 1 mL/s, confirm whether a laminar flow exists in the device. The density of the fluid is 1,060 kg/m^3 and viscosity is 10.23 g/m.s.

4.3 A parallel plate flow chamber ($56 \times 24 \times 0.28$ mm) is a common device used to test the effect of shear stress on various cells. For this system and for Newtonian fluids, show that $\tau_w = \mu \dfrac{6Q}{bh^2}$ where Q is the volumetric flow rate, b is the width of the chamber, and h is the height of the chamber. What is the flow rate required to expose the cells to 15 dyne/cm^2? Viscosity is 1.6 times water. If cells are seeded and their average height is 15 mm, then what is the actual shear stress on the cells?

4.4 A group of bioengineers are evaluating the adhesive strength of cells on a material using a parallel flow chamber of $7.6 \times 3.8 \times 0.02$ cm^3. What is the shear stress expressed on the wall, if the volumetric flow rate is 2 mL/min? Viscosity of the flow fluid is 0.007g/cm/s at 37°C.

4.5 Cordeiro et al. [9] built a flow chamber to study the behavior of adhering microcapsules exposed to wall shear stresses. The device consists of a poly-methylmethacrylate (PMMA) block ($120 \times 60 \times 8$ mm) in which a groove in the bottom part forms a channel of a 5-mm width, a 18-mm length, and a 0.4-mm depth. The bottom of the channel is closed with a thin microscope glass slide (thickness = 170 μm) with dimensions 24×24 mm and an optical quality surface. The device also had a slowly diverging area at the entrance and converging area at the exit to facilitate the development of the flow. Since in vivo shear stress values range up to 10–20 N/m^2, they want to know the fluid flow rate. Calculate the fluid flow rates needed to obtain the shear stresses.

4.6 Ujiie [10] performed pulsatile flow studies using pulse cycles of 5 seconds in 3-mm circular tubes. For the pulsatile condition, the mean flow velocity varied from 0.4 to 0.9 m/s. The solution density is 1.044 g/cm^3 and viscosity is 1.14 $\times 10^{-2}$ g/cm.s.

   (a) Calculate the range of Reynolds numbers for the pulsatile flow. Comment on the type of flow

   (b) Calculate the Woomersly number. Comment on the type of flow (10 points).

4.7   An approach to minimizing space in flow systems is to use MEMS-based min-
       iature pulsatile pumps. A group of researchers are interested in testing the
       influence of viscous forces and inertial forces under pulsatile flow for which
       they have developed a 50-mm-deep and 100-mm-wide rectangular channel.
       The effect of frequency on these channels is calculated using phosphate buffer
       saline solutions, which have a density and viscosity similar to water. Calcu-
       late the effect of 1-Hz and 1-kHz frequency on the Strouhal number and the
       Womersely parameter for a mean volumetric flow rate of 50 mL/min. Which
       factors are dominating? Is the flow laminar or turbulent?

4.8   Consider a 100-mm radius arteriole that has a Poiseuille parabolic flow
       profile.

       (a) What percentage of the total flow rate Q_{total} is within the center of a
           10-mm radius?

       (b) What percentage of the total cross-sectional area is within the center of a
           10-mm radius of the same vessel?

       (c) Make a graph with radius as the x-axis in 10-mm increments up to 100
           mm. Plot both $[Q(0\rightarrow r)/Q_{total}] \times 100\%$ and $[S(0\rightarrow r) / S_{total}] \times 100\%$ on
           the y-axis of the same graph.

4.9   A Lilly-type pneumotachometer is a device that measures lung volumes from
       flow measurements. It contains a fine wire mesh screen that introduces resis-
       tance to flow. For a patient, the nursing staff was measuring the lung volume
       so that they can provide an appropriate amount of anesthetic.

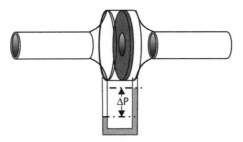

       (a) If the pressure drop is measured to be 50 Pa and the flow resistance is 0.2
           kPa.s/L for air at 25°C and 1 atm, then what is the flow rate?

       (b) When they wanted to give an anesthetic whose viscosity is twice that of
           air, they used the same pneumotachometer to measure the flow rate. What
           is the error introduced in the flow rate?

       (c) If another peumotachometer is used to measure the outflow, what is the
           error introduced due to the change in temperature from 25°C to 37°C?
           Viscosity of the gas at 37°C is 1.3 times that at 25°C.

4.10   In humans, the right lung is divided in three lobes perfused by different pul-
       monary arteries. Assume that the right upper lobe receives a blood flow (Q_{UL})
       of 1.4 L/min, the middle lobe a blood flow (Q_{ML}) of 0.6 L/min, and the lower
       lobe a blood flow (Q_{LL}) of 0.9 L/min. The pressures (P_{PA}) at the entrance of
       the three lobes are 15 mmHg and the pressures (P_{PV}) at the exit of the three
       lobes are 5 mmHg.

(a) Calculate the resistances of the three lobes in CGS units.

(b) If they are considered to be in parallel, what is the net resistance to flow?

4.11   Calculate the flow resistance in a 6-mm capillary. In the capillary, the Fahraeus-Lindqvist effect dominates, hence you need to use the actual viscosity. Assume the central-blood viscosity to be 8 cP and the cell-free plasma layer thickness to be 1 mm. If you were to ignore the Fahraeus-Lindqvist effect and calculate the flow resistance using viscosity of blood to be 3 cP, what is the change in resistance?

4.12   A cardiac patient undergoes bypass surgery in which an artery is replaced by one that is 10% larger in diameter. What is the percentage change in blood flow in the artery, assuming all other factors remain the same?

4.13   In Example 4.2, if there the air is flowing in the upward direction at 1 m/s, then how would it affect the settling velocity? If a person has 15 breaths per minute, will the particle settle or come out? How would you alter the settling properties?

4.14   In Example 4.5, if the flow is pulsatile (1-Hz frequency), how would the wall shear stress change? What is the St number? What information is required to calculate the pressure drop during a cycle of systole and diastole?

4.15   One of the areas of toxicology is understanding how a workplace affects the health of a person, particularly those working in locations where the process emits particle into the air. In one of the study sites, the process emits 20-mm rectangular particles (density = 2.0 gm/cm^3) into the air. A person working in that area breaths at 16 breath/min and has a 25-cm-long trachea of a 25-mm diameter. The tidal volume is 6 L/min. Will the particle settle in the bottom of the trachea:

(a) When the air flow rate is neglected?

(b) When the air flow rate is included?

4.16   If blood is flowing at 3 L/min in a 20-mm artery at an average pressure of 110 mmHg, what is the drag force experienced by a white blood cell of a 15-μm diameter? The density of the white blood cell is 1.10 gm/cm^3. If the thickness of the cell is 5 μm, then what is the change in the drag coefficient?

4.17   A person is 6 feet tall and his heart is located 2 feet from the top of the head. If the flow resistance to the top of the head and bottom of the foot are 5 mmHg, and arterial pressure near the heart is 110 mmHg, calculate the pressure in a location if the person is standing (orthostatic position) or if the person is laying flat (supine position). If the pressure in the venous pressure near the hear is 75 mmHg, which way would blood flow?

4.18   Blood is flowing through a blood vessel of 6-mm in diameter at an average velocity of 50 cm/s. Let the mean pressure be 100 mmHg. Assume blood density to be 1.056 gm/cm^3 and viscosity to be 3 cP.

(a) If the blood were to enter a region of stenosis where the diameter of the blood vessel is only 3 mm, what would be the approximate pressure at the site of narrowing?

(b) If the blood enters an aneurysm region instead with a diameter of 10 mm, determine the pressure in the aneursym. If for conditions of vigorous exercise, the velocity of blood upstream of the aneurysm were four times the normal value (200 cm/s), what pressure would develop at the aneurysm?

(c) Is the flow in the narrowed region laminar or turbulent?

4.19   There are two blood vessels, one of which is four times the diameter of the other. Determine the ratio of blood flow volume in the two vessels, assuming that both vessels are the same length, both have identical blood pressure gradients along their lengths, and both are identical in all other ways.

4.20   The simplest patient infusion system is that of gravity flow from an intravenous (IV) bag. A 500-ml IV bag containing a 0.9% saline solution (assume aqueous properties) is connected to a vein in the forearm of a patient. Venous pressure in the forearm is 0 mmHg (gauge pressure). The IV bag is placed on a stand such that the entrance to the tube leaving the IV bag is exactly 1m above the vein into which the IV fluid enters. The length of the IV bag is 30 cm. The IV is fed through an 18-gauge catheter (internal diameter = 0.953 mm) and the total length of the tube is 2m. Calculate the flow rate of the IV fluid. Estimate the time needed to empty the bag. If you assume laminar flow, show whether the assumption is correct or not.

4.21   Use the general mechanical energy balance equation to estimate the steady-state rate of blood infusion (in mL/min) from a 500-mL bag hung 1.00m above a patient. Assume the blood behaves as a Newtonian fluid with a viscosity of 3.0 cP (3.0310^{-3} Pa.s). Assume the line connecting the bag to the needle has a 4.00-mm inside diameter (I.D.) and is 1.40m long. Assume the needle has a 1.00-mm I.D. and is 60 mm long. Neglect the velocity of the blood at the inlet of the tube, and losses at the connection between the tubing and the needle. A spreadsheet is recommended to perform the calculations for this problem and possibly a trial-and-error approach.

4.22   A syringe pump is used to continuously delivery plasma (density 1.05 gm/cm^3) to a patient when using hypodermic needles. If the plasma is injected steadily at 5 mL/s, how fast should the plunger be advanced if the leakage in the plunger clearance is negligible, or if leakage is 15% of the needle flow?

4.23   Upon examination of a patient, the left ventricular pressure (systolic/end diastolic- mmHg) was determined to be 180/30 with a cardiac output 3.5 (l/min). The aortic pressure was 130/70 (mmHg) with the mean left atrial pressure of 40 mmHg. The heart rate was 100 bpm with a system to diastole time ratio of 2:3. The LV-aorta mean pressure difference was 65 mmHg and LA-LV mean pressure difference was 10 mmHg.

(a) Calculate the aortic and mitral valve orifice areas for the patient assuming the ideal orifice equation.

(b) However, if the actual flow rate is 50% of the ideal flow rate in the mitral valve and 70% in the aortic valve, then what are the diameters of mitral valve and aortic valve?

4.24   The coronary arteries (~6-mm diameter and ~5 cm in length) supply the heart muscle with oxygenated blood and nutrients. When they are obstructed, for example 50% by area, not enough blood gets to the heart. Bypass surgery is used to create a parallel flow around the obstruction and restore blood flow. Before the bypass, the pressure P_{in} at the beginning of the coronary arteries is 90 mmHg, the pressure P_{out} at the end of the coronary arteries is 83 mmHg, and the blood flow through the coronaries is 40 mL/min. Density and viscosity (μ) of blood are 1,060 kg/m^3 and 0.03 Poise (=0.003 kg/m.s), respectively.

(a) What is the coronary arterial resistance?

(b) If the resistance of the bypass channel is 25% of the coronary arterial resistance, what is the combined resistance of the coronary arteries and the bypass channel?

(c) Assuming that the pressures P_{in} and P_{out} do not change when the bypass is added, what is the blood flow to the heart with the bypass in place (in mL/min)?

(d) Are the flows laminar in obstructed and unobstructed arteries?

4.25   During a period of relative inactivity, Janet is losing −300 kJ of enthalpy every hour. This is due to the difference in metabolic and digestive waste products leaving the body and the raw materials ingested and breathed into the body. Janet is also losing heat to the surroundings given by $Q = hA\,(T_s - T_0)$, where A is the body surface area (roughly 1.8 m^2), T_s is the skin temperature (normally 34.2°C), T_0 is the temperature of the body surroundings, and h is the heat transfer coefficient. Assume that the typical values of h for the human body are 8 kJ/(m^{2}h°C) when fully clothed, a slight breeze blowing condition, and 64 kJ/(m^{2}h°C) when nude and immersed in water.

(a) Assuming Janet's body as a continuous system at steady state, write an energy balance on the body.

(b) Calculate the surrounding temperature for which the energy balance is satisfied (i.e., at which Janet would feel neither hot nor cold) when Janet is clothed and when Janet is immersed in water wearing a bikini.

(c) Comment on why Janet feels colder on a windy day than on a day when the temperature is the same but there is no wind.

4.26   In order to perform a transplant operation, a surgeon wants to reduce the patient's body temperature to 30°C. This is to be done by routing the blood through a chiller. Assume that the blood leaving the chiller is cooled to 25°C, how long will this take for a 70-kg patient?

4.27   A micro-organism is growing aerobically on glucose. When the overall growth is written as glucose consumption kinetics, the following stoichiometry is obtained:

$C_6H_{12}O_6 + 1.473O_2 + 0.782NH_3 \rightarrow 0.909C_{4.4}H_{7.3}N_{0.86}O_{1.2} + 3.854H_2O + 2CO_2$

Based on this equation, determine the respiratory coefficient quotient for this organism. Using the Weir equation, determine the energy expenditure.

4.28   A person is performing a physical work at a rate of 250 N.m/s. Assume all the physiological work is derived from protein metabolism in the muscle. The breathing rate is 15 breaths/minute and each breath has a volume of 0.5L. Assume air contains a 21% volume fraction of oxygen.

(a) If the efficiency of conversion of physiological work to physical work is 25%, what is the amount of oxygen required?

(b) Is there sufficient oxygen?

(c) If the breathing rate is increased to 20, what volume of air is required to achieve the desired work?

4.29   Consider the model of the heart as a pump. The right side takes blood returning from the veins at almost 0 mmHg and discharges blood toward the lungs at 15 mmHg. The left side takes blood in from the lungs at 6 mmHg and discharges it to the systemic circulation at about 90 mmHg. Assuming the steady-state operation, apply the macroscopic energy balance over the heart to estimate the rate of work done by the heart. Assume viscous losses l_V are small.

4.30   Verkerke et al. [11] performed numerical simulation of a pulsatile catheter (PUCA) pump which is a type of LVAD that could be mounted outside quickly. It consists of a single port membrane pump and a valved catheter, which is introduced into an easily accessible artery and positioned with its distal tip in the left ventricle. The pump aspirates blood from the left ventricle during systole (thus unloading it) and ejects into the ascending aorta during diastole, which ensures an adequate blood supply to the coronary arteries. To provide all organs with an adequate pulsatile blood flow, a pump output of 5 L/min is required. What is the pressure in the pump required for the aspiration and ejection when the pump is synchronized with a 70 beats per minute heart and bears a 100% load? 60% load?

The following data are known. The mean aortic pressure is 65 mmHg, and mean left ventricular pressure is 38 mmHg. Flow resistances caused by diameter changes from the membrane pump to the catheter during aspiration Kc = 1.0 and during ejection Kc = 0.5. Mean flow resistance in the valve during aspiration appeared to be 3.0, and during ejection, 3.2. Furthermore, $fV^2/2g$ can be approximated by $0.096 - 0.0157*\log_{10}(Re)$. The catheter is 0.5m long.

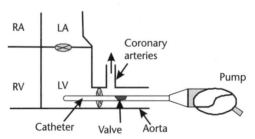

4.31   The Jarvik 2000 heart is a compact axial flow impeller pump positioned inside the ventricle with an outflow graft for anastomosis to the descending aorta. The pump rotor contains the permanent magnet of a brushless direct current motor and mounts the impeller blades and according to Westaby et al., the noise is very minimal—a problem with other pumps. The adult model measures 2.5 cm in diameter by 5.5 cm in length and is connected through a 16-mm diameter graft that is 15 cm long. The weight is 85g with a displacement volume of 25 mL. The adult pump is designed to achieve 5 L/min against 100 mmHg pressure at speeds of 10,000. The smaller pediatric version pumps achieve up to 3 L/min. The pediatric device measures 1.4 cm in diameter by 5 cm in length and is connected through a 10-mm diameter graft that is 10 cm long; the weight is 18g, and the displacement volume is 5 mL. If the manufactures says that for optimal mobilization of the patient and freedom to move, power can be delivered by two 12-V DC batteries (~10 W-h) for several hours. What is that several hours for an adult size? An infant size?

Make sure you include the frictional loss in the calculation [12].

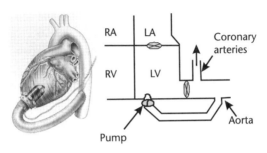

4.32   The Heartmate III [13] is an improved miniature centrifugal pump from its two previous LVAD models Heartmate 1 and II (made by Thoratec Corporation, Pleasanton, California). It is supposed to have improved performance and an extended life. In vitro testing shows that at 135-mmHg pressure, pump capacity is 7 L/m with an hydraulic efficiency of 30% at 4,800 rpm.

(a) Calculate the power put into the fluid.

(b) Calculate the energy needed for 4-hour functioning of the unit.

(c) If Heartmate II (an axial flow pump) delivers 4 L/m at a 100-mmHg differential at 10,000 rpm, did the new design meet the intended goal? Show the calculations.

4.33   A Canadian patient was transplanted with a Novacor left ventricular assist system (LVAS, WorldHeart Corp.) in Japan and had to be transported back to Canada for a heart transplant [14]. LVAS is connected in parallel to the natural circulation and takes blood from the left ventricle and returning it to the ascending aorta. The pump typically bears the entire workload of the left ventricle, but during exercise, native ventricular ejection via the aortic valve could occur. Prior to departure, the patient had 112 beats per minute and an output of 6.7L per minute. To minimize the power consumption, the thresholds were set for a pump output of 4.0L, a minimum stroke volume of 40 mL, and a maximum residual volume of 15 mL. If a two-cell pack of Li-ion D-cell batteries are used (total 30 W-h), was the power sufficient for 21.5 hours of travel time? By decreasing the capacity, how much power did they save during the journey? The diastolic pressure to be 80 mmHg, goes a pulse pressure of

40 mmHg. The mean arterial pressure is calculated using the equation $P_{mean} = (P_{systolic} + 2P_{diastolic})/3$ with the assumption that the duration in each beat to be 2/3 diastole and 1/3 systole.

4.34 AbioCor (ABIOMED Inc., Danvers, Massachusetts) is a totally implantable artificial heart that shows pulsatile flow. Two polyurethane blood pump chambers with a 60-mL stroke volume produce 8 L/min of flow. A German tourist suffered hemodynamic stability and had to be transplanted with AbioCor BVS 5000 total heart [15]. The aircraft flew at 40,000 feet with a cabin pressure of 8,000 ft where the flow rate decreased to 4 L/m despite the set target of 5 L/m (at the sea level) during 15.5 hours of travel. With this decrease, if power calculations are performed, what is the change in efficiency? Could you explain why the flow rate decreased?

4.35 A biventricular assist device (BiVAD) is one that supports both the right ventricle (pumping chamber of the heart) and the left ventrical. If the frictional factor is 30% more in the synthetic grafts and a surgeon uses three different lengths 10 cm, 20 cm, and 30 cm long and a 16-mm graft for three different patients, what happens to the flow rate in each case? If the pumps have to maintain the same flow rate, how would it affect their efficiency? Use the pressure values from a normal person.

References

[1] Casson, N., "A Flow Equation for Pigment Oil Suspensions of Printing Ink Type," in Mill, C. C., (ed.), *Rheology of Dispersed Systems*, Oxford, U.K.: Pergamon Press, 1959, pp. 84–104.

[2] Fournier R. L., *Basic Transport Phenomena in Biomedical Engineering*, 2nd ed., London, U.K.: Taylor and Francis, 2006.

[3] Thurston, G. B., "Plasma Release-Cell Layering Theory for Blood Flow," *Biorheology*, Vol. 26, 1989, pp. 199–214.

[4] White, F. M., *Fluid Mechanics*, 10th ed., New York: McGraw-Hill, 2002.

[5] Murray, C. D., "The Physiological Principle of Minimum Work. I. The Vascular System and the Cost of Blood Volume," *Proceedings of the National Academy of Sciences*, Vol. 12, 1926, pp. 207–214.

[6] Taber, L. A., "An Optimization Principle for Vascular Radius Including Smooth Muscle Tone," *Biophysical J.*, Vol. 74, 1998, pp. 109–114.

[7] Weibel, E. R., *Morphometry of the Human Lung*, New York: Springer-Verlag, 1963.

[8] Leondes, C. T., *Biomechanical Systems: Techniques and Applications, Volume II: Cardiovascular Techniques*, Boca Raton, FL: CRC Press, 2000.

[9] Cordeiro, A. L., et al., "Effect of Shear Stress on Adhering Polyelectrolyte Capsules," *J. Colloid Interface Sci.*, Vol. 280, No. 1, 2004, pp. 68–75.

[10] Ujiie, H., et al., "Hemodynamic Study of the Anterior Communicating Artery," *Stroke*, Vol. 27, 1996, pp. 2086–2094.

[11] Verkerke, G. J., et al., "Numerical Simulation of the Pulsating Catheter Pump: A Left Ventricular Assist Device," *Artificial Organs*, Vol. 23, No. 10, 1999, p. 924–931

[12] Westaby, S., et al., "Jarvik 2000 Heart: Potential for Bridge to Myocyte Recovery," *Circulation*, Vol. 98, No. 15, 1998, pp. 1568–1574.

[13]   Loree, H. M., K. Bourque, and D. B. Gernes, "The Heartmate III: Design and In Vivo Studies of a Maglev Centrifugal Left Ventricular Assist Device," *Artificial Organs,* Vol. 25, No.
5, 2001, pp. 386–391.

[14]   Haddad, M., et al., "Intercontinental LVAS Patient Transport," *The Annals of Thoracic
Surgery,* Vol. 78, No. 5, 2004, pp. 1818–1820.

[15]   Popapov, E. V., et al., "Transcontinental Transport of a Patient with an AbioMed BVS 5000
BVAD," *The Annals of Thoracic Surgery,* Vol. 77, No. 4, 2004, pp. 1428–1430.

Selected Bibliography

Chakrabarti, S. K., *The Theory and Practice of Hydrodynamics and Vibration,* New York:, World
Scientific Publishing Company, 2002.

Chandran, K. B., *Cardiovascular Biomechanics,* New York: New York University Press, 1992.

Fahraeus, R., "The Suspension Stability of Blood," *Physiology Reviews,* Vol. 9, 1929, pp.
241–274.

Fox, R. W., A. T. McDonald, and P. J. Pritchard, *Introduction to Fluid Mechanics,* New York:
John Wiley & Sons, 2005.

Fung, Y. C., *Biomechanics: Mechanical Properties of Living Tissues,* New York: Springer-Verlag,
1993.

Pedley, T. J., R. C. Schroter, and M. F. Sudlow, "Gas Flow and Mixing in the Airways," in West,
J. B., (ed.), *Bioengineering Aspects of the Lung,* New York: Marcel Dekker, 1977, pp. 163–265.

Saleh, J., *Fluid Flow Handbook,* New York: McGraw-Hill, 2002.

Vogel, S., *Life in Moving Fluids,* Princeton, NJ: Princeton University Press, 1994.

CHAPTER 5

Biomechanics

5.1 Overview

Previous chapters dealt with the movement of fluids and embedded molecules. Various structures in the body also move while performing day-to-day activity. These movements depend upon highly proficient neural control of motor neurons and muscles. Movement of body structures requires force generation by muscular system. Furthermore, many activities result in exerting pressure, tension, and shear forces to the tissues. Understanding force distribution is essential to obtaining more insight into the functioning of various tissues, the effect of load and overload on specific structures of living systems, the development of prosthetic devices, and the modeling and simulation of human movement. Furthermore, to incorporate safety features, for example, designing seat belts in automobiles, one has to understand the impact responses, injury tolerances, and injury mechanisms of the human body. Biomechanics deals with examining the forces acting upon and within a biological structure and the effects produced by these forces. Biomechanics is a broad field with many applications and can be subdivided into four areas: anthropometry (i.e., measurement of humans), gait analysis, stress analysis, and impact biomechanics. Some of the applications of each area are listed in Table 5.1.

Using the fundamentals of kinematics, one can study the positions, angles, velocities, and accelerations of body segments and joints during regular walking. Conservation of momentum and energy principles, discussed in Chapter 4, are used to assess the load and stress distribution. Complexity arises because the nature of solids is different than fluids. In solids, effects on individual elements have to be considered, unlike fluids, which are dealt with continuous streams of fluid without a beginning or end. Furthermore, most of the tissues are not simple solids and possess viscous and elastic characteristics. This chapter introduces the basic elements of biomechanics. First, the forces acting in various segments are described, assuming that body parts are rigid elements. Then, the concepts of elasticity and viscoelasticity are introduced. Finally, the conservation of energy principle is discussed in the context of injury prevention.

Table 5.1 Uses of Biomechanical Analysis

Area	Application
Anthropometry	Archeological surveys, human growth, populational variation, nutritional supplementation, design of tools, furniture, and equipment used in a workplace
Gait analysis	Identification of underlying causes for walking abnormalities in patients with neuromuscular problems (cerebral palsy, stroke, head injury), designing prosthetic components, improving athletic performance and rehabilitation engineering, and tools for persons with physical disabilities
Stress analysis	Prosthesis devices (heart valves and artificial heart), injury criterion for various tissues and organs, load distribution, anthropometry data, biomaterial selection, and rehabilitation engineering
Impact biomechanics	Development of safety devices, aircraft ejection seats, underwater rescue procedures, land mine detonation, pedestrian impacts, infant carriers, seat belt systems, air bag systems, child safety seats, and helmets and other protective gear in sports

5.2   Equations of Motion

5.2.1   Center of Mass

The mass of an object is a measure of the amount of matter in an object. The center of mass is a unique point at which the average of the mass factored by their distances is considered to be concentrated. The center of mass and center of gravity (CG) are used synonymously in a uniform gravitational field to represent the unique point in an object or system which can be used to describe the system's response to external forces. An object's CG is not always located physically inside of the object. In the human population, each person has a different weight and height. Locating the CG is of interest because, mechanically, a body behaves as though all of its mass were concentrated at the CG. Every time the body changes posture, its weight distribution and CG locations change.

The CG for the human body can be located either using a reaction board or segmental method. A reaction board consists of a long rigid board supported on pointed structures [Figure 5.1(a)]. Under each end of the board, there is a weighing scale. The CG location is measured using the conservation of momentum principle (see Section 5.2.2).

The segmental method is a more popular procedure for determining center of mass of the total body, as it can be used under dynamic situations to predict the body's response during walking, the joint moment of force, and the design of the workspace interface (crash test dummies). In the segmental method, the center of mass of individual segments is determined. Based on bony prominent landmarks, Dempster [1] grouped the body into 15 segments consisting of the foot, shank (leg), thigh, pelvis, thorax, hand, forearm, upper arm, and head [Figure 5.1(b)]. Based on the assumption that body segments are considered to be rigid bodies with frictionless hinges for the purposes of describing the motion of the body, body segments can be represented by geometric solids (cones, cylinders, spheres), and mass is uniformly distributed in a given segment. Although no joint in the human body is a simple single-axis hinge joint, the method can be individualized by capturing im-

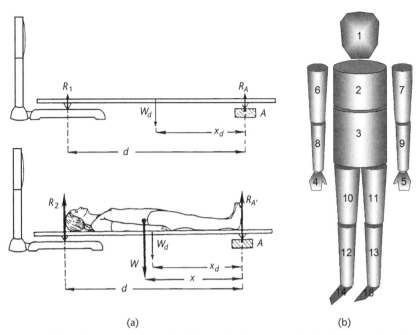

(a)   (b)

Figure 5.1 Analysis of the center of gravity: (a) reaction board method and (b) segmental method.

ages of the individuals while performing various tasks, resulting in a more accurate analysis.

Apart from mass, evaluating the size of each segment and the width is important when designing tools and furniture for people use since everyone is unique. Males have different overall measurements than women. Within males or females, some could be overweight, some could be tall, or some could be short. For children measurements would be smaller. Anthropometry is the general term used for the analysis of size, mass, shape, and inertial properties of the human body. Knowing the thickness and height of a leg is required to develop a prosthetic leg. Size measurement (using a regular measuring tape or calipers of different sizes) is also useful in determining subcutaneous adipose tissue. A design standard is to develop tools to fit the fifth to the 95th percentiles of the population with an ability to adjust the parameter. An ergonomic chair is a common example where the height and inclination can be adjusted to an individual's comfort level.

5.2.2 Newton's Laws of Motion

Force pushes or pulls to alter the body motion through direct mechanical contact, through the force of gravity or the electromagnetic field in some cases. The four primary sources of force (or load) relevant to biomedical applications are: gravity or weight of the body, muscles, externally applied resistance, and friction. The body could move to a new distance or rotate due to the action of the force. The consequences of forces acting on joints are joint compression, joint distraction, and pressure on body tissues. To understand the impact of an action on a joint, the net external force acting on any joint must be obtained. A diagram that depicts all the forces acting upon the object (body) of interest is drawn; this is called the free-body diagram. A force at an angle to a reference axis is resolved into vertical and horizontal components (Figure 5.2) in two dimensions using trigonometric relations. Alternatively, the magnitude and the direction of the resultant force are obtained by

$$F = \sqrt{F_x^2 + F_y^2} \text{ and } \tan\theta = \frac{F_y}{F_x}$$

Effective force is the component of the force in the direction of the displacement or that doing work on the object. Displacement is the distance along its path that the point of application of the force moves while the force is being applied. A positive sign is assigned to a force pointing to the right or up and a negative sign is assigned to a force if it is pointing left or down by convention. The force on an object in contact with a surface can also be resolved into a component perpendicular to the surface at a given point (the normal force), and a component parallel to the surface (the tangential force). The normal force is generally associated with the force that the surface of one body exerts on the surface of another body in the absence of any frictional forces between the two surfaces. The normal force is of special interest when dealing with friction. In particular, to keep a body on an inclined plane moving, one must overcome a frictional force, F_F, given by

$$F_F = \mu_K F_N$$

where μ_K is the coefficient of kinetic friction.

5.2.2.1   Newton's First Law of Inertia

A body will maintain a state of rest or constant velocity unless acted on by an external force that changes that state. This is called *Newton's first law of inertia* and the resistance that an object offers to changes in its state of motion is called *inertia, I*. For a linear motion, inertia is dependent upon the mass only. The greater the inertia, the more resistant the object is to a change in velocity.

Apart from linear motion, there is a rotational motion in many body segments at different joints. For example, a person trying to pick up an object from the floor

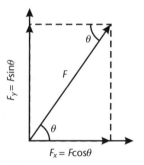

Figure 5.2   Cartesian components of force *F*.

has to rotate his or her arm while picking up that object. In an angular motion (Figure 5.3), the angular velocity (ω) is the angle (θ) turned per second. The direction of ω for a rotating rigid body is found by curling the fingers of the right hand in the direction of the rotation. The thumb points to the direction of ω. The angular acceleration is the rate of change of angular velocity ($\alpha = d\omega/dt$) and the units are radians/s^2 where 1 radian is $180°/\pi$.

The angular momentum is the tendency of a rotating object to keep rotating at the same speed around the same axis of rotation. The moment of inertia is the name given to rotational inertia experienced during an angular motion. For symmetric objects, the moment of inertia is calculated using the relation

$$I\left[\text{kgm}^2\right] = m\left[\text{kg}\right]r^2\left[\text{m}\right]^2 \tag{5.1}$$

where m is the mass of the particle and r is the distance from the center of mass of the particle to the axis of rotation. For nonsymmetric objects, the moment of inertia is calculated with the modification to (5.1).

$$I = mk^2\left[\text{kg.m}^2\right] \tag{5.2}$$

where k is the radius of gyration from the axis of rotation, that is, the distance from the axis of rotation to the point of center of mass (Figure 5.3). A rotating body will continue in a state of uniform angular motion unless acted on by an external torque (i.e., the rotary effect created by an applied force). Torque (T) is the moment of a force, that is, product of a force magnitude (F) and its perpendicular distance (d) from the direction of force to the point or axis of rotation.

$$T[\text{N.m}] = F[\text{N.}]d[\text{m}] \tag{5.3}$$

The perpendicular distance between the line of action of the force and the axis of rotation is also referred to as the *moment arm*. Similar to force, torque is characterized by both magnitude and direction; a positive sign is used if it is counterclockwise and negative sign is used if it is clockwise. In Figure 5.3, T is positive

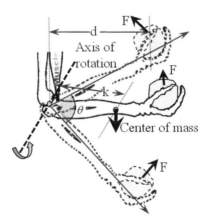

Figure 5.3   Angular motion in the forearm.

when the object is lifted up. The center of gravity is the point about which the sum of torques produced by the weights of the body segments is equal to zero. A pair of equal, oppositely directed forces that act on opposite sides of an axis of rotation to produce torque is known as a force couple (see Table 5.2).

Newton's second law of acceleration states that a force applied to a body causes an acceleration (*a*) of that body of a magnitude proportional to the force, in the direction of the force, and inversely proportional to the body's mass. This law, discussed in Chapter 4 (Section 4.4.2), is also applicable to linear motion of the body segments. In angular motion, Newton's second law states that a torque will produce an angular acceleration of a body that is directly proportional to the magnitude of the torque and in the direction of the applied torque, but inversely proportional to the moment of inertia of the body, that is,

$$T = I\alpha \qquad\qquad (5.4)$$

Table 5.2 Comparison of Linear Motion Versus Angular Motion

	Linear	*Angular*
Inertia	$I = m$	$I = mr^2$ or mk^2
Newton's second law	$F = ma$	$T = I\alpha$
Momentum	$M = mv$	$H = I\omega$
Kinetic energy	$KE = \dfrac{1}{2}mv^2$	$KE = \dfrac{1}{2}I\omega^2$
Work	$W = Fd$	$W = T\theta$
Power	$P = Fv$	$P = T\varpi$

EXAMPLE 5.1

The knee extensors insert on the tibia at an angle of 30° at a distance of 3 cm from the axis of rotation at the knee. How much force must the knee extensors exert to produce an angular acceleration at the knee of 1 rad/s², given that a mass of the lower leg and foot is 4.5 kg and the radius of gyration is 0.23m?

Solution: This is the case of angular motion. The produced torque is proportional to the angular acceleration determined by the momentum of inertia. Hence,

$$T = I\alpha = mk^2\alpha = 4.5[\text{kg}] * 0.23^2[\text{m}]^2 * 1\ [\text{rad/s}^2] = 0.238\ \text{N-m}$$

A force is applied by the knee extensor muscle to produce 0.238 N-m torque.

$$F * 0.03[\text{m}] = 0.238\ \text{N-m} \rightarrow F = 7.935\text{N}$$

However, the muscle is attached at 30° to the tibia. Hence, the actual force muscle has to apply

$$F_m = 7.935/\sin 30 = 15.9\text{N}$$

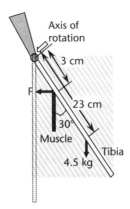

5.2.2.2 Application of the Conservation of the Momentum Principle to Solids

If the body is completely motionless, it is said to be in static equilibrium. Then the resultant force in a joint is obtained using the principle of static equilibrium, which states that for a body to be in a state of static equilibrium:

- The sum of all horizontal forces and sum of all horizontal moments must be zero.
- The sum of all vertical forces and sum of all vertical moments must be zero.
- The sum of all torques must be zero.

Consider the reaction board shown in Figure 5.1(a) with the person in position on the reaction board. The equation of static equilibrium for horizontal moments about a reference axis A becomes

$$\sum M_x = R_2 d - Wx - W_d x_d = 0 \qquad (5.5)$$

where R_2 is the weight scale reading, W is the weight of the person, x is the axis A to the CG of the person's body, x_d is the distance from axis A to the CG of the board, and W_d is the weight of the board. When the person is not loaded, the board is in static equilibrium, and horizontal moments about a reference axis A become

$$\sum M_x = R_1 d - W_d x_d = 0 \qquad (5.6)$$

where R_1 is the weight scale reading without the person. Solving (5.5) and (5.6), the location of the CG with respect to axis A can be obtained as

$$x = \frac{(R_2 - R_1)}{W} d \qquad (5.7)$$

Thus, knowing the weight scale readings and the weight of the person, the CG of that person in the supine position can be determined. However, the CG changes with every posture of the body motion and the reaction board cannot be applied.

For an object in motion (i.e., dynamics conditions), the sum of the external forces is equal to the mass of the object times its acceleration (i.e., Newton's second law). To emphasize the importance of the kinematics in a dynamics problem, it is essential to draw both the free body diagram and the kinetic diagram (sometimes referred to as the Hibbeler diagram). The combination of a free-body diagram and a kinetic diagram is a pictorial representation of Newton's second law. Then the equations of motion associated with the object of interest are written directly using these diagrams by considering the appropriate components. Steps involved in constructing a free body diagram and a kinetic diagram are as follows:

1. Identify and isolate the object or group of objects to focus on as the body.
2. Sketch the body "free" of its surroundings. The body could be represented by a single point located at the body's center of mass.
3. Draw the magnitude and the direction of those forces acting directly on the body.
4. Except for rotational problems, the forces can be sketched as though they were acting through a single point at the center of mass of the body. It is useful to draw the force vectors with their tails at the center of mass.
5. Do not include any forces that the body exerts on its surroundings; they do not act on the body. However, there is always an equal reaction force acting on the body.
6. For a compound body, do not include any internal forces acting between the body's subparts, since these internal forces come in action-reaction pairs that cancel each other out.
7. Choose a convenient coordinate system and sketch it on the free-body diagram.
8. Place the mass times acceleration vector for that body on the kinetic diagram, which is not for statics problems.

EXAMPLE 5.2

A physical therapist is applying cervical traction to a patient. The traction force is developed by the therapist leaning backwards while maintaining straight arms, and the system is in static equilibrium. The therapist's height is 1.6m, the therapist's mass is 60 kg, and the therapist's body proportions from the feet are as follows: the level of the glenohumeral joint (~shoulder height) = 85% of the total body height; the level of center of mass = 59% of the total body height; and the level of the hip joint = 52% of the total body height. Find the amount of traction force.

Solution: This problem is solved by setting up a moment equation about the heel. Draw the free body diagram as shown in the figure.

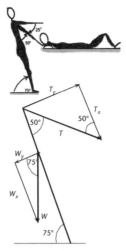

Convert the therapist's mass to a force:

$$F = mg = 60 * 9.81 = 588.6N$$

and it is acting downward.

The distance from the heel to the center of mass is

$$d_W = 59\% \text{ of } 1.6m = 0.944m$$

The distance from the shoulder to the heel is

$$d_T = 85\% \text{ of } 1.6m = 1.36m$$

Establishing momentum equation about the heel,

$$T_y d_T = W_y d_W$$
$$T\sin50 * d_T = W*\cos75 * d_W$$
$$T\sin50 * 1.36 = 588.6*\cos75 * 0.944$$
$$T = 138N$$

Newton's *third law of reaction* states that for every action there is an equal and opposite reaction. An example is when a person is walking. A ground reaction force (GRF) is generated to be equal in magnitude and opposite in direction to the force that the body exerts on the supporting surface through the foot. Although this is partly attenuated by joint structures and soft tissues, considerable force is transmitted to the bones of the lower limb. Hence, there is also an equal and opposite force between adjacent bones at a joint caused by the weight and inertial forces of the two body segments. This is normally called the joint reaction force and is useful in force analysis. Understanding these forces during walking alone is important in a number of applications and is discussed in Section 5.2.6. When two objects are in rotary motion, for torque exerted by an object on another, there is an equal and opposite torque exerted by the latter object on the former. The work done by a constant torque T is the product of the torque and the angle through which it turns. The muscle power (i.e., the rate of doing work) is defined as the scalar product of the joint torque and the segment's angular velocity.

5.2.3 Leverage

Forces arising due to various activities of the body are transferred to other anatomical structures. The amount of force transferred or required to perform an activity is a function of the framework of the anatomical structure and the location of joints. The concept levers are used to understand how body forces act. A lever is a simple mechanical device used to produce a turning motion about an axis with an application of the torque. The bones act as the levers about which the muscular system generates the movements. Lever components include:

- The fulcrum, which is the point of support, or axis, about which a lever may be made to rotate;
- The force arm, which is the distance from the fulcrum to the point of application of the effort;
- The resistance arm, which is the distance from the fulcrum to the weight on which force is acting.

The ratio of the force arm to the resistance arm is called a mechanical advantage. If the mechanical advantage is greater than one, more load can be lifted than the applied effort. If mechanical advantage is less than one, less load can be lifted than the applied effort. Levers are classified into three types [Figure 5.4(a)] based on the location of the fulcrum, effort, and resistance:

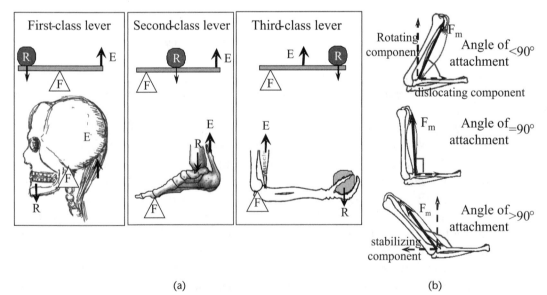

(a)   (b)

Figure 5.4 Leverage in musculoskeletal system. (a) Different classes of lever showing the location where E is the effort, R is the resistance, and F is the fulcrum. Also shown are examples from different parts of the body. (b) Resolving forces in bone-muscle actions.

1. A *first-class lever* is used to gain either force or distance, depending on the relative length of force and resistance arm. It is frequently observed in the body for maintaining posture and balance. Examples include the atlanto-occipital joint, where the head is balanced by the neck extensor muscle force and intervertebral joints, where the trunk is balanced by extensor spinae muscle forces.

2. A *second-class lever* is the applied force and resistance on the same side of axis, with the resistance closer to the axis. It provides a force advantage such that large weights can be supported or moved by small forces. An example is a person standing on his or her toe.

3. A *third-class lever* is the force and resistance on the same side of the axis, but with the applied force closer to the axis. This is more common in the body, and most muscle-bone lever systems are of the third class. Based on the application that a musculoskeletal system is performing, the lever classification is subjected to change.

The angle at which a muscle pulls on a bone affects the mechanical effectiveness of the muscle bone lever system [Figure 5.4(b)]. The force of muscular tension is resolved into two force components: one perpendicular to the attached bone and one parallel to the bone. Only the component of the muscle force acting perpendicular to the bone actually causes the bone to rotate about the joint center. The angle of maximum mechanical advantage for any muscle is the angle at which the most rotary force can be produced. The angle at the elbow at which the maximum flexion torque is produced is approximately 80°. The greater the perpendicular distance between the line of action of the muscle and the joint center (force arm distance), the greater the torque produced by the muscle at the joint.

The way that muscles are arranged in the body plays an important role in how they function. Flexion muscles are arranged to decrease the angle between articulating bones, whereas extensors are arranged to increase the angle between articulating joints. The velocity of contraction of muscles is influenced by the intrinsic speed of the shortening (the rate of conformational changes in the cross-bridge) number of sarcomeres in series. Factors such as the length of muscle influence the force of contraction. At lengths greater than their resting length, they develop tension or force. This force is passive, since it exists whether or not the muscle is active. The passive force acts in a direction from the muscle's points of attachment toward its center.

5.2.4   Impulse-Momentum Relation

The tolerance of various tissues to forces depends on the magnitude as well as the duration of forces acting on them. The sum of all forces acting during a certain period is termed the impulse. If the force is constant, then the product of force and time is the impulse. When the force is not uniform, a normal practice is to plot the changes in force as a function of time, and impulse is the area under the curve. Consider a person jumping. If the force generated against the floor is plotted over the jump time (Figure 5.5), then the shaded area represents the impulse generated. The impulse of a force is a vector quantity. To determine net impulse acting on the

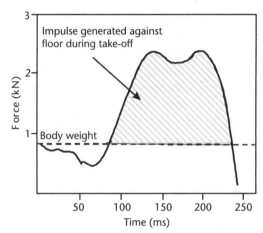

Figure 5.5   Impulse generation curve.

body, the force due to body weight is subtracted or the time axis is moved up to the corresponding force value. A similar curve can be developed for landing and other scenarios.

When an impulse acts on a system, the result is a change in the system's total momentum. The conservation of linear momentum is obeyed only if the impulse of the resultant force is zero. Hence, momentum will also increase when the magnitude of the impulse is increased. Since this increase may not change the mass, the result will be a corresponding increase in the magnitude of the velocity. Therefore, the change in resulting momentum is usually associated with a change in the velocity. An increase in velocity could be achieved by applying a large force for a short period of time or a small force for a long period of time. Typically, muscles contract more forcefully over short periods of time than when they are required to contract more slowly for long periods of time. Suppose a tennis player is playing tennis and wants to serve a ball that travels too fast (or an ace) so that the opponent has very little time to react. The ball should have the largest possible velocity, that is, the largest possible momentum, when it leaves the tennis racket. Since the ball would move slowly prior to being hit by the racket, it can be considered to have a negligible momentum. To change this momentum to a very large momentum toward the other side of the net, the player has to apply the largest possible impulse on the ball. Since impulse depends directly on the force applied and the duration of the force, the player should apply the largest possible force for the longest duration. Swinging the racket harder will hit the ball harder or apply maximum force. If the ball is hit at twice the force, it will impart at twice the impulse. However, the amount of force that can be applied depends on the strength of person and body coordination. Alternatively, the impulse on the ball can be increased by increasing the time that the racket exerts its force on the ball, a process called following through. If the ball is hit at twice the duration, the impulse of the ball will also increase twice. Thus, the greatest impulse is developed or a change in momentum is maximum if the ball is hit hard and followed through. In general, the linear impulse-momentum relation states that in the area under the force time (F-t) the curve is equal to the change in linear momentum, that is,

$$\int F(t)dt = m\Delta V \tag{5.8}$$

where dt is the duration of the force(s) and ΔV is the change in linear velocity. If $F(t)$ is idealized as a step function of duration Δt, then

$$F(t)\Delta t = m\Delta V$$

Using Newton's second law,

$$\Delta t = \frac{\Delta v}{a} = \frac{impulse}{force} \tag{5.9}$$

The impulse-momentum relationship is an important means of determining what motion results from an applied force. Impulse-momentum is also important in absorbing the generated forces. For example, when an athlete reaches out with the glove to catch a ball and as the ball contacts the glove, the arms flex inward to increase the time of force application. Unskilled players often keep the glove in the same place at impact and all the forces are applied over one instant, which could lead to loss of control of the ball or pain. This concept is discussed in Section 5.5.1.

EXAMPLE 5.3

A 59.5-kg volleyball player with one prosthetic leg is trying to block a ball for which he has to jump. During the jump, his feet are in contact with the ground for 500 ms. If his ground vertical reaction force during the jump is 1,000N and his total body velocity at touchdown is −0.50 m/s, determine his total body vertical velocity at jump takeoff. Assume that the applied force is constant during the duration.
   Solution: From the impulse-momentum relation,

$$\int_0^t F(t)dt = m(V_2 - V_1) \rightarrow (F_g - F_w)t = m(V_2 - V_1)$$

$$(1,000[N] - 59.5[kg]*9.81[m/s^2])(0.5[s]) = 59.5[kg](V_2 - (-0.5)[m/s]))$$

$$V_2 = 3m/s$$

5.2.5   Gait Analysis (Motion Analysis)

While a person walks, each foot repeats the same pattern of movement: (1) the foot comes on the ground, (2) the foot rests on the ground, (3) the foot lifts off the ground, and (4) the foot swings through the air. This is called a gait cycle. During this cycle, legs experience varying forces. These forces are necessary to support the entire body, to perform the desired motion, and to sustain the impact from the foot-floor contact. Gait analysis is a systematic study of the motion of an individual.

Gait analysis is useful in identifying the causes of walking disorders in individual patients with neuromuscular problems (cerebral palsy, stroke, head injury, or multiple joint disease), in aiding the treatment approach, or in the follow-up of the treatment. Gait analysis is also useful in understanding diseases that cause disabilities and in designing prosthetic components. For example, amputations cause a loss of limb accompanied with the deficit of many important physical functions. The newly developed prosthetic designs make some compensation on the amputated part and improve the travel ability and life independence of the patients. However, the design of the prosthetic device should ensure that the new prosthetic material is comfortably transferring the forces to the adjacent joint during various activities, for example, walking with different speeds on various surfaces.

A multitude of measurements can be made while a person is moving, and collected data is useful in determining the best course of treatment. The most commonly measured parameters to describe in an observational gait pattern include:

- *Step length* [Figure 5.6(a)], the distance between the point of initial contact of one foot and the point of initial contact of the opposite foot;
- *Stride length,* the distance between successive points of initial contact of the same foot.
- *Cadence* or walking rate (steps per minute);
- *Velocity,* the product of cadence and step length (free speed refers to the individual's comfortable walking speed; since individuals walk at different speeds depending on the situation, normal velocity values are somewhat arbitrary);
- *Walking base,* the sum of the perpendicular distances from the points of initial contact of the right and left feet to the line of forward progression;

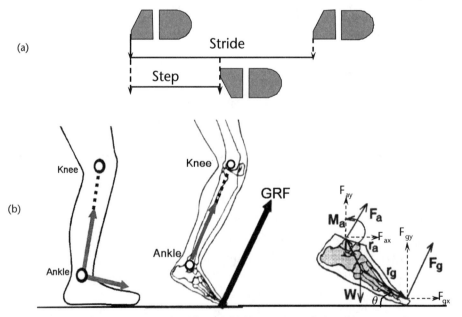

Figure 5.6   Gait analysis: (a) step and stride and (b) kinematic analysis using two markers.

- *Foot angle* or toe out, an angle between the line of progression and a line drawn between the midpoints of the calcaneus and the second metatarsal head.

In a normal gait, right and left step lengths are similar, as are right and left stride lengths. Table 5.3 shows the generic values of these parameters. While analyzing the forces and the moments during a full range of body motion, a person is as ked to walk or run over a force plate. Force exerted by the body on the ground is determined. Based on Newton's third law, the body would experience an equal and opposite reaction force (Figure 5.6). This is referred to as the ground reaction force (GFR), F_g.

There are treadmills equipped with force sensors located in their beds that allow one to measure vertical GRF and the center of pressure for complete, consecutive, multiple foot strikes during walking or running. Markers are placed over bony landmarks while imaging to define joint centers and relevant axes of rotation. These markers typically reflect the infrared light, which is picked up by cameras in a direct line of sight to the marker. Digitizing images translates the location of anatomical landmarks into coordinates.

There are two marker sets commonly used in gait analysis: the Cleveland Clinic marker set and the Helen Hayes marker set, originally developed at the Helen Hayes Hospital. However, marker locations vary from the type of analysis and the hospital where the analysis is performed. The Helen Hayes marker set is more popular, and it is also called the simplified marker set, as it does not require the time-consuming static data collection step. In the Helen Hayes marker set, 15 markers are used. A refinement to the Helen Hayes marker set uses a static trial to better define the knee and ankle centers and axes of rotation. This additional static trial gives more repeatable kinematic and kinetic measures.

Consider Figure 5.6(b), where two markers are placed, one on ankle and the other at the knee. When the person starts walking on a force plate, one could record the ground reaction force, F_g, and changes in the linear positions in those markers. By filming the motion and knowing the time, one could calculate the velocity and acceleration at those markers in addition to the angular velocity of the joints. Starting at the terminal segment of the foot, Newton's laws of motion are applied to determine the net ankle force, F_a, applied to the adjacent lower leg.

Table 5.3 Mean Stride Analysis Variables

	Males	*Females*
Step Length (cm)	79	66
Stride Length (cm)	158	132
Cadence (steps/min)	117 (60–132)	117 (60–132)
Velocity (m/sec)	1.54	1.31
Walking Base (cm)	8.1	7.1
Foot angle (degree)	7	6

Source: [2, 3].

$$\text{Sum of the forces in the } x\text{-direction} \rightarrow \sum F_x = ma_x = F_{ax} + F_{gx} \qquad (5.10)$$

$$\text{Sum of the forces in the } y\text{-direction} \rightarrow \sum F_y = ma_y = F_{ay} + F_{gy} - W \qquad (5.11)$$

where W is the weight of the ankle. Similarly,

$$\text{Sum of all the moments} \rightarrow \sum M = I\alpha = M_a + \left| r_g F_g \right| + \left| r_a F_a \right| \qquad (5.12)$$

where r_a and r_g are the perpendicular distances to the respective forces from the axis of rotation (i.e., the CG of the foot). Using landmarks corresponding to the segmental CG, one can determine these distances and calculate the ankle moment, M_a, caused mainly by muscles and ligaments. After determining F_a and M_a, similar analysis is performed at the knee joint using that landmark, from which F_k and M_k are determined. Similar analysis is carried out to all the necessary segments for each action. Plotting force values at various time points (similar to Figure 5.4), one could determine the impulse generated at each joint and the motion generated, which is used for subsequent therapies. Calculating values at each joint for every move manually is time-consuming. With recent advances in computational tools, sophisticated software packages are available that can analyze and display the geometry of movement as the feet travel through space and the forces that cause movement (kinetics). When a person is walking, the analysis of the musculoskeletal system, including dynamic changes in muscle length and moments, can be performed in three-dimensional space. Furthermore, the animation can also be built for better visualization. For three-dimensional joint kinematics and kinetics, an accurate representation of the underlying anatomical structure is necessary so that an anatomical coordinate system can be properly defined. However, one has to be careful about the possible landmark movement caused by skin movement, which could lead to inaccuracies in the joint center location.

EXAMPLE 5.4

A 100-kg football player was injured in the knee. To understand how he is recovering, gait analysis was performed. After attaching the markers, he was asked to walk on a force platform, which was videotaped. Analysis of the film showed that his leg accelerated at -25 m/s² in the horizontal direction and 140 m/s² in the vertical direction. Furthermore, the ground reaction forces were determined to be 2,000N in the vertical direction and -500N in the horizontal direction. If the mass of the leg is determined to 5 kg, determine the knee joint forces.

Solution: x-direction: Using (5.10),

$$5[\text{kg}] * -25[\text{m/s}^2] = F_{kx} + (-500)[N] \rightarrow F_{kx} = 375N$$

y-direction: Using (5.11),

$$5[\text{kg}] * 140[\text{m/s}^2] = F_{ky} + 2,000[N] + 5[\text{kg}] * 9.81[\text{m/s}^2]$$
$$F_{ky} = -1,349N$$

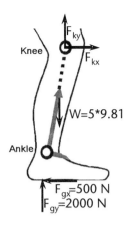

5.3   Ideal Stress-Strain Characteristics

The forces (or load) transferred across each joint are calculated assuming that each segment is a rigid body. In reality, each segment is not a rigid body and there is elastic storage and recovery. For example, tensile loading produces an elongation and a narrowing of a structure (Figure 5.6), whereas compressive forces on a structure tend to shorten and widen. In this section, we will discuss the stress-strain behavior in relevance to biomedical applications.

5.3.1   Structural Parameters and Material Parameters

To understand the load-bearing capacities of materials, and compare material characteristics, two terminologies are used: stress and strain. If F is the applied force on a cross-sectional area of A, then stress, σ, is given by

$$\sigma = F/A \tag{5.13}$$

The deformation of a material to an applied force can be expressed either as an absolute change in length (ΔL) or as strain that is defined as the change in length per unit initial length ($\Delta L/L_i$). Strain is a dimensionless unit and is expressed as a percentage, while shear strain is expressed in radians. Strength of a material is defined as the ultimate stress (σ_u) at which failure occurs.

In the body, tissues come in different geometric dimensions as well as different densities. For example, cortical (compact) bone has very high density relative to cancellous (trabecular or spongy) bone. When a force is applied on two tissues of different cross sections, stresses developed in a tissue of small cross section are significantly higher than in a tissue with a larger cross section. Hence, when comparing two elements, geometric parameters play a significant role. Strength defined structurally is the ultimate load (or force) at which failure of the system occurs. Consider the lower leg bone tibia and its corresponding fibula that support the upper body weight [Figure 5.7(a)]. When the stress strain curves are plotted for various loads, they show identical characteristics (i.e., the tibia and fibula have similar material strength per unit bone volume). However, the tibia has a larger cross section relative to the fibula. Hence, for any given stress, the force experienced by the tibia will be higher than the fibula. In addition, the failure of fibula will occur at a much lower force than the tibia. Under such a circumstance the tibia is said to have a higher structural strength than the fibula. Hence, while evaluating the material characteristics, structural differences are important.

5.3.2 Axial Stress and Strain

If a material is pulled along a symmetric axis of the material with a small force (called a tensile force), a temporary deformation occurs in the material due to the elastic displacements of molecules [Figure 5.7(b)]. Removal of the force will gradually result in the recovery of this extension. British scientist Robert Hooke stated that for small strains, stress is proportional to strain. This is called *Hooke's law* of linear elasticity. In one dimension,

$$\sigma = E * \varepsilon \tag{5.14}$$

where E is called the Young's modulus with the units of force per unit area. Cancellous (or spongy) bone has a modulus of 1 GPa, whereas a cortical (or compact) bone has a modulus of 18 GPa (Table 5.4). The material returns to its original size only to a certain applied force. This region is called the elastic region [Figure 5.8(a)] and its stress-strain behavior proceeds in a linear fashion from point A to point B. Hooke's law is also stated in terms of applied force (F) and observed displacement (x) from its initial position as

$$F = kx \tag{5.15}$$

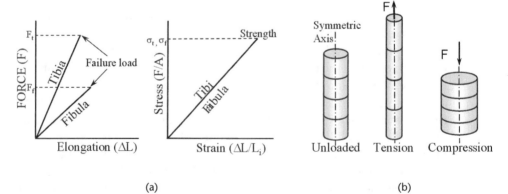

(a)   (b)

Figure 5.7 Tensile and compressive properties of materials: (a) the difference between the force-elongation curve and the stress-strain curve, and (b) the effect of the axial tensile and compressive loading on cylindrical materials.

Table 5.4 Material Properties Materials Used in Biomedical Applications

Material	σ_{yield} (MPa)	$\sigma_{ultimate}$ (MPa)	E (GPa)
Stainless steel	700	850	180
Cobalt alloy	490	700	200
Titanium alloy	1,100	1,250	110
Bone	85	120	18
PMMA (fixative)		35	5
UHMWPE (bearing)	14	27	1
Patellar ligament		58	

Source: [4].

where *k* is called the stiffness (or spring constant) with the units of force per unit length. Stiffness is affected by the structure's geometry (length and cross-sectional area), unlike the elastic modulus. In general, materials with large *k* values can balance larger forces than materials with low *k* values. By convention, a positive sign is used for tensile loading and a negative sign is used for compressive loading.

If the material is stressed beyond point B yielding or plastic deformation is observed in the material and the stress corresponding to point B is referred to as the yield strength, σ_{yield}, of the material. If material is stressed up to point C and then the load is removed, the sample does not return its original size at point A. Instead, the elastic component of deformation occurs along the line C-C', but the plastic portion does not. At C' the unloaded specimen is permanently elongated at a distance A-C' and by convention the specimen is said to have a permanent strain, C', expressed in percent. Loading beyond point C continues to plastically deform the sample until the ultimate tensile strength (σ_u) is reached and failure occurs. Stiffness or modulus is an important material parameter relating the degree of deformation to the applied load. The area under the stress-strain curve reflects the energy stored by the material prior to failure. The slope of the plot in the linear elastic region is the modulus and reflects the material's stiffness.

Stress-strain behavior of a few structural materials is compared in Figure 5.8(b). Metal has the highest stiffness (elastic modulus) and at stresses beyond its yield point exhibits typical ductile behavior (large plastic deformations before failure) in the nonelastic region. Glass also has a much higher modulus than bone; however, as a material it readily undergoes brittle failure, having no discernible nonelastic (plastic) region. Bone has a much lower modulus than either metal or glass, but in terms of failure mechanics, bone behaves similar to glass, fracturing in a brittle (low-deformation) mode. Brittle materials fracture into two or more pieces that have very little permanent plastic deformation and therefore have the potential to be nicely pieced together into the original prefracture conformation. In the case of bone, this behavior facilitates accurate anatomical fracture reduction and reconstruction using internal fixation. A fractured metal plate, in contrast, does not conform to its original prefractured shape, owing to the permanent plastic deformation that occurred prior to fracture. However, the ductile behavior of the

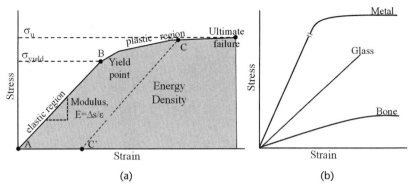

Figure 5.8 Stress-strain behavior: (a) different regions and (b) a comparison to other structural materials.

plate is desirable for internal fixation purposes, allowing the surgeon, within limits, to plastically contour the plate to bone without incurring brittle fracture.

Maximal tensile forces or compressive forces in the structure are generated on a plane perpendicular to the applied load (i.e., the normal force). Consequently, failure typically occurs along this plane. As in pure tension, maximal stresses occur on a plane perpendicular to the applied load; however, the stress distribution and resultant fracture mechanics in compressive failure are often very complicated, particularly for an anisotropic material such as bone.

5.3.2.1 Stresses Due to Pressure

Many conduits in the body transport fluids under pressure. For example, blood vessels carry liquids under pressure. In these cases, the material comprising the vessel is subjected to pressure loading, and hence stresses, from all directions within the conduit. The normal stresses resulting from this pressure are functions of the radius of the element under consideration, the shape of the pressure vessel (i.e., cylinder, or sphere), and the applied pressure. The method of analyses can be grouped into thick-walled pressure vessels and thin-walled pressure vessels. Generally, a pressure vessel is considered to be thin-walled if its radius r is larger than 10 times its wall thickness. For most engineering applications, the thin-wall pressure vessel can be used and is discussed later. For thick-walled pressure vessels, equations are called the Lame's equation and detailed derivations can be obtained from textbooks related to the strength of the materials.

Consider a cylindrical pressure vessel with a wall thickness, δ, and an inner radius, r (Figure 5.9). The fluid within this vessel has a gauge pressure of P and is flowing in the z-direction. Two types of normal stresses are generated based on the direction:

- Longitudinal (or axial) stress and, if the end of the vessel is closed, stress experienced by the end cap;
- Hoop stress, σ_h, the stress in a pipe wall acting circumferentially in a plane perpendicular to the longitudinal axis of the cylindrical vessel.

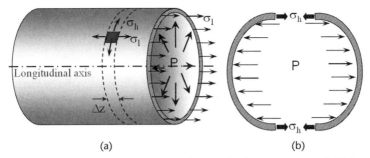

Figure 5.9   Stresses in pressure vessels: (a) longitudinal shear stress and (b) hoop stress in closed conduits.

Applying the conservation of momentum principle, described in Chapter 4 in (4.17), for equilibrium in the z-direction,

$$\sum F_z = 0, \text{ that is, } \sigma_l.\pi\left(r_o^2 - r_i^2\right) = p.\pi r_i^2$$

Substituting for $r_o = r_i + \delta$ and simplification gives

$$\sigma_l.\pi\left(2r_i\delta + \delta^2\right) = p.\pi r^2$$

With the assumption of a thin wall, t is small and t^2 is negligible

$$\sigma_l.2\pi r\delta = P.\pi r_i^2$$

Rearranging

$$\sigma_l = \frac{P.r_i}{2\delta} \tag{5.16}$$

For equilibrium in the radial direction,

$$\sum F_r = 0, \text{ that is, } 2.\sigma_h.\delta.dz = p.2r_i.dz$$

Rearranging

$$\sigma_h = \frac{P.r_i}{\delta} \tag{5.17}$$

Comparing (5.16) with (5.17), hoop stress is twice the longitudinal stress. Similar equations are derived for spherical pressure vessels (for example, bladder and alveoli) where hoop stress is the same as the longitudinal stress.

> ### EXAMPLE 5.5
>
> The aorta of a male patient had an inner radius of 13 mm and was 2.2 mm thick in the diastolic state. It was 50 cm long and expanded due to the pumping of the heart. When the heart valve opened in the systolic phase, 70 mL of blood was discharged. Half of this blood was initially stored in the aorta, expanding its wall to some inner radius. Assume the diastolic pressure and systolic pressure to be 80 mmHg and 130 mmHg, respectively, and the heart rate to be 72 beats per minute. Calculate the systolic radius and the wall thickness. What is the stress in the wall in the systolic state? How much energy is stored in the elastic wall? What is the average Reynolds number in the aorta? Is the flow laminar or turbulent?
>
> Solution:
>
> (a) Volume of the lumen in diastole = $\pi * (13 * 10^{-3})^2 * = 0.5 = 0.265*10^{-3}$ m^3
>
> Increase in volume due to storage of blood = $70/2 = 35$ mL $= 0.035$ m^3
>
> Volume of the lumen in systole = $(0.265 + 0.035) * 10^{-3} = 0.3 * 10^{-3}$ m$^3 = \pi * (r_s)^2 * 0.5$
>
> Systolic radius = 13.8 mm
>
> (b) Cross-sectional area in diastole = $\pi * [((13 + 2.2) * 10^{-3})^2 - (13 * 10^{-3})^2] = 194.9 * 10^{-3}$ m^2
>
> Cross-sectional area in systole = $194.9 * 10^{-3}$ m$^2 = \pi * [((13.8 + t_s) * 10^{-3})^2 - (13.8 * 10^{-3})^2]$ $t_s = 2$ mm
>
> Hoop stress, $\sigma_s = P_s R_s / t_s = 130$ mmHg $* (133$ Pa/mmHg$) * 0.0138$m$/0.002 = 0.113$ MPa
>
> (c) Energy stored in the elastic wall = $P_{mean} \Delta V$
>
> $P_{mean} = (P_{systolic} + 2P_{diastolic})/3 = (130 + 2 * 80)/3 = 96.67$ mmHg $= 96.96 * 133$ Pa $= 12.9$ $\Delta V = 35$ mL $= 35 * 10^{-6}$ m^3
>
> $P_{mean}\Delta V = 12.9 * 10^3 * 35 * 10^{-6} = 0.45$J
>
> (d) Volumetric flow rate = $70*10^{-6}$ (m^3/beat)$/(60$ sec$/72$ beats$) = 84.1*10^{-6}$ (m^3/s)
>
> Velocity = volumetric flow rate/average cross-sectional area $= 84.1 * 10^{-6}/(\pi * ((13 + 13.8)/2 * 10^{-3})^2] = 0.149$ m/s
>
> Density of blood (assume), $\rho = 1.1$ gm/cc $= 1,100$ kg/m^3
>
> Viscosity (assume), $\mu = 3$ cP $= 3 * 10^{-3}$ kg/m.s
>
> From (5.2), $N_{Re} = \rho \Delta v / \mu$ $= 1,100 * (2 * 0.0134) * 0.149/3 * 10^{-3} = 1,465$
>
> Hence, the flow is laminar.

Compliance is an expression used in biological materials to indicate their elastic nature. This concept is applied to blood vessels, the heart, lungs, and practically any other situation involving a closed structure. Compliance is defined as the change in volume for a given change in pressure. The reciprocal of compliance is elastance. For example, the lungs expand and contract during each breath. The pressure within and around the lungs changes equally at the same time. If a pressure-volume curve of the tissue is drawn, then the slope of the line is the compliance of the tissue. In the majority of the tissues, the pressure-volume relation is not linear due to the heterogeneous nature of tissues. For example, in a normal healthy lung at a low volume, relatively little negative pressure outside (or positive pressure inside) is applied to the lung expansion. However, lung compliance decreases with increasing volume. Therefore, as the lung increases in size, more pressure must be applied to get the same increase in volume. Hence, local compliance is calculated. Compliance can also change in various disease states; compliance decreases in scarring and increases in fibrous tissues. For example, fibrotic lungs are stiffer and require a large pressure to maintain a moderate volume. Such lungs would have a very low compliance value. However, in emphysema, where many alveolar walls are lost, the lungs become loose and floppy. Only a small pressure difference is necessary to maintain a large volume and the compliance value is significantly high.

The units of compliance in the above definition are volume per unit pressure [mL/mmHg] and depend on the volume of the vessel. To avoid the dependency on dimension, an alternative definition of compliance is the ratio of change in volume per unit volume of the vessel to the change in pressure. Then the units of compliance are inverse pressure units and that of elastance will be similar to elastic modulus.

EXAMPLE 5.6

A surgeon is inserting a 5-cm-long Palmaz-Schatz stent inside a 10-mm diameter arterial wall. However, during insertion it is crimped to a 4-mm diameter to easily navigate through the vessel and is mounted on a balloon. Using a pressure of 10 atm in the balloon, it is expanded to the final size. What is the compliance of the entire unit (i.e., stent and the balloon) during mounting?

Solution: Initial volume of the vessel, $V = \pi r^2 * L = \pi 0.2^2 * 5 [cm^3] = 0.628$ mL

Final volume of the vessel, $V = \pi r^2 * L = \pi 0.5^2 * 5 [cm^3] = 3.927$ mL

Change in volume = $3.927 - 0.628 = 3.299$ mL

$$\text{Compliance} = \frac{\Delta V / V}{\Delta P} = \frac{3.299 [mL] / 0.628 [mL]}{10 [atm]} = 0.525 \text{ atm}^{-1}$$

5.3.2.2   Composite Materials

Most biological materials are made up of multiple components and can be considered as composites, where different materials coexist as individual components but work together to give unique material properties. For example, tendons and ligaments consist of many matrix elements including collagen. Bone is a composite comprising a polymer matrix reinforced with inorganic salts; the polymeric matrix is mainly collagen (nearly 30% dry weight) and inorganic salts of calcium such as carbonates and phosphates (70% dry weight). The inorganic component provides the strength whereas the polymeric matrix contributes to the ductility. Extracellular matrix elements in connective tissue are also mixtures of elastin, collagen, and proteoglycans. Composites were first devised to provide higher levels of performance than individual materials alone. One advantage of composite materials is that two or more materials could be combined to take advantage of the characteristics of each material. The major attractions of composites are that they can be made stiffer, stronger, or tougher to different loading conditions and also can be lighter than their constituent single components. Biological substitutes in many applications can also be developed by blending (not dissolving) two or more components.

Consider a composite made up of two materials: one material (the matrix or binder) surrounds and binds together a cluster of fibers or fragments of a much stronger material (the reinforcement). Properties of the composites depend on the individual material properties and the amount and geometrical arrangement of the constituents (i.e., volume fraction) in a parallel [Figure 5.10(a)] or in a series [Figure 5.10(b)] arrangement. In general, elastic properties can be calculated with the following assumptions: (1) Hooke's law is applicable (i.e., stresses are below the elastic limit), (2) both phases are continuously in contact, and (3) there is no slip between the two phases. In order to account for directionality in mixing two components, two limiting cases are discussed later for the elastic properties of the two-phase composite systems.

5.3.2.3 Young's Modulus of a Composite Material of Two Components in a Parallel Configuration

When a force F is applied, the stress in the fiber and the stress in the matrix are not the same, as the cross sections of the two components are different. However, the axial strain (ε) must be the same (known as the isostrain condition) in both the components, with the assumption that there is no slip between the two phases. Otherwise, holes would appear in the ends of the composite as we stretched it. Then stresses can be calculated knowing the elastic modulus of each material.

For material 1, axial stress $= E_1\varepsilon$
For material 2, axial stress $= E_2\varepsilon$

To design the composite one has to evaluate the composition of each component. For this purpose, volume fraction for material $1 = \phi = \dfrac{V_1}{V_{total}} = \dfrac{A_1}{A_{total}}$ as the length of the fibers are same.

Thus, the volume fraction $= 1 = \phi = \dfrac{V_2}{V_{total}} = \dfrac{A_2}{A_{total}}$

For the entire material,

$$A_1 E_1 \varepsilon + A_2 E_2 \varepsilon \tag{5.18}$$

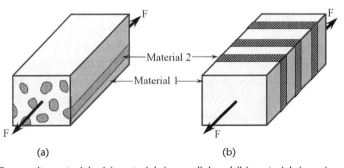

Figure 5.10 Composite materials: (a) materials in parallel and (b) materials in series.

$$\frac{A_1 E_1 \varepsilon + A_2 E_2 \varepsilon}{A_{total}} = \varphi E_1 \varepsilon + (1 - \varphi) E_2 \varepsilon \tag{5.19}$$

The combined modulus of elasticity is given by

$$E = \phi E_1 + (1 - \phi) E_2 \tag{5.20}$$

Thus, by selecting two components with known properties, composites with desired properties can be obtained by altering the volume of each component. A similar analysis can be extended to composites with more components. In the above derivation, the volume fraction of fibers is used, but while manufacturing the weight fraction is easy to measure. Knowing the densities of each component, mass fractions can be converted to volume fractions.

5.3.2.4   Young's Modulus of a Composite Material of Two Components in a Series Configuration

When a force F is applied, the stress in the fiber and the stress in the matrix are the same as the cross sections of the two components are same (known as the isostress conditition). If the force is in the transverse direction of the same composite, the components are now in an isostrain situation. If the two components have varying elastic properties, then each will experience a different strain. The strains can be calculated knowing elastic modulus of each material.

For material 1, axial strain, $\varepsilon_1 = \sigma / E_1$
For material 2, axial strain, $\varepsilon_2 = \sigma / E_2$

The strain in the composite will be the volume average of the strain in each material.

The volume fraction for material $1 = \phi = \dfrac{V_1}{V_{total}} = \dfrac{L_1}{L_{total}}$ as the cross-sectional area is constant. Thus, the volume fraction equals $1 = \phi = \dfrac{V_2}{V_{total}} = \dfrac{L_2}{L_{total}}$

For the entire material, the axial deformation equals $L_1 \dfrac{\sigma}{E_1} + L_2 \dfrac{\sigma}{E_2}$

and the axial strain equals $\dfrac{L_1 \sigma}{L_{total} E_1} + \dfrac{L_2 \sigma}{L_{total} E_2} = \phi E_1 \varepsilon + (1 - \phi) E_2 \varepsilon$

The combined modulus of elasticity equals $\dfrac{\sigma}{\phi \sigma / E_1 + (1 - \phi) \sigma / E_2}$

$$E = \frac{E_1 E_2}{\phi E_2 + (1 - \phi) E_1} \tag{5.21}$$

It is also important to determine the effective stiffness, which is an average measure of the stiffness of the material, taking into account the properties of all phases of the heterogeneous media and their interaction. The effective stiffness, K_{eff}, properties can be used in the analysis of a structure composed of the composite material by

$$K_{eff} = \frac{A_{total} E_{combined}}{L_{total}} \qquad (5.22)$$

EXAMPLE 5.7

A bioengineer is designing a new artificial ligament using a composite structure having a total cross-sectional area of 4 cm^2 and a length of 10 cm. Also, each structure is made up of material 1 (Young's modulus of E = 600 MPa) and material 2 (Young's modulus of E = 10 MPa), in proportions of 80% and 20%, respectively. In the first design, constituents are arranged in parallel, while in design 2, constituents are arranged in series. A tensile axial load of 1,000N is applied to each structure. For each design, determine the tensile force and the tensile stress in each constituent along the cross-section, the strain and axial extension in each constituent, the total extension, the effective modulus, and the total effective stiffness. Which design more closely resembles a real tendon, which has a Young's modulus of 400 MPa? Which one would you choose?

Solution:
Total axial force = 1,000N
Total axial stress = total force/total area = 1,000/0.0004 =2.5 MPa
(a) Material in parallels
From (5.19), total axial stress = $\phi E_1 \varepsilon + (1 - \phi)E_2 \varepsilon$
Hence, $\varepsilon = \dfrac{2.5[\text{MPa}]}{0.8 * 600[\text{MPa}] + (1 - 0.8) * 10[\text{MPa}]} = 0.0052$
Force in Material 1 = 0.0052 * 0.004 [m^2] * 0.8 * 600 * 10^6 [Pa] = 995.9N
Force in Material 2 = 1,000 – 995.9= 4.1N
$\varepsilon_1 = \varepsilon_2 = \varepsilon = 0.0052$
Total$_{extension}$= Material 1$_{extension}$ = Material 2$_{extension}$ = 0.0052 * 0.10[m] = 0.52 mm
From (5.20), effective modulus E = 0.8 * 600 [MPa] + (1 – 0.8)10[MPa] = 482 MPa
From (5.22), $K_{eff} = \dfrac{0.004 * 482 * 10^6 [\text{Pa}]}{0.1[\text{m}]} = 1{,}928$ kN/m
(b) Materials in series
Force in Material 1 = Force in Material 2 = 1,000N
Stress in Material 1 = Stress in Material 2 = 2.5 MPa
$\varepsilon_1 = \sigma/E_1 = 2.5[\text{MPa}]/600[\text{MPa}] = 0.0042$
$\varepsilon_2 = \sigma/E_2 = 2.5[\text{MPa}]/10[\text{MPa}] = 0.25$
Material 1$_{extension}$ = 0.0042 * 0.8 * 0.10 [m] = 0.33 mm
Material 2$_{extension}$ = 0.25 * 0.2 * 0.10 [m] = 5 mm
Total$_{extension}$ = 5.33 mm
Total strain = 5.33/100 = 0.053
From (5.121), effective modulus
$E = \dfrac{600[\text{MPa}] * 10[\text{MPa}]}{0.8 * 10[\text{MPa}] + (1 - 0.8)600[\text{MPa}]} = 46.9$ MPa
From (5.22)
$K_{eff} = \dfrac{0.004 * 46.9 * 10^6 [\text{Pa}]}{0.1[\text{m}]} = 188.8$ kN/m

Since the effective modulus of the composite in parallel is close to the modulus of the natural tendon, a composite with a parallel configuration needs to be selected.

5.3.3   Shear Stress

When a shearing stress is applied to a liquid, it suffers a shear deformation and, as long as the shear is applied, it continues to shear. When the stress is removed, the shearing stops, but does not recover. However, when a shearing stress is applied to a solid, it suffers a shear deformation. When the shearing stress is removed, if the solid is elastic, the deformation recovers. For example, square or rectangular solids (Figure 5.11) become parallelograms under shear loading. Shear strain is the measure of the amount of angular deformation, α, in response to the application of shear force. Hence, shear loading is described graphically by a load-angle-change plot instead of a load-deformation plot as in the case of tension and compression. A graph of shear stress versus shear strain will normally exhibit the same characteristics as the graph of axial stress versus axial strain (Figure 5.11). Hence, stiffness, yielding, and elastic and nonelastic behavior are interpreted similarly from the plots regardless of loading mode. Within the elastic region, Hooke's law for shear stress is

$$\tau_{xy} = G\gamma_{xy} \tag{5.23}$$

where G is the shear modulus or modulus of rigidity with the units similar to shear stress [Pa], τ_{xy} is the shear stress in the x-y-plane, and γ_{xy} is the shear strain in the xy-plane. By convention, shear stress is considered positive when a pair of shear stresses acting on opposite sides of the element to produce a counterclockwise torque. The coordinate directions chosen to analyze a structure are usually based on the shape of the structure. As a result, the direct and shear stress components are associated with these directions. For example, to analyze a bar, one always chooses one of the coordinate directions along the bar's axis [Figure 5.11(b)].

5.3.3.1   Shear Stress in Combination with Axial Stresses

Consider a plane cut through a rectangular bar at an angle θ. Acting on this plane will be both an axial stress σ_θ and a shear stress τ_θ. By multiplying these stresses by the appropriate cross-sectional areas, forces on each surface can be determined. Then, using the conservation of momentum at static equilibrium conditions,

$$\sum F_x = -\sigma_x(A\cos\theta) - \tau_{xy}(A\sin\theta) - \tau_\theta(A\sin\theta) + \sigma_\theta(A\cos\theta) = 0 \tag{5.24}$$

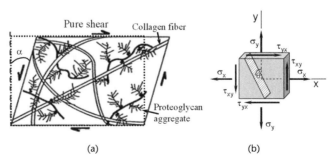

(a)   (b)

Figure 5.11   Shear stress: (a) pure shear stress effect on cartilage and (b) stresses in a coordinate system.

$$\sum F_y = -\sigma_y (A\sin\theta) - \tau_{xy}(A\cos\theta) + \tau_\theta(A\cos\theta) + \sigma_\theta(A\sin\theta) = 0 \qquad (5.25)$$

Solving (5.20) and (5.21) simultaneously for σ_θ and τ_θ using trigonometric relations,

$$\sigma_\theta = \frac{\sigma_x + \sigma_y}{2} + \frac{\sigma_x - \sigma_y}{2}\cos 2\theta - \tau_{xy}\sin 2\theta \qquad (5.26)$$

$$\tau_\theta = \frac{\sigma_x - \sigma_y}{2}\sin 2\theta - \tau_{xy}\cos 2\theta \qquad (5.27)$$

Equations (5.26) and (5.27) are referred to as transformation equations. There exist a couple of particular angles where the normal axial stresses are either at the maximum or at the minimum (Figure 5.12). They are known as the *principal stresses*, and the planes at which they occur are known as the principal planes. The angle θ_p is found by setting τ_{xy} to zero in (5.24),

$$\tan 2\theta_p = \frac{2\tau_{xy}}{\sigma_x - \sigma_y} \qquad (5.28)$$

θ_p defines the *principal directions* where the only stresses are normal stresses. The stresses are found from the original stresses (expressed in the x,y,z directions) via

$$\sigma_{max} = \frac{\sigma_x + \sigma_y}{2} + \sqrt{\left(\frac{\sigma_x - \sigma_y}{2}\right)^2 + \tau_{xy}^2} \qquad (5.29)$$

$$\sigma_{min} = \frac{\sigma_x + \sigma_y}{2} - \sqrt{\left(\frac{\sigma_x - \sigma_y}{2}\right)^2 + \tau_{xy}^2} \qquad (5.30)$$

Furthermore, the maximum shear stress is equal to one-half the difference between the two principal stresses,

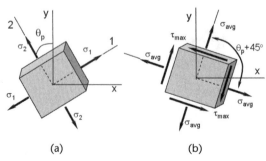

(a)   (b)

Figure 5.12   (a) Principal stresses and (b) maximum shear stress.

$$\tau_{max} = \sqrt{\left(\frac{\sigma_x - \sigma_y}{2}\right)^2 + \tau_{xy}^2} = \frac{\sigma_{max} - \sigma_{min}}{2} \qquad (5.31)$$

θ_s, where the maximum shear stress occurs, can be obtained by solving the shear stress transformation equation,

$$\tan 2\theta_s = \frac{\sigma_x - \sigma_y}{2\tau_{xy}} \text{ and } \theta_s = \theta_p \pm 45° \qquad (5.32)$$

5.3.4   Bending

Consider a beam attached to the wall at one end (Figure 5.13) and a force applied at the unattached end. Forces on the spine are frequently in this situation. The applied force will bend the beam with a tensile force maximum on the convex surface and with a compressive force maximum on the concave side. Between the two surfaces (i.e., concave and convex through the cross-section of the member), there is a continuous gradient of stress distribution from tension to compression. An imaginary longitudinal plane corresponding to the transition from tension to compression, approximately in the center and normal to applied force, is designated the neutral surface. Along this surface there is theoretically no tensile or compressive load on the material. Another designation is the neutral axis, which is the line formed by the intersection of the neutral surface with a cross-section of the beam, perpendicular to its longitudinal axis. The bending of the beam distorts the shape of the beam, which can be described as an arc of an imaginary circle. If r is the radius of the neutral axis and θ is the angle that describes the arc segment, then the length of the beam along the neutral axis is given by $r\theta$. Length of the beam at height y from the neutral axis can be calculated using the relation

$$L = (r + y)\theta$$

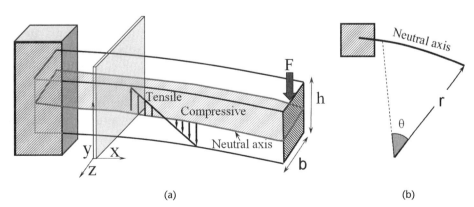

(a)   (b)

Figure 5.13   Bending moment in a cantilever: (a) stress distribution and (b) schematic appearance of the beam.

Axial strain ε_x at height y from the neutral axis $= \varepsilon_x = \dfrac{(r+y)\theta - r\theta}{r\theta} = \dfrac{y}{r}$

Axial stress σ_x at height y from the neutral axis $= \sigma_x = E\dfrac{y}{r}$

The beam is in static equilibrium. Hence, the forces across the section are entirely longitudinal and the total compressive forces must balance the total tensile forces. Applying the conservation of momentum principle in the x-direction,

$$\sum F_x = 0$$

The force on each element of cross-section dA is $\sigma_x dA$ and the sum of forces on the entire cross-section can be obtained by integration, that is,

$$\int \sigma_x dA = \frac{E}{r}\int y\, dA = 0$$

This indicates that the neutral axis (where $y = 0$ by definition) must pass through the centroid of the cross-sectional area. If M_b is the bending moment developed, the equilibrium of moments in the z-direction is

$$M_b - \int y\sigma_x dA = 0$$

Substituting for σ_x,

$$M_b = \frac{E}{r}\int y^2 dA$$

Replacing $\int y^2 dA$ with I_{zz} leads to

$$M_b = \frac{E}{r}I_{zz} \tag{5.33}$$

I_{zz} is commonly referred as the area moment of inertia about the z-axis. In a bending mode of loading, strength and stiffness are dependent on the cross-sectional area as in tension and compression and on the arrangement or distribution of mass about the neutral axis. The strength parameter is the *area moment of inertia* (*I*) and helps in evaluating the strength of a specific shape or geometry under conditions of bending. The area moment of inertia takes into account the fact that in bending, a structure gets stronger (and stiffer) as its mass is moved further from its neutral axis. In structural engineering, this concept is adapted by the frequent use of I-beams, which provide a maximum resistance to bending with a minimum weight. Similarly, the tubular shape of bone is appropriately designed to uniformly resist bending in all directions and has its mass located circumferentially at a distance from the neutral axis, thus providing a high area moment of inertia and high resistance to bending. Equations for calculating the area moment of inertia for few commonly utilized cross-sections are summarized in Table 5.5.

EXAMPLE 5.8

A company is investigating a new plate for use in a bone fracture. They have a plate that is 1 cm thick and 3 cm wide. They want to know whether Case A or Case B orientation is better for resisting bending. Could you help them?

Solution: For a rectangular cross-section, $I = bh^3/12$

For Case A, $I = 3 * 1^3/12 = 0.25$ cm^4

For Case B, $I = 1 * 3^3/12 = 2.25$ cm^4

Since the moment of inertia for Case B is significantly higher, the developed stresses and deflection are significantly less. Hence, orientation according to Case B is preferred.

Substituting for (E/r) in (5.33) in terms of σ_x gives the expression

$$M_b = \frac{\sigma_x}{y} I_{zz}$$

Rearranging, the variation in axial stress throughout the cross-section is calculated as

Table 5.5 Formula for Moment of Inertia of Common Shapes

Cross-Sectional Shape	Area Moments of Inertia (I)	Polar Moments of Inertia (J)
Square of side length a	$\frac{1}{12}a^4$	$\frac{1}{6}a^4$
Circle with radius r	$\frac{\pi}{4}r^4$	$\frac{\pi}{2}r^4$
Circular tube with inner radius r_i and outer radius r_o	$\frac{\pi}{4}\left(r_o^4 - r_i^4\right)$	$\frac{\pi}{2}\left(r_o^4 - r_i^4\right)$
Ellipse with a minor axis of $2a$ and major axis $2b$	$\frac{\pi}{4}ab^3$	$\frac{\pi}{4}ab\left(a^2 + b^2\right)$
Equilateral triangle with side length a	$\frac{\sqrt{3}}{96}a^4$	$\frac{\sqrt{3}}{80}a^4$
Rectangle of b breadth and h height	$\frac{1}{12}bh^3$	$0.1406bh^3$

$$\sigma_x = \frac{M_b y}{I_{zz}} \qquad (5.34)$$

As axial stress depends on the height from the neutral axis (y), the greatest positive (tensile) stress is in the top surface of the beam. The greatest negative (compressive) stress is in the bottom surface of the beam, which can be determined by

$$\left(\sigma_x\right)_{y=y_2} = -\frac{M_b y_2}{I_{zz}}$$

The quantities I_{zz}/y_1 and I_{zz}/y_2 are referred to as *section modulii*. Equation (5.34) also states that stress does not depend on elastic modulus. However, axial strain depends on a modulus. The equation for axial strain can be obtained by substituting Hooke's law

$$\varepsilon_x = \frac{M_b y}{EI_{zz}} \qquad (5.35)$$

The product EI_{zz} in the denominator is called the flexural rigidity. The length (L) of a structure does not affect its flexural rigidity in bending. However, the length will affect the peak moment, the peak stress, and the maximum deflection under bending. To determine flexural rigidity in biological samples, three-point bending experiments are performed. In this case, the moment and the axial stress are greatest at the point of application of the middle load. For a small deflection in the Y-direction relative to the length of the cantilever with a point load (in the y-direction), the maximum beam deflection can be obtained using the expression

$$\delta_{max} = \frac{FL^3}{3EI} \qquad (5.36)$$

The assumption in deriving this expression is that the plane section remains linear. Equation (5.36) cannot be used for large deflections and nonlinear equations are utilized.

5.3.5 Torsion

Torsional loading is a geometric variation of shear loading and acts to twist a structure about the neutral axis. Torsion produces maximum shear stresses over the entire surface, and these stresses are proportional to the distance from the neutral axis. The magnitude of deformation is measured in terms of the shear angle. Maximum shear stresses are on planes perpendicular and parallel to the neutral axis, but maximum tensile and compressive stresses are on a diagonal to the neutral axis. A type of bone fracture called a spiral fracture is attributed to torsional loading. The location of a crack initiation and the direction of its propagation are dependent on the inherent strength of the material and on the magnitude of the imposed stresses within the material. It has been suggested that for a bone subjected to pure torsional loading, failure begins with a crack initiation in a shear mode, followed by a crack propagation generally along the line of a maximum tensile stress perpendicular to the neutral axis.

Consider a cylindrical object (Figure 5.14) subject to equal and opposite torque (T) at both ends. To illustrate the effect of torsion, a square is drawn the surface. With the application of torsion on the cylinder, the circular symmetry of the object is maintained, which implies in turn that plane cross-sections remain plane, without warping. However, the square element shows shear-type deformation twisted by an angle θ. The shear stress varies from zero in the neutral axis to a maximum at the outside surface of the cylindrical structure. On the contrary, if a diamond-shaped schematic is drawn on the surface of the cylinder, then the diamond-shaped element experiences a deformation analogous to simple tension and compression; it elongates and narrows, with maximum tensile stresses acting on a plane perpendicular to the axis of elongation and maximum compressive stresses orthogonal to that.

Performing a momentum balance similar to bending moment derivation, the shear stress (τ) due to torsion is given by

$$\tau = \frac{Tr}{J} \tag{5.37}$$

where J is the polar moment of inertia of the cross-section. Similar to the area moment of inertia, the polar moment of inertia is a measure of resistance offered by an object to torsion. The polar momentum of inertia is defined by

$$J = \int r^2 dA$$

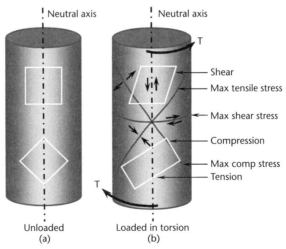

Figure 5.14 Effect of torsional loading cylindrical material: (a) before application and (b) after application and stress distribution.

The polar moments of inertia for common cross-sections are summarized in Table 5.5. Using Hooke's law, the shear stress is related to the shear strain (i.e., the rotation of the cross sections). The corresponding shear strain (γ) is given by

$$\gamma = \frac{Tr}{GJ} \tag{5.38}$$

The product GJ is called the *torsional rigidity*. Typically, the failure plane of a brittle shaft under torsion is often at a $45°$ angle with respect to the shaft's axis.

EXAMPLE 5.9

A skier with a mass of 65 kg catches the edge of the right ski while making a turn. The torque developed during the accident is 15 Nm. The minimum cross-section of the tibia is approximated by a tube of elliptical shape with an outer major axis of 20 mm and a minor axis of 16 mm. The inner major axis is 18 mm and the minor axis is 14.5 mm. If the average ultimate shear strength is 60 MPa, determine whether the tibia will fracture or not.

Solution: $T = 1$ Nm

$c = 0.01$ m (outer radius of the major axis)

J = polar moment of inertia of the elliptical tube = $J_{out} - J_{in}$

Using the relation for J from Table 5.5 for elliptical geometry,

$$J = \frac{\pi}{4} * 10 * 8 \left(10^2 + 8^2 \right) - \frac{\pi}{4} * 9 * 7.25 \left(9^2 + 7.25^2 \right) = 3,460 \text{ mm}^4 = 3,460 * 10^{-12} \text{m}^4$$

Using (5.37), $\tau_{xy} = \dfrac{Tc}{J} = \dfrac{15 * 0.01}{3,460 * 10^{-12}} = 43.35$ MPa

Since it is below the ultimate shear strength, the tibia would not break.

5.4   Nonidealities in Stress-Strain Characterization

5.4.1   Failure Under Combined Loading

Tension, compression, shear, bending, and torsion represent simple and pure modes of loading. In reality, all loading patterns are observed in the body when activities are performed. Injuries encountered in the body are more commonly a product of a combination of the aforementioned modes exposed and short-term and long-term loadings and movements combined. Furthermore, the mode of loading is determined by the direction of the load application and the fact that in the case of traumatic injuries (e.g., automobile) there is virtually no constraint on applied load orientation (or magnitude). In addition, all materials show anisotropic mechanical behavior owing to their anatomical adaptations. For example, bone exhibits greater strength when subjected to tension directed longitudinally versus tension directed transversely. Under tensile loading, the orientation of the fracture line is perpendicular to the long axis of the bone. Unlike failure in tension, a compressive failure in the bone does not always proceed along the theoretic perpendicular plane of maximum stress, but rather once a crack is initiated, it may propagate obliquely across the thickness following the line of the maximum shear stress. The tearing strength of the aorta is higher in the longitudinal direction than in the transverse direction.

A thorough knowledge of the complex biomechanical function of the body structures and joints is essential to make accurate clinical diagnoses and decisions regarding injury treatments. Techniques such as a universal force-moment sensor, which can mimic different loading conditions, have been developed for a few joints. However, evaluating the material properties and predicting failure under combined loading for various components are difficult due to problems associated with the repeatability of results. Computational techniques such as the finite element analysis (FEA) are used in conjunction with experiments for a better understanding of the stress and deformation characteristics. For detailed description of the FEA, refer to [5].

5.4.2 Viscoelastic Characteristics

The behavior of metals such as steel or aluminum and quartz do not deviate much from the linear elasticity at room temperature within a small strain range. However, all other materials deviate from Hooke's law (5.14) in various ways, for example, by exhibiting viscous-like and elastic characteristics. A perfectly elastic material stores all the energy created by the deformation forces so that on the removal of the forces, material can return to its original dimensions independent of time. In contrast to elastic materials, a Newtonian viscous fluid under shear stress obeys

$$\sigma = \mu \frac{d\varepsilon}{dt} \tag{5.39}$$

with μ as the viscosity, and this fluid dissipates energy as heat or in the drag of the fluid that is exuded and absorbed during loading and unloading. There are materials whose response to a deforming load combines both viscous and elastic qualities; that property is called *viscoelasticity*. The relationship between stress and strain depends on time. Anelastic solids represent a subset of viscoelastic materials that possess a unique equilibrium configuration and ultimately recover fully after the removal of a transient load. Synthetic polymers, wood, and all human tissues (cells, cartilage, bone, muscle, tendon, and ligament) as well as metals at high temperature display significant viscoelastic effects. As an example, consider the behavior of a soft tissue such as cartilage or ligament subjected to a constant compressive strain (Figure 5.15). When the tissue is subjected to a constant deformation [point B in Figure 5.15(b)] and is maintained at that state, its stress varies over time. Initially, stress increases with the efflux of water, followed by a time-dependent reducing stress phase where the fluid within the cartilage redistributes. Immediately after loading occurs, the small permeability of the extracellular network impedes an instantaneous fluid flow through the matrix. With time, the load causes the fluid to be driven away from the loaded site, through pores in the matrix. This fluid flow also explains another feature of viscoelasticity (i.e., that the biomechanical behavior

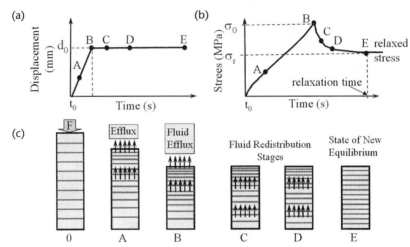

Figure 5.15 Viscoelastic behavior in a soft tissue: (a) application of strain and holding, (b) alteration in stress, and (c) schematic representation of structural rearrangement with strain.

of the tissue is dependent on the strain rate and on the time after the stress application). For instance, when the strain and stress are applied rapidly, the slope of the stress-strain curve will be steeper than when they are applied slowly.

5.4.3 Dynamic Loading

Many biological materials possess time and history dependent mechanical properties. The loading and unloading of a specimen yield different paths, forming a hysteresis loop that represents the energy lost as a result of a nonconservative or dissipative process. For example, the medial collateral ligament (MCL) heals by itself after an injury, remodeling the tissue, whereas an anterior cruciate ligament (ACL) does not heal and requires surgical intervention. Furthermore, the loading pattern can be classified into two types: static loading and dynamic loading. For example, static loading occurs during standing and sitting, whereas dynamic loading occurs during walking and running. Since the above-mentioned methods do repetitive loading-unloading, they are not suitable for evaluation during dynamic loading. This requires the assessment of the responses to a cyclic loading at a wide range of frequencies. In dynamic tests, a cyclic or sinusoidal stress is commonly produced for the determination of the behavior during dynamic loading. The peak sustainable load under cyclic conditions is significantly lower than for constant loading. The fatigue behavior of a material is typically represented on a plot of the peak stress per cycle (σ) versus the number of cycles (N) to failure. Under cyclic loading, tissues can quickly settle into a steady-state response called the endurance limit or preconditioning [see the dotted line in Figure 5.16(a)]. Sometimes, there is also some slow secondary creep of the steady-state cycle. However, this is generally negligible compared with the differences in response over the first few cycles of loading. Furthermore, the stress-strain curve during loading is essentially different from that during unloading, and this feature is called hysteresis [Figure 5.16(b)]. The respective dimensions of the hysteresis loops of the stress-strain curves may

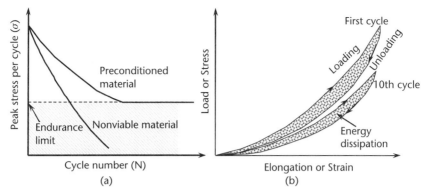

Figure 5.16   Influence of cyclical loading: (a) fatigue properties and (b) hysteresis curve.

change dramatically during the first few cycles of loading and unloading. After preconditioning, the hysteresis loop during loading and unloading will remain essentially unchanged. A material can be cycled virtually endlessly at stresses below this limit.

To describe the general response of a system, a model must allow for details of loading history. One of the approaches is using the *Boltzmann superposition principle*. Boltzmann proposed that the long-term deformation produced by two or more forces is independent of the order and the times at which forces are applied, but the long-term effect of two or more forces applied at different times is equal to that produced applying them simultaneously.

5.5   Energy Conservation

5.5.1   Conservation of Energy

The conservation of energy principle described in Section 5.4.4 is also useful in assessing various interactions in biomechanics. For biomechanical applications in linear motion, (5.25) can be simplified to

$$\frac{V_1^2}{2g} + z_1 = \frac{V_2^2}{2g} + z_2 + h_w \tag{5.40}$$

where h_w is the heat loss associated with energy loss due to friction or the action performed by the activity. Multiplying h_w by the weight, the energy loss by the system can be calculated. Consider the case of a wheelchair standing on a hill of height h. Initially when it is at rest, all its energy is potential energy. As it accelerates down the hill, some of its potential energy is converted into kinetic energy and some is used to overcome friction. This energy used to overcome friction is not lost but converted into heat. At the bottom of the hill the energy will be purely kinetic, with the assumption that the potential energy is zero.

A person weighing 70 kg is skiing from the top of a hill that is 50m tall and has a slope of 1 in 20 measured along the slope. He starts at a speed of 1 m/s, after which he skies down to the bottom of the hill, where his speed increased to 25 m/s. How much energy has he lost?
Solution:

$$\frac{1^2[\text{m/s}]^2}{2*9.81[\text{m/s}^2]} + 50[\text{m}] = \frac{25^2[\text{m/s}]^2}{2*9.81[\text{m/s}^2]} + 0 + h_w$$

$$= h_w = 18.196[\text{m}]$$

Energy lost = $mgh_w = 70[\text{kg}]*9.81[\text{m/s}^2]*18.196[\text{m}] = 12,495\text{J}$

A simplified version of conservation of energy principle is the conservation of mechanical energy where the frictional loss is neglected. The sum of the kinetic energy, ΔE_{KE}, and the potential energy, ΔE_{PE}, is referred as the total mechanical energy, ΔE_{ME}. If the system is isolated, with only conservative forces acting on its parts, the total mechanical energy of the system is a constant. The mechanical energy can be transformed between its kinetic and potential forms, but cannot be destroyed. Rather, the energy is transmitted from one form to another. Any change in the kinetic energy will cause a corresponding change in the potential energy and vice versa. For objects in rotary motion, kinetic energy is calculated using angular velocity. Potential energy is a form of stored energy and is a consequence of the work done by a force. Examples of forces that are associated with potential energy are the gravitational and the electromagnetic fields and, in biomechanics, a spring. For a body moving under the influence of a force F, the change in potential energy is given by

$$\Delta E_{PE} = -\int_1^2 F \cdot ds \qquad (5.41)$$

where 1 and 2 represent the initial and final positions of the body, respectively. Equation (5.41) states that the performance of work requires the expenditure of energy. This relation is commonly referred to as the *work-energy relation* and can be applied to kinetic energy as well.

$$\Delta E_{KE} = \int_i^f F \cdot ds$$

Consider a body of mass m being accelerated by a compressed spring. The compressive force exerted by a spring within the linear range is given by (5.15),

$$\Delta E_{KE} = \int_i^f - \quad = \frac{1}{2}\left(\begin{smallmatrix} 2 \\ 2 \end{smallmatrix} - \begin{smallmatrix} 2 \\ 1 \end{smallmatrix} \right)$$

If the initial position is considered $x_1 = 0$ (the spring's equilibrium position and where its potential energy is defined to be zero) and set $x_2 = x$, then the above equation becomes

$$\Delta E_{PE} = \frac{1}{2}kx^2 \qquad (5.42)$$

The work-kinetic energy relation is used to analyze the motion of particles. However, the work-kinetic energy relation is difficult to apply to many particle systems and limited in application to mechanical process involving motion only.

5.5.2   Energy Absorption

When two objects collide, they exert equal and opposite forces on each other (i.e., both experience the same force). For example, when a person lands from a jump, the GRF returns the force of the landing body back to the body. Since the contact areas associated with the collision are same for both the objects involved in a collision, the impulse applied to each object and the change in momentum experienced by each object are the same. Hence the sum of the linear momentums of the two objects before the collision is the same as the sum of the linear momentum after the collision, that is,

$$m_1 v_{i1} + m_2 v_{i2} = m_1 v_{f1} + m_2 v_{f2} \qquad (5.43)$$

There are two primary types of collisions: perfectly *elastic* and perfectly *inelastic*. Inelastic collisions occur when the objects involved in the collision stick together. The sum of their kinetic energies is reduced, compared with the initial value, because a part of it has changed into internal energy. Elastic collisions occur only between extremely hard objects and are of great interest in biomechanics. Objects in elastic collisions exchange momentum by "bouncing" from each other during the collision process (like billiard balls). The elasticity of a collision is indicated by the coefficient of restitution, k_r

$$k_r = \frac{\text{speed after collision}}{\text{speed before collision}}$$

For a perfect elastic collision, k_r is 1, but for most collisions in human movement k_r is less than 1. As the speed of collision increases, k_r tends to decrease. The kinetic energy is also conserved in elastic collisions. For a two-body system,

$$\frac{m_1 v_{i1}^2}{2} + \frac{m_2 v_{i2}^2}{2} = \frac{m_1 v_{f1}^2}{2} + \frac{m_2 v_{f2}^2}{2}$$

$$v_{1f} = \frac{m_1 - m_2}{m_1 + m_2} v_{i1} \text{ and } v_{2f} = \frac{2m_1}{m_1 + m_2} v_{i1} \qquad (5.44)$$

Due to the impulse-momentum relationship, the body develops a certain impulse that has to be dissipated. If the time over which the velocity is decreased is very small, then the force will be very large. If the time over which the velocity is decreased is very large, then the magnitude of the forces will be small. Therefore, one should try to increase the time component of the landing or catch as much as possible in order to decrease the force per unit time.

EXAMPLE 5.11

A 90-kg person wanted to try skiing. He went to the top of a Killington Ski Resort, which is located at a height of 140m. Being new to the sport and unable to control his skis, he quickly finds himself going very fast through the icy (negligible friction) hill. At the base of the hill he is unable to stop. He skis across an icy deck, through the open doors of the ski lodge, and across the concrete floor of the rental shop a distance of 15m before coming to rest against the back wall of the shop. The deck and lodge are on level ground at the base of the hill. If the unfortunate skier experiences an impulse of 4,500 Ns during the collision, what is the coefficient of kinetic friction between the skis and the concrete?

Solution: First, the velocity of the person before colliding with the wall needs to be determined. From impulse-moment relation,

$$\int_0^t F(t)dt = m(V_2 - V_1) \rightarrow 4,500 \ [\text{Ns}] = 90[\text{kg}]V_2[\text{m/s}] \ \text{or} \ V_2 = 50[\text{m/s}]$$

From (5.61), $0 + 140[\text{m}] = \dfrac{50^2[\text{m/s}]^2}{2*9.81[\text{m/s}^2]} + 0 + h_L = 12.579\text{m}$

Since the skier had to dissipate this energy by friction on concrete (15m), the coefficient of friction is 12.579/15 = 0.839.

The number and pattern of cracks formed in a bone depend largely on the rate at which load is applied. High velocity impacts release a large amount of energy on fracture resulting in the fragmentation of bone and soft tissue injury. Fractures from high-energy injuries have a higher rate of healing complications due to the severity of soft tissue injuries. Since bone has a larger elastic modulus (or stiffness), it can absorb more energy to failure if the load is applied more rapidly (i.e., it is stiffer and tougher). A single crack, however, has a finite threshold energy for initiation and a finite capacity to dissipate stored or applied energy. Thus, under conditions of a high loading rate, if the stored energy in the structure exceeds that which can be dissipated via the formation of one crack, multiple cracks will form and energetically less favorable fracture mechanisms may initiate. Stated in another way, bone has a finite capacity to absorb energy that increases significantly with load rate. Fractures are arbitrarily grouped into three general categories based on the energy required to produce them: low-energy, high-energy (e.g., fractures observed following automobile trauma), and very high energy.

Many factors determine the occurrence of an injury to an anatomical segment of the body. Most important are the mass and size of the anatomical segment, in addition to the age and sex of the person. Strength depends on a person's medical history and health. If the injury is not a bone fracture but soft tissue damage or an internal injury, there may be no precise onset loading but a gradual worsening of injury with increasing load. To compare severity of injuries in various anatomical segments, the American Medical Association (http://www.ama-assn.org/) and the American Association for Automotive Medicine (http://www.aaam.org/) jointly developed a universally accepted injury scaling procedure called the Abbreviated Injury Scale (AIS). According to AIS, AIS 1 is a minor injury, AIS 2 is a moderate injury, AIS 3 is a serious injury, AIS 4 is a severe injury, AIS 5 is a critical injury, and AIS 6 is a maximum and virtually unsurvivable injury. Anatomical segments addressed by the AIS include external, head, neck, thorax, abdomen/pelvic contents, spine, and extremities/bony pelvis. These injury scales are used for patient care as well as designing safety equipment for injury prevention.

For each anatomical segment, injury tolerances are investigated and tolerance curves for the injury risk have been developed. The tolerance curve primarily relates the injury outcome on the magnitude and duration of the applied force. For example, head injury tolerance curves reflect that the risk for brain injury depends on the magnitude and the duration of the applied force. A high magnitude of force will cause injury if it exceeds the tolerance limits. Using these tolerance curves, the safety of the occupant can be improved. A basic design strategy is to absorb energy without producing forces that exceed injury tolerance. The best shock-absorbing technique is based on receiving a small force spread over a long period of time or force spread over a large surface area. Other approaches used to protect body parts in the case of impact loads are energy-absorbing soft pads such as airbags and energy-shunting hard shells such as body armor and bulletproof vests and combined systems.

The principle of an energy-absorbing device is attenuating the impact force by means of a shock-absorbing material [Figure 5.17(a)]. This is achieved by using a highly elastic (or low stiffness) material. An example is inserting a silicone rubber that can undergo considerable deformation into the helmet or gymnasium mats. The area under the load-extension curve corresponds to the energy absorbed by the material while undergoing a deformation. The area enclosed by the stress-strain curves during loading and unloading is a measure of the amount of energy dissipation per unit volume (also referred as the energy density) of that material in that material [Figure 5.8(b)]. Furthermore, the area under the hysteresis loop is also the energy dissipated during each cycle. If the energy dissipated is lower, then a more

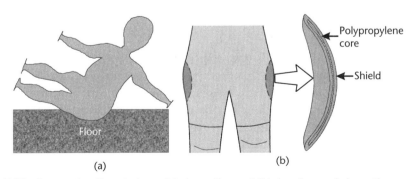

(a)   (b)

Figure 5.17   Energy shunting strategy: (a) absorption and (b) shunting and absorption.

elastic response is observed in the tissue. The hysteresis energy may dissipate as heat or in the drag of the fluid that is exuded and absorbed during loading and unloading, respectively. The use of airbags in automobiles and a fluid-containing pad system are also based on energy-absorbing mechanisms. An energy-shunting configuration is used to distribute impact loads away from the vulnerable area to the surrounding soft tissue. For example, hip protectors and wrist guards use hard shells that shunt energy away from the joints, which are more vulnerable to injury.

A number of organizations mandate the required standards in designing, developing, and testing safety equipment. The National Operating Committee on Standards for Athletic Equipment fosters and encourages the dissemination of information on research findings on athletic equipment, injury data, and other closely related areas of inquiry. In terms of automobile safety, the Federal Motor Vehicle Safety Standard (FMVSS) has established various testing protocols. Furthermore, in testing to develop protective gear during a crash, biomechanical analyses are carried out using artificial bodies or joints (popularly known as dummies). Dummies are developed to mimic few physiological joint structures, ligaments, muscle response, and bone response.

Problems

5.1   Consider an arm, abducted 20° from the surface, of a person of 150 pounds with a 20-pound weight (as shown in the figure). The angle of insertion of the abductor muscle is 25° and 2 inches from point A. The arm weight is typically 5% of the body weight and the center of gravity is four-ninths of the total length from the proximal end. What is the force generated by the muscle?

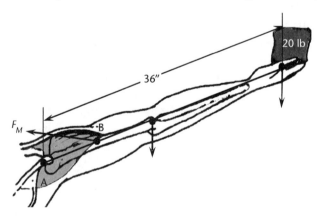

5.2   Ali wants to carry some weight in his hand maintaining a static position, as shown in the figure. The weight of his forearm and the hand is 3 kg with a center of mass at 21 cm from the elbow joint. The weight is at 29 cm from the elbow joint. The elbow flexion muscle is attached at 3 cm from the elbow joint and can exert a maximum force of 2,500N. Calculate the mass of the heaviest ball that Ali can hold statically in that position.

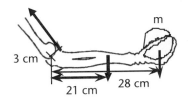

5.3   Angie is carrying 70N at 36.5 cm from the elbow joint. The weight of her fore-
       arm and hand is 20N with a radius of gyration of 18 cm. She has a complete
       lesion of the musculocutaneous nerve, so only the brachioradialis can produce
       a flexion when the elbow is 90° of flexion. The brachioradialis is attached to
       the forearm at a distance of 26 cm from the elbow joint, and the angle of inser-
       tion is such that the vertical distance at the elbow joint is 7.8 cm.

       (a) Identify the class of lever operating during the isometric elbow flexion.

       (b) Calculate the resistance torque and the effort force.

       (c) Calculate the resistance joint force and its angle of application.

5.4   Sayed is carrying a 50-pound load in his hand. His forearm is at a 45° angle to
       the ground and his upper arm is at a 30° angle to the ground. The distance of
       the weight from the elbow joint is 14 inches and the distance from the elbow
       to the shoulder joint distance is 12 inches. Ignoring the weights of forearm and
       upper arm, (a) draw the free body diagram, and (b) determine the moment
       exerted about the elbow joint and about the shoulder joint.

5.5   The illustration shows the abdominal skeletal structure of person (upper body
       weight = 250 pounds) standing on the right lower extremity. The right hip ab-
       ductor muscles must contract to counteract the force of gravity and keep the
       pelvis level. Note that the hip abductor muscle force is acting at an angle of
       70° relative to the horizontal plane.

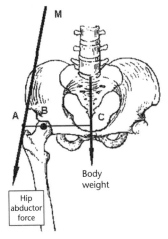

       (a) How much force must the right hip abductor muscles generate to keep the
           pelvis level?

       (b) If the muscle contraction $\Delta L/L = 10\%$, and the average cross-sectional area
           is 14 cm^2, what is the elastic modulus of the muscle (assume Hooke's
           law)?

(c) Is Hooke's law a valid assumption? Why?

(d) A cane is used as an assist device to compensate if the muscles are not strong enough to counter the moments produced by gravity. How does that work? Give a two-line explanation or a free body diagram in which AB is 5 cm, and BC is 12 cm.

5.6   In the 1970s, hip prostheses were designed with short femur necks to reduce the torque experienced about the vertex of the prosthesis. As shown in the figure, the torque of the hip abductors must equal the torque of the body weight in order to have equilibrium. For a normal length of the femur neck, the distance (d_1) is 6 cm and the shortened neck of the femur is 3 cm. The distance from the weight is 12 cm and the weight of person is 800N.

(a) What class lever is involved?

(b) What are the forces required for the equilibrium in each case?

(c) What are the magnitude and direction of force on the head of the femur in each example?

5.7   The most common cause of knee pain is the patellofemoral pain syndrome (PFPS), which is caused by imbalances in the forces controlling patellar (knee cap) tracking during knee flexion and extension. The quadriceps muscles straighten the knee by pulling at the patellar tendon via the patella. In patients with PFPS, a quadriceps resistance force is applied to evaluate the amount of discomfort and to determine if pain is correlated with the angle of the knee joint. A correlation used in determining the patellofemoral compression force (F_{PFC}) is

$$F_{PFC} = \sqrt{F_{QT}^2 + F_{PT}^2 + 2F_{QT}F_{PT}\cos\frac{(180-\alpha)-\theta}{2}}$$

where F_{QT} is the quadriceps tendon force, F_{PT} is the patella tendon force, α is the knee joint angle, and θ is the patellar tendon angle.

(a) Typically, F_{QT} is equal with F_{PT} and is nearly 410N. Determine the PFCF for 30° and 90° of knee flexion by using the graphical method of determining resultant forces. Note that when α is 30°, θ is 18°; and when α is 90°, θ is 1°.

(b) Determine the joint angle where you would expect the patient to have the least discomfort.

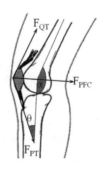

5.8   In Problem 5.7, Yan is interested in measuring the difference in the developed patellofemoral compression force in a healthy individual and in an individual who had a knee operation. Both the subjects are told to sit with their knee extended at 45° to the ground ($\theta = 15°$). A force of 100N is applied at the ankle joint which is 40 cm from the knee joint. The patellar tendon force is calculated using the equation

$$F_{PT} = \frac{L\left(F + F_{LW} / 2\right)}{2}$$

where F is the applied force at the ankle, F_{LW} is the force due to the lower leg mass, and L is the distance from the ankle joint. Assume that both the subjects have a similar lower leg weight (5 kg). The healthy patient develops an equal amount of F_{QT}; however, the patient who had the knee operation develops an F_{QT} that is twice the F_{PT}. Calculate the developed compressive force in the two cases. Comment on the direction of the compressive force.

5.9   Several muscles in the lower leg have common insertions. One example is the triceps surae—a combination of the two heads of the gastrocnemius and the soleus—which inserts into the posterior aspect of the calcaneus (heel bone) at different angles. Using the right lower extremity as our model (and in reference to the right horizontal), the medial gastrocnemius inserts at 100°, the lateral gastrocnemius inserts at 70°, and the soleus inserts at 85°. What are the magnitude and direction of the resultant (identified with respect to the horizontal) under the condition where all three muscles exert a force of 1,000N?

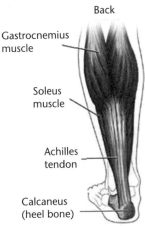

Back

Gastrocnemius
muscle

Soleus
muscle

Achilles
tendon

Calcaneus
(heel bone)

5.10   Based on the gait analysis experiments, the following information is available: the vertical component of ground reaction (F_z) is 150 pounds, the horizontal component of ground reaction (F_x) is 15 pounds, $b = 3.2$ mm, $d = 2.6$ mm, the inclination of the tibia in the sagital plane, θ is 10° and the Achilles tendon force, Mt, is assumed to be parallel to the tibial longitudinal axis at a distance a is 25.4 mm. Calculate: (a) the joint reaction force (Rx, Rz) and (b) the

Achilles tendon force, *Mt* [personal communication with Dr. Noshar Lagrana, Rutgers University, 2002].

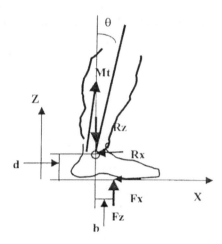

5.11   Arul is sitting with his knee flexed to 45° while maintaining resistance at the ankle area (for dimensions, select one from your group). The distance of insertion is 1 inch.

(a) Compute the force of quadriceps muscle if the angle of the patellar tendon insertion at this angle is 20°. Provide a free-body diagram.

(b) Compute the force of quadriceps muscle if the knee is fully extended and the angle of patellar tendon insertion is 35°. Provide a free-body diagram.

(c) Which is more tiring to maintain: at 45° or in full extension? Compare with computed results.

(d) Compare if there is no patella at all (the angle of insertion is 10° in both cases).

(e) How important is the patella? Why?

5.12   Hari weighs 150 pounds. With his 2-ft-long arm (the weight is 5% of the total body weight), he tries to hit a tennis ball using a 0.3-pound (CG is 6 inches from the handle end) and 8-inch tennis racket. The ball travels delivering a force of 120 pounds. Assume that all movements occur in the frontal plane and the arm is acting at 10 inches from the joint.

(a) Compute the amount of force needed to drive the ball 120 pounds back. The distance of insertion is 2 inches. The angle of insertion is 20°, supposing that the point of impact occurs at a 90° arm abduction.

(b) What are the advantages and disadvantages of using implements like the racket?

(c) In what way can you improve the technique in stroke to avoid or diminish the disadvantage?

5.13   During the ground contact (stance) phase of running, a peak force of 1,500N is applied to the foot, at a location 11 cm in front of the ankle joint. (a) What will be the corresponding deformation in the Achilles tendon, if it is oriented as shown in the figure and has a cross-sectional area of 85 mm², a length of

280 mm, and a Young's modulus of 700 MPa? (b) If the modulus of the Achilles tendon is 150 MPa, determine the tensile force *FA* required to cause the strain in the tendon to equal its yield strain of 0.2 m/m [6].

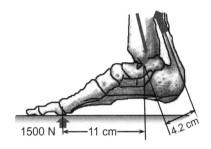

5.14   If a person's forearm has a cross-sectional area of 2.5 cm² and an effective mass of 3 kg, can normally withstand a maximum compression of 16×10^7 Pa (N/m²), and comes to rest at constant deceleration in 2 ms, what is the maximum speed it could have been going before it hit?

5.15   A nursing aide is being assessed for low back pain in your clinic. One of her main tasks is bed making. She reports that she often leans over the bed to tuck sheets in on the far side. You want to analyze the forces and moments at the L5 lumbar disc seen in her technique compared to the "proper" technique. Using the information given in the table, answer the following:

(a) How large is the bending moment on the L5 disc when the patient is making the bed while bending over? While standing upright?

(b) How large must the erector spinae muscle force be to counteract the bending moment for both cases?

(c) How large will the compressive and shear force on the L5 disc be in the two cases? Assume that the line of application of the erector spinae muscle force is perpendicular to the disc, independent of its degree of inclination [personal communication with Dr. Michele Grimm, Wayne State University, 2004].

Weight of head and trunk	340N
Weight of arms	60N
Moment of arms	
Bending over: arms c-of-g	0.67m
Bending over: body c-of-g	0.31m
Upright: arms c-of-g	0.18m
Upright: body c-of-g	0.17m
Erector spinae	0.05m
Angle of inclination of L5 disc	
Bending over	70°
Upright	30°

5.16   A child holds a 10-kg mass in front of herself at a distance of 25 cm from her chest. The mass of the child's body segments and their moment arms about the lumbar spine are given in the table.

Body Part	Mass (kg)	Moment Arm Length (cm)
Head	4	5
Torso	30	2
Arm	2	15
Anterior chest	N/A	7

The pair of erector spinae muscles is symmetrically distributed on either side of the spine. They act at 1 cm from the center of rotation of the spine in the anteroposterior direction. What is the required force generated by the erector spinae muscles in order to allow the child to carry this weight? If the muscles have a maximum intensity of 1,000 kPa and the cross-sectional area of each is 5 cm², what is the maximum possible load that can be generated by this child's erector spinae? Assuming that they are the only muscles acting, will the child be physically able to lift this weight [personal communication with Dr. Michele Grimm, Wayne State University, 2004]?

5.17   A parade organizer is being sued by some of the "waving marchers" for repetitive motion injuries. You are serving as a consultant on the case and have been asked to estimate the forces and moments at the elbow. You could assume that the waving motion is: pure curvilinear motion, only in the coronal plane, and rotation of the forearm at the elbow about the axis of the humerus (arm). The mass of the hand and forearm of one of the plaintiffs is 2.5 kg, the length of the forearm/hand is 0.5m, and the moment of inertia about the elbow is 0.0127 N-m-s². Calculate: (a) the segment angle at the above five time points; (b) the angular velocity and acceleration of the forearm/hand complex, given the above data; and (c) the force and moment at the elbow after 10.2 seconds. State all assumptions that you make [personal communication with Dr. Michele Grimm, Wayne State University, 2004].

Time (s)	X_1 (m)	Y_1 (m)	X_2 (m)	Y_2 (m)
10	0.05	0.02	0.225	0.14
10.1	0.045	0.04	0.20	0.185
10.2	0.04	0.055	0.15	0.237
10.3	0.03	0.063	0.10	0.264
10.4	0.02	0.069	0.05	0.279

5.18   Hinne weighs 100 kg and rides a 10-kg bike to work everyday. One day he is in a hurry to get to work and increases the velocity from 0 to 10 m/s. What impulse was required?

5.19   Heidi is an athletic person and weighs 50 kg. She wants to hit the volleyball, for which she jumps with an average vertical ground reaction force of 750N for 0.5 second. What is Heidi's vertical takeoff velocity?

5.20   In a fight between Evander Holyfield and Mike Tyson, Evander swings his arm at an angular velocity of 5 rad/s. Assume that Evander Holyfield's arm weighs 5 kg and the radius of gyration is 23 cm. If Mike Tyson has to stop the motion, what average amount of force must be applied over a period of 0.2 second by the elbow flexors inserting at an average perpendicular distance of 1.5 cm from the axis of rotation?

5.21   If a pitched ball with a mass of 1 kg reaches a catcher's glove traveling at a velocity of 30 m/s:

(a) How much momentum does the ball have?

(b) How much impulse is required to stop the ball?

(c) If the ball is in contact with the catcher's glove for 0.5 second, how much force is applied by the glove?

5.22   A soccer ball (4.17N) was traveling at 7.62 m/s until it contacted with the head of a player and was sent traveling in the opposite direction at 12.8 m/s. If the ball was in contact with the player's head for 22.7 milliseconds, what was the average force applied to the ball?

5.23   Professor Loosehead (65.0 kg) is running at 5.00 m/s along a sidewalk. Being an absentminded professor, his mind wanders. Just before he is about to cross an intersection, he snaps to attention and realizes that he is in danger of being struck by a car. What was his momentum while his mind was wandering? Naturally, Professor Loosehead tries to stop quickly to avoid the accident. Assume a contact time per step of 0.3 second and that he places his leg at 30° from the vertical, the maximum resultant force he can apply to the ground is three times his body weight, and the force is applied like a square wave. How many steps does he need to come to a complete stop? Answer with an integer (1, 2, 3, and so on), but support your answer with calculations.

5.24   (a) Calculate the stress and strain in a typical femur (bone in the thigh) while standing. Assume that the cross-sectional area of the femur can be approximated by a circle with a diameter of 2.5 cm. Assume that the average individual weighs 150 pounds. The modulus of the elasticity of bone is about 17 GPa. Give your answer in metric units. Don't forget that you have two legs.

(b) If the compressive strength of bone is 170 MPa, how much weight would the individual in part (a) have to be carrying to have a compressive failure in the femur? Assume now that the individual is walking and therefore all of the weight is occasionally on one leg. Ignore any dynamic forces.

(c) Using the numbers in part (a), if the individual wanted to be an inch taller permanently, how much weight would he have to hang from his ankles to make himself stretch that much? Assume that the "body" is 6 feet long and that the body can be completely modeled as bone. Would the resulting stress be greater than the failure strength of bone [personal communication with Dr. Noshar Langrana, Rutgers University, 2002]?

5.25   In a biomedical laboratory, a 2-mm-thick rectangular plate (10 mm wide and 50 mm long) was tested by pulling axially with a load of 1,000N. If the observed extension is 70 μm, then:

(a) What are the axial stress and axial strain?

(b) What is the modulus of elasticity?

(c) What material was evaluated in this test? Justify your answer.

(d) Will this axial stress cause failure? If yes, why? If not, why not?

5.26   Consider a square bar with a cross-sectional area of 0.01 m². Equal and opposite 100N forces are applied. Determine the shear stress τ and normal stress σ acting on the planes defined by $\theta = 0°$, $\theta = 45°$, and $\theta = 90°$.

5.27   The bladder can be considered as a thin-walled sphere. Show that the hoop stress and longitudinal stress have same magnitude in this case. Determine the stresses when the internal pressure is 40 mmHg. Assume the thickness of the bladder to be 4 mm and the diameter to be 9 cm.

5.28   A 10-cm-long tubular steel of internal diameter of 10 mm and a wall thickness of 1 mm is considered for a stent application. The possible loads include axial (F), torsional (T_x), and internal pressure (P) loads.

(a) Calculate the tensile stress of the tube when $F = 300N$ applied along the x-axis. If that force is at an angle of 30° to the x-axis then, what is the tensile stress?

(b) Calculate the shear stress when $T_x = 12$ N-m.

(c) For a 6,000-Pa internal pressure, calculate the hoop stress and axial stress.

(d) If the ultimate stress that the steel can withstand is 700 MP, how would you prove that your material can perform under the above conditions?

5.29   In a typical human aortic stent of a 10-cm length and an average diameter of 2.5 cm, a 4% increase in diameter is measured for a pulse pressure of 40 mmHg. Compute the compliance of the aortic stent.

5.30   An artery of an internal diameter of 1 cm and a wall thickness of 2 mm at an end diastolic pressure of 85 mmHg goes through a 8% increase in the diameter for a pulse pressure of 45 mmHg. Assume that the arterial wall is thin walled and made of linear isotropic elastic material. Compute: (a) the mean arterial pressure using the equation $P_{mean} = (P_{systolic} + 2P_{diastolic})/3$; (b) the circumferential stress and the elastic modulus of the vessel at the mean arterial pressure; and (c) the elastic modulus at the mean arterial pressure.

5.31   In an experiment, a 0.5-m-long aorta of a human cadaver of an inner diameter of 18 mm and a wall thickness of 2 mm is used as the sample. Experiments were conducted at physiological conditions including the pulsatile flow of blood, that is, the diastolic pressure (P_d) and the systolic pressure (P_s) are 83 mmHg and 120 mmHg, respectively. In the systolic phase(s) the blood volume of 60 mL is discharged (assuming 70 heartbeats per minute). Half of this blood is initially stored in the aorta, expanding its wall to some inner radius $2R_s$. The mean pressure can be estimated by the relation, $P_{mean} = (P_{systolic} + 2P_{diastolic})/3$.

(a) Calculate systolic radius and the compliance of the aorta.

(b) Calculate the hoop stress using and neglecting the changes in the wall thickness.

(c) Assuming the Poiseuille equation, determine the pressure drop over the length of the aorta. Is your laminar flow assumption valid? Show the calculations.

(d) How much energy is stored in the elastic wall?

(e) Suppose that 70% of the cross-sectional area was blocked, but the experiment was conducted at the same volumetric flow rate of 60 mL/stroke, what would be the flow velocity and the Reynolds number at the constriction? The density and viscosity of the liquid used in the experiment are 1.050 g/cm^3 and 1.1 cP, respectively.

5.32   Daku is trying to develop a new extracorporeal tubing for hemodialysis. Consider a blood flow through a 6-mm tubing; the goal is to deliver 5L per minute. Using a 3-cP viscosity and a 1.056-g/cm^3 density, calculate the shear stress developed at the wall. If the red blood cell wall break stress is 30 dyne/cm^2, will the cell be alive or dead?

5.33   Stella is evaluating a material for use as a single compression plate in repairing a long bone fracture. She needs to determine how the rigidity and strength of the fixation is affected by the orientation of the plate. She prepares a plate that is 3 mm thick and 18 mm wide. She wants to try two orientations: one when the load is applied perpendicular to the thickness, and one when the load is applied perpendicular to the width. She assumes that the plate will be subjected to pure bending M, and that the bone will initially contribute negligibly to load bearing.

(a) Which orientation has the highest area moment of inertia?

(b) Assume that the plate is made of stainless steel that has a yield stress of 800 MPa. For each of the two orientations, determine the magnitude of the bending moment M required to initiate yielding.

5.34   Every year, spinal cord injuries due to motor vehicle accidents cost around $3 billion. A group of researchers are trying to understand the properties of the cervical spine and also the forces involved in causing these injuries so that appropriate safety measures can be incorporated in designing automobiles. One of the first things to calculate is the compressive stress due to the head weight.

(a) If the weight of the head is 60 pounds and the average spine area is 10 cm², what is the compressive stress experienced by the spine?

(b) If the maximum compressive force that a spine can sustain is calculated using Euler's buckling theory, what is the critical load (P_r)?

(c) At critical load, what pure bending moment is required at the free end (as shown in the figure) to introduce a 5° bend in the spine?

(d) What assumptions (list only four) would you try to eliminate to increase the accuracy of calculations?

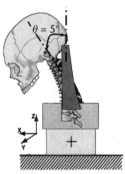

5.35 Suppose that the femur mid-shaft is approximately tubular with a 25-mm outside diameter and a 6-mm wall thickness. Consider the relative strength of the bone and of the nails used for fracture fixation.

(a) How much of a moment can an intact femur withstand?

(b) How much of a moment can a solid cylindrical stainless steel nail that is 9 mm in diameter withstand?

(c) How much of a moment can a tubular cylindrical stainless steel nail that has a 9-mm outside diameter and a 1-mm wall thickness withstand?

(d) It is claimed a cloverleaf nail can withstand 23 Nm. A patient with a broken femur was treated with such a nail. One day he stood on one leg to put on his pants. He felt the nail bend, and returned to the doctor. Was the nail at fault?

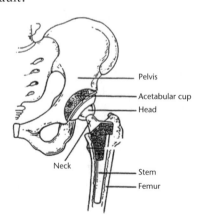

5.36   There has been an increased interest in using translucent fiber reinforced composites in prosthodontics. Narva et al. [7] evaluated electrical-glass fiber (15 to 17 μm thick, the elastic modulus is 73 GPa, and the density of the fiber is 2.54 g/cm^3) and reinforced urethane dimethacrylate (the elastic modulus is 3.9 GPa and the density of the polymer is 1.19 g/cm^3) composites as a possibility. They tested for flexural rigidity of these composites using cylindrical unidirectionally reinforced samples that were 2 mm in diameter and 60 mm long. A one-way constant deflection test was performed by fixing the sample. Applying a load of 40N at a 11-mm span from the fixed end showed a deflection of 1 mm. An independent analysis of the fiber content in the composite showed 53.5 % by wt. Do you think these results are consistent? If not, why not? State your reasons and show your calculations.

5.37   Suppose a tendon is modeled as a cylindrical structure (see the figure) of length L, a cross-sectional area A, and a modulus of elasticity E, which is subject to equal and opposite axial loads of magnitude F. The amount of the tendon extension under the load P is ΔL.

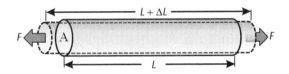

   (a) What is the stiffness k of the tendon (the slope of the force-deflection curve) in terms of A, E, and L?

   (b) For a given magnitude of P, how would the extension ΔL be affected if A is doubled? If E is doubled? If L is doubled?

   (c) Let the symbol W represent body weight in Newtons. Assume that the tendon cross-sectional area A is proportional to $2W/3$, the tendon length L is proportional to $W/3$, and the tendon modulus of elasticity E does not vary with W. How would the tendon stiffness k then vary with W?

5.38   A common technique for replacing a torn anterior cruciate ligament involves constructing a graft from a removed section of the patient's patellar tendon. A surgeon is interested in understanding how much of the patellar tendon can by removed before its strength is compromised. The surgeon initially considers the common but strenuous activity of rising from a chair. He conducts a free-body diagram analysis, which indicates that for the 700N patient, the tensile force in each of the right and left patellar tendons during this activity is approximately 410N. The surgeon also considers the published data on the adult patellar tendon [4], which are as follows: the average length is 30 mm, the average cross-sectional area is 74 mm^2 (and does not vary substantially along its length), the elastic modulus is 400 MPa, and the ultimate stress is 58 MPa.

   (a) What is the axial stress in an intact tendon under 410N of axial force and the "factor of safety" (the ratio of ultimate stress divided by working stress) associated with this activity?

(b) For the conditions described in part (a), determine the tendon's elongation (in meters), axial strain, and axial stiffness (in Newtons per meter).

(c) During ACL replacement surgery, the cross-sectional area of the tendon is typically reduced by one-third. What effect does this have on the maximum axial load that the tendon can maintain without rupture?

(d) What is the maximum amount that the tendon cross-sectional area can be reduced, before we would predict it to fail when rising from a chair?

References

[1]   Dempster, W. T., *Space Requirements of the Seated Operator*, WADC Technical Report 55-159, Wright-Patterson Air Force Base, OH, 1956.

[2]   Murray, M. P., A. B. Drought, and R. C. Kory, "Walking Pattern of Normal Men," *Journal of Bone and Joint Surgery*, Vol. 46A, 1966, pp. 335–362.

[3]   Murray, M. P., R. C. Kory, and S. B. Sepic, "Walking Pattern of Normal Women," *Archives of Physical Medicine and Rehabilitation*, Vol. 51, 1970, pp. 637–650.

[4]   Burstein, A. H., and T. M. Wright, *Fundamentals of Orthopaedic Biomechanics*, Baltimore, MD: Williams and Wilkins, 1994.

[5]   MacDonald, B. J., *Practical Stress Analysis with Finite Elements*, Dublin, Ireland: Glasnevin Publishing, 2007.

[6]   KIN402, Mechanical Behavior of Tissues, 2003.

[7]   Narva, K. K., L. V. Lassila, and P. K. Vallittu, "Fatigue Resistance and Stiffness of Glass Fiber-Reinforced Urettiane Dimeltiacrylate Composite," *Journal of Prosthetic Dentistry*, Vol. 91, No. 2, 2004, pp. 158–163.

Selected Bibliography

Benzel, E. C., *Biomechanics of Spine Stabilization*, Rolling Meadows, IL: American Association of Neurological Surgeons, 2001.

Berger, S. A., W. Goldsmith, and E. R. Lewis, *Introduction to Bioengineering*, New York: Oxford University Press, 2001.

Enoka, R. M., *Neuromechanics of Human Movement*, Champaign, IL: Human Kinetics Publishers, 2001.

Ferry, J. D., *Viscoelastic Properties of Polymers*, New York: John Wiley & Sons, 1980.

Freivalds, A., *Biomechanics of the Upper Limbs; Mechanics, Modeling and Musculoskeletal Injuries*, Boca Raton, FL: CRC Press, 2004.

Fung, Y. C., *Biomechanics: Mechanical Properties of Living Tissues*, New York: Springer, 1999.

Hibbeler, R. C., *Mechanics of Materials*, 6th ed., Upper Saddle River, NJ: Prentice-Hall, 2004.

Humphrey, J. D., *Cardiovascular Solid Mechanics*, New York: Springer, 2002.

King, A. I, "Fundamentals of Impact Biomechanics: Part I—Biomechanics of the Head, Neck, and Thorax," *Annual Review of Biomedical Engineering*, Vol. 2, 2000, pp. 55–81.

Nigg, B. M., and W. Herzog, *Biomechanics of the Musculoskeletal System*, 2nd ed., New York: John Wiley & Sons, 1999.

Whiting, W. C., *Biomechanics of Musculoskeletal Injury*, 2nd rev. ed., Champaign, IL: Human Kinetics Publishers, 2008.

Whittle, M. W., *Gait Analysis: An Introduction*, 4th ed., Boston, MA: Butterworth-Heinemann, 2007.

Zatsiorsky, V. M., *Kinematics of Human Motion,* Champaign, IL: Human Kinetics Publishers. 1997.

CHAPTER 6

Biomaterials

6.1 Overview

Various materials are utilized in a wide spectrum of medical applications (Figure 6.1). Biomaterials form the critical components of every biomedical application. Cotton pads are used to stop bleeding of minor cuts within a few minutes, polymeric catheters are used to feed nutrients or therapeutic molecules, stainless steel surgical tools are used to operate on a person, and composite total hip replacement devices are used with the intention of lasting for 10 to 15 years or the lifetime of the patient. The extent of material interaction with the body varies with the type of application, either with different cells, tissues, or extracellular fluids in addition to the duration. Apart from understanding the mechanical property requirements (discussed in Chapter 5), the design of many biomedical devices requires the processing of materials to the required dimensions and the global acceptance of the material by the surrounding tissues and fluids of the body and by the body as a whole. Processing and fabrication of components could affect the dimensional stability, surface characteristics, and surface chemistry/physics. In many cases, the blood interactions required within a device create fluid eddies and tissue damage that are favorable to blood clotting with grave clinical consequences in the absence of an anticoagulant.

Advancements in material and manufacturing technologies have endowed products with much greater reliability and reproducibility. Polymeric materials are used as contact lenses and biodegradable sutures; contact lenses are designed to interact with tissues minimally for a few hours to weeks, whereas sutures are designed to slowly resorb over a few weeks to months and continuously exposed to the wound healing process. As the basis for biomaterials selection design advances, tailoring biomaterial properties for individual applications becomes a practical means of improving implant safety and efficacy. This chapter aims to establish a fundamental understanding of and appreciation for the selection, development, and performance of the many types of biomaterials used in medicine and medical devices. First, different types of biomaterials utilized in medical applications and their material characteristics with the focus on common problems and importance of surface topographies are discussed. Next, the basic biological interactions such as blood protein adsorption, inflammatory reaction, and infection are discussed. Finally, the use of biomaterials in the context of tissue engineering is discussed.

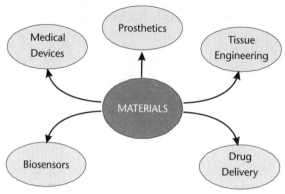

Figure 6.1 Uses of biomaterials.

However, interactions with cells and various cellular processes are discussed in Chapter 7.

6.2 Types of Biomaterials

In general, a biomaterial is defined as a nonviable material, used in a medical device that is intended to interact with biological systems [1]. A material that performs with an appropriate host response in a specific application is termed as a *biocompatible* material. Currently used biomaterials can be grouped into five categories: metals and alloys, ceramics, polymers, composites, and biological materials.

6.2.1 Metals and Alloys

Metals and their alloys (blended with other metallic or nonmetallic substances to improve a specific quality) are used as biomaterials because of their strength and toughness. Most important in this class are stainless steel, cobalt-chromium alloy, titanium, nickel, aluminum, zinc, and their alloys. Metals and alloys are utilized in: replacement devices for joints such as hips and knees; internal fixation devices such as bone plates, screws, and pins for hard tissues and stents for the opening of blood vessels [Figure 6.2(a)], the esophagus, and urinary tracks. After severe trauma in the head region, the surgical removal of part of the cranium is often the only possibility to save the patient. This leads to large defects and material is necessary to protect the brain mechanically. Rebuilding the contours with long-term stability, the reduction of patient complaints, the easy handling of the implants, and the possibility of the oncological follow-up with qualified imaging methods are the basic and most important requirements. The best available material for this particular purpose is pure titanium, which is used in the manufacturing of the required implants [Figure 6.2(b)]. Dental applications employ gold and its alloys. Tantalum is used in special cases such as wire sutures in some plastic and neurosurgical applications. Alloys of platinum and other noncorrosive metals in that group are used in pacemakers as conducting leads and other components. Nickel containing stainless steel has been used in early hip implants for its good strength, ability to work hard, and pitting

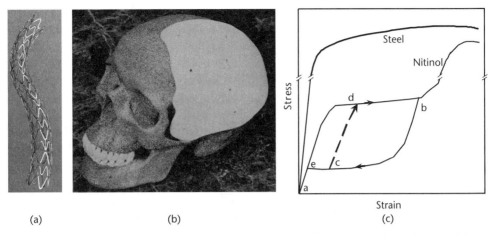

(a)   (b)   (c)

Figure 6.2 Metals and alloys. (a) Stainless steel stent with a polymer coating (TAXUS Express[2] Atom Paclitaxel-eluting coronary stent system, Boston Scientific). (b) MEDPOR TITAN cranial temporal implant. (Used with permission from Porex Surgical, Inc.) (c) Stress-strain behavior of SMAs.

corrosion resistance. Cobalt alloys are used in both cast condition and wrought condition; the wrought condition provides superior mechanical and chemical properties due to finer grain sizes and a more homogenous microstructure. The ability to form an inert thin oxide layer on the surface of aluminum and titanium is also advantageous in many applications.

While the widely used implant metals are generally biocompatible, they may release ions. For example, stainless steels could potential release Ni^{2+}, Cr^{3+}, and Cr^{6+} ions on a long-term basis into the body, due to which their use is restricted to temporary devices. Further, the linear elastic deformation (following Hooke's law, discussed in Chapter 5) of many metals including stainless steel cobalt-based alloys is limited to nearly 1–2% strain. This limits usage in many medical devices where broader elastic range is necessary. For example, stainless steel arch-wires have been employed as a corrective measure for misaligned teeth. Owing to the limited extensibility and tensile properties of these wires, considerable forces are applied to teeth, which can cause a great deal of discomfort. When the teeth succumb to the corrective forces applied, the stainless steel wire has to be retensioned. Repeated visitation to the orthodontist may be needed for retensioning every 3 to 4 weeks in the initial stages of treatment.

There is a special class of alloys, called *shape memory alloys* (SMAs), which have the ability to return to a predetermined shape with thermal changes, and deformation of more than 10% strain can be elastically recovered. The most common SMA is that of nickel and titanium, called Nitinol. SMAs have also been formed by blending nickel and titanium with other components such as copper, hafnium, palladium, and zirconium. Further, niobium-aluminum, nickel-iron-zinc-aluminum, titanium-niobium, gold-cadmium, iron-zinc-copper-aluminum, uranium-niobium, iron-manganese-silicon, copper-aluminum-iron, and zirconium-copper-zinc also behave as SMAs. When an SMA is cold or below its transformation temperature, it has a very low yield strength and can be deformed easily into any new shape. When the material is heated above its transformation temperature it undergoes a change in crystal structure, which causes it to return to its original shape. The stress-strain behavior of nitinol [Figure 6.2(c)] shows non-Hookean behavior, unlike stainless steel. This property of thermally induced elastic recovery can be used to change a small volume to a larger one. When coronary artery disease reduces the cross-section of the artery and consequently reduces blood flow to the heart muscle, a stent made of SMA can be used to dilate or support a blocked conduit. A stent of a given size, typically larger than the vessel where it is deployed (point a) is crimped into a delivery state (point b, smaller diameter), then packaged, sterilized, and shipped. The stent is introduced either in conjunction with a dilation balloon or by itself in the deformed shape by traveling through the arteries with the stent contained in a catheter. After insertion to the target site, the stent is released from the constraining catheter into the vessel, the device is triggered by heat from the body and will expand from point "b" until movement is stopped by impingement with the vessel (point c). Because the stent did not expand to its preset shape, it continues to exert a low outward force, opening the vessel lumen and reinstating blood flow. However, it will resist recoil pressure or other external compressive forces because of stiffness required large loads (path c to d), which are substantially steeper (stiffer) than the unloading line (towards point e). If the SMA encounters any resistance during this transformation, it can generate large constant forces. Vena-cava filters have a long record of successful in vivo application.

SMAs are also used in applications such as healing broken bones, and teeth alignment. Owing to their elastic properties and extendibility, the level of discomfort can be reduced significantly as the SMA applies a continuous, gentle pressure over a longer period. Visits to the orthodontist are reduced. However, nitinol requires controlled processing to achieve optimal shape memory and superelastic properties. Surface processing is also required in order to promote optimal corrosion resistance and biocompatibility. Nitinol, like titanium and stainless steel, is a self-passivating material (i.e., it forms a stable surface oxide layer that protects the base material from general corrosion). The corrosion resistance of properly treated nitinol competes with the titanium or other common implant materials.

6.2.2 Ceramics

Ceramics are inorganic, nonmetallic, refractory, polycrystalline compounds. The chemical compositions of ceramics vary considerably from simple compounds to a mixture of many complex phases. Ceramics are generally used for their hardness, high wear resistance, high modulus (stiffness), high compressive strength, low coefficient of friction, good electrical insulation, and dielectric properties. Except for their brittle behavior and low tensile strength, ceramics are ideally suited for a variety of implant applications such as artificial joints, implantable electronic sensors, and teeth as well as bone bonding surfaces. Electronic components of cochlear implants are typically housed in a ceramic casing, whereby the ceramic protects the electronics and also acts to conduct the device's radio-frequency waves. Bioceramics (ceramics used in biomedical applications) are classified into three categories:

1. *Bio-inert or nonabsorbables.* This class of ceramics is compatible to body fluid, lacks interactions with the surrounding tissue. These ceramics are

nontoxic, noncarcinogenic, nonallergic, and relatively noninflammatory. They offer excellent corrosion resistance as they are made of the metal oxides. Examples include alumina (Al_2O_3), zirconia (ZrO_2), silicone nitride, and pyrolytic carbon. Alumina and zirconia can also be polished to a high surface finish, which make them ideal candidates for wear application articulating surfaces in hip and knee joints, where they operate against materials such as ultrahigh molecular weight polyethylene (UHMWPE). Other applications for alumina encompass porous coatings for femoral stems and porous alumina spacers. Several oxides such as MgO, CaO, and Y_2O_3 dissolve in the zirconia crystal structure and stabilize the temperature-dependent crystal structure changes. Obtained zirconia by such treatment is called partially stabilized zirconia (PSZ). PSZ has higher flexural strength and fracture toughness as well as lower Young's modulus and the ability to be polished to a superior surface finish compared to alumina. Although PSZ has less hardness relative to alumina, femoral heads made of PSZ are more predominant in clinical applications.

Ceramics such as lead zirconate titanate (PZT) that can convert mechanical energy into an electrical charge or vice versa are used in a range of medical implantable components, including sensors, actuators, motors, micropumps, ultrasonic generators, transducers in ultrasound applications, and valves for precision metering.

Low-textured pyrolytic carbon is commonly used in mechanical heart valves [Figure 6.3(a)] and is the most popular material for this application. Pyrolytic carbon has been alloyed with silicon carbide to maintain consistent hardness. Apart from good strength, wear resistance, and durability, pyrolytic carbon is thromboresistance (i.e., resist blood clotting). Pyrolytic carbon closely matches the mechanical characteristics of the finger bones. Hence, pyrolytic carbon is also used in implants for small orthopedic joints such as the carpometacarpal implant, which relieves basal thumb joint pain associated with arthritis.

2. *Surface-reactive ceramics.* When these ceramics are placed within the human body, they interact with the surrounding tissue through an ion exchange reaction at the surface, for example replacing Na^+ with H^+. These favorable interactions result in the direct chemical bonding with the host biological tissue. Examples include hydroxyapatite and nonporous glasses.

(a)   (b)

Figure 6.3   Bioceramics: (a) St. Jude's bileaflet valves and (b) porous ceramic.

Hydroxyapatite is a naturally occurring ceramic mineral present as a structural component of bone. Hydroxyapatite ($Ca_{10}(PO_4)_6(OH)_2$) does not break down under physiological conditions unlike the other calcium phosphates, which are considered bioresorbable. It is stable at physiological pH and actively takes part in forming strong chemical bonds with the surrounding bone. This property is exploited for rapid bone repair after major trauma or surgery. While its mechanical properties have been found to be unsuitable for load-bearing applications such as orthopedics, hydroxyapatite is used as a coating on materials such as titanium and titanium alloys. Hydroxyapatite contributes the bioactive properties, while the metallic component bears the load. Such coatings are typically applied by plasma spraying at high processing temperatures. Careful control of processing parameters is necessary to prevent thermal decomposition of hydroxyapatite into other soluble calcium phosphates.

Silicate-based materials with calcium and phosphate in proportions mimicking natural bone composition have also been developed. After implantation in bone tissue, these materials resist bonded to bone due to the surface reaction. Various kinds of bioactive glasses and glass-ceramics with different properties such as high mechanical strength, and high machinability have been developed. Bioactive glass and ceravital (additionally contains oxides of magnesium and potassium) are the most prominent. Bioactive glass is shown to bond to bone as well as soft tissue. They are used in nonload-bearing applications such as in the middle ear, and alveolar ridge maintenance implants.

3. *Resorbable ceramics.* This class of ceramics degrades upon implantation to the host. Examples include a variety of phosphates (calcium, tricalcium, aluminum-calcium, zinc sulfate-calcium), oxides (zinc-calcium-phosphorous, ferric-calcium-phosphorus), corals (mostly calcium carbonate), and plaster of Paris (calcium sulfate dihydrate). They are used in implants where endogenous tissues gradually infiltrate as the implants degrade and are absorbed by the body. Degradation could be caused by physiologic environment such as pH, physical disintegration such as grain boundary attack, or biological processes such as phagocytosis. They are useful in the replacement or repair of damaged bone by trauma or disease, coating of metal implants to promote bone in-growth, repair and fusion of vertebrae, repair of herniated disks, repair of maxillofacial and dental defects, and drug delivery.

Calcium phosphate compounds are abundant in living systems. Apatites constitute the principal inorganic phase in normal calcified tissues such as enamel, dentin, and bone. There are several calcium phosphate ceramics that are considered biocompatible and of these, most are resorbable when exposed to physiological environments. The calcium phosphates attain their biocompatibility through the release of Ca^{2+} and $(PO_4)^{-3}$ ions to the surrounding tissue after implantation, and the ion release rate is determined by the composition, structure, porosity, and other factors, and thus can be controlled somewhat by manipulation of the material during processing. The in vivo response to these materials is also a variable. The order of solubility for some of the materials is tetracalcium phosphate $(Ca_4P_2O_9)$ > amorphous calcium phosphate > alpha-tricalcium phosphate $(\alpha\text{-}Ca_3(PO_4)_2)$ > beta-tricalcium phosphate $(\beta\text{-}Ca_3(PO_4)_2)$ >> hydroxyapatite (most stable form).

6.2.3   Polymers

Polymers are the most versatile class of biomaterials, being extensively applied in medicine. The property of polymers depend on the monomers they are made of and the how the monomers are organized. Compared with metals and ceramics, polymers offer the advantage of cost-effective synthesis in desirable compositions with a wide variety of chemical structures with appropriate physical, interfacial, and biomimetic properties. Further, they are easy to work with. Traditionally, metals or ceramics are chosen for hard-tissue applications, and polymers are selected for soft-tissue applications. However, the very strength of a rigid metallic implant used in bone fixation can lead to problems with stress shielding, whereas a bioabsorbable polymer implant can increase ultimate bone strength by slowly transferring load to the bone as it heals. Biopolymers (polymers used in biomedical application) are usually used for their flexibility and stability, but have also been used for low friction articulating surfaces. Some of the common applications are listed in Table 6.1 and Figure 6.4. For drug delivery, the specific properties of various degradable systems can be tailored to achieve optimal release kinetics of the drug or active agent. The basic design criteria for biopolymers call for compounds that are biocompatible, have manufacturing compatibility, sterilizability, chemical resistance, rigidity, and are capable of controlled stability or degradation in response to biological conditions. Biopolymers can be classified according to the following criteria:

- Natural, synthetic, or a combination of both (semisynthetic);
- Degradable or nondegradable;
- Structural or nonstructural.

Naturally derived polymers are abundant and usually biodegradable. For example, polysaccharide chitin is the second most abundant natural polymer in the world after cellulose. The principal disadvantage of natural polymers lies in the development of reproducible production methods, because their structural complexity often renders modification and purification difficult. Additionally, significant batch-to-batch variations occur because of their preparation in living organisms.

Synthetic polymers are available in a wide variety of compositions with readily adjusted properties. Most synthetic polymers are linear homopolymers and are defined by monomer chemistry, stereochemistry, and a degree of polydispersity. Copolymers (multimers) are comprised of more than one monomer, which can be arranged in an alternating pattern, ordered in blocks, or randomly distributed. Processing, copolymerization, and blending provide means of optimizing property of a polymer to a required application. However, the primary difficulty is the general lack of biocompatibility of the majority of synthetic materials, although

Table 6.1   Commonly Used Polymers

Type	Example	Representative Areas of Application
	Polyethylene [UHMWP] polypropylene	Hip replacement
	Polyethylene oxides	Drug delivery systems
	Poly(urethanes)	Artificial organs, cosmetic implants, adhesives, catheters, tubing, and other blood accessing devices
	Silicone (polydimethylsiloxane)	Adhesives, cosmetic applications, nondegradable drug delivery devices, hydrocephalus shunts, catheters, tubing, and other blood accessing devices and parts in many implants, soft lithography, and microfluidic devices
	Poly(methacrylates)	Intraocular lenses and cement spacers for orthopedic prostheses, contact lenses, dental fillings
	Aromatic polyesters (nylon and Dacron)	Permanent vascular grafts
	Poly(ethylene vinyl acetate)	Implantable biosensors, nondegradable drug delivery devices
	Poly(vinyl alcohol)	Ophthalmic formulations, contact lenses
	Polystyrene	Tissue culture plastic
Synthetic nondegradable	Fluoropolymers such as polytetrafluoroethylene (Teflon), polyvinylidene fluoride	Vascular grafts, tubings, and fittings
	Poly(glycolic acid), poly(lactic acid), and copolymers	Barrier membranes, drug delivery, guided tissue regeneration (in dental applications), orthopedic applications, stents, staples, sutures, tissue engineering
	Poly(caprolactone)	Long-term drug delivery, orthopedic applications, staples, stents
	Poly(hydroxybutyrate), poly(hydroxyvalerate), and three copolymers	Long-term drug delivery, orthopedic applications, stents, sutures
	Polydioxanone	Fracture fixation in nonload-bearing bones, sutures, wound clip
	Polyanhydrides	Drug delivery
	Polycyanoacrylates	Adhesives, drug delivery
	Poly(amino acids) and pseudo-poly(amino acids)	Drug delivery, tissue engineering, orthopedic applications
	Poly(ortho ester)	Drug delivery, stents
	Poly(phosphazenes)	Blood contacting devices, drug delivery, skeletal reconstruction
Synthetic degradable	Poly(propylene fumarate)	Orthopedic applications
	Collagen	Artificial skin, coatings to improve cellular adhesion, drug delivery, guided tissue regeneration in dental applications, orthopedic applications, soft tissue augmentation, tissue engineering, scaffold for reconstruction of blood vessels, wound closure
	Gelatin	Capsule coating for oral drug delivery, hemorrhage arrester
	Fibrinogen and fibrin	Tissue sealant
	Cellulose	Adhesion barrier, hemostat
	Chitin, chitosan, alginate	Drug delivery, encapsulation of cells, sutures, wound dressings, tissue engineering
Natural	Starch and amylose	Drug delivery, soft tissue engineering

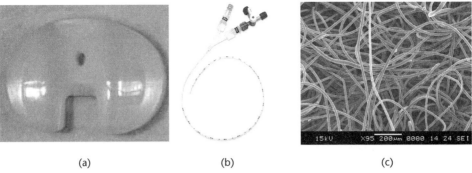

(a)   (b)   (c)

Figure 6.4   Polymers: (a) UHMWPE hip, (b) polyurethane catheters, and (c) polyglycolic matrix.

poly(ethylene oxide) and poly(lactic-co-glycolic acid) are notable exceptions. Synthetic polymers are often associated with inflammatory reactions, which limit their use to solid, unmoving, impermeable devices. Most commonly used synthetic polymers for catheters, tubings, contact lenses, and hip joints include polyurethane, polyethylene, polypropylene, polyimides, polytetrafluoroethylene, and polymethylmethacrylates, poly (vinyl chloride), polyethylene, polypropylene, and polystyrene; and its copolymers with acrylonitrile and butadiene, polyesters, polyamides or nylons, polyfluorocarbons, polyurethanes, natural and synthetic rubbers, silicone polymers, polyacetal, polysulfone, and polycarbonates.

A majority of the polymers made of carbon-carbon backbone alone tend to resist degradation. However, if polymers contain other atoms in the backbones, they tend to be biodegradable. Using this basic concept, one can synthesize polymers to a desirable degradation rate by the addition of chemical linkages such as anhydrides, esters, or amide bonds, among others. For example, copolymers of poly(ethylene oxide) and poly(butylene terephthalate) have been developed. These materials are subject to both hydrolysis (via ester bonds) and oxidation (via ether bonds). One could control the degradation rate by adjusting the poly(ethylene oxide) molecular weight and content.

Hydrogels are also polymers that can swell without dissolving when placed in water or other biological fluids. At equilibrium, hydrogels typically comprise of 60–90% fluid and only 10–30% polymer. The structural changes are reversible and repeatable upon additional changes in the external environment. Hydrogels are attractive because of their high water content, tissue-like mechanical properties, and ability to be polymerized in vivo under physiological conditions. The ability to polymerize the hydrogel transdermally would minimize the invasive procedure that is normally required for plastic and orthoscopic surgery. The injectable hydrogel would allow implantation through a large needle or arthroscopic instrument. The key to this method is the ability to inject the gel into the joint, control the shape of the gel, and be able to polymerize the gel beneath the skin.

6.2.4   Biological Materials

Materials include natural skin, arteries, veins, cord blood vessel, excised diseased, and defective tissues. Tissue-derived substances or modified extracellular matrix components such as collagen have also been utilized in various applications. Matrices obtained from xenogeneic tissues have been used in tissue replacement applications. For example, heart valves recovered from animals are used to replace a diseased heart valve. These are called bioprosthetic valves and they mimic the functional properties of native valves to be replaced. The most commonly used animal sources are porcine (pigs) and bovine (cows). Tissues are typically conditioned by a cross-linking solution such as glutaraldehyde prior to usage. Other biologically derived tissues include acellular dermis and small intestinal submucosa (SIS), a dense connective tissue [Figure 6.5(a)] harvested from the small intestine.

Plant-derived substances such as cotton are used as a wound dressing material [Figure 6.5(b)]. Cotton is composed of cellulose, a polymer made of glucose monomers. Qualities of cotton include ability to balance the fluid due to high absorbancy, ease of application and removal, bacterial protection, air permeability, and mechanical characteristics to accommodate movement that are needed for bandages. The woven cotton has strength that remains unchanged whether it is dry or wet, whereas a nonwoven material is weaker when wet. By combining woven cotton with nonwoven cotton, cotton gauze dressings are made where woven cotton contributes its high strength in tension ensures that the finished product is strong and the nonwoven material has very high absorption capacity. Gauze is also manufactured after mixing fibers of cotton, rayon, polyester or a combination of these fibers. Most woven products are a fine or coarse cotton mesh, depending on the thread count per inch. Fine mesh cotton gauze is frequently used for packing, such as normal saline wet-to-moist dressings. Coarse mesh cotton gauze, such as a normal saline wet-to-dry dressing, is used for nonselective debriding (i.e., the removal of dead/dying tissue and debris).

6.2.5 Composites

In many applications, a single biomaterial may not provide all the necessary properties. For example, one of the major problems in orthopedic surgery is the mismatch of stiffness between the bone and metallic or ceramic implant as the modulii of

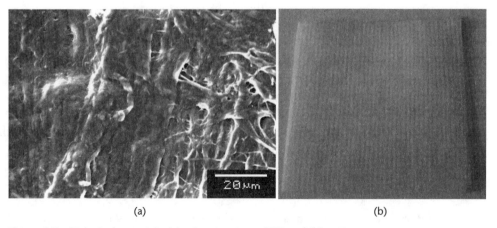

(a)   (b)

Figure 6.5   Biological materials: (a) microstructure of SIS and (b) cotton gauze.

metals and ceramics are very high (Table 5.4). To overcome some of the limitations, composites are developed by combining two or more components. As discussed in Chapter 5, in a composite, one component forms the continuous phase in which other components are embedded as discontinuous inclusions in the shape of platelets, fibers, or particles.

A popular concept is using polymers as the matrix component in combination with continuous carbon fibers to form reinforced composite materials. Polymer-based composites possess a wide spectrum of properties, which allow them to be used in a diverse range of medical applications. For example, titanium mesh is used to reconstruct cranifacial defects where contours suitable to the patient have to be formed. However, titanium mesh exhibits many sharp points when cut and the edges can make insertion difficult. To minimize sharp edges when the implant is cut, the titanium mesh is embedded on both sides with high-density polyethylene. A composite of poly(ether ether ketone) (PEEK) polymer and short carbon fibers is made to increase the strength of the natural unfilled polymer significantly. The two components are processed into filaments, brought together into bundles, and shaped into a rod. Carbon fiber loading of 30–35% by weight increases the material's modulus from 4 to nearly 18 GPa and its tensile strength from 100 to 230 MPa. With a stiffness close to that of cortical bone, carbon-fiber-reinforced PEEK composites are used in applications for which stress shielding may have a critical effect on the lifespan of an implant. For example, hip stems are made from carbon-filled PEEK compounds that demonstrate elastic properties similar to the surrounding bone and that reduce the effects of stress shielding. Further, the modulus of PEEK can be varied to suite to the requirement. This adaptability reduces stress concentrations that can be transferred to the bone and stimulates the healing process. Some of the composite biomaterials include dental composites (acrylic polymer matrix with inclusions of inorganics such as quartz, barium glass, and colloidal silica) and orthopedic components (high density polyethylene matrix with inclusions of carbon fibers). While forming the composite, the process used should produce good bonding strength between two phases. For example, when carbon fiber-reinforced ultrahigh molecular weight polyethylene (UHMWPE) was employed clinically in tibial components, it failed catastrophically. UHMWPE failures have been attributed to poor bonding strength between the carbon fiber and the UHMWPE matrix.

Biodegradable composites can also be formed for use as bioactive matrices to guide and support tissue in-growth. Composites are prepared using polyhydroxy-butyrate (PHB), a naturally occurring β-hydroxyacid linear polyester, and as much as 30% by volume of either hydroxyapatite (HA) or tricalcium phosphate (TCP). One of the goals is to achieve a reasonably homogeneous distribution of the HA/TCP particles in the PHB matrix, as this uniformity would provide an anchoring mechanism when the materials would be employed as part of an implant. The composites are manufactured through a compounding and compression molding process. It is observed that microhardness increased with an increase in bioceramic content for both the HA ad TCP compounds.

6.3   Material Characteristics

6.3.1   Mechanical Performance

The starting point in selecting an appropriate biomaterial for an application is the identification of the loading conditions (discussed in Chapter 5) under which the material has to perform. For example, a hip prosthesis must be strong and rigid as it is exposed to high load bearing conditions. For hip prosthesis, the most commonly selected biomaterials are ceramics as they possess high load bearing capacity. However, brittle materials exhibit high ultimate stress and high modulus and a low breaking strain and therefore are not as tough (strong ductile materials have a moderate ultimate stress but large ultimate strain). Thus, using ceramics is restricted to designs involving limited tensile loading, and no impact loading conditions. High elastic modulus of alumina and zireconia also limits their effectiveness as bone interface materials. In total hip joint arthroplasty where both articulating surfaces of the joint are replaced, the most widely used implant configuration [Figure 6.6(a)] includes a metal component articulating against a polymeric component fabricated from UHMWPE. Although other bearing couples such as metal-on-metal or ceramic-on-ceramic total joint replacements have been investigated, metal-on-UHMWPE total joint replacements provide better care for degenerative joint disorders. Commonly used metal alloys are made of cobalt, chromium, and molybdenum. Metal-on-UHMWPE total joint arthroplasty is a popular treatment modality providing remarkable restoration of mobility for patients with disabilities.

Other physical properties also need to be considered. The dialysis membrane needs specified permeability, the articular cup of the hip joint must have high lubricity, and the intraocular lens has clarity and refraction requirements. Optimizing the mechanical properties of biomaterials is an important step in determining their clinical performance. Forming composites with good engineering design provides an opportunity for long-term clinical survival prosthesis.

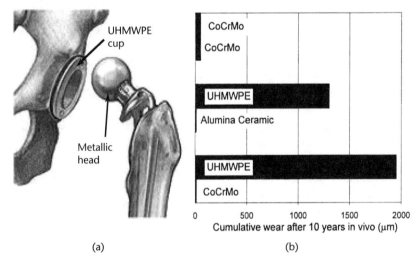

(a)   (b)

Figure 6.6   Total hip replacement: (a) femoral head/cup prosthesis and (b) wear behavior of different head and cup combinations [2].

6.3.2   Mechanical Durability

Apart from suitable mechanical performance, mechanical integrity and wear resistance of a biomaterial are vital to its continuing success. For example, polymer behavior is strongly influenced by temperature, moisture absorption, and other environmental factors. The use of these materials for biomedical applications in a surrounding with varying environmental conditions, needs design procedures, which take into account the loading history and effect on thermomechanical characteristics, in order to obtain a reliable prediction of the long-term behavior. The material could deteriorate physically or chemically, resulting in reduced mechanical properties. Understanding the mechanical durability under the intended loading conditions is important, particularly for long-term implantable devices such as total hip joint replacements, which need to function effectively over periods of 10 years or more.

Durability of the prosthetic device is greatly influenced by the process of cyclic loading in which they are present. Implant materials must have a high degree of fatigue resistance to perform over the long term. Otherwise products will not perform well or for very long. By reducing wear, the tribologist (who deals with friction, heat, wear, bearings, and lubrication) prevents the failure of prosthetic components within the body. Wear depends upon the nature and geometry of the interacting surfaces, the environment in which they interact and the mechanical load (static, dynamic, or impact type).

There are two main types of wear: mechanical and chemical. Mechanical wear involves processes that may be associated with friction, abrasion, impact and fatigue. Chemical wear arises from an attack of the surface by reactive compounds and the subsequent removal of the products of reaction by mechanical action. Mechanical wear results when surfaces produce local mechanical damage, unwanted loss of material, and the resultant generation of wear particles. Fatigue wear occurs as a result of repetitive stressing of a bearing material. Wear at the interface of two components is grouped into abrasive and adhesive wear. Abrasive wear occurs when a surface roughness cuts or plows into the opposing surface, particularly when the two surface materials have different hardnesses and the harder material cuts into the softer material. Adhesive wear occurs when bonding of microcontacts exceeds the inherent strength of either material. The weaker material may then be torn off and adhere to the stronger material. Other factors in wear include surface roughness, material hardness, contact areas, and loads applied.

For example, in the popular head-cup design of the total hip prosthesis and knee prosthesis [Figure 6.6(a)], UHMWPE is the weaker component. The production of large number of particles with UHMWPE is a major factor limiting the life of prosthetic joints. If UHMWPE is replaced by another alloy, the number of wear particles produced decreases significantly. Cells in the immune system sense the presence of wear debris, which leads to aseptic loosening of the hip prosthesis (discussed in Section 5.4.4). To aid in the development of load-bearing devices, multidirectional simulators that mimic the intended loading condition have also been developed. To understand the wear mechanisms, wear rates, wear debris morphology, and wear-surface morphology, these simulators are operated under the appropriate conditions. The wear rate of UHMWPE against alumina ceramic is two to three times less than the wear rate of UHMWPE against stainless steel. The wear rate of UHMWPE against alumina ceramic is about 20 times lower compared to polyethylene against Co-Cr-Mo alloy. However, alumina ceramic exhibits a brittle tendency and is sensitive to microstructural flaws.

There are a number of endogenous factors and exogenous factors that affect the mechanical durability. *Endogenous factors* include the starting raw material, the manufacturing process, the thickness of the device, the rupture energy of the material, sterilization methods, packaging conditions, and aging. For example, the wear resistance of UHMWPE is affected by sterilization techniques. Gamma irradiation in air breaks molecular bonds in the UHMWPE chains, giving rise to free radicals. Oxygen present in the environment combines with these free radicals, leading to subsurface oxidation. An increase in oxidation increases fatigue cracking and delamination. Components that have been on the shelf for less than a year before implantation show decreased in vivo oxidation and better in vivo performance than those with longer shelf lives. Laboratory wear studies have shown increased wear rates in polyethylene gamma irradiated in the air compared to nonirradiated material. *Endogenous factors* include design of bearing components, level of conformity, alignment, surface conditioning, and modularity. For example, the wear rate of certain polymer materials has been demonstrated to be highly dependent on the motion-path or crossings, which occur during the relative motion of the impinging surfaces. The wear rates of hip implants using UHMWPE are related to the crossings of the wear path.

6.3.3 Corrosion and Degradation

Corrosion and degradation are two processes that determine the nature and quantity of wear debris and ions produced during the lifetime of implants. Corrosion occurs due to the interaction between a solid material and its chemical environment, which leads to a loss of substance from the material, a change in its structural characteristics, or a loss of structural integrity. These processes take place at the surface of the implant and the surface of wear debris produced by physical processes. The resistance of a device to chemical or structural degradation (biostability), and the nature of the reactions that occur at the biological interface determine the corrosion characteristics of biomaterials. Ceramics are extremely inert to corrosion. The implant material is only available for chemical reaction at the surface.

Corrosion of metals is widely investigated and there are many types of corrosion. Pitting corrosion arises from a breakdown of protective film. Frequently the pitting is initiated at inclusions, indicating the importance of clean metals. Fretting corrosion typically arises from micromotions occurring at the tissue/implant interface in addition to the action of mechanical loading and electrochemical oxidation. The main mechanical factors are the contact pressure, slip amplitude at the interface, and the frequency of movement, the latter depending on the age and activity of the patient. Osteoporosis affects the mechanical properties of bone and thus influences the stiffness of the contact. As a result, oxide particles accumulate and act as abrasive particles between the rubbing surfaces. This leads to the degradation of the implant and loosening of the prosthesis.

There is always a concern that the corrosion of metals and alloys that the implant is made of could occur in the wet, salty surroundings of the human body, which could release toxins into the systemic circulation. A large number of metals and alloys, including aluminum, copper, zinc, iron and carbon steels, silver, nickel, and magnesium are discarded for usage as they are too reactive in the body. When stainless steel was introduced into general engineering as a new corrosion-resistant material, it was soon utilized in surgery. However, many implant metals corrode in vivo and release varying amounts of ions into the local tissues and corrosion products are formed that might be systemically transported to distant sites and accumulate in the organs. For example, when implanted, Co-Cr-Mo alloys release Co, Cr, and Mo ions to host tissues, and over time the level of metal ions may become significant to cause toxicity problems. The local and systemic consequences of such release include metabolic, immunological, and carcinogenic effects. Released metal ions may contribute to inflammatory and hypersensitivity reactions, changes in tissue remodeling processes that lead to aseptic loosening of the implant. Also, cobalt shows signs of causing anemia by inhibiting iron from being absorbed into the blood stream. Ulcers and central nervous system disturbances have been detected as a result of chromium. Aluminum present in some implant materials may cause epileptic effects.

Surface treatment techniques to form a passivating oxide layer are used to minimize corrosive effects. The oxide layers at the surface of CoCr alloys and stainless steel provide chemical stability and thus good protection from further corrosion. Presence of oxide layers show protective effects with minimal release of ionic or corrosion byproduct residue into the surrounding tissue. However, the periodic removal and reformation of the passive oxide film under fretting conditions can lead to a significant increase in corrosion and in the rate of formation of wear fragments. To reduce the amount of corrosion, better quality materials such as titanium, fiber-reinforced composites, and ceramics are selected.

Polymers are generally stable in the physiological environment. However, in some cases it is the additives added to improve the property of the polymer that cause concerns. For example, additives (plasticizers, stabilizers) added while manufacturing polyvinyl chloride tubing could leach and cause toxic effects. Another factor is the use of degradable polymers in tissue engineering and controlled drug delivery. The problem is to formulate materials that degrade in a controlled manner with tissue-acceptable degradation products. The chemical resistance and physical property of polymers are based on the type and arrangement of atoms in the polymer chain. In general, polymers that are hydrophilic and contain hydrolyzable linkages are most likely to suffer from degradation. On the other hand, polymers that are hydrophobic or do not contain hydrolysis-susceptible bonds are much less prone to environmental attack. Polymers containing carbon-fluorine bond such as poly(tetra fluoro ethylene) have a much wider range of chemical resistance than polymers containing carbon-hydrogen bonds such as poly(ethylene). However, hydrophobic polymers with no reactive sites may degrade slightly by the action of many enzymes present in the body. Degradation of the polymer could result in swelling, softening, discoloration, or delamination in the case of composites.

Degeneration of an implant could occur due to indirect interactions with the body. For example, glutaraldehyde cross-linked bioprosthetic heart valves fail due to degeneration mediated by the deposition of calcium containing salts. Calcification or mineralization of a tissue can be a normal or abnormal process. Nearly 99% of calcium entering the body is deposited in bones and teeth. The remaining 1% of calcium is dissolved in the blood. When a disorder affects the balance between calcium and other minerals or hormones, calcium can be deposited in other parts of the body, such as arteries, kidneys, lungs, brain, or prosthetic materials, which results in tissue hardening. The failure of certain implants is frequently caused by the deposition of calcium phosphate or other calcium-containing compounds. These deposits increase with time causing the failure of bioprosthetic device [3].

Calcification may occur on the surface of an implant (extrinsic calcification), where it is often associated with attached tissue or cells, or within structural components (intrinsic calcification). Pathological calcification can be either *dystrophic* (deposition of calcium salts in damaged or diseased tissues or biomaterials in individuals with normal calcium metabolism) or *metastatic* (deposition of calcium salts in previously normal tissues, as a result of deranged mineral metabolism such as elevated blood calcium level). In general, the determinants of biomaterial mineralization includes factors related to both the host metabolism (in presence of abnormal mineral metabolism, calcification associated with biomaterials or injured tissues is enhanced) and implant structure and chemistry.

6.3.4   Surface Roughness

Manufacturing plays an important role on the surface finish of various devices. The finest irregularities of a surface generally results from a particular production process or material condition. Imperfections at the surface lead to variation in the surface topography, which influences the surface interactions. Surface texture is the combination of fairly short deviations of a surface from the nominal surface. To avoid tissue damage by disturbing fluid flow, many applications including materials for orthopedic implants and contact lenses need a smooth surface. If cell-surface interactions are desired, as in polymer scaffolds useful in tissue regeneration, surface roughness enhances the cell in-growth and tissue integration within the implants. In such cases, careful attention must be paid to the thrombogenicity of the inner surface. A number of defects could occur due to the quality of the raw material used during the process or during the finishing stages such as cleaning the edges so that no dross or burrs are attached. More uniform distribution of plastic deformation during tube drawing is critical to achieve better mechanical properties. Better controlled dimensional tolerances and surface texture are important to ensure smooth cutting. Minimizing drawing contaminations and more effective surface finishing techniques for removing surface contaminations and oxides are paramount to eliminate corrosion concerns and to improve the biocompatibility of the final devices.

Surface texture includes waviness, lay, and roughness. Waviness includes the more widely spaced deviations of a surface from its nominal shape. Lay refers to the predominant direction of the surface texture. Lay is determined by the particular production method and geometry used. Turning, milling, drilling, grinding, and other machining processes usually produce a surface that has lay (striations or peaks and valleys in the direction that the tool was drawn across the surface). Other processes such as sand casting and grit blasting produce surfaces with no characteristic direction. Sometimes these surfaces are said to have a nondirectional, particulate, or protuberant lay. Lay is important for optical properties of a surface.

A number of techniques are available to assess the surface roughness. A popular method is atomic force microscopy (AFM) which uses a small probe attached at the end of a cantilever to probe surfaces. AFM can be operated either in contact mode or in noncontact mode. In contact mode, also known as repulsive mode, an AFM tip makes soft "physical contact" with the sample. The tip is attached to the end of a cantilever with a low spring constant, lower than the effective spring constant holding the atoms of the sample together. As the scanner gently traces the tip across the sample (or the sample under the tip), the contact force causes the cantilever to bend to accommodate changes in topography. In constant-height mode, the spatial variation of the cantilever deflection can be used directly to generate the topographic data set because the height of the scanner is fixed as it scans.

Noncontact AFM is one of several vibrating cantilever techniques in which an AFM cantilever is vibrated near the surface of a sample. The spacing between the tip and the sample for NC-AFM is on the order of tens to hundreds of angstroms. NC-AFM is desirable because it provides a means for measuring sample topography with little or no contact between the tip and the sample. Like contact AFM, noncontact AFM can be used to measure the topography of insulators and semiconductors as well as electrical conductors. The total force between the tip and the sample in the noncontact regime is very low, generally about 10–12N. This low force is advantageous for studying soft or elastic samples. A further advantage is that samples like silicon wafers are not contaminated through contact with the tip. AFM with receptor molecule tips are also available. Apart from the contact angle measurements and atomic force microscopy techniques described above, a number of techniques have been developed to assess the surface characteristics of a material.

Scanning electron microscopy (SEM) uses a focused beam of high-energy electrons to scan the surface of a material. The beam interacts with the material, which generates a variety of signals (such as secondary electrons, backscattered electrons, X-rays) each of which can be used to characterize the material. It is typically used in analyzing the surface characteristics of materials. Transmission electron microscopy (TEM) uses a highly focused beam, which bombards a thin sample to generate transmittable electrons. The transmitted electron signal is magnified by a series of electromagnetic lenses and observed through electron diffraction or direct electron imaging. Using electron diffraction patterns, one can determine the crystallographic structure of the material. It is typically used in structural and compositional analyses and high-resolution imaging. Scanning tunneling microscopy (STM) uses piezoelectric translators that can bring sharp metallic crystalline tips with several angstroms of the surface. The change in the tunneling current established by a bias voltage between the tip and surface permits the density and energy maps of electron states. It is typically used to obtain atomic and molecular resolution images of conductive and semiconductive materials.

Conoscopy is a novel polarization interferometric technique, which provides depth information using quasi-monochromatic and spacially incoherent light. It is based on the interference behavior of doubly refractive crystals under convergent, polarized light. Since it eliminates the sensitivity and stability problems associated with the LASER-beam, it can be used for a wide range of measurements. Conoscopic devices include range finders, surface profilers, and roughness gauges.

X-ray photoelectron spectroscopy (XPS) also called electron spectroscopy for chemical analysis (ESCA) is generally operated in the 10^{-9} millibar pressure. The energy of the photoelectrons leaving from the solid sample is determined using a concentric hemispherical analyzer and this gives a spectrum with a series of photoelectron peaks. The binding energy of the peaks is characteristic of each element. The peak areas can be used (with appropriate sensitivity factors) to determine the composition of the materials surface. The shape of each peak and the binding energy can be slightly altered by the chemical state of the emitting atom. XPS provides chemical bonding information about all the elements except hydrogen or helium.

Optical or stylus-based profilometry uses a tip, which is dragged across the surface, similar to AFM. The difference is it does not have a feedback loop to regulate force. It can be used analyze the chemical and mechanical properties. There is also a penetration based mechanical testing such as indentation techniques.

6.3.5   Sterilization Techniques

All materials used within the body or placed in contact with corporeal fluids must be sterilized to prevent the introduction of harmful organisms into the body. According to the Center for Disease Control and Prevention, "sterilization means the use of a physical or chemical procedure to destroy all microbial life, including highly resistant bacterial end ospores" (http://www.cdc.gov/ncidod/dhqp/bp_sterilization_medDevices.html) and "disinfection means the use of a chemical procedure that eliminates virtually all recognized pathogenic micro-organisms but not necessarily all microbial forms (e.g., bacterial endospores) on inanimate objects" [4]. Many choices are available for the sterilization process, including exposure to autoclaving, dry heat, ultraviolet and gamma radiation, and ethylene oxide. In general, sterilization processes alter the structure or function of the macromolecules within the pathogenic micro-organism, leading to its death or the inability to reproduce. For example, when ethylene oxide (EtO) is used for sterilizing, it replaces hydrogen atoms on molecules needed to sustain life, and stops their normal life-supporting functions. However, no single sterilization process is capable of sterilizing all medical devices. Biomaterials can be attacked by the same mechanisms, and different forms of sterilization may result in different adverse effects leading to changes in the physical, mechanical, and biological properties of biomaterials. Some of the commonly used techniques are described in the following sections.

6.3.5.1   Sterilization by Heat

Both dry heat and moist heat are used for sterilization. Dry heat is applied by infrared radiation or incineration typically 134°C. Moist heat is generally applied using a pressurized steam (121 kPa) at 121°C. The process is carried out in a pressure vessel called the autoclave, designed to withstand the elevated temperature and pressure. Exposures of 20 minutes usually result in complete sterilization. Steam sterilization is efficient, reliable, rapid, relatively simple, and does not leave toxic residues. Steam sterilization is widely used in hospitals for sterilizing heat-resistant surgical equipments and intravenous fluids. However, it is not the predominant method in the commercial sterilization of medical devices due to the difficulties in autoclaving-packaged products. Also, heat sterilization is applicable only if damage by heat and/or moisture is not a problem. The high temperature, humidity, and pressure used during the steam sterilization process can lead to hydrolysis, softening, or degradation of many biopolymers. Hence, it may not be suitable for the sterilization of many polymers.

6.3.5.2 Sterilization by Gases

The most widely used gas in medical applications to sterilize heat sensitive items is ethylene oxide (EtO). The EtO sterilization procedure involves:

1. Creating a vacuum in the sterilization vessel;
2. Injecting EtO at a desired concentration (typically around 600–1,200 mg/L);
3. Maintaining the device at the desired conditions (typically 30–50°C and 40% to 90% humidity for 2 to 8 hours);
4. Evacuating the chamber several times to remove residual EtO.

Furthermore, aeration is usually required after removal from the chamber, with aeration time ranging from 2 hours to 2 weeks, depending on the device and packaging. EtO has advantages such as low processing temperatures, high penetration, and compatibility with a wide range of materials. The main disadvantage of EtO relates to its flammability, explosive nature, residual toxicity (EtO and its secondary products, i.e., ethylene chlorohydrin and ethylene glycol may remain in the medical devices after sterilization), and suspected carcinogenicity of the gas and residuals in the product and manufacturing environment. To reduce its toxic effects and flammability, EtO is usually mixed with inert gases, such as fluorinated hydrocarbons and carbon dioxide. Furthermore, to prevent residual toxicity, after the sterilization by EtO it is essential to allow the residual gas to dissipate from the material to acceptable levels prior to its usage. Methods for rapidly removing absorbed EtO are based on exposing the materials to repeated air washings, exposing the load to forced air circulation in a closed chamber, or placing the exposed material in well-ventilated areas for long periods. In order to overcome some of these drawbacks, more strict regulations have been established concerning EtO sterilization, such as the American National Standard ANSI/AAMI ST27-1988 (guidelines for industrial ethylene oxide sterilization of medical devices). Gas plasmas are increasingly being employed to sterilize the surfaces of medical components or devices. This method uses less-toxic materials than does EtO processing and can be more cost-effective than irradiation.

6.3.5.3 Sterilization by Irradiation

Radiation sterilization utilizes ionizing radiation generated by gamma rays (see Chapter 8 for details on radiation), an electron beam, X-rays, or ultraviolet rays (220–280-nm range) to sterilize medical devices. Gamma irradiation is the most popular form of radiation sterilization. Materials to be sterilized are exposed to gamma rays (generated typically from a cobalt-60 isotope source) for a defined period of time. Gamma radiation, despite the dangers associated to the ionizing radiation source, is a rapid, residue-free method that can be monitored due to the availability of various tools. Gamma rays have a high penetrating power (up to 50 cm). They are widely used to sterilize medical products such as surgical sutures, metallic bone implants, and knee and hip prostheses, among other materials. However, gamma radiation cannot be used to sterilize some types of polymers because of their inherent sensitivities or susceptibilities. Further, properties and performance can be negatively affected in these cases because of material degradation or cross-linking.

Electron beam sterilization uses a generator (between 10^{-13}J and 20×10^{-13}J) to produce a beam of high energy electrons that destroys organisms. The electrons in the beam have a much lower penetrating power (penetrate ~5 cm of the sample), but higher dose rates than gamma rays. Similar to gamma rays, products at the edges of the pack are subjected to higher doses than products at the center to insure that full sterilization is achieved. Dosage for either process is measured in Megarad (Mrad) and as a general rule a radiation dose of around 2.5 Mrad will sterilize clean articles in air. The higher dose rates and shorter times used for E-beam sterilization can slightly improve the dosage to produce substantial damage due to the reduced exposure to oxygen during the process.

6.3.5.4 Sterilization by Solution

In spite of the fact that ethanol is not a sterilizing agent, but a good disinfectant (due to its dehydration action and protein coagulation effect, which destroy membranes and denature proteins), this solvent does not act on the endospores of many bacteria, limiting its use as a surface-sterilizing agent.

6.4 Physiological Responses to Biomaterials

Biomaterials are utilized to repair, assist, or replace living tissue or organs that are functioning below an acceptable level. When an artificial devices is used, the biocompatibility of that material is influenced by a number of factors, including the material characteristics discussed in the previous section (the form and design of the implant, the dynamics or movement of the device in situ, the resistance of the device to chemical or structural degradation), the toxicity of the materials employed, and the nature of the reactions that occur at the biological interface. For example, saline bags routinely used for intravenous infusion made of PVC cause concern about the possible toxic effects of phthalate plasticizers often used to soften PVC. Hence PVC alternative materials (e.g., polypropylene, polyethylene vinyl acetate) that are not only clear, resilient, and autoclavable but also do not leach toxic chemicals into the body are necessary.

The determination of the biocompatibility of materials involves detailed characterization of the material (e.g., bulk and surface chemical composition; density; porosity; and mechanical, electrical, and degradation properties) and extensive testing, first at the protein/cell/tissue, then in animal models, and ultimately in human clinical trials. Before the introduction of a new material into clinical testing, several tests have to prove its safety and usefulness. These tests have to be meaningful and appropriate. The design and use of biocompatibility testing protocols is provided by a variety of regulatory organizations (described in Chapter 11), including the American Society for Testing and Materials International (ASTM); the International Standards Organization (ISO standard 10993 "Biological Evaluation of Medical Devices"); the Food and Drug Administration (FDA); and the Medical Device Directive (93/42/EEC) in the European Union. Documentation of biocompatibility tests are required by law in the United States and other countries, and are used to ensure that biomedical devices and their constituent materials are safe and effective under intended use conditions.

6.4.1   Toxicity Analysis

The mechanisms by which substances are rendered toxic are varied and complex. It may be due to direct chemical toxicity, accumulation of products from wear, corrosion or degradation, and excess inflammatory response. Biomaterials should be carefully evaluated and studied for toxicity in vitro before being implanted. Such testing must include the intact material, as well as the degradation and wear products, which might be produced during function. Several in vitro (a Latin word means in glass and refers to processes and experiments occurring outside an organism in an artificial environment) tests are used to screen materials, their components, and or leachable/soluble/degradation products for cytotoxic, genotoxic, immunological, and hemolytic effects. *Cytotoxicity* is the extent to which the material kills cells in cell cultures (described in Chapter 7). Measuring cytotoxicity in vitro includes using standard cell lines, which are available from various biological suppliers. The *direct contact test* is the simplest test where cells are placed on top of the material to be tested. In the *agar diffusion test*, a layer of agar is placed between the cells and the sample. In the *elution test*, the material is soaked in fluid and the fluid is applied to the cell culture.

Medical devices that are implanted in the cardiovascular system offer an even more challenging task for creating specific testing standards. Apart from the material properties, hemocompatibility for implantable devices is dependent on the shape, function, and location of the implant. For example, many of the vascular graft materials are hemolytic although in vitro tests prove otherwise. Then, the hemolysis assay has little predictive value for problems encountered in clinical use for grafts.

Systemic toxicity refers to toxicity at some distance from the site of initial usage. Animal models are used to evaluate material-host tissue interactions and to predict how the device or prototype works at the systemic level. Ultimately, the safety and effectiveness of the device must be evaluated in humans prior to widespread use. At each stage, biocompatibility data must be correlated with material properties and with manufacturing, sterilization, packaging, storage, and other handling procedures that also may influence test outcomes.

6.4.2 Surface Adhesion

In general, when a material is used in biomedical applications, the surface of that material is exposed to a mixture of small solute molecules, proteins, and cells. Plasma contains more than 100 identified proteins with specific functions and varying biologic properties. The most abundant proteins in blood are albumin, fibrinogen, and IgG. These get adsorbed to the surface of the implant instantaneously following exposure to the systemic circulation. The continuous interaction of blood with artificial contact surfaces can lead to a substantial damage of blood cells and plasma factors. Routinely used blood-contacting devices such as needles, cannulae, and blood containers all have very different blood compatibility (termed as hemocompatibility) requirements. For example, a needle may reside in the bloodstream for only a short time and the primary concern for a needle would be hemolysis, the destruction of red blood cells as a result of chemical interaction with the needle material. However, a cannula may be implanted for a much longer time and the primary concern is thrombogenicity, or clotting, which can be caused not only by chemical interaction, but also by the flow rate of the blood. Localization of the formed blood clot near the brain could potentially cause a stroke with the blockade of blood flow.

Surface properties of the biomaterial could be different from the bulk material properties, uniquely reactive, readily contaminated, and mobile (i.e., can change depending on environment). The character of the adsorbed protein layer that mediates subsequent events is thought to be dependent on the properties of the substrate surface such as roughness, chemistry of molecules, inhomogenous surfaces, crystallinity or disorder, and hydrophobicity (wettability). All affect how the material would interact with the biological component.

Understanding biomaterial surface structure and its relationship to biological performance is important in utilizing the biomaterial for any biomedical application and to develop surface modification strategies for biomaterials, if needed. Water spreads on or wets some solids and does not others. Surface energy and work function determine the wettability and ultimately blood compatibility. Figure 6.7(a) shows some of the possible wetting behaviors of a drop of liquid placed on a horizontal, solid surface (the remainder of the surface is covered with air, so two fluids are present). Case A represents a liquid, which wets a solid surface well (e.g., water on a very clean copper). The angle θ shown is the angle between the edge of the liquid surface and the solid surface, measured inside the liquid. This angle is called the *contact angle* and is a measure of the quality of wetting. The contact angle describes the shape of the drop at the liquid-solid-vapor three-phase line while in thermodynamic equilibrium. For perfect wetting, in which the liquid spreads as a thin film over the surface of the solid, θ is zero. Case C represents the case of no wetting. If there were zero wetting, θ would be 180°. However, the gravity force on the drop flattens the drop, so that 180° angle is never observed. If gravity is not present, the drop shape is spherical assuming. In general, a liquid is said to wet a surface if θ is less than 90° and does not wet if θ is more than 90°. Values of θ less than 20° are considered strong wetting, and values of θ greater than 140° are strong nonwetting. This might represent water on teflon or mercury on clean glass.

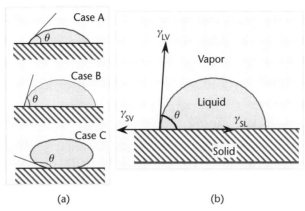

Figure 6.7 (a) Types of contact angles. (b) Forces acting when a liquid is in contact with a solid surface.

Several techniques are available to measure the contact angle [5]. Using contact angle measurements, surface energy can be calculated. While using contact angle measurements, assumptions are that:

- The surface is absolutely smooth and homogenous and does not contain any contaminants.
- The solid surface is rigid and nondeformable.
- The solid surface is stationary and does not reorient with the application of the liquid drop.
- The surface tension of the liquid is known and remains constant.
- The solid and the liquid phases do not react.
- The vapors of the liquid does not adsorb to the solid.

The strength of adhesion to a solid surface can be measured directly using suitable force measurements, or can be estimated from a value of the thermodynamic *adhesive work*, W_A (the negative of the Gibbs energy of adhesion). In a system [Figure 6.7(b)] where a liquid (L) drop adheres to a solid (S), based on the mechanical equilibrium of a liquid under the action of interfacial forces adhesive work is defined as

$$W_A = W_{SL} = \gamma_{SV} + \gamma_{LV} - \gamma_{SL} + \pi_e \tag{6.1}$$

where γ_{SL} represents the interfacial tension between the solid and the liquid, γ_{SV} and γ_{LV} are, respectively, the surface tension of the solid and the surface tension of the liquid in equilibrium with their own vapors. Also π_e is the equilibrium spreading pressure of the liquid's vapor over the solid surface, which is typically very small and negligible. Using (6.1) to determine W_A is difficult since only γ_{LV} can be measured with any confidence. Those involving the solid cannot be measured independently. Alternatively, expression for $\cos\theta$ can be written from the definition of cosine in trigonometry [Figure 6.7(b)] as

$$\cos\theta = \frac{\gamma_{SV} - \gamma_{SL}}{\gamma_{LV}} \tag{6.2}$$

Rearranging

$$\gamma_{SV} = \gamma_{SL} + \gamma_{LV}\cos\theta \tag{6.3}$$

where θ is the equilibrium contact angle. Equation (6.3) is called Young's equation although Young did not prove this equation and suggests that by knowing γ_{SV}, γ_{LV}, and θ, γ_{SL} can be calculated. Substituting into (6.1)

$$W_A = W_{SL} = \gamma_{SL} + \gamma_{LV}\cos\theta + \gamma_{LV} - \gamma_{SL}$$

Hence,

$$W_A = \gamma_{LV}(1 + \cos\theta) \tag{6.4}$$

To determine W_A, the contact angle of various liquids whose γ_{LV} are known (see Table 6.2) are measured on a given surface. Then $1/\gamma_{LV}$ is plotted against $(1 + \cos\theta)$ and from the slope W_A is determined. The plot used to determine W_A by this technique is called Zeisman plot. Typically, higher surface tension corresponds to a lower contact angle.

EXAMPLE 6.1

Yan and Lin [6] reported a modification strategy to improve the polyurethane films with dimethyol propionic acid PUDPA. They measured the contact angles (in degrees) of two different dry PUDPA films with different test liquids (see the table). Using these values, calculate the adhesive work of the two membranes.

Polymer	Water	Ethylene Glycol	Glycerol	Formamide
PUDPA1	98	73.3	93.3	83.1
PUDPA2	82.7	65.3	82.7	77.7

Solution: Obtain the γ_{LV} from Table 6.2 and compute the following.

			PUDPA1				PUDPA2			
	γLV	1/gLV	Degree	Radians	Cos(θ)	1+Cos(θ)	Degree	Radians	Cos(θ)	1+Cos(θ)
Water	72.8	0.0137	98	1.710	–0.139	0.861	82.7	1.443	0.127	
Ethylene glycol	48.3	0.0207	73.3	1.279	0.287	1.287	65.3	1.140	0.418	1.418
Glycerol	63.4	0.0158	93.3	1.628	–0.058	0.942	82.7	1.443	0.127	1.127
Formamide	58.2	0.0172	83.1	1.450	0.120	1.120	77.7	1.356	0.213	1.213

Then plot these using the Young equation and determine the slope to obtain the adhesive function.

As shown in the graph, W_A for PUDPA1 is 69.885 dyne/cm and for PUDPA2 it is 62.504 dyne/cm.

Table 6.2   Liquid Contact Angles

Liquid	γ_{LV} (dyne/cm)	γ_{LV}^d (dyne/cm)	γ_{LV}^p (dyne/cm)
Water	72.8	21.81	50.98
Fibrinogen	65.0	24.70	40.30
Albumin	65.0	31.38	33.62
Glycerol	63.4	37.21	26.21
Formamide	58.2	39.44	18.66
Diiodomethane	50.46	50.46	0
Human blood plasma	50.5	11.00	39.50
Ethylene glycol	48.3	29.27	19.01
Human blood	47.5	11.2	36.30
Tetrabromoethane	47.5	42.12	5.38
α-Bromonaphthalene	44.6	31.70	12.89
4-Octanol	27.5	7.4	20.1

Most of the values were obtained at 20°C and from [7] and others from [8].

6.4.2.1   Determine the Components of Surface Energy

The intermolecular attraction that is responsible for surface energy, γ, results from a variety of intermolecular forces whose contribution to the total surface energy is additive. The majority of these forces are functions of the particular chemical nature of a certain material. A common approach to treating solid surface energies is that of expressing any surface tension (usually against air) as a sum of:

- Polar or nondispersive component, γ^p (hydrogen bonding);
- Dispersive component, γ^d (e.g., van der Waals forces present in all systems regardless of their chemical nature).

Hence, the surface energy of any system can be described by

$$\gamma = \gamma^d + \gamma^p \tag{6.5}$$

In order to prove functionalization, one has to find dispersion (γ^d) and polar components (γ^p) of surface energy. The interfacial tension between the liquid and solid phases is then expressed in terms of the two components for each phase as follows:

$$\gamma_{SL} = \gamma_{SV} + \gamma_{LV} - 2\left[\left(\gamma_{SV}^d \gamma_{LV}^d\right)^{1/2} + \left(\gamma_{SV}^p \gamma_{LV}^p\right)^{1/2}\right]$$

Substituting into (6.1),

$$W_A = \gamma_{SV} + \gamma_{LV} - \left(\gamma_{SV} + \gamma_{LV} - 2\left[\left(\gamma_{SV}^d \gamma_{LV}^d\right)^{1/2} + \left(\gamma_{SV}^p \gamma_{LV}^p\right)^{1/2}\right]\right)$$

Simplifying,

$$W_A = 2\left(\gamma_{SV}^d \gamma_{LV}^d\right)^{1/2} + 2\left(\gamma_{SV}^p \gamma_{LV}^p\right)^{1/2} \qquad (6.6)$$

γ_{LV}^d and γ_{LV}^p for a few of the compounds are given in Table 6.2. There are two unknowns in (6.6), γ_{LV}^d and γ_{LV}^p, the components of the solid surface energy. Rearranging (6.6)

$$\frac{W_A}{2\left(\gamma_{LV}^d\right)^{1/2}} = \left(\gamma_{SV}^d\right)^{1/2} + \left(\gamma_{SV}^p\right)^{1/2} \left(\frac{\gamma_{LV}^p}{\gamma_{LV}^d}\right)^{1/2} \qquad (6.7)$$

Contact angle data from at least two liquids of different polarity are measured from which W_A is calculated. Then, $\dfrac{W_A}{2\left(\gamma_{LV}^d\right)^{1/2}}$ is plotted against $\left(\dfrac{\gamma_{LV}^p}{\gamma_{LV}^d}\right)^{1/2}$ and from the slope and intercept values of the linear line γ_{LV}^d and γ_{LV}^p can be calculated using. Then the net surface tension γ_{SV} is calculated using the relation

$$\gamma_{SV} = \left(\gamma_{SV}^d\right) + \left(\gamma_{SV}^p\right) \qquad (6.8)$$

EXAMPLE 6.2

Using the W_A obtained in the Example 6.1, calculate the γ_{LV}^d, γ_{LV}^p, and the γ_{SV}.

Polymer	γ_{LV}^d (dyne/cm)	γ_{LV}^p (dyne/cm)	γ_{SV} (dyne/cm)
PUDPA1	19.3	2.3	21.6
PUDPA2	9.1	14.2	23.3

Solution: Using the W_A values, calculate the x and y coordinates. From the slope and intercept, dispersion and polar components can be obtained.

To understand whether the plasma proteins attach and which one has a preference over the other, interaction of serum albumin and serum fibrinogen with the material surfaces are evaluated. The preferential binding of fibrinogen over albumin indicates the possibility of thrombus formation as albumin shows thromboresistant properties. The interfacial tension, Γ, between the surface of the biomaterial and the absorbed protein is mainly determined by

$$\Gamma_{ij} = \left[\left(\gamma_S^d\right)_i^{1/2} - \left(\gamma_S^d\right)_j^{1/2}\right]^2 + \left[\left(\gamma_S^p\right)_i^{1/2} - \left(\gamma_S^p\right)_j^{1/2}\right]^2 \qquad (6.9)$$

where i represents the biomaterial and j the kind of protein analyzed, either albumin or fibrinogen. Once the Γ_{ij} for the two proteins are calculated, their ratio is determined, from which thrombogenecity can be estimated. (Typically γ_S^d ranges from 10–50 dyne/cm.)

EXAMPLE 6.3

Low temperature isotropic carbon is one of the biomaterials with the potential to develop a mechanical heart valve. However, its hemocompatibility has to be improved for long-term application. Wang et al. [9] suggested depositing a layer of titanium dioxide film to improve the hemocompatibility using the following data. Do you agree with their suggestion? Show the calculations. From Table 6.2, $\gamma_{Albumin}^d = 31.38$ dyne/cm and $\gamma_{Albumin}^p$ dyne/cm.

Solution: Equation (6.9) for this case is rewritten as

$$\Gamma_{LTI-Albumin} = \left[\left(\gamma_{LTI}^d\right)^{1/2} - \left(\gamma_{Albumin}^d\right)^{1/2}\right]^2 + \left[\left(\gamma_{LTI}^p\right)^{1/2} - \left(\gamma_{Albumin}^p\right)^{1/2}\right]^2$$

$$\Gamma_{LTI-Albumin} = \left[(30.19)^{1/2} - (31.38)^{1/2}\right]^2 + \left[(26.73)^{1/2} - (33.62)^{1/2}\right]^2 = 6.35 \text{ dyne/cm}$$

Material	γ_S^d (dyne/cm)	γ_S^p (dyne/cm)
LTI-carbon	30.19	10.76
TiO$_2$ layered	26.73	6.50

Applying this interaction to other conditions, obtain other values and calculate the rations.

Material	γ_S^d (dyne/cm)	γ_S^p (dyne/cm)	$\Gamma_{Albumin}$ (dyne/cm)	$\Gamma_{Fibrinogen}$ (dyne/cm)	$\dfrac{\Gamma_{Albumin}}{\Gamma_{Fibrinogen}}$
LTI-carbon	30.19	10.76	6.35	9.75	0.65
TiO$_2$ layered	26.73	6.50	10.72	14.43	0.743

The results indicate that the interfacial tension of LTIC with water, blood, and proteins is greater than that for titanium oxide films, implying that the interaction of blood components with LTIC is stronger than with a TiO$_2$ film. Since TiO$_2$ has a higher ratio than uncoated LTI-carbon, probably, it is more thromboresistant.

When blood or plasma comes in contact with electron-conducting materials, another possible reaction apart from adsorption is the charge transfer reaction. The charge transfer reaction with one of the proteins in the blood could lead to the formation of a blood clot. Although there are many proteins in the blood, the formation of a blood clot on a biomaterial is correlated with fibrinogen undergoing charge transfer reactions with the surface of the biomaterial; then fibrinogen decomposes to fibrin monomers and fibrinopeptides. The charge transfer of fibrinogen, decomposition, and formation of a thrombus are related to the composition and semiconductor properties of the biomaterials. It is proven that fibrinogen has an electronic structure similar to an intrinsic semiconductor with a band gap of 1.8 eV. In order to inhibit the transfer of the charge carrier from fibrinogen to the biomaterial, the work function of the film must be reduced.

Tests to determine the interactions between the surface and the blood proteins have been developed. Fibrinogen adsorption and albumin adsorption are evaluated to determine the relative thrombogenic potential of a material. To some extent, synthetic materials adsorbing less fibrinogen could attach fewer platelets and thereby exhibit improved blood compatibility. Fibrinogen adsorption from plasma to biomaterials surfaces passes through a maximum when studied as a function of adsorption time, plasma concentration, or column height in narrow spaces. Leo Vroman first observed that adsorption from plasma or serum occurred through a complex series of adsorption-displacement steps in which low molecular weight proteins arriving first at a surface are displaced by relatively higher MW proteins arriving later. However, certain proteins, such as albumin, are observed to be relatively resistant to displacement at hydrophobic surfaces whereas other proteins such as kininogen, readily displaces fibrinogen. These are called the Vroman effect. Subsequent studies have demonstrated that the Vroman effect is a general phenomenon of plasma proteins and reflects the competitive adsorption of plasma proteins for a limited number of surface sites. Exact molecular mechanisms underlying the Vroman effect process have not yet been resolved. However, there are no specific "mixing rules" stipulating how interfacial behavior of complex protein mixtures can be deduced from the behavior of single-protein solutions.

6.4.3   Blood-Material Interactions

A general sequence of interactions leading to the formation of blood clot (thrombus) is shown in Figure 6.8. Following contact with blood, most if not all artificial surfaces quickly acquire a layer of blood proteins. The interaction begins by a few molecules present in the plasma forming a layer, which could be activated or passivated by the underlying synthetic biomaterial surface. The concentration of the adsorbate and the conformation of the constituent proteins determine the adsorption and activation of subsequently interacting proteins and other cells present in blood. These interactions occur reversibly and irreversibly between surfaces and blood components. The composition of the adsorbed layer varies with time in a complex manner and depends in part on the physicochemical properties of the substrate surface. There is also a redistribution of proteins according to the relative biochemical and electrical affinity of each protein for a specific biomaterial surface.

At least 12 plasma proteins interact in a cascade of reactions leading to blood clotting. They are designated as clotting factors distinguished with a suffix of Roman numerals made in order of discovery. An example is factor VIII, commonly called the von Willebrand factor. Each enzyme is present in an inactive form referred to as zymogen. During the reaction, the inactive factors become enzymatically activated due to the proteolytic cleavage. A cascade of interactions occurs with a series of positive and negative feedback loops, which control the activation process. These interactions lead to the production of thrombin, which then converts soluble fibrinogen into fibrin. Formed fibrin then forms a clot with the interactions of platelets and red blood cells. Since the sequence involves a series of

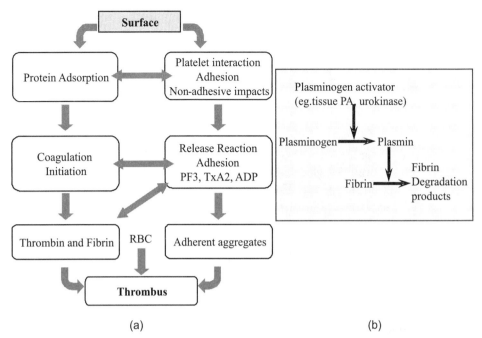

Figure 6.8   Interaction of blood components with a surface: (a) coagulation pathway and (b) fibrinolytic pathway.

steps, and since one enzyme can activate many substrate molecules, the reactions are quickly amplified to generate significant amounts of thrombin and subsequent platelet activation, form fibrin, and stop bleeding.

6.4.3.1   Platelet Interaction to Artificial Surfaces

Platelets are the smallest (diameter 2–4 mm) anuclear cells present in human blood and play an important role in the formation of blood clot. The number varies from 150,000 to 300,000/mL blood (~0.3% of blood volume). Platelets serve three primary functions in blood clot formation:

1. *Adhering to the surface.* Platelets interact with the adsorbed proteins present on the synthetic surface through the mediation of the von Willebrand factor and platelet membrane glycoproteins. Since platelets interact predominantly with the bound proteins and not the prosthetic material itself, the concentration of surface ligands presented by the bound proteins affect short and long term functionality of the biomaterial.
2. *Releasing and providing molecules required for many of the reactions for the processes of the coagulation cascade.* After adherence, platelets undergo morphological changes, from discoid cells to spherical cells with tentacles or pseodopodia. Granules within the cells release a number of components including platelet factor 4 (PF4), β thromboglobulin (bTG), ADP, Ca^{2+}, and fibrinogen.
3. *Recruiting, activating other platelets and attaching to them to form aggregates.* Activated platelet-surface phospholipids tend to recruit and stimulate

other platelets. Increased platelet activity stimulates previously inactive coagulation factors, which are always present and circulating in normal blood. Platelet aggregation requires the proteolytic enzyme thrombin, which is generated locally. Thrombin also converts fibrinogen to fibrin, which stabilizes the platelet aggregation. Membrane phospholipids are metabolized to generate thromboxane A2 (TxA2). TxA2, ADP, and thrombin recruit additional circulating platelets into an enlarging platelet aggregate and forming blood clot. Activation of these clotting factors results in the initiation of a series of reactions or pathways. The coagulation cascade continues, which ultimately leads to the formation of an insoluble, stabilized fibrin clot. The formation of a thrombus takes approximately 12–16 seconds in a normal individual.

In vitro blood-coagulation tests are developed and used to evaluate the blood compatibility of materials. Typically, tests compare the weight of the thrombus formed, the amount of unclotted blood, and the reduction in platelet count of the blood exposed clot formation rate on a surface to that of a control material. These data are used to calculate a relative index whereby materials can be rated quantitatively as to the rate of clot formation on their surface. However, the relationships between material surface properties, the initially deposited protein layer, cell-surface interactions, later events (which could be of clinical consequence for the evaluation of biocompatibility), and the design of blood-contacting cardiovascular devices, are not well understood. Hence, the experiment results have to be carefully performed.

6.4.3.2   Fibrinolytic Sequence

In the natural process of wound healing, the clot or thrombus dissolves from the repaired site through the degradation of the fibrin mesh with the action of many proteins. This is termed as fibrinolytic sequence [Figure 6.8(b)] and helps restore blood flow following thrombus formation and facilitates the healing process. The fibrinolytic sequence is also involved in tissue repair and macrophage function. Plasminogen, an inactive protein produced in the liver and circulating in the blood, plays a central role in the fibrinolysis. Plasminogen adheres to a fibrin clot, being incorporated into the mesh during polymerization. Plasminogen is converted into an active form called plasmin by the actions of plasminogen activators, which may be present in blood or released from tissues. Plasmin then digests the fibrin clot, releasing soluble fibrin-fibrinogen digestion products (FDP) into circulating blood. FDPs (e.g., fibrin D-D dimer fragment) may be assayed for in vivo fibrinolysis.

Plasmin also binds and degrades many matrix proteins including fibronectin, von Willebrand factor, thrombospondin, and laminin. The proteolytic network of susceptible matrix proteins is further extended to include the collagens and elastin by the ability of plasmin to activate certain matrix metalloproteinases (MMPs), which, in turn, can activate other proteins. Also, certain growth factors, cytokines, and chemokines can be released, activated, and/or degraded by plasmin. The activity of plasmin is regulated by limiting the conversion of plasminogen to plasmin by the action of plasminogen-activator inhibitory proteins or directly by antiplasmin proteins.

6.4.4   Inflammatory Response

The immune system protects the host against infectious agents by identifying the pathogens as foreign materials. Tissue damage by a wide range of harmful stimuli including mechanical trauma, tissue necrosis, and infection causes a protective vascular connective tissue reaction called *inflammation*. The purpose of inflammation is to destroy (or contain) the damaging agent, initiate the repair processes, and return the damaged tissue to useful function. The inflammatory response to surgically implanted biomaterials is one of the major obstacles to their deployment. Based on the time sequence of response, inflammation is grouped into *acute* and *chronic inflammation*.

The initial response of the body to an infection or trauma is called the *acute* inflammatory response (Figure 6.9). *Acute* response is nonspecific and is the first line of defense of the body against infectious agents. *Acute* response consists of a coordinated local and systemic mobilization of immune, endocrine, and neurological mediators. Typical symptoms of acute inflammation are redness caused by hyperemia (excess accumulation of blood), swelling caused by fluid exudation, increased local temperature caused by hyperemia, pain, and loss of function. The outcome of acute inflammatory response depends on a number of factors, including the degree of toxicity to the cells in the tissue, the reason for triggering the inflammation, infection of the material, and the status of the tissue prior to implantation. In a healthy response, the inflammatory response becomes activated and clears the pathogen (in the event of infection). Fragile capillary buds proliferate and grow new blood vessels, and also connect with existing capillaries. With

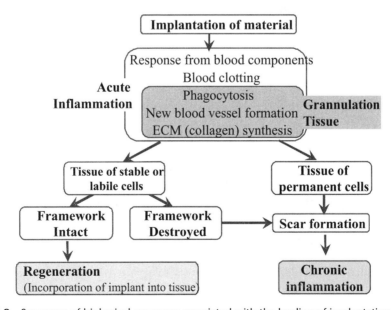

Figure 6.9   Sequence of biological processes associated with the healing of implantation site.

the restoration of oxygen supply and nutrients, fibroblasts produce collagen fibers that span wound, epithelium regenerates and proliferates beneath the clot. Wound contraction occurs to pull closer together ends of damaged tissue, and temporary healing is acheived. Upon initial healing, the wound remodels over an extended period of time where matrix elements are broken down and resynthesized with precise orientation and mechanical properties. This leads to a complete regeneration of the tissue.

If the framework is destroyed or large number of cells is damaged, acute inflammation may *resolve* or *heal by scarring* and may progress to chronic inflammation. Sometimes, it is common for a mixed acute and chronic response to coexist. Acute inflammation itself can damage otherwise healthy cells, which could then further stimulate inflammation. A number of local or systemic signaling molecules are released to which the immune system responds by recruiting more cellular components. These cellular components release cytotoxic or degradative molecules to destroy the material. Persistent activation of the immune system is deleterious: the released products can destroy the neighboring healthy tissue and chronic macrophage activation culminates in the fibrotic scar tissue formation, where increased amounts of fibers are present. Scar tissue is not as vascularized as original tissue and it is not as flexible, elastic, or strong as the original tissue. Scar tissue formed in a muscular organ such as the heart or bladder may inhibit function. Scar tissue may adhere to adjacent organs, which could lead to loss of function of ligaments and tendons. Formation of scar tissue near a joint compromises the joint mobility.

Consider the case of total hip replacement prosthetic devices. Debris created by the wear of implant-articulating surfaces is attacked by macrophages. Macrophages can respond directly to wear particles released from the surface. Phagocytosis of wear particles by macrophage induces signaling mechanisms, which release inflammatory mediators [Figure 6.6(b)]; integration of signaling events triggers a dynamic process of cytoskeletal rearrangement accompanied by membrane remodeling at the cell surface that leads to engulfment. The time scale for these interactions begins within seconds of implantation and can continue through many years. Some fibroblasts may also phagocytose wear debris. The mechanism of uptake and dealing with wear debris is identical to the way that phagocytic cells ingest and kill bacteria.

Macrophages gather and ingest foreign substances and introduce these substances (antigens) to T cells and B cells of the immune system for appropriate action. However, the immune response is not effective against the plastic particles and cells may continue to respond. Chronic activation of macrophages occurs through the components of *adaptive immune response* (i.e., T cells and B cells). For example, activated T cells produce stimulatory molecules, which influence a number of immune compartments including the activation of more innate immune cells. However, T cells are normally quiescent and cannot directly recognize intact foreign materials; their activation and subsequent clonal expansion is dependent on proper presentation by antigen presenting cells (APCs that includes macrophages and dendritic cells). This continuous response is deleterious as the released cellular mediators affect cells present in the surrounding tissue including bone such as fibrobalsts, osteoblasts, and osteoclasts. Activation of osteoclasts in the bone may lead to bone resorption, which eventually leads to bone loss (osteolysis) next to the implant. Bone loss contributes to the loosening of implants. Thus, immune response can affect the performance of the graft.

6.4.5 Infection

Biomaterial-associated infection is the second most common cause of implant failure. The infecting micro-organisms are either introduced during implantation of a prosthesis device or are carried to the biomaterial surface by a temporary infection of the patient. Although sterilization techniques (described in Section 6.3.5) are used to prevent bacterial, fungal, and viral contamination, implanted biomaterials are more vulnerable to microbial colonization. Surfaces well colonized by healthy tissue cells tend to resist infection by virtue of cell membranes and eukaryotic extracellular polysaccharides. The high incidence of infection occurs because host defenses in the vicinity of the prosthesis are generally impaired in the absence of cells. Micro-organisms arriving at the surface, through introduction with the device or transient infection, are not eliminated by the immune system. Infection of a prosthetic implant will almost always result in reoperation, amputation, or death. Combined rates of death or amputation from infected cardiac, abdominal, and extremity vascular prosthesis may exceed 30%.

Adhesion-mediated infections develop that are resistant to antibiotics and host defenses and tend to persist until the biomaterial or foreign body is removed. Clinical retrievals of prosthetic devices indicate that a few bacterial species such as *Staphylococcus epidermidis* and *Staphylococcus aureus*, dominate biomaterial-centered infections. However, other bacterial species including *Escherichia coli*, *Pseudomonas aeruginosa*, *Proteus microbilis*, *b-hemolytic streptococci*, and *enterococci* have also been isolated. Bacteria are surface-adherent organisms and 99% of their biomass exists on surfaces rather than in floating forms. The hemodynamic interactions required within a device create fluid eddies and tissue damage that are favorable to clotting cascades and the initial events of microbial adhesion. Surface disruption by wear, corrosion, trauma, toxins, biosystem chemical degradation, or bacterial mechanisms establishes appropriate environmental conditions for opportunistic bacteria within a microenvironment.

6.4.5.1 Theories of Bacterial Adhesion

When biomaterials are introduced into a patient, their surface may become an adhesive site for bacterial colonization. The process of bacterial adhesion is biochemically parallel, and may be competitive or mutually exclusive to the process of tissue integration. A number of interactions may occur depending on the specificities of the bacteria or biomaterial surface characteristics. The process of bacterial colonization can be broadly classified into attachment, adhesion, agregation, and dispersion at the substratum surface (Figure 6.10). A bacterium swims towards the prosthetic material using polar flagella and forms random loose attachments to the prosthetic material. Initial attachment depends on the general long-range physical characteristics of the bacterium, the fluid interface, and the substratum. Subsequent to attachment, specific irreversible adhesion could occur as a time-dependent chemical process involving chemical binding, hydrophobic interactions, and interactions

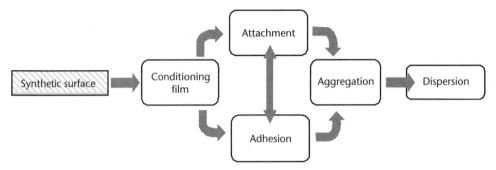

Figure 6.10 Steps involved with the bacterial interaction with the biomaterial surface [10].

through specific recognition molecules present on the surface of the bacterium. Adhesive molecules interact with surfaces at distances greater than 15 nm, bridging repulsive forces.

Surfaces provide an interface for the concentration of charged particles, molecules, and nutrients from mineral or organic sources, or they themselves may be metabolized. When environmental conditions are favorable, bacteria propagate and secrete polysaccharides to the extracellular environment. These polysaccharides form a slime, which is a loose amorphous material composed of a range of low and high molecular weight polymers associated in large part through ionic interactions. Extracellular polysaccharide is an important factor in the development and persistence of biomaterial-centered infections. This complex exopolysccaride is believed to act as an ion-exchange resin for enhanced nutrition, to interfere with phagocytosis, to influence response to antibodies, and to function in later stages of surface adhesion, aggregation, and polymicrobial interaction. Micro-organisms in colonies on surfaces form layers two to hundreds of organisms thick composed of cellular material, extracellular polysaccharides, environmental adsorbates, and debris. This 3D surface composite is called the biofilm or slime. Biofilm acts as a protective barrier from antimicrobial agents and host defenses. The biofilm mode of growth offers enhanced protection against natural host defenses and antibiotic therapy. Further, the microzones, metal ions required by pathogenic bacteria are not lost by diffusion and may be shielded from host protein-binding complexes. When environmental conditions become unfavorable, some bacteria detach and swim away to find a new surface in a more favorable environment. To avoid these complications, the design of many implants requires, in addition to general biocompatibility, that a sector of the prosthesis or artificial organ be colonized by host cells to provide secure fixation or integration.

6.5 Tissue Engineering

Biodegradable porous scaffolds are used as templates (or scaffolds) in tissue engineering, an alternative way to generate replaceable tissue parts, which minimize the complications presented by bioinert materials without cells. From a biomaterials point of view, the scaffolds are synthesized from materials that support the growth of cells and that are biodegradable. The porous scaffold may be used directly, or after for cell growth in vitro (Figure 6.11) that is degraded by the growing cells prior to implantation. The scaffold can also be formulated to contain bioactive agents for more rapid tissue growth or compatibility. Regenerated tissues outside the body are also useful in toxicology tests, and understanding the progression of a pathological condition that is not clearly understood. Scaffolds of required shape are generated, which guide and support the in-growth of cells. While cells grow and synthesize their own matrix at the site of grafting or in vitro, the scaffold will degrade leaving only the necessary healthy tissue in a topologically required form.

In many applications, where tissue functionality is critical, developing tissues outside the body via cell colonization prior to transplantation is the preferred *modus operandi*. For example, porous leaky materials cannot be grafted without cell colonization in the cardiovascular systems. The absence of cellular components typically leads to blood clotting, an overgrowth of cells that clog blood flow, and to the subsequent failure of the graft. Another example is the liver cells, the metabolic hub of the body. If liver has to be replaced, the replaced tissue should have the ability to provide critical metabolic functions. For cells to colonize the scaffolds successfully, various molecular events have to be carefully programmed to regulate cellular bioactivity such as adhesion, spreading, migration, proliferation, and functionality, which are discussed in Chapter 7. Here, the importance of scaffold characteristics and processing techniques are discussed.

6.5.1   Material Selection

The selection of biocompatible materials for scaffold formation includes considerations that go beyond nontoxicity to bioactivity as it relates to interacting with and, in time, being integrated into the biological environment as well as other tailored

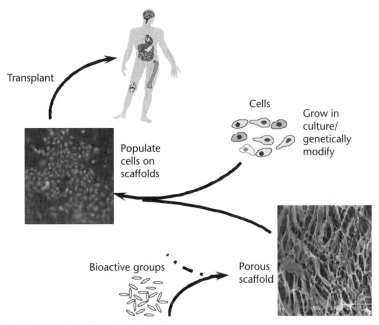

Figure 6.11   Concept of engineering tissue/organ.

properties depending on the specific application. Further, the degradation products define the biocompatibility of the material used, not necessarily the material itself. Scaffolds generated from natural and synthetic polymers or biological materials from xenogeneic sources have been used. The batch-to-batch variation in the properties of nature of biological materials limits the wide spread usage in clinical settings. Thus, synthesizing scaffolds using various polymer processing techniques (described in Section 6.5.2) has gained more attention. The general criteria for selecting polymers are to match the mechanical properties and the time of degradation to the needs of the application. Synthetic and natural polymer-based scaffolds have been formed and used with and without prior cell-seeded configurations. Ideally, tissue engineering scaffolds should meet at least the following requirements:

1. Polymer and its degraded products should not either show cytotoxicity or initiate inflammation.
2. The processing of material into a 3D scaffold should allow the formation of required pore sizes and pore orientations and should be reproducible.
3. Surface should permit appropriate cellular in-growth.
4. Scaffold should provide a mechanical property similar to the native tissue to be replaced.
5. The scaffold should have interconnected open pores and high porosity to provide adequate space for cellular proliferation, transport of nutrients/metabolic waste and de novo deposition of the matrix components.
6. Scaffold should resorb gradually and should be replaced by cells and cellular components completely without leaving any material at the end of the tissue growth.
7. Degradation rate should be amendable so that cell growth kinetics and biomaterial degradation rate can be synchronized to ensure no space restriction due to slower degradation or the loss of structural support due to faster degradation.

Among different classes of biodegradable synthetic polymers, polyesters such as polylactide (PLA), polyglycolide (PGA), polycaprolactone (PCL), and their copolymers have been extensively employed as biomaterials due to their favorable biodegradability and tunable mechanical properties. For example, PGA was used to develop as a first synthetic absorbable suture. PLA products including tissue screws, tacks, and suture anchors, as well as systems for meniscus and cartilage repair. These polymers degrade by hydrolysis (i.e., nonenzymatically); their degradation rates and mechanical properties can be altered via co- and graft-polymerization techniques and processing conditions. For example, a block copolymer of caprolactone with glycolide, offering reduced stiffness compared with pure PGA, can be generated. Degradation of these materials yields the corresponding hydroxy acids. However, formed degradation products are relatively strong acids, which raise concerns about possible inflammation; also, the scaffold microenvironment may not be ideal for tissue growth. Further, scaffolds may show structural instability due to swelling during degradation. Synthetic polymers do not possess a surface chemistry that is familiar to cells (i.e., extracellular matrix elements). Hence, many synthetic polymers lack biological activity suitable for cell colonization. Tests have been conducted to improve these limitations via copolymerization and grafting signaling molecules, or immobilizing bioactive molecules into the synthetic system. For example, grafting a small peptide arginine-glycine-aspartic acid (RGD) onto polymers is an approach taken to incorporate bioregulation of matrix elements. The use of RGD is based on the understanding that the majority of cellular communication with outside takes place via integrins, a family of transmembrane receptors. A number of new materials are developed by abstracting good design from the natural world. Synthetic poly(amino acids) were developed, however, high crystallinity makes them difficult to process and results in relatively slow degradation. Further, concerns related to immune response towards polymers with more than three amino acids in the chain also makes them inappropriate for tissue regeneration. Modified "pseudo" poly(amino acids) have been synthesized by using a tyrosine derivative. Tyrosine-derived polycarbonates are high-strength materials that may be useful as orthopedic implants.

Involvement of an extracellular matrix (ECM) in tissue remodeling under pathological conditions and their role in diverse molecular mechanisms have been extensively studied. Based on this concept, purified components such as collagen (and gelatin, which is a denatured collagen) and glycosaminoglycans (GAGs) have been investigated for generating scaffolds and tissues. Other natural polymers such as alginates, chitosan, and their various combinations have also been used as scaffolding materials. A commonly used system is collagen/GAGs. Collagen/GAG-based skin equivalents are in clinical use. However, using GAG and collagen components together may not be suitable for applications where only one component exhibits the required biological function. Due to the restricted processing characteristics of GAGs, an approach to using GAGs alone is to form an ionic complex with chitosan by electrostatic interactions. However, weak mechanical strength, inadequate tailorability options in altering mechanical and degradation properties limit their usage. Blending both synthetic and natural polymers is also an option to obtain polymers with tailorable mechanical and biological properties.

6.5.2 Scaffold Formation Techniques

The 3D scaffolds provide physical cues of porous structures, mechanical strength to guide cell colonization, and chemical cues for cell-binding sites to support cell attachment and spreading. Unlike prosthetic materials, these porous materials require a different set of processing conditions. For example, porous structures with an optimum pore size range for supporting cell ingrowth for a majority of the mature cell types (with few exceptions) is in the range of 100–$200\ \mu$m. Many cells are unable to completely colonize scaffolds with pore sizes larger than $300\ \mu$m due to the difficulty in crossing large bridging distances. In addition to pore size, the topography of scaffold surfaces influences spreading characteristics and activity of cells. Scaffold formation techniques can be grouped into two categories: additive processes and subtractive processes.

Additive processes include self-assembled monolayer techniques and free form fabrication where matrices are assembled using fundamental building blocks. Self-assembled monolayers (SAMs) can be prepared using different types of molecules and different substrates. Self-assembled 2D layers of proteins are of particular interest for the fabrication of biomaterials. Such layers play important roles in biological mineralization by controlling the size, orientation, and morphology of inorganic crystals at the surface of the protein layer. This system is broadly studied and widely believed to serve as the platform of choice to develop a variety of biological technologies.

Rapid prototyping is the name given to a host of related technologies that are used to fabricate physical objects directly from digital data sources. These methods are unique in that they add and bond materials in layers to form objects. Such systems are also called solid freeform fabrication, layered manufacturing, stereolithography, selective laser sintering, fused deposition modeling, laminated object manufacturing, inkjet-based systems, and 3D printing. These names are often used as synonyms for the entire field of rapid prototyping; however, each of these technologies has its strengths and weaknesses. The general process begins with developing a computer-generated model using computer-aided design (CAD) software. Potentially, one could use digital images obtained by computer tomography or magnetic resonance imaging (discussed in Chapter 9) scans to create a customized CAD model. This CAD model is then expressed as a series of cross-sectional layers. The data is then implemented to the rapid prototyping machine, which builds the desired object a layer at a time, similar to laser printing (Figure 6.12). The material could be self-adhesive or one could use a small quantity of an adhesive. Once one layer is completely traced, it is lowered by a small distance and a second layer is traced right on top of the first. The adhesive property of the materials used cause layers to bond to one another and eventually form a complete, three-dimensional object after many such layers are formed. Objects can be made from multiple materials as composites, or materials can even be varied in a controlled fashion at any location in an object. Although objects can be formed with any geometric complexity or intricacy without the need for elaborate final assembly, few limitations include manipulating the porosity of the scaffold, and the type of material that can be printed.

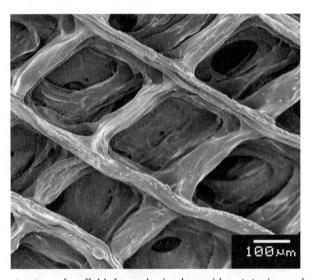

Figure 6.12 Microstructure of scaffolds formed using by rapid prototyping and chitosan solution.

Subtractive processes include the particulate-leaching technique, controlled rate freezing and lyophilization technique, and laser ablation techniques. In these processes, materials are removed from the assembled matrices to obtain porous structures.

Particulate leaching is a process whereby the internal architecture is determined by embedding a high density of particles such as salt crystals into a dissolved polymer or ceramic. The dissolved mixture is then poured into a mould and treated under heat and pressure to form the external shape. The particles are subsequently leached out to leave pores. Although pores of uniform size can be obtained by using particles of same size, nonporous layers are formed during different steps of the generation process, which make controlling the pore size difficult. Due to the transport limitations of the leaching technique, obtaining thick samples is cumbersome in addition to the difficulty in uniformly distributing the particles within the matrix. Scaffolds produced by solvent-casting particulate leaching cannot guarantee the interconnection of pores because this is dependent on whether the adjacent salt particles are in contact.

The controlled rate freezing and lyophilization technique involves introducing phase changes in a homogeneous polymer solution by freezing solvent into crystals followed by sublimation of solid crystals into vapors (Figure 6.13). This is the method of choice while forming scaffolds from natural polymers such as collagen, which typically dissolve in acidic water. One advantage of the process is that it is carried out at low temperatures and avoids the loss of biological activity of proteins or other conjugates due to thermal denaturation. Hence, bioactive molecules can also be incorporated during the fabrication process. Also, custom shaped scaffolds can be easily synthesized by freezing polymer solutions in appropriate molds. Although pore sizes can be controlled to a certain degree by altering the initial freezing temperature, the pore sizes show distribution, and controlling pore geometry is difficult.

There are a number of other techniques based on the textile industry to form porous structures. These include electrospinning, fiber spinning, and extrusion

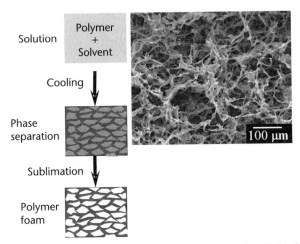

Figure 6.13 Controlled rate freezing and drying. Microstructure of scaffolds formed using a mixture of chitosan and gelatin.

techniques. All synthetic polymers can be melted or processed by conventional means such as injection molding, compression molding, and extrusion. In addition, alterations in pressure of the system to create porous structures have also been used to form 3D scaffolds.

Problems

6.1   What are the advantages and disadvantages of using metals as biomaterials?

6.2   What is the major advantage of using ceramics as a biomaterial?

6.3   List several physical properties that are characteristics of polymers. What are the major problems of using polymers as biomaterials?

6.4   What should be the criteria in selecting a particular material for biomedical applications?

6.5   You are trying to develop a new coating for a prosthetic application. What primary interactions will you consider if you were to test the compatibility of the material for the application? Write only the five tests.

6.6   Magnesia partially stabilized zircona bioceramic exhibits high mechanical strength, excellent corrosion resistance and minor host response. Thus, it is suitable for orthopedic and dental applications. However, low adherence with the body fluids limits its use. Hao et al. [11] investigated using a CO_2 laser to alter the surface structure. They report the following contact angle data.

Test Liquid	Glycerol	Formamide	Etheneglycol	Polyglycol E-200	Polyglycolycol 15-200
Untreated	79	73	61	53	35
1.6 kW/cm^2 CO_2 laser treated	40	36	29	35	19

Determine whether surface energy changed significantly to make the surface more wettable.

6.7   Hao et al. [12] studied the modifications of a 316LS stainless steel surface following the irradiation of a CO_2 laser and a high-power diode laser (HPDL). They report the following contact angle measurements in various liquids. Using these values, they calculated the adhesive work towards SBF to be 72.5 mJ/m^2 and SBF+BSA to be 54.9 mJ/m^2. Check their calculations and compare the effect of the surface treatment.

Test Liquid	Distilled Water	Glycerol	Simulated Body Fluid (SBF)	SBF+BSA
Untreated	89.0 ± 1.3	73.8 ± 0.8	82.9 ± 0.8	61.4 ± 0.2
CO_2 laser	77.8 ± 1.9	67.9 ± 1.1	73.2 ± 1.6	57.1 ± 0.4
HPDL laser	83.1 ± 0.3	70.5 ± 0.5	76.4 ± 1.3	59.4 ± 1.1

6.8   How does the inflammation influence the biomaterial?

6.9   A research group worked on using a novel surface treatment method to enhance surface hydrophilicity. How would they know if the surface was indeed chemically modified by the plasma treatment?

6.10   BioX has developed a new prosthetic device but they do not know how to sterilize it. They want a summary of different available options so that they can compare methods. Please summarize different available options while highlighting the deciding factor.

6.11   What differentiates biomaterial-centered infections from the common infections?

6.12   State two main differences between material selections for manufacturing a prosthetic device and tissue engineering scaffold.

6.13   Describe different processing techniques available for generating porous structures.

References

[1]   Williams, D. F., *Definitions in Biomaterials*, New York: Elsevier, 1987.

[2]   Protek, AG, Technical brochure, Switzerland, 1994.

[3]   Schoen, F. J., and R. J. Levy, "Tissue Heart Valves: Current Challenges and Future Research Perspectives," *J. Biomed. Mater. Res.*, Vol. 47, 1999, pp. 439–465.

[4]   The Center for Disease Control and Prevention, http://www.cdc.gov/ncidod/dhqp/bp_sterilization_medDevices.html.

[5]   Ratner, B. D., et al., *Biomaterials Science*, 2nd ed., San Diego, CA: Academic Press, 2004.

[6]   Yan, J. M., and H. T. Lin, "Wettability and Protein Adsorption on HTPB-Based Polyurethane Films," *Journal of Membrane Science*, Vol. 187, No. 1–2, 2001, pp. 159–169.

[7]   Kaelble, J., "A Surface Energy Analysis of Bioadhesion," *Polymer*, Vol. 18, 1977, p. 475.

[8]   Lawrence, J., et al., "A Two-Stage Ceramic Tile Grout Sealing Process Using a High Power Diode Laser-Grout Development and Materials Characteristic," *Optics and Laser Technology*, Vol. 30, No. 3, 1998, pp. 205–214.

[9]   Wang, X. H., et al., "Improvement of Blood Compatibility of Artificial Heart Valves Via Titanium Oxide Film Coated on Low Temperature Isotropic Carbon," *Surface and Coatings Technology*, Vol. 128–129, 2000, pp. 36–42.

[10]   Gristina, A. G., "Biomaterial-Centered Infection: Microbial Adhesion Versus Tissue Integration," *Science*, Vol. 237, No. 4822, 1987, pp. 1588–1595.

[11]   Hao, L., and J. Lawrence, "CO_2 Laser Induced Microstructural Features in a Magnesia Partially Stabilised Zirconia Bioceramic and the Effects Thereof on Wettability Characteristics," *Materials Science and Engineering: A Structural Materials: Properties, Microstructure and Processing*, Vol. 364, 2003, pp. 171–181.

[12]   Hao, L., J. Lawrence, and L. Li, "The Wettability Modification of Biograde Stainless Steel in Contact with Simulated Physiological Liquids by the Means of Laser Irridation," *Applied Surface Science*, Vol. 247, No. 1–4, 2005, pp. 453–457.

Selected Bibliography

Dee, K. C., et al., *An Introduction to Tissue-Biomaterial Interactions*, New York: John Wiley & Sons, 2002.

Lanze, R., R. Langer, and J. Vacanti, *Principles of Tissue Engineering,* 3rd ed., San Diego, CA: Academic Press, 2007.

Lawrence, J., "Identification of the Principal Elements Governing the Wettability Characteristics of Ordinary Portland Cement Following High Power Diode Laser Surface Treatment," *Materials Science and Engineering A*, Vol. 356, No. 1–2, 2003, p. 15.

Ma, P. X., and J. Elisseeff, (eds.), *Scaffolding in Tissue Engineering*, Boca Raton, FL: CRC Press. 2005.

Park, J. B., and J. D. Bronzino, (eds.), *Biomaterials: Principles and Applications*, Boca Raton, FL: CRC Press, 2002.

Ramakrishna, S., et al., *An Introduction to Biocomposites,* London, U.K.: Imperial College Press, 2004.

Saltzman, M. W., *Tissue Engineering: Engineering Principles for the Design of Replacement Organs and Tissues*, Oxford, U.K.: Oxford University Press, 2004.

Schweitzer, P. A., *Corrosion of Polymers and Elastomers,* 2nd ed, Boca Raton, FL: CRC Press, 2006.

Vadgama, P., (ed.), *Surfaces and Interfaces for Biomaterials,* London, U.K.: Woodhead Publishing Limited, 2005.

Yoneyama, T., and S. Miyazaki, (eds.), *Shape Memory Alloys for Biomedical Applications* London, U.K.: Woodhead Publishing Limited, 2008.

CHAPTER 7

Cellular Engineering

7.1 Overview

Cellular activities in the body depend on coordinated signaling mechanism(s) from the extracellular environment. Upon receiving a signal, some cells secrete regulatory molecules such as hormones, neurotransmitters, digestive enzymes, or antibodies into their surrounding extracellular fluid. Each event is highly regulated with a precise signaling pattern. For example, the secretion of specific antibodies in the presence of a foreign material is an important immune response mechanism. If antibodies are expressed without any foreign body, it could lead to autoimmune diseases. To understand and utilize the cellular activity, technology has been developed to harvest cell populations of interest, culture them in a regulated environment, and expose them to various conditions. Evaluating cellular interactions in an isolated environment minimize the complexities of studying in the entire body. Exploring cellular activity outside the body is referred to as an in vitro (Latin for "in glass") system as opposed to an in vivo (Latin for "in the organism") system. A basic understanding of cellular interactions and processes is important for various applications (Figure 7.1).

Significant advances have been made in successfully culturing many mammalian cells with optimum environment. Cells can be cultured in large quantities, which can be utilized to relate function and pathology to a genomic sequence, to regenerate an artificial tissue, to detect a biological signal, to test the toxicity of a chemical, or to collect cell-derived products for use as therapeutic molecules. Developing therapeutic molecules such as antibodies and anti-inflammatory products by genetically engineering a proliferating cell population has offered the possibility of targeted treatment. Furthermore, the intracellular metabolic fluxes can be assessed by the measurement of extracellular metabolite concentrations in combination with the stoichiometry of intracellular reactions. Using the metabolic flux analysis, one could manipulate a metabolic pathway to produce a desired product in that pathway. Success in bone marrow transplantation and other cellular therapies to treat various diseases has opened up opportunities to address critical medical needs. In addition, some cells and tissues can be preserved without compromising their proliferative capacity, offering an easy access to eternal source. In this chapter, basic cellular interactions and cellular processes are described. Furthermore, how

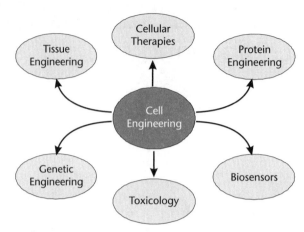

Figure 7.1   Uses of cellular engineering.

various parameters that define the cellular environment are manipulated and modeled is discussed.

7.2   Cellular Interactions

7.2.1   Cell Culture Microenvironment

A cell culture environment is designed to mimic human homeostasis that is critical for survival, growth, and the possible expression of the specific function of cells. A basic culture medium can be formulated to control the pH, temperature, and osmolarity within a narrow range of physiologically acceptable conditions. Since cells need nutrients for energy and metabolism, the medium also provides carbon and nitrogen sources. In addition, several small molecules and critical vitamins that are essential for cell viability are incorporated into the medium. Depending on the tissue of origin, each type of cell will have different requirements for culture. Cells will respond in different ways to signals from their environment. Hence, more sophisticated microenvironments can be obtained by altering four major components (Table 7.1) in addition to this basic medium.

Polystyrene is the most commonly used substrate for cell culture as it is easy to mold into a required shape, relatively inexpensive, and transparent (or optically clear). Furthermore, the hydrophobic (nonwettable) surface of polystyrene to which cells have difficulty attaching can be easily modified by a variety of chemical (e.g., sulfuric acid) and physical (e.g., corona discharge, gas-plasma or irradiation) methods. Commercial tissue culture plasticware is made of polystyrene that has been exposed to an ionized gas generated by an electric arc. The ionized gas reacts with the polystyrene surface and changes the uncharged hydrophobic surface into a more ionic hydrophilic surface. When serum is added to a culture medium, some of the serum proteins adsorb to the surface of the treated plastic and cells readily attach to these adsorbed proteins. In particular, fibronectin and vitronectin found in the serum are important for cell attachment. When using a serum-free medium, the plastic tissue culture dishes are sometimes coated with ECM molecules (e.g.,

Table 7.1 Components of Cell Culture Microenvironment

Component	In Vivo Environment	In Vitro Strategy for Cultured Cells
Soluble factors	Cells are exposed to numerous systemic factors in circulation and those produced locally	Add serum, bioextracts, purified hormones, and growth factors
Matrix elements	Cells are surrounded by complex extracellular matrix elements	Add purified ECM proteins and culture cells in a three-dimensional matrix
Influence of other cells	Presence of other cells regulate via secreted molecules (homotypic or heterotypic)	Add growth factors, adjust cell seeding density, and coculture with other cell types
Physical factors	Cells are subjected to numerous physical forces	Expose cells to physical forces, fluid shear stress, mechanical stretch, and low oxygen tension

fibrinogen) prior to seeding the cells. For some cell types, it is not necessary to coat the dish with ECM proteins as the cells will attach and spread on the charged surface of the plastic. Once attached, many cells will begin to synthesize and secrete their own insoluble ECMs onto the plastic surface. Glass surfaces can also be used instead of plastic and promote cell adhesion by the same mechanisms.

Cells are continuously exposed to diverse stimuli ranging from soluble factors to signaling molecules on neighboring cells [Figure 7.2(a)]. By adjusting the concentration of some of the basal media constituents or by adding new components discovered during a disease state, it is possible to effect significant changes in the cultured cells. For example, physicochemical modifications of ECM are recognized by the cells and modify cellular functions which, in turn, modify the ECM itself. In an effort to produce a tissue-engineered blood vessel, vascular endothelial cells are cultured on the inner lumen of a tube-like structure that is surrounded by fibroblasts and smooth muscle cells in concentric circles. Although this complex culture regenerates a reasonable blood vessel, formed structures have very low strength and burst open at very low fluid pressures than those seen for in vivo. If vitamin C is added to the culture, the resulting vessel shows improved strength and can withstand a tenfold increase in fluid pressure. Vitamin C, among other activities, is an important cofactor for a component that cross-links collagen fibers, and this may

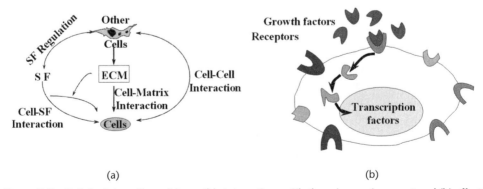

(a)   (b)

Figure 7.2 Cellular interactions: (a) possible interactions with the microenvironment and (b) effect of growth factor interaction.

be the explanation for the improvement. While culturing, it is important to provide extracellular signals appropriate for the cell in order to achieve an optimal function or proliferative response. ECM molecules also interact with soluble factors, altering their bioactivity and transmitting signals across the cell membrane to cytoplasmic molecules that, in turn, initiate cascades of events involving cytoskeleton and nucleus. These "nuclear events" affect specific gene expression that reciprocally regulates ECM structures and contents.

7.2.2 Cell-Soluble Factor Interactions

Soluble factors play an important role in various functions including cell growth and the secretion of other soluble factors. In general, soluble factors are proteins, show very high specificity in interactions, are members of a family of related proteins, and are pleiotropic (i.e., are able to mediate a variety of actions on different cells). *Autocrine signaling* is a way that cells self-regulate their extracellular environment by secreting soluble factors, which in turn affects the way that the cell functions. Some cells regulate their growth by this signaling. *Paracrine signaling* occurs when soluble factors released by cells are taken up by neighboring cells. Paracrine signals are short-lived with local effects and play an important role in early development. *Hormonal signaling* occurs due to the released signaling molecule from the endocrine glands being collected and distributed via bloodstream. *Synaptic signaling* occurs through the nervous system via neurons producing neurotransmitters close to the target cells. Synaptic signals persist briefly. The serum provides a good source of soluble factors, of which only a few are known such as vitamins and hormones. As a serum is formed, the platelets in the whole blood release their soluble factors, many of which are called the *wound healing growth factors*. With the advent of gene cloning and recombinant DNA technology, many of the genes encoding growth factors have been identified, and these genes are used to produce large amounts of highly purified growth factors that are added to the cell culture medium. Some soluble factors that play an important role in influencing the immune system are referred to as cytokines.

Soluble factors typically transduce the signal into the nucleus by binding to other proteins, for which the name *receptors* is in common use. During these interactions, soluble factors are referred to as *ligands*. Small hormones such as steroid hormones readily diffuse across the plasma membrane and bind with cytosolic receptors or directly enter the nucleus and then bind to intranuclear receptors. These activated receptors act like transcription factors, leading to the production of mRNA and the protein synthesis or an increased rate of secretion of specific molecules. These types of soluble factors are called *first messengers* because they do not depend on other molecular interactions that may be required to achieve the end results. However, many soluble factors do not readily diffuse across the plasma membrane. They exert their influence on intracellular processes by interaction with cell surface receptors [Figure 7.2(b)]. In these cases, binding activates the receptors, which initiate a sequence of biochemical reactions inside the cell to produce other messengers that will transfer the signal to alter necessary function. These are called *second messengers,* which include cyclic adenosine mono phosphate (cAMP), cyclic guanosine monophosphate (cGMP), diacylglycerol (DAG), inositol triphosphate

(IP3), and Ca^{2+} ions. Once activated, receptors could function as ion channels (Chapter 4), enzymes (i.e., protein kinases: transfer a phosphate group from ATP to a substrate), or G-protein (i.e., guanine triphosphate binding proteins) activators. An understanding of ligand (growth factor or cytokine, hormones, drugs, and toxins) interactions with the respective receptor is critical to numerous research areas including embryonic development, gene therapy, separation of cell populations, inflammation, designing new ligands, and regulating cellular activity. The influence on physiological changes depends on the type of receptor-ligand interactions.

Some soluble factors referred to as *enzymes* catalyze specific chemical reactions converting a substrate to a product. Specificity relies on the virtue of the enzymes' exquisite three-dimensional structure, and enzymes recognize differences of a single atom in the substrate's chemical structure. Enzymes are involved in many functions including: (1) metabolism (see Section 7.3.3) inside a cell or outside a cell, (2) synthesizing many chemicals needed by the cell including other enzymes, and (3) degrading other proteins that have served their purpose, thus recycling the components (see Section 7.3.4). Enzymes are also involved in processes such as blood clotting (see Chapter 6). Enzyme-substrate interactions are treated similarly to receptor-ligand interactions. Furthermore, antibody-antigen interactions (see Chapter 9) are also treated similarly. Antibodies are immune system-related proteins called immunoglobulins, produced primarily in response to the entry of unfamiliar materials into the body.

Since receptor-ligand interactions regulate many events necessary for maintaining homeostasis, understanding these interactions, engineering their production, and predicting the cellular response are of significant importance. Receptor-ligand binding, the nature of enzyme-catalyzed reactions, and engineering antibodies with high affinity to a desired target are important to develop new products and therapies. Furthermore, the magnitude of the cellular response that results from receptor-ligand binding may not be directly proportional to the ligand concentration. Hence, one has to understand the dosage requirements to generate a necessary signal. Evaluating binding characteristics when one receptor binds to one ligand is described later.

7.2.2.1   Single Site Binding

Binding occurs when a ligand and a receptor collide due to diffusion and when the collision has the correct orientation. During the interaction the ligand remains bound to the receptor for a finite period of time, attributed to the attractive force between them. The law of mass action is the origin for quantitative treatment of chemical equilibrium in receptor-ligand interactions. It is obtained from either a thermodynamic point of view as the driving force of a chemical reaction is directly proportional to the active masses of each of the reactants or a kinetic point of view as the rate of a chemical reaction is directly proportional to the active masses of each of the reactants. The assumptions of law of mass action are that:

- The driving force is directly proportional to the active mass of a reactant (all receptors are equally accessible to the ligands).

- The driving force of each mole of each reactant acts independently of any other (binding does not alter the ligand or receptor).

These interactions are reversible and at chemical equilibrium, the total driving force(s) of the reactant(s) is equal to the total driving force(s) of the product(s).

$$R + L \underset{k_{off}}{\overset{k_{on}}{\rightleftharpoons}} R \cdot L$$

This type of interaction is commonly referred as a *noncooperative interaction*. The rate of association is:

Number of binding events per unit of time = k_{on}[Ligand][Receptor]

The probability of dissociation is the same at every instant of time. The receptor does not know how long it has been bound to the ligand. The rate of dissociation is:

Number of dissociation events per unit of time = k_{off} [ligand • receptor]

After dissociation, the ligand and receptor are the same as at they were before binding. If either the ligand or receptor is chemically modified, then the binding does not follow the law of mass action. The time rate of change of the ligand/receptor complexes

$$\frac{d[R \cdot L]}{dt} = k_{on}[R][L] - k_{off}[R \cdot L] \tag{7.1}$$

From this expression, a time-dependent analytical solution can be obtained if rate constants are available. When equilibrium is reached, that is, the rate at which new ligand•receptor complexes are formed equals the rate at which the ligand•receptor complexes dissociate, then,

$$\frac{d[R \cdot L]}{dt} = 0 \tag{7.2}$$

Rearrange (7.1) to define the equilibrium dissociation constant K_D.

$$K_D = \frac{k_{off}}{k_{on}} = \frac{[R][L]}{[R \cdot L]}$$

The K_D has a meaning that is easy to understand. From the K_D value, free energy in the ligand/receptor binding can be calculated using

$$\Delta G^0 = RT \ln K_D \tag{7.3}$$

where T is the temperature and R is the gas constant. A small K_D value corresponds to a large negative free energy (i.e., the reaction is thermodynamically favored). If the receptors have a high affinity for the ligand, the K_D will be low, as it will take a low concentration of ligand to bind half the receptors. K_D spans a wide range from 10^{-6} (low affinity) to 10^{-15} M (high affinity or tight binding). However, (7.1) is not of practical significance as it requires the concentration of unoccupied receptor, $[R]$, at any given time. Alternatively, assuming that the receptors do not allow for more than one affinity state or states of partial binding, the fractional receptor occupancy (fraction of all receptors that are bound to ligand) at equilibrium as a function of ligand concentration can be determined. The ligand bound receptor is called, *Bound* (B) and the unbound ligand is called *Free*, L_0. Both of these are measurable experimentally. Substituting these measurable terms into (7.2),

$$\frac{L_0 [R]}{B} = K_D \tag{7.4}$$

Each cell has a specific number of receptors, for a specific ligand, at a given physiological state. Receptors are either free or bound to the ligand. In other words,

Total receptors = free receptors + receptors bound to the ligand

For any given cell, the total receptors present can be termed as the maximal number of binding sites, B_{max}. Hence,

$$B_{max} = [R] + B \tag{7.5}$$

Hence (7.4) can be rewritten as

$$\frac{L_o (B_{max} - B)}{B} = K_D \tag{7.6}$$

This can be rearranged to obtain

$$B = \frac{L_0 . B_{max}}{K_D + L_0} \tag{7.7}$$

It is possible to estimate the B_{max} and K_D from (7.7). Plotting (7.7) gives a hyperbolic curve, which is a typical response of many biological reactions. This type of equation is popularly referred to as the Michaelis-Menten equation and is often valid for many reactions including transport of substrates, enzyme-catalyzed reactions, and other drug-related effects. In enzyme-catalyzed reactions, (7.7) is rewritten as

$$V = \frac{S.V_{max}}{K_M + S} \tag{7.8}$$

where V is the reaction velocity, S is the substrate concentration, V_{max} is the maximum reaction velocity, and K_M is the Michaelis-Menten constant (defined as the concentration at which the rate of the enzyme reaction is half of V_{max}).

A common approach used to determine the rate constants is linearizing (7.7) and plotting the coordinates. There are a variety of algebraically equivalent ways to linearize (7.7) and plot, including the Lineweaver-Burk (or the double reciprocal equation), Eadie-Hofstee, Wolff, and Scatchard-Rosenthal plots. With perfect data, all yield identical answers, yet each is affected more by different types of experimental error. A commonly used method in receptor-ligand interactions is the Scatchard plot (Figure 7.3), where (7.7) is linearized as

$$\frac{B}{L_0} = \left(-\frac{1}{K_D} \right) B + \frac{B_{max}}{K_D}$$

Then the X-axis is specific binding (B) and the Y-axis is the ratio of specific binding divided by free-to-bind ligand concentration (B/L_0). From the plot, K_D is the negative reciprocal of the slope and B_{max} is the X intercept. The Lineweaver-Burk plot and the Eadie-Hofstee plot are more common enzyme-catalyzed reactions.

EXAMPLE 7.1

A bioengineering group is interested in generating a novel estrogen receptor similar to the native form of the estrogen receptor-alpha (ER-α). They successfully construct a mutant and produce a mutated ER-α. They perform experiments using surface plasmon resonance biosensors to determine the affinity for the estrogen, and the data is given below. Calculate K_D and determine whether the mutant is as good as the native ER-α or if there is a need for a new design.

Bound estrogen (nM)	1	0.8	0.6	0.4	0.2	0.05
Free estrogen in the native ER-α mixture (nM)	5.2	1.38	0.67	0.27	0.1	0.025
Free estrogen (nM) in the mutant ER-α mixture (nM)	15.35	4.12	1.8	0.85	0.39	0.08

Solution: Using the given data, calculate the ratio of bound-to-free ligand and plot the Scatchard plots.

| | Native | | Mutant | |
B (nM)	L_0 (nM)	B/L_0	L_0 (nM)	B/L_0
1.00	5.20	0.19	15.35	0.07
0.80	1.38	0.58	4.12	0.19
0.60	0.67	0.90	1.80	0.33
0.40	0.27	1.48	0.85	0.47
0.20	0.10	2.00	0.39	0.51
0.05	0.03	2.00	0.08	0.63

Plot B/L_0 and B for both samples. On the plot, fit linear regression lines and obtain the slopes and intercepts. From the slopes, the K_D value for a native ER-α is 0.415 nM and for the mutant, it is 1.672 nM. Since K_D for the mutant is nearly four times that of the native, it has four times less affinity for the ligand. Hence, there is a need for a new mutant design.

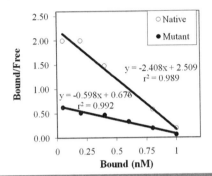

7.2.2.2   Multiple Site Binding

Many receptor-ligand interactions and enzyme-catalyzed reactions are complex. For example, some molecules have two or more identical ligand binding sites. Binding at one site may be independent or dependent on the binding at the second site. When binding at the second site is not independent, it is called cooperative binding and it could be positive (i.e., subsequent binding is enhanced) or negative (i.e., subsequent binding is diminished). One is often an active site, and the others are control sites that can increase or decrease the affinity or activity of the active site. Such a protein is referred to as an allosteric protein, indicating that there are other sites besides the active site. In these scenarios, the B_{max} and K_D values determined by the Scatchard plot are likely to be far from their true values, although

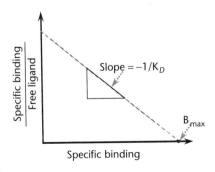

Figure 7.3   Scatchard plot.

many times data analysis by nonlinear regression could correlate well with experimental results. Cooperative interaction between ligand and receptor can be written in a series of interactions

$$R + L \underset{k1_{off}}{\overset{k1_{on}}{\rightleftharpoons}} R \cdot L \rightarrow R \cdot L + L \underset{k2_{off}}{\overset{k2_{on}}{\rightleftharpoons}} R \cdot L_2 \rightarrow R \cdot L_{n-1} + L \underset{k_{n,off}}{\overset{k_{n,on}}{\rightleftharpoons}} R \cdot L_n$$

Interaction of one ligand with N types of binding sites can be represented as

$$R_j + L \underset{k_{joff}}{\overset{k_{jon}}{\rightleftharpoons}} R_j \cdot L; \quad j = 1...n$$

or they can also be a combination of all three of these schemes and can be not well explained by a simple bimolecular reaction such as (7.1). In addition, the concentration of receptors on a cell and the dissociation constant could modify in many physiological and pathophysiological situations; the receptor concentration can reflect functional receptor modifications, and the dissociation constant can reflect genetic alterations of the receptor. Consider the transport of O_2 in the body, which occurs through a binding to a transporter (nearly 97% of O_2 is carried by hemoglobin inside red blood cells) along with the dissolution in plasma (1.5–3% of O_2). Each hemoglobin molecule has four heme groups, each of which binds a molecule of O_2 in a reversible reaction to form oxyhemoglobin (HbO_2).

$$Hb + 4O_2 \Leftrightarrow Hb(O_2)_4$$

This reaction assumes that the subunits of Hb are fully cooperative, that is, if one subunit binds oxygen, they all must bind oxygen, or if one releases oxygen, they must all release oxygen. However, this assumption of total cooperativity produces a very steep curve and does not agree with the experimental results. Consider a general binding interaction of Hb with oxygen:

$$Hb + nO_2 \Leftrightarrow Hb(O_2)_n \text{ or } [P] + n[L] \Leftrightarrow [P:L_n]$$

where $[P]$ is the unliganded protein (hemoglobin), $[L]$ is the ligand (oxygen), and $[P:L]$ is the concentration of the protein:ligand complex (hemoglobin with oxygen bound). At any equilibrium condition, the rate constant is

$$K_D = \frac{[P][L]^n}{[P:L_n]} \tag{7.9}$$

Defining the fraction, Y, of protein that is in complex with a ligand (e.g., hemoglobin molecules that are saturated) is:

$$Y = \frac{[protein : ligand \quad complex]}{[Total \quad protein \quad conc.]} = \frac{[P : L_n]}{[P] + [P : L_n]} \quad (7.10)$$

Substitute (7.9) for $[P:L_n]$ into (7.10) and simplification results in

$$Y = \frac{[L]^n}{K_D + [L]^n} \quad (7.11)$$

Equation (7.11) is popularly called the Hill equation and it is used to analyze the binding equilibria in ligand-receptor interactions where there could be cooperativity. When Y is 0.5 (for 50% saturation), (7.11) reduces to

$$K_D = [L_{50}]^n$$

where L_{50} is the ligand concentration at 50% saturation. Hence, (7.11) is normally rewritten as

$$Y = \frac{[L]^n}{[L_{50}]^n + [L]^n} \quad (7.12)$$

The Hill coefficient (n), described as an *interaction coefficient*, reflects cooperativity rather than estimating the number of binding sites. A similar equation can be written for enzyme-catalyzed reactions. Furthermore, (7.12) can be rearranged to get

$$\left(\frac{Y}{1 - Y} \right) = \frac{[L]^n}{[L_{50}]^n}$$

Plotting the above equation gives a nonlinear curve (Figure 7.4). Taking logarithms on both sides and plot gives the linear form of Hill equation.

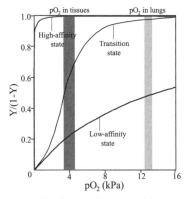

Figure 7.4   Characteristics of oxygen binding with hemoglobin.

$$\ln\left(\frac{Y}{1-Y}\right) = n\ln[L] - n\ln[L_{50}] \tag{7.13}$$

The exponent, n, is the slope of the line. Plotting saturation versus oxygen pressure (or concentration), apparent cooperativity can be determined from the slope. For Hb the slope is 2.8, that is, Hb is *partially* cooperative as the 4 subunits only partially cooperate to behave like 2.8 completely cooperative subunits.

EXAMPLE 7.2

Suppose you get the following data for a type of hemoglobin that you have isolated from a new animal species:
(a) What is the Hill coefficient? What is the P_{50}?
(b) Describe the cooperativity and affinity of this new hemoglobin as compared to normal human hemoglobin.

Y% (HbO$_2$/Hb)	pO$_2$ (mmHg)
25.1	39.8
6.3	15.8
1.6	6.3
0.4	2.5

Solution: Using the given data calculate the X- and Y-axes in (7.13).

Y%	pO$_2$ (mmHg)	Y	$\frac{Y}{1-Y}$	$\ln\left(\frac{Y}{1-Y}\right)$	$\ln(pO_2)$
25.1	39.8	0.251	0.3351	−1.093	3.684
6.3	15.8	0.063	0.0672	−2.700	2.760
1.6	6.3	0.016	0.0163	−4.119	1.841
0.4	2.5	0.004	0.004	−5.517	0.916

(a) Plot $\ln(Y/(1-Y))$ and $\ln(pO_2)$ and fit linear regression lines and obtain the slopes and intercepts. From the slope, Hill coefficient is 1.593.
From the intercept, $n*\ln(p_{50}) = 7.022$.
Hence, $P_{50} = 82.12$ mmHg
(b) The cooperativity of the new hemoglobin is positive but less than the normal hemoglobin.

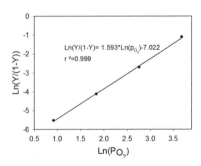

In addition to providing a measure of the affinity of the ligand for the receptor, the Hill plot could provide information on number of binding sites. However, the conditions under which the Hill coefficient provides an accurate estimate of the number of binding sites only when an extreme positive cooperativity is present between the binding of the first and subsequent ligand molecules. The shelf-life of donated blood is only about 40 days. Considering the small margin of surplus, a shortage is likely in an event requiring massive transfusion. For these reasons, substitutes for allogeneic donor blood, primarily the red blood cell component, is an actively investigated area. The prospect of developing a substitute for red blood cells has been advanced for well over 100 years from fluorocarbon, modified hemoglobin, encapsulated hemoglobin, synthetic heme, recombinant hemoglobin, and others. The properties with respect to storage form, stability, physiologic characteristics, toxicity, half-life, and interference with laboratory tests should be the same. As we gain a greater appreciation of the physiologic derangement associated with shock, ischemia, reperfusion injury, and tissue metabolism, further enhancements to the formulation of hemoglobin-based red blood cell substitutes can be anticipated.

7.2.2.3   Inhibition of Binding

Molecular structures within soluble factors dictate the function of soluble factors and anything that significantly changes the structure of a soluble factor would change its activity. Hence, extremes of pH and high temperature can denature the protein and inhibit the activity of the soluble factor in an irreversible fashion, unless it could refold properly. However, a typical approach is to add a small molecule with the knowledge of the binding characteristics and protein structure, which interacts with the soluble factor to induce responses (called agonists) or inhibit response (called antagonists) of the soluble factor. These molecules are either reversible (i.e., the removal of the inhibitor restores soluble factor activity) or irreversible (i.e., the inhibitor permanently inactivates the soluble factor). Reversible inhibitors can be classified as competitive, uncompetitive, noncompetitive, or mixed, according to their effects on K_D and B_{max}. These different effects result from the inhibitor binding to the receptor alone, the receptor-ligand complex, or to both. Enzyme inhibitors can also irreversibly inactivate enzymes, usually by covalently modifying residues at the active site. The type of inhibitor can be discerned by studying the rate of kinetics as a function of the inhibitor concentration. These reactions follow exponential decay functions and are usually saturable. Below saturation, they follow first-order kinetics with respect to inhibitor. Different types of inhibitions produce Lineweaver-Burke and Eadie-Hofstee plots that vary in distinctive ways with inhibitor concentration. The method of nonlinear regression of the enzyme kinetics data to the rate equations above can yield accurate estimates of the dissociation constants. The reader is referred to [1] for a detailed discussion of inhibitory action.

7.2.3   Cell-Matrix Interactions

Cells that must attach to a substrate for survival are known as anchorage-dependent cells. The majority of cells, except some blood cells, are anchorage-dependent, and cell attachment regulates a number of cellular processes including cell shape, migration, growth, and death. Cell adhesion and spreading involves many cellular activities such as production of adhesive proteins and cytoskeleton transport. The substratum characteristics shine through the adsorbed proteins toward adhering and spreading cells. The general dogma is that most cells attach to the ECM (Figure 7.5) and communicate (both mechanically and chemically) via cell adhesion receptors, particularly integrins that can bind to an ECM containing a specific polypeptide sequence of arginine-glycine-aspartic acid (RGD). After binding, receptors communicate signals across the cell wall with the help of a focal adhesion (FA) complex. The FA complex is comprised of many molecules including focal adhesion kinase (FAK), vinculin, talin, and paxillin. The FA complex interacts directly with cytoskeletal actin and changes its polymerization state. Actin reorganizes and redistributes, modulates the structure of the cell cytoskeleton, and alters cell shape and characteristics. Hence, cell morphology which reflects the organization of the intracellular cytoskeletal network is partially determined by the interaction between the cells and their surrounding ECMs. One approach to ensure proper adhesion of cells to extracellular substrate is conjugating polypetides containing the RGD sequence. There are a few other peptides including arginine–glutamic acid–aspartic acid–valine (REDV) (from fibronectin), tyrosine–isoleucine–glycine–serine–arginine (YIGSR) (from laminin), and andisoleucine–lysine–valine–alanine–valine (IKVAV) (from laminin) that are explored for promoting cell adhesion.

Mechanical stresses are either transmitted to cells from the ECM or are generated within the contractile cytoskeleton of cells and exerted on their ECM adhesions. Hence, understanding the force generated in cell-matrix interactions is important. The net force generated can be calculated knowing the number of receptor-ligand bonds formed and the force generated in each bond. An approach taken to quantify the force generated is to deposit a matrix of interest on a tissue culture plastic and evaluated cell interactions. The adherent cells on a surface are analyzed for the force of attachment based on the force required to detach them under an applied shear stress. Similar to (7.1), bonds formed with matrix elements are expressed as

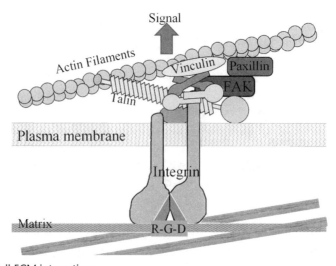

Figure 7.5 Cell-ECM interactions.

$$\frac{dn_b}{dt} = k_f n_L n_R - k_r n_b \tag{7.14}$$

where n_b is the number of bonds formed, n_L is the number of ligands, n_R is the number of receptors, k_f is the forward rate constant, and k_r is the reverse rate constant. American scientist George I Bell suggested that the rate constants should be treated as functions of forces. By assuming that the association rate is unaffected by applied force, the reverse rate constant, k_r, (and thus the affinity) will vary exponentially with the balance of forces. The equation is written as

$$k_r = k_{r0} Exp\left[\frac{\gamma F}{N_C K_B T}\right] \tag{7.15}$$

where k_{r0} is the unstressed reverse rate constant in the absence of force, γ is a length parameter defined as the bond interaction distance, F is the applied force, N_C is the number of receptor-ligand complexes, K_B is the Boltzmann's constant, and T is the temperature. In terms of a chemical reaction, the bond interaction distance can be thought of as a length scale for the transition of the reaction from a bound state to unbound state. To extend this model to adhesion of cells to surfaces (and to other cells), an assumption is that all bonds are stressed equally. Bell proposed the empirical equation for the adhesion strength per bond basis of an interaction as:

$$f = \frac{F}{n_b} \approx 0.7 \frac{K_B T}{\gamma} \ln\frac{[L]}{K_D^{0}} \tag{7.16}$$

where f is the adhesion strength per bond, K_D^{0} is the unstressed dissociation rate constant, and $[L]$ is the concentration of a free ligand. If a stationary cell of a radius r is exposed to a laminar flow of a velocity V, the force F experienced by the cell can be determined using the conservation of momentum principle. Assuming that the adherent cell is in mechanical equilibrium, that is, has no net force acting on and no net torque acting around its center of mass, the adhesive strength can be determined using (7.16) and the conservation of momentum equation, knowing the velocity of the fluid stream at which cells detach from a surface. However, a major limitation is generating reliable experimental data relevant to the body system and the inherent complexities of the adhesive process. Furthermore, cell membranes are viscoelastic in nature. Alternative models have been proposed using principles of mechanics and thermodynamics, although they too suffer from the complexity associated with the experimental data generation.

A simple approach to understanding the cell interaction with a material is quantifying the extent of cell adhesion to the surface by: (1) suspending a known concentration of cells over the surface, (2) incubating the cells in the culture medium for some period of time, (3) removing loosely adherent cells under controlled conditions, and (4) quantifying the cells that remain on the surface or the number of cells that are retrieved in the washes. Cell spreading and migration are also used to indicate adhesion strength. However, these cellular processes exhibit complex dependencies on adhesive strength and do not provide direct or sensitive measurements.

To better understand signaling patterns involved in changing the behavior of cells during disease and development, presenting ECMs in their physiological form is important. Many implantation studies using biocompatible materials have shown that the microarchitecture of the materials is the primary determinant in the foreign body response. Mechanical and biochemical properties of the ECM in the body are different from the two-dimensional (2D) rigid tissue culture plastic (or glass) surface often used in cell culture. On a 2D substrate, cells are restricted to spread and attach to a prefixed ECM coating on the tissue culture plastic. Hence, the effects of biophysical properties of the matrix that provide a spatiotemporal effect in the body are not part of the effect. The mechanical nature and composition of the substrate upon which they grow are critical for cellular activity. Porous matrices of ECM with a different rigidity show that cells respond differently and the matrix stiffness is an important factor in cellular behavior. Many cell types attach differently to three-dimensional (3D) matrices than to the 2D culture. Focal adhesions appear distinct in 3D from 2D and are called 3D matrix adhesions to distinguish them from their 2D counterparts. Such differences in cell adhesion between 2D and 3D cause different signal transduction and subsequent alteration in cellular rearrangement. For example, rigid films of collagen promote the cell spreading and growth of hepatocytes (found in the liver), whereas collagen hydrogels promote a rounded morphology and differentiated phenotype of the same cells. This could be via the response of tractional forces between cells and the substrate; a scaffold should be able to withstand cell contractile forces. Thus, ECM interactions with the cell are chemical as well as physical, unlike soluble ligands, which are chemical only.

7.2.4 Cell-Cell Interactions

Direct interactions between cells are also important in the development and function of tissues. Several different types of stable cell-cell junctions are critical to the maintenance and function of the epithelial barrier. However, some cell-cell interactions are transient, for example, the interactions between cells of the immune system and the interactions that direct leukocytes to sites of tissue inflammation. Interactions between the same types of cells are called homotypic interactions and interactions between two different cell types are termed as heterotypic interactions. Cell-cell contact sites provide important spatial cues to generate cell polarity and enable cells to sense and respond to adjacent cells. Junctional complexes at cell-cell contact sites are also control points for regulating solute flow across cell monolayers (through tight junctions) and from one cell to another (through gap junctions).

During in vitro cell culture, cell seeding density needs to be optimized for proper growth. Most normal diploid cells, with the exception of immortal cell lines, require a minimum cell seeding density for successful growth in vitro. When cells proliferate to cover the entire dish, they reach a state of confluency where cell-cell interactions are maximal. Many normal diploid cells such as fibroblasts respond to the maximum interactions by slowing or stopping their growth, called the state of contact inhibition. Heterotypic interactions are also important in cell culture. One cell type is used to promote the proliferation and/or induce a specific function of a different cell type. Both cell types may be allowed to proliferate or it is possible to preferentially limit the proliferation of one cell type. For example, if one of the cell types is subjected to lethal irradiation or treated with mitomycin C (a DNA cross-linker), the cells will remain attached and viable but do not proliferate. The cell type that is inhibited to proliferate is often celled the feeder layer. If another cell type is cocultured with this feeder layer, they proliferate with the nourishment produced by the feeder cells. Coculturing stem cells with feeder layers has been the strategy to proliferate them without differentiation.

Cell-cell interaction is mediated by transmembrane receptors, which can be grouped into four major types:

- *Cadherins* are Ca^{2+}-dependent adhesion molecules that are thought to be the primary mediators of adhesion between the cells of vertebrate animals. Cadherin-mediated cell adhesion is thought to be homotypic (i.e., one cadherin binds to another in the extracellular space).
- *Selectins,* which are carbohydrate binding proteins, recognize oligosaccharides exposed on the cell surface and, in the presence of calcium, bind to specific oligosaccharides on another cell.
- *Integrins* usually bind cells to matrix. However, they also bind cells to cells, and binding is calcium dependent. Binding is from an integrin to a specific ligand on the target cell. Binding may involve actin filaments, but is not associated with a cell junction. Integrins contain an alpha and a beta subunit. Most cell-to-cell interactions involve integrins with an alpha and a beta-2 subunit. Examples are integrins on white blood cells that allow tighter binding to endothelial cells before they migrate out of the blood stream to tissue.
- The *immunoglobulin* (Ig) *superfamily* contains structural domains similar to immunoglobulins. Noncalcium dependent cell-cell binding belongs to the Ig superfamily. Examples of the Ig family include neural cell adhesion molecules and intercellular adhesion molecules.

7.3   Cellular Processes

Using the interactions from the extracellular environment, cells carry out numerous processes ranging from migration to metabolism. Many of these processes follow complicated cascades of biochemical reactions. For example, cells dividing in culture undergo a dramatic sequence of morphological changes, characterized by cytoskeletal disassembly as cells round up, by a redistribution of actin, myosins and other cytoplasmic and surface molecules into the cleavage furrow, before daughter cells finally separate at the mid-body. Nevertheless, individual cells receive multiple signals, yet they rapidly process and integrate these signals to produce few possible actions. For example, during the branching of a new capillary, an individual cell proliferates, while neighboring cells differentiate or program to die. Understanding cellular processes such as migration, proliferation, metabolism, and intracellular degradation is important in a number of biomedical applications.

7.3.1 Migration

Cell migration is important in a number of physiological and pathological processes such as embryo development, immune response, tumor metastasis (abnormal movement of cancer cells), and wound healing. During immune response or inflammation, white blood cells migrate through the tissue space to the location where the host defense is necessary. During the wound healing of skin, blood enters the wound, and signaling processes are enabled that give rise to the activation of clotting cascades and aggregation of macrophages and leucocytes. Furthermore, skin cells migrate to close the gap. During cancerous tumor growth, endothelial cells migrate to form new vascular networks. Cells move either in association with the surface of a material or through an ensemble of other cells. Cell migration is believed to be regulated through classical signal transduction pathways. Some of the motility factors such as platelet-derived growth factor, vascular endothelial growth factor, hepatocyte growth factor, and fibroblast growth factor bind to their receptors on the surface and induce migration stimulatory signals that result in the reorganization of the cellular cytoskeletal architecture and stimulation of the motility machinery of the cell resulting in cell migration.

The mechanics of cell migration are largely addressed in tissue culture systems by controlling the culturing conditions and using a large number of similar cell types. From these studies, the migration of slow-moving adherent cells such as fibroblasts (involved in wound healing) and endothelial cells (involved in new blood vessel formation) over a 2D substrate is proposed to occur in five steps:

1. Cells polarize and extend protrusions in the direction of migration and protrusions are driven by actin polymerization.
2. Extensions are stabilized by adhesion to the ECM or adjacent cells via transmembrane receptors such as integrins and adhesions serve as traction sites for migration.
3. Cells contract from both ends in towards the nucleus exerting forces (or tractions) against the substrate.
4. The rear of the cell is released from the substrate. The opposing forces act together to move the cell body forward through contraction of the cytoplasm.
5. Membrane receptors from the rear are recycled to the front of the cell by endocytosis and/or forward-directed movement on the cell surface. If the release of adhesive bonds is hindered, cell migration decreases. Hence, while selecting a biomaterial where cell colonization is expected, one has to consider the release effect.

These steps are not obvious in fast-moving cells such as neutrophils, which glide over the substratum. Neutrophils are activated in response to soluble factors (commonly referred to as chemotactic factors in this literature) recognized by receptors on their plasma membranes. Migration into inflamed tissues occurs in three steps: (1) *rolling* along the vasculature (mediated through transient interactions between selectin proteins and their carbohydrate ligands), (2) *activation* of both neutrophils and endothelial cells and a high affinity interaction between integrins and glycoproteins of immunoglobulin superfamily, and (3) *extravasation* (crawling along the endothelium, diapedesis, and migration into tissue) in response to a chemoattractant gradient. Somatic cells migrating in vivo show large single protrusions and highly directed migration, in contrast to the multiple small protrusions they display on planar substrates, and cancer cells can modify their morphology and nature of migration in response to environmental changes.

7.3.1.1 Measuring Cell Migration

Migratory behavior of all cells depends on their microenvironment. *Chemokinesis* is the induction of random, nondirectional migration in response to a ligand without any oriented cues. To describe the random motility of a cell in the x-direction, a flux of the form similar to Fick's first law (Chapter 2) is assumed, that is,

$$J_{random} = -D_n \frac{dn}{dx} \tag{7.17}$$

where D_n is the cell random-motility coefficient and n is the cell density (similar to concentration of a chemical). Random motility in 3D can be obtained by

$$J_{random} = -D_n \left(\frac{dn}{dx} + \frac{dn}{dy} + \frac{dn}{dz} \right) = -D_n \nabla n \tag{7.18}$$

where ∇ is the gradient operator. However, cell migration primarily occurs due to the presence of ligands secreted by certain cell types in addition to the concentration of binding domains in the extracellular matrix. Chemotaxis (from the Greek *taxis* "to arrange") describes the directed migration of cells towards a positive gradient of soluble chemoattractant, whereas haptotaxis is the directed migration of cells along a gradient of anchored substrate (laminin and fibronectin). Analogous to random motility, chemotactic flux can also be written as

$$J_{chemo} = \chi(c)n\nabla c$$

where $\chi(c)$ is a chemotactic function and c is the concentration of the soluble chemoattractant. Chemotactic function is dependent on the receptor-ligand interactions. Similarly, the influence of haptotaxis is modeled as

$$J_{hapto} = \rho_0 n \nabla f$$

where $\rho_0 > 0$ is the (constant) haptotactic coefficient and f is the concentration of the substrate-bound haptotactic factor. When all three factors (chemokinesis, chemotaxis, and haptotaxis) are considered to assess the migratory pattern, the mathematical model is called a continuous model. The three contributions to the total cell flux J_n, are given by

$$J_n = J_{random} + J_{chemo} + J_{hapto}$$

$$J_n = -D_n \nabla n + \chi(c)n\nabla c + \rho_0 n \nabla f \tag{7.19}$$

Neglecting the cell growth or death, the conservation equation for cell density n is written as

$$\frac{\partial n}{\partial t} = \frac{dJ_{n,x}}{dx} + \frac{dJ_{n,y}}{dy} + \frac{dJ_{n,z}}{dz} \rightarrow \frac{\partial n}{\partial t} = \nabla \cdot J_n$$

Substituting for J_n,

$$\frac{\partial n}{\partial t} = -D_n \nabla^2 n + \nabla \cdot (\chi(c)n\nabla c) + \nabla \cdot (\rho_0 n \nabla f) \tag{7.20}$$

This is called the continuous model of cell migration. Equation (7.20) can be simplified based on the application and solved using numerical integration techniques. However, obtaining reliable biological data to validate the model is very difficult. Segregating haptotactixis from chemotaxis is not always possible. Furthermore, one has to consider the changes in the concentration of the chemoattractant or the changes in the haptotactic agent; there could be a decay of the chemoattractant or a degradation of the matrix elements. The model needs improvement to account for the effect of the mechanical property of the substrate.

Cell migration studies are performed on individual cells [Figure 7.6(a)] as well as on groups of cells [Figure 7.6(b)]. As the latter depends on the net sum of the motions of the former, the detailed study of individual cell trajectories can usually reveal greater insights into cell motion behavior. Additionally, mathematical models have been developed to relate a summation of individual cell paths to population

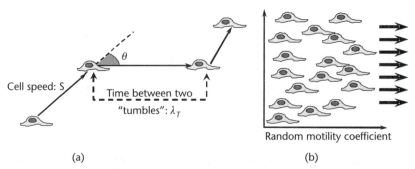

(a)   (b)

Figure 7.6 Modeling cell migration: (a) random walk model in 2D and (b) population-based description.

movements. To observe the movements of a small number of cells, they are seeded on the surface and condition of interest and cell motion is tracked using time-lapse videomicroscopy tracks as a function of time elapsed. To increase the accuracy, agarose gels are used and the leading front of the cell movement beneath the gel is monitored. Alternatively, a number of cells are seeded and collective movements of group of cells are observed. To increase the accuracy, the number of cells migrating through a filter (called the filter chamber assay) or a membrane is measured. Microporous membrane inserts are widely used for cell migration, and invasion assays using Boyden chambers are a normal practice. Both methods enable the estimation of intrinsic cell motility parameters such as random motility coefficient and persistence time. *Persistence time (P)* is the average time period between significant changes in the direction of movement.

$$P = \frac{1}{\lambda_T (1 - \cos\phi)} \tag{7.21}$$

where λ_T is the time between two changes in direction of the cell and ϕ is the change in direction [Figure 7.6(a)]. A single cell moving through an isotropic environment follows a straight path over short time intervals, yet exhibit random movement over long time intervals. Overall, this cell motion is characterized as a persistent random walk model, which is extensively utilized in statistical physics. This persistent random walk model for describing cell migration is written as

$$\langle d^2 \rangle = n_d \langle S^2 \rangle P[t - P(1 - \exp(-t/P))] \tag{7.22}$$

where $<d^2>$ is the mean squared displacement over a cell population, n_d is the number of dimensions in which the cells are migrating, and S is the characteristic speed of a cell (measure of centroid displacement per time). Obtained values are fit to the model so that speed and persistence time are determined. However, current methods of analysis are time-consuming and tedious, involving cotton swabbing of nonmigrated cells on the top side of insert, manual staining, and counting. From the persistence time and the characteristic speed of a cell, random motility coefficient can be obtained using $D = S^2 P/2$.

EXAMPLE 7.3

In an experiment using mature fibroblasts in tissue culture plastic surface, the persistent time is determined to be 30 minutes and the speed is 12 μm/hr. What is the displacement of a cell after 20 hours? Determine the random motility coefficient.

Solution: $n_d = 2$, $S = 12$ μm/hr, $P = 0.5$ hour

When $t = 20$ hours, what is d?

$$\langle d^2 \rangle = 2 * 144 [\mu\text{m/hr}]^2 * 0.5[\text{hr}](20[\text{hr}] - 0.5[\text{hr}](1 - \exp(-20/0.5))) = 2,808 \ \mu\text{m}^2$$

Hence, $d = 53$ μm

$$D = S^2 P/2 \rightarrow 12^2 * 0.5/2 = 36 \ \mu\text{m}^2/\text{hr} = 1 \times 10^{-10} \text{ cm}^2/\text{s}$$

Comparing this value to the diffusion coefficient of albumin (Example 2.6) suggests that the random motility coefficient is nearly 1,000 times less.

7.3.2 Proliferation and Differentiation

After the fertilization of an egg and sperm is complete, one division occurs to form a zygote. The newly formed cells of the zygote begin to differentiate (i.e., irreversible changes in the pattern of gene expression and cell function begin to occur). It is the state of differentiation that dictates whether and under what circumstances cell proliferation (also referred to as cell growth) may occur. Most cells in the body are fully differentiated to perform a specific function; they are called mature cells. These cells lose their ability to make copies of themselves, and they do not change their identity. However, few cells in various mature tissues have been discovered to possess self-renewal properties to enrich their population and differentiate to other cell types. These are called adult stem cells. There is an increased interest in understanding what causes stem cells to be special due to the possibility of treating various disorders and developing biomedical devices. The best-known stem cell therapy is the bone marrow transplant, which is used to treat leukemia and other types of cancer.

Stem cells were first discovered to exist during embryonic development. Eggs that are successfully fertilized are allowed to divide for 3–6 days [Figure 7.7(a)], forming tiny balls of nearly 100 cells called blastocysts. The blastocysts are like tennis balls, solid on the outside but hollow in the middle. By cracking the ball, cells are collected and placed in a dish containing an appropriate in vivo environment to grow them to make exact copies of themselves. This technology was first established in 1981 in mouse embryonic stem cells, which were demonstrated to be cells with the ability to develop into all somatic cell lineages. In 1998, human embryonic stem cells (hESCs) were derived from blastocysts obtained from in vitro fertilized human eggs.

Blood retrieved from the umbilical cord following the birth of a baby is a rich source of stem cells that produces all other blood cells. Like donated bone marrow, umbilical cord blood is used to treat various disorders that affect the blood and immune system, leukemia and certain cancers, and some inherited disorders. There is a high degree of tolerance to sibling mismatches, suggesting that the same cell source can be utilized for treating several members of a given family. Because umbilical cord blood lacks well-developed immune cells, there is less chance that the transplanted cells will attack the recipient's body, a problem called graft versus host disease, typically encountered when cells from a matched donor are used. There is also some evidence of using these cells to treat genetic disorders such as sickle cell anemia. However, the amount of cells available per sample is limited, and the proliferative potential to other cell types is not completely understood.

Much of the regenerative medicine and tissue engineering embodies using stem cells that have the potential to differentiate to various mature cells. Two important factors need to be considered: the degree of potency and the plasticity of stem cells. The degree of potency refers to the number of lineages that can be obtained

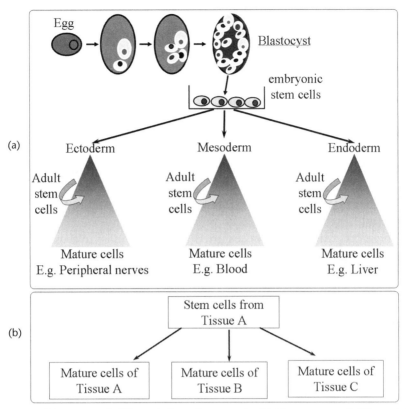

Figure 7.7   Proliferation and differentiation of stem cells. (a) The process of differentiation. (b) Plasticity of stem cells.

assuming a hierarchical structure of differentiation (i.e., the more mature the cell, the less ability it has to form other cells). As a cell matures, it begins to differentiate or head down the path toward a specific cell type. For example, neural stem cells are cells that can self-renew but will only become nervous system cells.

Cells that have a more limited potential are called multipotent. Adult and embryonic stem cells differ in the number and type of differentiated cells types that they can become. Embryonic stem cells can become all cell types of the body that are described as pluripotent or totipotent cells. Adult stem cells are generally limited to differentiating into various cell types of their tissue of origin and are described as multipotent cells. However, some evidence suggests that adult stem cell plasticity may exist, increasing the number of cell types that a given adult stem cell can become.

Plasticity refers to the possibility of obtaining cells for one type of tissue from another tissue source [Figure 7.7(b)], for example, obtaining cardiac myocytes from neural cells or the ability of bone marrow cells to restore liver function by differentiating into liver cells after transplantation.

In general, differentiated cells proliferate in response to specific extracellular signals including soluble factors, ECM, and neighboring cells. Very often, the same factors also regulate differentiation. Cell proliferation and differentiation are vigorous during embryonic and neonatal development, but also occur in many, but not all, tissues of adult organisms. Cell proliferation and differentiation are intriguingly complex processes, and there is no single molecular switch that controls either process. In a normal resting cell, the intracellular signaling proteins and genes that are normally activated by extracellular growth factors are inactive, which leads to no proliferation of cells. When the normal cell is stimulated by an extracellular growth factor, these signaling proteins and genes become active and the cell proliferates. Nonproliferating cells may remain viable for long periods of time and, in some cases, may be very active metabolically. For example, adult hepatocytes and striated muscle cells are proliferatively quiescent but metabolically very active.

Whether proliferation is continual or is induced in a cell, it proceeds via an ordered set of events termed the cell cycle. The cell cycle culminates in the formation of two daughter cells. Resting cells may enter either a reversible or irreversible growth arrested state, and differentiated cells enter an irreversible growth state. Therefore, differentiated cells are often said to exist in a unique growth state, G_D. Cells capable of resting in a nondividing state are said to be in a quiescent state, G_0. The cell cycle is divided into two functional phases, the S (DNA synthetic) and M (mitosis) phases, and two preparatory gap phases, G_1 and G_2 [Figure 7.8(a)]. A successful cell cycle requires the orderly and unidirectional transition from one cell cycle phase to the next. The passage of the cell cycle checkpoints ultimately requires the activation of intracellular enzymes known as cyclin-dependent kinases. They initiate one or more cycles of cell division when there is a need to maintain or replace tissue, and they terminate dividing when the necessary growth is complete. In G_1, protein and RNA synthesis are active. If conditions are permissive for subsequent cell division, cells quickly move into the S period when new DNA is synthesized and DNA replication occurs. Under normal circumstances, the time

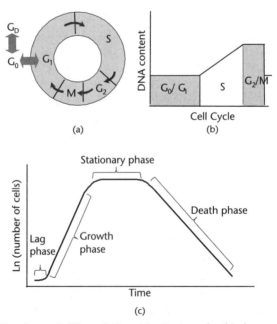

Figure 7.8 Cell proliferation and differentiation: (a) mitotic cycle, (b) changes in DNA content during cell proliferation, and (c) different phases observed during in vitro cell culture.

taken for a mammalian cell to complete S [Figure 7.8(b)] phase is about 8 hours. Another gap (G_2) follows when the newly duplicated chromosomes condense. In the M period, the chromosomes divide into two sets, and the cell forms two nuclei and then divides into two daughter cells. The length of the M phase is about 1 hour and is also normally invariant. When normal cells differentiate, typically with a gain in the properties required for organ or tissue functions, they generally lose the capacity to continue cell division. G_1 and G_2 phases are required for the synthesis of cellular constituents needed to support the following phase and ultimately to complete cell division. The length of G_2 phase is about 2 hours. The length of G_1 phase is highly variable and can range from about 6 hours to several days or longer. Nondividing cells are not considered to be in the cell cycle.

7.3.2.1 Calculating Cell Proliferation

Cells cultured in vitro undergo four major phases of growth [Figure 7.8(c)]. In the lag phase or induction phase, cell division does not occur. The length of the lag phase is mostly attributed to the type and age of the cell, cell density, and culture parameters such as nutrient concentration. In the growth phase, the cell number follows an exponential trend as division takes place. The growth rate increases with the cell density. Lack of nutrients and overpopulation push the micro-organism into a stationary phase where no further increase in cell density is to be expected. This stage is followed by a decrease in the number of viable cells as a result of toxic waste buildup and a depletion of nutrients.

Growth-phase classification is very important for control or optimization purposes. Hence, it is commonly assumed that all cells are in the growth phase and cells double with every cell cycle (i.e., the daughter cells themselves divide upon completion of the next cell cycle). The most prominent parameter for analyzing cell proliferation is the measurement of DNA content or synthesis as a specific marker for replication. In measuring DNA synthesis, labeled DNA precursors ([3H]-Thymidine or 5-bromo-2′-deoxyuridine [BrdU]) are added to cells (or animals), and their incorporation into genomic DNA is quantified following incubation and sample preparation. Incorporation of the labeled precursor into DNA is directly proportional to the rate of cell division occurring in the sample. Exponential growth is expressed as

$$X = X_0 \exp(\mu t) \tag{7.23}$$

where X is the number of cells (or number of cells per unit volume) at time t, C_0 is the initial number of cells seeded, and μ is called the specific growth rate constant with day^{-1} units. While utilizing bioreactors for large-scale production, (7.23) is rewritten in the differential form as

$$\frac{dX}{dt} = \mu X \tag{7.24}$$

To determine the value of μ, a linear form of (7.23) is used where $\ln(X/X_0)$ is plotted against time and the slope corresponds to μ. A smaller μ value indicates that the cell cycle is longer.

EXAMPLE 7.4

To understand the effect of newly created material, Stella seeded 10,000 cells in a tissue culture plate. To measure the growth rate, she added ^{3}H thymidine, a nontoxic radioactive substance, to the culture and read the incorporated radiation (counts per minute) 24 hours later. The radiation reading was 5,000 counts per minute (cpm) which correlates to the amount of DNA synthesized. To calibrate the amount of cpm to number of cells, she recounted the cells after 24 hours and found that the total cell number is 20,000 per culture plate. If she adds a specific soluble factor to the medium under the same conditions (10,000 cells/well over 24 hours), the incorporated radiation reads 8,000 cpm. What is the specific growth rate constant under the new conditions? List few assumptions used during these calculations.

Solution:

X_0 is 10,000 cells and $X(t = 24$ hours): 20,000 cells

Hence, a growth rate of 10,000 new cells/day corresponds to 5,000 cpm (i.e., 0.5 cpm/cells in 24 hours).

If the cpm count is 8,000,

Number of newly proliferated cells = 8,000/0.5 =16,000 cells

Hence,

$X(t = 24) = X_0 + 16,000 = 26,000$

From (7.23), 26,000 = 10,000 exp(24μ)
Hence, $\mu = 0.040$ hr^{-1}
The assumptions are that the growth rate is directly proportional to ^{3}H thymidine uptake and is taken up only during cell division and is in excess so that it will not run out. The plating efficiency is 100% (i.e., all the 10,000 cells are capable of dividing). cpm = 0 when there is no proliferation (i.e., the background signal is negligible).

For comparison purposes, the time interval t_d required to double the population is sometimes calculated. Nevertheless, (7.23) is applicable for cells in the exponential growth phase. A lag phase or time required to adjust to a new environment can be included by correcting time, that is,

$$X = X_0 \exp(\mu(t - t_{lag})) \tag{7.25}$$

However, if only a fraction (f) of the cells are dividing and cells have different phases, then the total number of cells after one cell cycle is given by

$$X = (1 - f)X_0 + 2fX_0 \tag{7.26}$$

where $(1 - f)$ is the fraction of cell that is not dividing. Nevertheless, μ depends on the quantity of the substrate. Hence, relating μ to the substrate concentration is necessary. For proliferation limited by one substrate, S, a widely used Monod equation (named afer French biologist Jacques Monod) is

$$\mu = \frac{\mu_{max} S}{K_s + S}$$

where μ_{max} is the maximum specific growth rate constant [hr^{-1}] and K_s is the Monod constant. This empirically derived equation for bacterial cultures is similar to the Michaelis-Menten model. The Monod equation does not consider the fact that cells may nee d a substrate (or may synthesize the product) even when they do not proliferate. Hence, more sophisticated proliferation models are developed combining receptor-ligand interactions.

7.3.2.2   Consideration of Cell Death

Cell death is part of the normal development and maturation cycle and is the component of many responses of cells to pathogens and to endogenous modulations, such as inflammation and disturbed blood supply. In the treatment of cancer, the major approach is the removal of the cancerous tissue and/or the induction of cell death in cancerous cells by radiation, toxic chemicals, and antibodies and/or cells of the immune system. On the other hand, cell death decreases the overall biopharmaceutical yield in large-scale cell cultures. To optimize cell viabilities and protein yields in culture, one has to develop strategies for controlling cell death. Cell death occurs in two general forms: necrosis and apoptosis. Necrosis results from immediate, extreme conditions that physically damage cells causing them to swell and rupture, releasing their cellular contents into the surrounding environment. Necrosis occurs without the consumption of energy, without DNA cleavage, and without the role of mitochondria. It is not regulated by any intracellular signaling pathway and varies from cell type to cell type. Necrosis can be controlled by optimizing environmental conditions in large scale cell cultures. Apoptosis (also referred to as a type of programmed cell death) is a regulated physiological process resulting from a nonlethal stimulus that activates a cellular cascade of events culminating in cell death leading to the removal of unnecessary, aged, or damaged cells. Apoptosis is important for removing viral-infected cells, cancer cells, cytotoxic T-cells to prevent self-attack, and cells with DNA abnormality. Apoptosis allows the formation of vital organs by removing excessive cells in the area, sloughing off inner lining of the uterus at the start of menstruation, and the formation of synapses between neurons in the nervous system. Apoptosis is energetically dependent and genetically regulated by the interplay of pro- and anti-apoptotic proteins. Although the apoptotic program is complex and the cellular events involve the activation of many signaling cascades, apoptosis is characterized by cell shrinkage, plasma membrane blebbing, and DNA fragmentation. Blebbing involves the shedding of membrane fragments from the cell in the form of apoptotic bodies that often include cytosolic and nuclear contents.

To account for cell death, (7.23) is rewritten as

$$X = X_0 \exp((\mu - k_d)t) \tag{7.27}$$

where k_d is the specific death rate constant. In the differential form, (7.24) reduces to

$$\frac{dX}{dt} = (\mu - k_d)X \tag{7.28}$$

Understanding the death rate helps in developing strategies for the quantitative improvement of cellular and biologic product productivity.

7.3.3 Metabolism

Metabolism is the sum of many individual chemical reactions by which cells process nutrient molecules and maintain a living state. The end-products of a catabolic process are often the substrates for anabolism or vice versa. For example, there are actually over 30 individual reactions that lead to breakdown of glucose, although the combustion of glucose is written in one equation. Each reaction is controlled by a different enzyme. The general sequences of these enzyme-controlled reactions are called metabolic pathways. Metabolic pathways may be linear, branched, cyclical, or a combination of these different systems.

To understand the functional determinants of cellular components, understanding changes in molecular fluxes through metabolic pathways is important as they integrate genetic and metabolic regulation. The measurement of metabolic fluxes allows observation of the functional output of the cell. Furthermore, one could utilize the metabolic pathways to generate: (1) an essential intermediate metabolic product by diverting the competing reactions at the genetic level, or (2) an aromatic amino acid that is difficult to process through traditional chemical processing techniques. The fermentation process has used this concept to produce ethanol from the various, less expensive, renewable carbon sources. Using the biochemical understanding of metabolic pathways, the production of secondary metabolites, or small molecules important for pharmaceutical and materials applications is also possible. The underlying goal is to improve the yield and productivity of native products synthesized by micro-organisms via alterations of competing metabolic reactions. Based on the required target, one can engineer the genetic sequence to increase selectivity of a biochemical reaction.

A challenge while making genetic changes and assessing performance is evaluating the complex metabolic networks. The structure of such networks can be derived from genomic or biochemical data for a large diversity of organisms. Mathematical analysis and modeling require kinetic parameters that are rarely available or not available at all. Structural analyses of the network require mainly reaction stoichiometries, which are oftentimes well known. This is one of the reasons that stoichiometric modeling or metabolic flux analysis (MFA) has become an essential tool and a promising approach for understanding the functionality of metabolic networks. By MFA, the intracellular metabolic fluxes are calculated by the measurement of few extracellular metabolite concentrations in combination with the stoichiometry of intracellular reactions. MFA is applied to evaluate the intracellular metabolic conditions and to identify key metabolic pathways or metabolites in the central metabolism. In MFA, mass balances over all the intracellular metabolites are used in combination with the stoichiometry of intracellular reactions to calculate the fluxes through the different branches of the network. The intracellular fluxes are calculated by combining measurements of extracellular metabolite concentrations, either with linear algebra or with linear optimization. However, in any given scenario, completing the material balance is a challenging task due to the possibility of multiple product formation.

7.3.4   Intracellular Degradation

Intracellular degradation of various materials in cells results from the action of a combination of multiple mechanisms. Understanding intracellular degradation is important to understanding the trafficking of metabolites during various disease states and gene therapy. The use of various viral (such as adenoviruses and retroviruses) and nonviral (such as synthetic lipids and polymers) delivery systems to deliver DNA for gene therapy applications (to correct genetic deficiencies or treat acquired diseases) has shown that the efficiency of the transfection of nonviral gene transfer is poor relative to viral vectors. Viral vectors have evolved mechanisms to attach to cells, cross cellular membranes, evade intracellular transport systems, and deliver their genomes into the appropriate subcellular compartment. However, viral vectors have several restrictions such as limited DNA-carrying capacity, a lack of target-cell specificity, immunogenicity, and, most importantly, the safety of the patient. Developing nonviral gene delivery carriers with a high transfection efficiency is important for safe use, improving target-cell specificity, and eliminating immunogenicity. Apart from concerns related to in the extracellular inactivation and initial favorable interactions with the cell surface, intracellular degradation is an important barrier. Hence, understanding the intracellular degradation mechanism is very important for novel designs. There are three major degradation pathways:

1. Lysosomes contain nearly 50 hydrolytic enzymes, including a variety of proteases known as cathepsins. Lysosomes have an internal pH of nearly 5. If genetic material is introduced into a cell, a major concern is the degradation of the foreign genetic material by the lysosomes. The internalization of the gene delivery system by endocytosis results in an endosomal trafficking process, which, if not escaped, leads to degradation of the therapeutic DNA in lysosomes. Hence, some strategies focus on evading the lysosomal degradation. Nevertheless, the products of this activity are monomers, which leave the lysosomes either via diffusion or with the aid of specialized transport systems. The building blocks formed during degradation are either reutilized for the biosynthesis of complex molecules or further degraded to provide metabolic energy.

2. Another degradation pathway involves two Ca^{2+}-dependent cytosolic enzymes (called Calpains) activated by increased intracellular calcium and ATP depletion. Calcium-dependent proteases have neutral pH optima and are dependent on Ca^{2+} ions for catalytic activity. Two molecular species are based on the requirements for Ca^{2+} ion concentration: μ-calpains and m-calpains. Although their role in trauma-induced proteolysis is not clear, they have been implicated in the muscle catabolism accompanying denervation and muscular dystrophy.

3. A third pathway involves proteasome catabolism regulated in part by the ubiquitin conjugation of targeted proteins; recent work suggests that the ubiquitin-proteasome pathway accounts for the majority of hypercatabolic changes. Ubiquitin is a small peptide of 76 amino acids, which may be the most conserved protein in evolution. It is present in all eukaryotes with a primary sequence perfectly conserved from insects to man. Ubiquitin participates in a variety of cellular processes such as the cell cycle control, DNA repair, ribosome biogenesis, and the stress response chaperone; in higher organisms, ubiquitin is also involved in the immune response. The ubiquitin molecule acts as a targeting label for proteolysis by covalently associating with proteins to be degraded. The ATP-dependent conjugation of the ubiquitin moiety with the target protein is carried out by enzymes called E2. The proteolysis proper is classically mediated by a cytosolic multicatalytic ATP-dependent complex called the proteasome. In eukaryotes, the proteasome complex often associates with other proteins to form even larger complexes.

7.4 Bioreactors

Most tissue cultures in the laboratory are performed on a small scale to prolifarate a small population needed for various experiments. At this scale, cells are usually grown in T flasks ranging from a 25 cm² to a 175 cm² surface area [Figure 7.9(a)]. The typical cell yields in a T175 flask range from 10^7 for adherent cells to 10^8 for

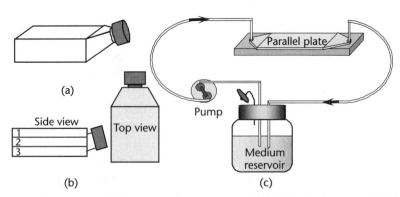

Figure 7.9 Scale-up of a cell culture using flat-geometry: (a) T-flask, (b) three-tiered flask, and (c) parallel plate reactor.

cells which grow in suspension. However, the exact yields vary depending on the type of cell used. It is not practicable to produce very large quantities of cells using T-flasks, due to a limited surface area and problems associated with gaseous exchange (particularly oxygen depletion and low solubility of oxygen in the water).

As more cellular products are commercialized, increasing demand for the consistent supply of clinical-grade material of tens of kilograms is necessary. Furthermore, tissue engineering strategies need improved mass transport throughout 3D porous matrices. Many parts of the body are exposed to stresses either due to the weight they carry (such as bone), the function they perform (such as bladder and cartilage), or the flow of fluid (lung and blood vessels). For example, cells colonizing the bladder are constantly under a mechanical strain as the bladder cyclically expands and deflates. Thus, it is important to grow the cells outside the body by exposing them to the same conditions that they are exposed to within the body. For this purpose, large-scale production is performed in devices referred to as bioreactors, which provide tightly controlled environmental (e.g., pH, temperature, nutrient supply, and waste removal) for cell growth. However, a number of parameters need to be considered while scaling up a cell culture process. These include problems associated with nutrient depletion, gaseous exchange (particularly oxygen), and the buildup of toxic byproducts such as lactic acid. Optimal fluid movement and chemical conditions within the bioreactor are essential for the proper growth and development of the cells and tissues. Nonuniformity in the bioreactor microenvironment can lead to undesirable differences in the cell growth rate. To maximize cell growth, one must adjust the temperature and provide sufficient aeration with gases such as oxygen and nitrogen.

7.4.1   Different Shapes of Bioreactors

Different types of bioreactors have been designed to regenerate tissues with the intention of improving the nutrient distribution while applying mechanical stimuli. Some of the configurations are described next.

7.4.1.1   Extend T-Flask Technology

Few designs emulate the routinely used T-flask technology in scaling up operations. There is a 500-cm^2 triple flask, a three-tiered chamber inoculated [Figure 7.9(b)] and fed like a traditional T-flask, with the same footprint as a T175 flask. There are a series of 632-cm^2 stackable chambers welded together as a single unit, with common fill and vent ports connected by channels to allow liquid and air flow between them. The mass transport of gases such as oxygen and ammonia is the greatest impediment to static cultures in large flasks. Cells grow best when the culture medium is constantly moving, allowing improved aeration at the surface. Cell growth is increased in response to mechanical stresses such as fluid shear (see Chapter 4), as compared to those grown under static culture conditions. Parallel plate reactors [Figure 7.9(c)] that apply controlled mechanical forces are used as model systems of tissue development under physiological loading conditions. The medium is perfused directly on the cells using a pump. The ease of monitoring flow distribution, the ease of varying the shear stress, and the ease of nondestructively observing cells under a light microscope are attractive features of parallel flow reactors. They are widely used to study cell migration, function, and proliferation in cardiovascular applications. However, the major drawback is related to the scale-up issues, specifically the total volume of media required to perfuse all the cells. The volume of the medium required is directly proportional to the total surface area on which the required number of adherent cells is growing. Although these limitations have been addressed by stacking the plates, it is not an attractive configuration if the focus is secreted products.

7.4.1.2   Roller Bottles

A gentle stirring or rocking may be ideal for cells that grow in suspension in a small scale, but many cells grow only on a substrate. Roller bottles [Figure 7.10(a)] are developed converting T-flasks to bottles whose entire inner surface is available for cell growth. A special apparatus slowly rotates (between 5 and 60 revolutions per hour) the bottles, and the cells are alternately bathed in the medium and exposed to the atmosphere, which in turn allows for a larger usable surface area and increased oxygenation. The associated mixing also prevents formation of concentration gradients of nutrients. Most roller bottles accommodate standard-sized ($\sim 1,050$ cm^2) bottles. Expanded surface roller bottles that use different patterns to obtain a larger surface ($\sim 4,200$ cm^2) are developed. However, the size of the roller bottles is a problem since they are difficult to handle inside the biological safety cabinet. Also, uniform cell seeding and the inability to routinely observe cell cultures under light microscopy are additional problems.

7.4.1.3   Spinner Flasks

For cells that have been engineered to grow in suspension, the simplest device is a spinner flask [Figure 7.10(b)]. It is a wide cylinder with ports for the addition and removal of cells and media and gases with CO_2 enriched air. To achieve a sufficient air exchange, tanks are constantly agitated by magnetic stir bars of different shapes and are either kept small enough for the headspace to provide ample oxygen or are equipped with an aeration device such as a sparger to provide a continuous flow of oxygen. Spinner flasks have been used for culturing cells inside porous 3D matrices. During seeding, cells are transported to and into the scaffold by convection. During culture, stirring enhances an external mass transfer but also generates turbulent eddies, which could be detrimental for the development of the tissue. Spinner flask systems designed to handle culture volumes of 1 to 12 liters are available. For large-scale operations, spinner flask technology is extended into large bioreactors where cells are kept in suspension by either a propeller in the base of the chamber vessel or by air bubbling through the culture vessel. However, excessive fluid-shear or turbulence damages cells. It is important to determine and control the fluid stress level inside the bioreactor.

For culturing adherent cells, microcarrier (30–200 μm in diameter) beads (either porous or smooth surface) have been developed. These offer a large surface area per unit volume onto which cells can attach and grow. This technology has a vastly expanded cell culture capacity. The surface properties of the beads should

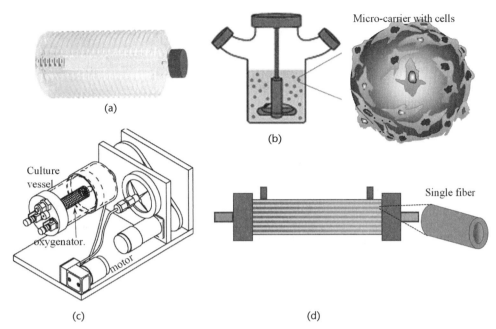

Figure 7.10 Scale-up of cell culture using different cylindrical configuration: (a) extended roller flask, (b) spinner flask, (c) rotating wall reactor, and (d) hollow fiber-reactor.

be such that cells can adhere and proliferate rapidly. Microcarriers are made from materials such as gelatin and dextran. The density of the beads should be slightly more than that of the culture medium (1.03–1.09 cm^3) to facilitate easy separation. However, the density should not exceed a certain limit to avoid settling.

7.4.1.4 Hollow Fiber Reactor

Several systems use porous fibers and membranes to trap cells and larger macro-molecules such as secreted proteins, while granting media and its nutrients free access; toxic metabolites are simultaneously diluted or eliminated as well. Among these systems are hollow fiber reactors [Figure 7.10(d)], available in a variety of configurations. Hollow fiber bioreactors are used to enhance mass transfer during the culture of highly metabolic and sensitive cell types such as liver cells. However, there are disadvantages when considering tissue engineering applications in which the cells are the desired product.

The retention of cells or enzymes within reactors allows the reactor to be con-tinuously perfused without concerns about washout due to process upsets. Fur-thermore, the reactor allows a high biomass density to be achieved. Membranes provide an in situ separation of cells/enzymes from the product, obviating the need for this step downstream. They can be used with relatively crude raw materials because particulates and other unwanted substances can be excluded from the compartment containing cells and enzymes. However, the culture environment is spatially inhomogeneous, and this creates potentially large concentration gradients of critical nutrients, as well as oxygen and pH. As modules are scaled up, the space between fibers is not typically kept constant, which would significantly change the oxygen transfer characteristics of the system. Thus, while the culture performance of the hollow fiber reactors might be adequate, the difficulties encountered in monitoring and controlling the culture environment suggest that a different type of reactor would be more appropriate for the expansion of primary cells.

7.4.1.5 Rotating Vessel Bioreactor

This was developed initially as a microgravity simulator to mimic and model effects of microgravity on cells in earth-based laboratory studies. It spins a fluid medium filled with cells to neutralize most of gravity's effects and encourage cells to grow in a natural manner [Figure 7.10(c)]. A rotating vessel bioreactor provides a dynamic culture environment to the constructs, with low shear stresses and high mass-transfer rates. Hence, they are very attractive in tissue engineering applications. The porous matrix containing cells remains in a state of free fall through the culture medium. Tissues grown in this manner often form 3D aggregates with similarities to the native tissue.

7.4.2 Different Modes of Operation

The most popular bioreactor employed for various applications is the stirred tank reactor. Strategies to improve yields in bioreactors include reducing or selectively removing toxic byproducts such as lactate and ammonia, reducing gas transfer limitations, and maintaining uniform reaction conditions without causing excessive shear damage. Bioreactors are operated in a batch, a fed-batch, or a perfusion mode.

7.4.2.1 Batch Reactor

Many antibodies, growth factors, vaccines, and antibiotics are produced in a single-batch reactor [Figure 7.11(a)] as they provide a number of advantages, including: (1) higher raw material conversion levels, resulting from a controlled growth period, (2) more flexibility with varying product/biological systems, (3) lower capital investment when compared to continuous processes for the same bioreactor volume, and (4) reduced risk of contamination due to a relatively brief growth period. Cells are inoculated into a fixed volume of medium. As the cells grow, they consume nutrients and release toxic metabolites into the surrounding medium. During this

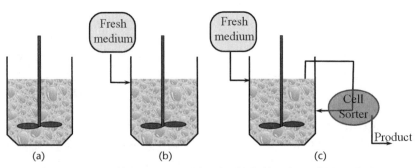

Figure 7.11   Different modes of bioreactors: (a) batch, (b) fed batch, and (c) perfusion.

dynamic reaction period, cells, substrates (including the nutrient salts and vitamins), and concentrations of the products vary with time. Proper mixing keeps the
differences in composition and temperature at acceptable levels. Cell growth ceases
due to an accumulation of toxic metabolites and a depletion of nutrients. The disadvantages include: (1) a decreased productivity due to time for filling, heating,
sterilizing, cooling, emptying, and cleaning the reactor, (2) a large expense incurred
in preparing several subcultures for inoculation, (3) an increased labor cost, and (4)
an increased focus on instrumentation due to frequent sterilization.

7.4.2.2   Fed-Batch Bioreactor

Eliminating the cell-growth restriction by a nutrient supply is replenishing depleted
nutrients to enable the production of cellular products for a prolonged period of
time. This is the principle of fed-batch operation [Figure 7.11(b)], which, in general, produces antibodies at a higher concentration due to the accumulation of
the product. A more frequently used method is initiating the reactor in the batch
mode until the growth-limiting substrate has been consumed. Then the substrate
is fed to the reactor as specified (batch) or is maintained by an extended culture
period (continuous). For a secondary metabolite production, in which cell growth
and product formation often occur in separate phases, the substrate is typically
added at a specified rate. This mode of operation provides an increased opportunity
for optimizing environmental conditions in regard to the growth phase or the
production and the age of the culture. The volume throughput per run is restricted
by the size of a bioreactor and is rate-limited by its prolonged operation time. To
compensate for the low-volume throughput, a much larger volume bioreactor is
often required. The controlled feeding of key nutrients is often performed and the
nutrient concentration in cultures is maintained at low levels.

7.4.2.3   Perfusion Bioreactors

In the perfusion mode, cells are retained by some type of retention device, whereas
the cell-free medium containing the product is removed continuously from the bioreactor. A fresh medium is added to maintain a constant volume [Figure 7.11(c)]. As
a result, nutrients are constantly replenished, and toxic metabolites are removed. By
comparison, a perfusion operation can produce a large volume of product from a
size-limited bioreactor on a continuous basis. However, the product concentration
is lower than that in a fed-batch operation due to the dilution effects resulting
from a constant medium replenishment. The specific metabolite production rate is
estimated from the expressions:

$$\frac{dC}{dt} = qX - pC \tag{7.29}$$

where C is the metabolite concentration, p is the normalized perfusion rate (1/day),
and q is the specific metabolite production rate. The perfusion rate (also referred
to as the dilution rate) is calculated as the ratio of volumetric flow of the media to

the volume of the reactor. The consumption of a substrate is expressed similarly to (7.26) with a negative sign to the production rate. For growth-limiting substrates, the Monod growth kinetic model is also used.

$$\frac{dC}{dt} = \frac{-k_{max}SX}{K_S + S} - pS \tag{7.30}$$

where k_{max} is the maximum substrate consumption rate, S is the substrate concentration, and K_s is the Monod growth constant. Perfusion bioreactors provide a number of advantages, including: (1) less nonproductive time expended in emptying, filling, and sterilizing the reactor, (2) reduced labor expense, (3) consistent product quality due to invariable operating parameters, (4) decreased toxicity risks to staff, due to automation, and (5) reduced stress on instruments due to sterilization. The disadvantages of perfusion bioreactors include: (1) higher investment costs in control and automation equipment and increased expenses for continuous sterilization of the medium, (2) minimal flexibility, since only slight variations in the process are possible (throughput, medium composition, oxygen concentration, and temperature), and (3) greater processing costs with continuous replenishment of nonsoluble components.

EXAMPLE 7.5

Thrombopoietin (TPO) is a hematopoietic growth factor that induces the proliferation of megakaryocytes and the differentiation of immature megakaryocytes. TPO is a potential therapeutic glycoprotein for the amelioration of thrombocytopenia associated with chemotherapy, irradiation, and bone marrow transplantation. To produce TPO, Sung et al. [2] transfected a Chinese hamster ovary (CHO) with the hTPO gene and cultured it in a 50-mL medium at 2×10^5 cells/mL. These cells can produce $4 \, \mu g/10^6$ cells per day of TPO. If an environmental condition supporting a specific growth rate constant is 0.39 day^{-1} and a death rate constant is 0.04 day^{-1}, how much TPO is generated in 3 days?

Solution:

To determine the production of TPO, we need to solve (7.28) and (7.29) simultaneously.

$$\frac{dX}{dt} = (0.39 - 0.04)X$$

$$\frac{dC_{TPO}}{dt} = 4X$$

Solving the two equations simultaneously in the Mathsolver with $X(t = 0) = 0.2 \times 10^6$ cells/mL.

$$X(t = 3 \text{ days}) = 0.571 \times 10^6 \text{ cells/mL}$$

$$C = 4.246 \, \mu g/mL$$

For 50 mL, $C = 0.213$ mg

In selecting a type of bioreactor and mode of operation, one should determine whether the goal is to obtain a secreted product or the cells themselves. Other issues to consider include the ease of sterilization and use, the need for automation, the expertise required, expenses, the space required, the time and speed of the operation, downstream processing needs, the process scalability and validation concerns, single or multiple vessels, and the batch or continuous product. Furthermore, using cell culture media that contain no serum or other components of animal origin is necessary. Serum is not preferred for many reasons, including: (1) possible introduction of contaminants and toxins such as viral particles, (2) a component variability inherent in biological sources, which affects both quality and performance of the final product, (3) a high cost as a raw material, and (4) interference with purification of the final product. Also, regulatory issues with the use of biological materials within the pharmaceutical industry are problematic; meeting these requirements raises the final production costs. For more details on bioreactors, refer to [3].

7.4.3   Downstream Processing

Once a product is obtained from a biosynthesis in bioreactors, the most time-consuming and cost-intensive factor is the purification of the product. Based on the physical and biochemical properties of the desired product and the level of contaminating molecules from the cultivation broth, a broad range of separation methods can be applied. If the supernatant is already cell-free and highly concentrated with product (as it often is with membrane systems), there is a significant reduction in downstream processing. However, for suspension cells cultured in larger batches, supernatants must be pumped out and clarified before they can be subjected to purification procedures. The purification process includes chromatographic and filtration technologies.

With purification technologies, the challenge is to recover the target protein in its active form and in high yields from other proteins found in the host organism, as well as from the many components of the medium in which the cells were grown. Time-dependent phenomena such as degradation, proteolysis, clipping, and cleaving, which would not be significant in a small scale because of a faster processing time, become the major problems in large-scale downstream processing. Hence, the combination and design of the single steps are of paramount importance for the economic process development. In the overall process design, the engineer has to consider different aspects such as the elimination of contaminants, process scalability, automation, the capacity of the production line, and regulatory compliance. Optimization of downstream processing technologies is considered the central element in the appropriate process design. For more details, see [4].

7.5   Preservation of Cells and Tissues

With rapid advances in the clinical diagnosis and treatment of diseases, there is an increasing need for using cell and organ transplantation to cure diseases and to correct genetic defects. Furthermore, more types of living cells, tissues, organs, and engineered tissues are required to be preserved. Preservation is important for modern medicine/healthcare and for many other areas:

1. The banking of a large quantity of living cells/tissues for genetic typing and matching between the recipients and donors to meet the increased clinical needs and sometimes the urgent needs (e.g., in a war or in events such as terrorist attack, or natural disasters);
2. Facilitating the transport of cells/tissues between different medical centers;
3. Allowing sufficient time for the screening of transmissible diseases (e.g., HIV) in donated cells/tissues before transplantation;
4. Engineered tissues needing to be successfully preserved before their practical use in any applications and commercialization;
5. The preservation of sperm and oocytes/eggs of endangered or transgenic species.

A fact that permits the long-term preservation of the living cells and tissues is that biological metabolism in living cells diminishes and eventually stops at low temperatures. Thus, for short-term storage (less than couple of days), cells and organs can be stored at 4°C. This is a clinically employed organ preservation technique and there are specific preservation solutions that minimize cellular swelling and membrane pump activity. Cell survival in cold storage depends on the cell type.

7.5.1 Long-Term Storage of Cells

For long-term storage (months to years), cells must be stored frozen at as low a temperature as possible; the colder the temperature, the longer the storage. Storage in liquid phase nitrogen (−196°C) allows the lowest possible storage temperature to be maintained with absolute consistency, but requires the use of large volumes (depth) of liquid nitrogen. There could be the possibility of cross-contamination by virus pathogens via the liquid nitrogen medium. For these reasons low temperature storage is commonly in the vapor phase nitrogen. For vapor phase nitrogen storage, cells present in small containers (called ampules) are positioned above a shallow reservoir of liquid nitrogen. Since a vertical temperature gradient exists in the vapor phase, the liquid level should be properly monitored.

While preserving, cooling the cells to required temperature plays a significant role in the survival. When cells are cooled too slowly (Figure 7.12), the extracellular environment freezes first, and extracelluar ice forms. Extracellular ice creates a chemical potential difference across the membrane of cells, creating an osmosis of water that dehydrates and shrinks the cell. The slower the cells are cooled, the longer the cells are dehydrated, causing irreparable damage called a solution injury. On the other hand, when cells are cooled too quickly, the cell retains water within the cell. Inside the cell water freezes and expands, as the density of ice is less than that of water. The abrasive ice crystals physically destroy the cell itself; this is called an intracellular ice injury. If the cooling rate is optimized properly, maximum survival is achievable where the total cell damage from both mechanisms is minimized. However, this is only a narrow range of a cooling rate, which permits a high viability of cryopreserved cells. Specific cell properties such as membrane permeability

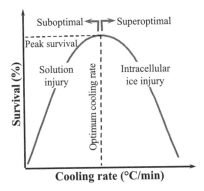

Figure 7.12   Effect of cooling rate on cells.

to water and initial intracellular water concentration will determine the precise rate of cooling. To reduce injury to cells during cryopreservation, cryoprotective agents (CPA) have been developed to both permeate the cell and to displace water to prevent intracellular ice crystal formation. CPAs are antifreeze components added to the solutions in which the cells are being frozen. Polge et al. [5] successfully cryopreserved mammalian cells in 1949 using glycerol as a CPA. Another commonly used CPA is dimethyl sulfoxide, which may not be appropriate for all cells, as it could induce differentiation.

It is important to control the freezing as well as thawing environment in order to maximize cell viability. Cryopreserved cells are extremely fragile and require gentle handling. Cryopreserved cells should be thawed for use as rapidly as possible, without letting the temperature exceed the physiological temperature range. Quick thawing prevents recrystallization (the tendency of small ice crystals to increase in size) of intracellular ice and minimizes the exposure time to high solute concentrations. After thawing, the cells are transferred to a flask and a fresh medium is added dropwise and slowly to prevent osmotic damage to cells. After adherent cells attach and spread, the medium is changed to remove the CPA. A thawed culture may require a certain amount of time before usage in a specific application.

Steps for obtaining increased cell numbers have been developed. For most cells, 1°C/min is used, which is performed in a controlled rate freezer. First, adherent cells are removed from the substrate by enzymatic treatment and resuspended as a suspension at high cell densities (at least 106 cells/mL) in a medium containing serum and a CPA. When glycerol is used, the cells should be allowed to rest in the solution containing CPA for 30 minutes at room temperature prior to cooling to ensure that the CPA permeates through the cell membrane without suffering from osmotic shock. To avoid the possible effect of the osmotic pressure induced by a sudden change of concentration of CPA, it might be effective to raise the concentration of CPA gradually. The major steps in the cryopreservation process include:

1. Adding CPAs (at a concentration of 10% v/v) to cells/tissues before cooling;
2. Cooling the cells/tissues to a low temperature (e.g., −196°C, liquid nitrogen temperature at pressure of 1 atm) at which the cells/tissues are stored;
3. Warming the cells/tissues;
4. Removing the CPAs from the cells/tissues after thawing.

7.5.2   Storage of Tissues

To cryopreserve successfully a large size tissue, one would rather avoid the deleterious effects of a cooling rate, similar to single cells. The power of the cold to destroy undesirable cells like cancer cells is used to surgically operate on cells in cryosurgery. In addition to the two types of cellular damage discussed earlier (solution injury and intracellular ice injury), exposure to the cold can eliminate blood circulation (called ischemic injury) in the undesirable cells, thereby preventing the delivery of nutrients and oxygen as well as the removal of wastes and CO_2; as blood contains a large amount of water, ice will readily form and grow throughout the interconnected tubes. One of the practices for the short-term storage of vascularized tissues is to purfuse the tissue with preservative solutions such as noncolloidal histidine-tryptophan-ketoglutarate (HTK) solution, hyperkalaemic N-Tris (hydroxymethyl) methyl-2-aminoethane sulphonate balanced (anionic buffer), or salt solution (CPTES). The key areas that need advancement are additives to preservation solutions, alternatives/adjuncts to preservation solutions, and optimum perfusion technology.

Cryopreserving tissues for the long term poses more complex scenarios, although equations developed for single cells are used to predict changes in tissue volume. Tissues normally contain more than one cell type, which could demand different freezing rates. Furthermore, the presence of an extracellular matrix and a large amount of water make the penetration of CPA into the cell more difficult. First, the microenvironment surrounding each cell needs to be equilibrated with the CPA and then the CPA can enter the cell. However, the prolonged exposure of cells to hypertonic conditions (caused by a high CPA content) and a high concentration of CPA at temperatures are also deleterious. An emerging technology is vitrification (glass formation), which allows cells to be preserved in their existing state without going through the dehydration that is achieved in slow cooling. In vitrification, high concentrations of CPAs (three to four times higher) are used with a cooling rate nearly 10,000 times faster by plunging them directly into liquid nitrogen. The rapid cooling is necessary for preventing the toxicity of the high levels of CPA at room temperature and achieving vitrification. The glass transition occurs even at a low cooling rate, if the concentration of CPA is high enough. However, introducing such a high concentration of CPA into cells, as well as into a tissue, is difficult unless new CPAs (low viscosity, low toxicity) are discovered. Alternatively, adding a pressure during a freezing process is made to reduce the concentration of CPA, which can vitrify.

The cryopreservation procedures that are currently in use for tissues can vary greatly from tissue bank to tissue bank. Furthermore, some of the current practices are protected by trade secrets. Although the American Association of Tissue Banks (AATB), a nonprofit organization, publishes standards for tissue banking, it does not recommend any specific cryopreservation procedures. Rather, the AATB imposes guidelines for establishing procedures and implementing standard operations in retrieval, processing, storage, and/or distribution of tissues that meets the needed quality control for safe clinical usage. Thus, individual tissue banks need to develop an optimal cryopreservation protocol for a particular tissue. Then the AATB can help the tissue bank in implementing the protocol and accrediting the tissue bank.

Problems

7.1   Among the neuroendocrine factors that affect the immune system, the pineal
        hormone melatonin appears to be an important one. Melatonin receptors have
        been described in human immune cells such as T-cells ($K_d = 240$ pM), human
        Th2 lymphocytes from bone marrow ($K_d = 350$ pM), platelets ($K_d = 4$ nM),
        neutrophils ($K_d = 132$ pM), granulocytes ($K_d = 2$ nM), and monocytes with a
        K_d around 23 pM). Arrange these cells in the order in which melatonin stimu-
        lation will be observed [6].

7.2   Growth hormone receptor (GHR) antagonists are a new class of drugs, whose
        development has been facilitated by an understanding of the molecular interac-
        tion between GH and its class 1 cytokine receptor. Pegvisomant (B2036-PEG)
        is a GHR antagonist developed for the treatment of acromegaly, a condition
        usually caused by excessive GH secretion. To determine the effectiveness of
        this drug, Ross [7] used a cell line and performed binding assays. Scatchard
        analysis revealed that the Ka for GH and B2036 was 0.36×10^9 and $0.32 \times$
        10^9 mol/L^{-1}, respectively; and the binding capacity was 4×10^5 and 5.5×10^5
        receptors/cell, respectively. If so, is the antagonist binding as good as the GH?
        Show calculations.

7.3   The peroxisome proliferator-activated receptor (PPAR) belongs to the nuclear
        receptor superfamily that plays an important role in the regulation of the stor-
        age and catabolism of fats. It is known that binding of ligands to PPAR-γ (one
        of the subtypes) induce a conformational change. Interaction of PPAR-γ on
        its affinity to various ligands can be modeled [8] by assuming single bind site
        model as $\dfrac{dR}{dt} = k_{on}L(R_{max} - R) - k_{off}R$ where R represents the response units, L
        is the concentration of the ligand, k_{on} is the association rate constant and k_{off}
        is the dissociation rate constant. From experiments, following data for two
        ligands were obtained. Using these values determine the rate constants and
        equilibrium dissociation constant. Which ligand is better to inhibit the action
        of PPAR-γ?

Ligand 1 [µM]	0.1	0.5	1	5	20
Response Unit	3	9	13.8	20.5	24
Ligand 2 [µM]	0.1	0.5	1	5	20
Response Unit	1	3	6.5	14	17

7.4   Glucocorticoids are secreted from the adrenal glands and act as a periph-
        eral effector of the hypothalamic-pituitaryadrenal (HPA)1 axis, playing an
        essential role not only in energy metabolism but also in stress response and
        homeostatic regulation. Glucocorticoids elicit hormone action via binding to
        their cognate receptor glucocorticoid receptor (GR), which is a member of the
        nuclear receptor superfamily and localizes in the cytoplasm as a latent species.
        The GR is composed of a modular structure with a number of functional do-
        mains, including AF-1 transactivation domain and AF-2. Specific mutations
        in the GR cause selective loss of GR dimerization and so prevent dimer-depen-
        dent transactivation. Such dissociated receptors retain the ability to oppose

other transcription factor function. One of the ligands is dexamethazone. To understand the importance of LBD, a new GR with a mutation in a tyrosine residue was introduced and binding affinities were measured [9]. Using these values, determine the change in the binding affinity.

Bound [fmol]	50	100	200	300	400
Free ligand with wild type GR [nmol]	0.5	1.3	3	6	10
Mutant [nmol]	0.8	2	5	10.5	20

7.5   Warfarin, an anticoagulant, is known to bind to albumin. There are two different binding sites on human serum albumin (HSA) as well as bovine serum albumin (BSA). To understand the difference in binding characteristics, Sun [10] measured the binding characteristics of Warfarin to two albumins and reported the following values.

$Warf_{free}$ (μM)	86.3	73.9	61.6	49.3	37	24.6	12.3	6.2	1.2	0.61	0.36	0.12
R_{BSA} (%)	4.87	4.74	4.52	3.9	3.68	2.63	1.54	1.01	0.7	0.64	0.48	0.39
R_{HSA} (%)	5.57	4.69	3.83	3.39	2.94	2.39	1.58	1.04	0.61	0.48	0.39	0.31

In this table R is the average number of moles of ligand bound per mole of albumin. The absolute amount of BSA and HSA used in the experiment are 2.28 nmol and 1.14 nmol, respectively. Determine the difference in binding characteristics of the two albumins with Warfarin using: (a) a Scatchard plot and a nonlinear regression and (b) the Hill equation with the assumption $Alb + nW \Leftrightarrow Alb(W)_n$.

7.6   Starting with the equilibrium expression for O_2 binding to myoglobin, $O_2 + MbO \leftrightarrow Mb\,O_2$, derive the equation $Y = \dfrac{pO_2}{pO_2 + P_{50}}$ where Y is the fractional saturation of myglobin.

7.7   The partial pressure of O_2 in the capillaries of active muscle is approximately 20 mmHg.

(a) Calculate the fractional saturation, Y, for myoglobin and hemoglobin.

(b) How do these values relate to the function of myoglobin in muscle?

7.8   A pharmacologist testing a new drug determines its effect on oxygen dissociation of hemoglobin. Under control conditions, when the drug is absent, the hemoglobin solution is 50% saturated at an oxygen tension of 30 mmHg, and it is 25% saturated at an oxygen tension of 20 mmHg. Under experimental conditions, with the drug present, the hemoglobin solution has a P50 of 40 mmHg. Predict the percent difference in saturation of control and experimental solutions at an oxygen tension of 83 mmHg. Please show your work and indicate your reasoning.

7.9   Calculate the amount of protein containing a bound ligand assuming a single binding site with a K_d of 10^{-6} M, a protein concentration of 10^{-6} M and a total ligand concentration of 10^{-6} M. How much is bound if the protein concentration is reduced to 10^{-9} M but the total ligand concentration is maintained at 10^{-6} M (i.e., a thousand-fold greater)? What does this tell you about the protein concentration required to perform a binding measurement?

How much will be bound if the protein concentration is kept at 10^{-9} M and the ligand concentration is increased to 10^{-3} M?

7.10 One of the methods to measure receptor-ligand interactions is an equilibrium dialysis experiment. If a protein is placed in a dialysis bag and then dialyzed against a solution containing a ligand, the ligand will distribute itself across the bag but the protein will remain within the bag. At equilibrium, the free ligand concentration will be identical inside and outside the bag; but an additional ligand will be found in the bag due to binding to the protein. Thus, the free ligand concentration is the concentration measured outside the bag. A single dialysis experiment will give one point in a plot and a series of dialyses are set up, each with the same protein concentration, but with different ligand concentrations.

In an equilibrium dialysis experiment the protein concentration inside the bag is 2×10^{-3} M and the equilibrium concentration of the ligand is 1.2×10^{-4} M inside and 1.00×10^{-4} outside. Calculate the fraction of the sites filled. What percentage of the sites is filled? Describe how you would proceed to determine the binding constant and number of sites for the protein. How would you determine if there was more than one binding site per protein? If there were more than one binding site per protein, how would you determine if they had different affinities for the ligand?

7.11 Assume a steady flow that is also used to estimate the average and maximum wall shear stress in dynamic flow regimes. Dynamic flow regimes have been employed without note of whether the applied frequencies are within limits for using the PPFC. Flow frequencies employed in the stimulation of cells generally remain below 10 Hz.

7.12 A desirable property for an artificial oxygen carrier is to match the properties of hemoglobin (MW is 128 kDa). It has an affinity (p50) of 26–28 mmHg at a concentration of 7.14 gHb/dL (4–8 mM heme). A circulation half-life is greater than 24 hours and a shelf life is 6 months. Do you think the artificial carrier is as good as hemoglobin?

7.13 A diver starts at a depth of 66 feet of water (3 atm) with a lung volume of 3.0L BTPS. If he ascends to 33 feet (2 atm) immediately without breathing, what will his lung volume be at 33 feet?

Enzyme	V_{max} mol/ (min.mg)	km (mM)
$B_{synthase}$	1.0	0.05
$C_{synthase}$	1.8	0.3
$D_{synthase}$	2.7	0.5

7.14 A reaction sequence in a tissue proceeds at an overall rate of 0.9 mol/min.mg of tissue. Included in the overall sequence is the following reaction scheme.

$$A \xrightarrow{\text{Bsynthase}} B \xrightarrow{\text{Csynthase}} C \xrightarrow{\text{Dsynthase}} D$$

The kinetic constants of the enzymes are given in the table. All other enzymes involved have maximum velocities of the order of 10 mol/(min.mg) or more.

(a) What are the intracellular concentrations of A, B, and C, assuming all reactions have the same rate?

(b) What is the rate limiting step in the process? Why?

7.15   A serine kinase has a histidine, a serine, and aspartic acid in the active site. When any of them are mutated to alanine, the reaction rate drops by several orders of magnitude. This serine kinase catalyzes the reaction of glucose with the help of ATP to glucose-6-phosphate at a velocity of 20 μM/min at [S] = 0.01M. The K_m value is 1.5×10^{-5} and k_{cat} is 190 s^{-1}. Assuming that this enzyme follows the Michaelis-Menton kinetics, what will the reaction velocity be at [S] = 1.0×10^{-5}M?

7.16   In the laboratory, you want to measure the adhesive strength of cells to a novel biomaterial in a parallel flow chamber. Bell [11] proposed a model for a spherical surface. Since the cells (shown in the figure) typically assume a hemispherical shape, these equations have to be modified. Modified equations for hemispherical shape can be obtained from [12]. Assuming a volumetric flow rate (Q):

(a) Calculate the net force exerted to a cell located at 3 cm using a force value from [11].

(b) Calculate the volumetric flow rate.

Assume r_p = 10–15 μm, h_1 = 5–10 μm. The dimensions for the parallel flow chamber can be obtained from [13].

7.17   Explain why knowing the spatial distribution of adhesion ligands at a surface in addition to the average total density is a key part of understanding cellular responses to ligand-modified biomaterials.

7.18   You have seeded 50,000 cells in a tissue culture plate. You added a nontoxic radioactive substance (^{3}H thymidine) to the culture and read the incorporated radiation (counts per minute) 24 hours later. The radiation reading is 5,000 counts per minute (cpm), which correlates to the amount of DNA synthesized. You have also recounted the cells after 24 hours, and found that the total cell number is 60,000 per culture plate. If you add a specific growth factor to the medium under the same conditions (50,000 cells/well over 24 hours), the incorporated radiation reads 800 cpm.

(a) What is the growth rate under the new conditions?

(b) In making these calculations, what assumptions did you make (you need to mention three)?

7.19   A group of engineers is trying to evaluate a new material for a biomedical application. They seeded 10,000 mammalian cells in a 24-well tissue culture plate and onto their new material inserted inside a 24-well plate. Assume that the areas are the same. To understand the growth rate, they added 5-bromo2'-deoxy-uridine (BrdU), which incorporates into the DNA in place of thymidine. To detect the amount of BrdU incorporated in the DNA, they detected using a fluorescently conjugated monoclonal antibody against BrdU after denaturation of the DNA (usually obtained by exposing the cells to acid, or heat). After 24 hours, the intensity reading was 1.5, which correlates to the amount of DNA synthesized. They also counted the cells after 24 hours and found that the total cell number was 20,000 per culture plate.

(a) If the count on the new material is 1.8 intensity units after 48 hours, is the growth rate higher or lower to the tissue culture?

(b) What are the assumptions?

7.20   Describe the impact of substrate concentration on growth and substrate utilization. Given a Monod curve, be able to determine the maximum specific growth rate and half saturation constant.

	A	B	C
Specific growth rate constant, μ (hr^{-1})	0.014	0.02	0.001
Extent adhesion (% of initial seeding)	70	40	10
Cell spreading (μm^2)	40	50	25

7.21   There are three biodegradable materials (A, B, and C) to be tested for a tissue engineering application. The primary requirement is to form a monolayer of endothelial cells on the surface. Using an initial seeding of 10^4 cells, the following parameters have been reported for these biomaterials. Using these values,

(a) Calculate the time required to generate a 1 cm^2 surface of the monolayer.

(b) Which material will you select for further analysis?

(c) If you wish to accelerate the process and reduce the time required for monolayer development, what parameters will you focus upon?

(d) What interactions will you consider if you were to test the scaffold in vivo? How would you evaluate these interactions?

7.22   By reading the published literature extensively, you have generated the following summary of the differences between the two major types of endothelial cells used for in vitro investigations:

	Umbilical Vein	Bovine Aortic
Growth rate	Slow	Fast
Number of population doublings before senescence	20	35
Strength of adhesion to the substrate	High	Low

Based on these observations, which cell type would you expect to be the fastest to align in the direction of shear when subjected to fluid shear stress? Give a rationale for your answers (20 points).

7.23   A bioreactor has an oxygen mass transfer rate of (ka) of 8 hr^{-1}. Hybridoma cells are inoculated into the bioreactor at an initial viable cell density of 100,000 cells/mL. The cell doubling time is 18 hours. The head space in the bioreactor contains 5% CO_2/95% air. The oxygen consumption rate of cells is 5×10^{-17} mol/ cell-s. Calculate the bulk oxygen concentration after 72 hours.

7.24   Cocultures of hepatocytes and fibroblasts have been used to build prototype bioartificial livers. A series of tests to measure the effect of fibroblasts on the expression of liver-specific functions by the hepatocytes was carried out.

A fixed number of hepatocytes (0.25×10^6 cells) were seeded onto surfaces in circular patterns of a 490-μm diameter. A variable number of fibroblasts were then seeded around the hepatocyte islands (fibroblasts will not attach on top of the hepatocyte islands, but only on the free spaces in-between). To keep the seeding density of fibroblasts (number of fibroblasts per unit area of free space available in-between hepatocyte islands) constant, the total area used for culturing the cells was decreased with decreasing fibroblast numbers according to the table. The secretion rate of urea was measured for each culture condition shown above, and the results found to fit a dose-response curve which can be described by

$$P = \frac{P_{max} N}{K_{0.5} + N} + P_0$$

where P_0 and P_{max} are the minimum and maximum urea secretion rates, respectively, N is the number of fibroblasts, and $K_{0.5}$ is the number of fibroblasts to obtain a half-maximal response. For the hepatocyte-fibroblast system, the parameter values are $P_0 = 23.7$ mg/day, $P_{max} = 230$ mg/day, and $K_{0.5} = 4.66 \times 10^3$ fibroblasts.

Experiment	A	B	C
Fibroblasts seeded	1.5×10^6	0.25×10^6	0.125×10^6
Center-to-center spacing between hepatocyte islands	1,230 μm	650 μm	560 μm

(a) Estimate the culture area and the P value for each experiment shown in the table above.

(b) What culture condition maximizes the urea secretion per hepatocyte? What culture condition maximizes the urea secretion per unit surface area of culture?

(c) In a prototype bioartificial liver device of given size and surface area available for culturing cells, which culture configuration would you use to maximize the clinical benefit [personal communication with Dr. Berthiaume, Harvard Medical School, 2010]?

7.25 The epidermal growth factor (EGF) is known to influence cell growth through the interaction with its receptor. As a way to inhibit cell growth in the treatment of psoriasis, a drug is tested and shows that it reduces the receptor level in keratinocytes by 50%. Estimate the decrease in the proliferation of keratinocytes compared to cultures without the drug. K_D and B_{max} for EGF-EGFR are known to be around 70 nM and of 5 μmol/μmol of receptor protein. List your assumptions. (Some assumptions are that the drug does not change the interactions of EGF-EGFR, the same number of cells is used, and single binding interaction is assumed.)

7.26 Fox [14] reported using Chinese hamster ovary (CHO) cells for the production of Interferon (IFN)-γ in a batch reactor. Approximately 250,000 cells/mL were seeded in 100 mL of a medium containing 20 mM glucose (MW = 180.16). These cells are known to have a lag phase of 12 hours and the following rate constants are known. If they are cultured for a total period of 96 hours, what is the concentration of IFN-γ and glucose? How many cells are present in the broth? Substrate consumption and cell growth are assumed to follow Monod growth kinetics for which following kinetic data is available.

Maximum specific growth rate constant, μ_{max} (hr^{-1})	0.036
Monod growth constant, K (mg/mL)	0.80
Maximum specific glucose consumption rate, S_{max} (mg/cell.hr)	2.4×10^{-7}
Monod growth constant, K_S (mg/mL)	11
Specific IFN-γ production rate, C_{IFN} (μg/cell.hr)	1.5×10^{-8}

7.27 Chinese hamster ovary (CHO) cells are used to produce Factor VIII in a continuous culture with cell recycle. The μ_{max} is 0.035 hr^{-1}. The dilution rate used is 1.5 times of the μ_{max}. Under the operating conditions cells have a finite death rate; thus, viable and dead cells are both seen in the culture. However, the cell separation device used preferentially recycles the viable cells with a 90% recovery and purging 10% of viable cells, whereas for the dead cells, the recovery is only 50% (i.e., purging 50% cells). The concentration of viable and dead cells in the outlet stream are 10^9 cells/L and 2×10^8 cells/L. The operation recycles 20% of the stream taken from the stirred tank bioreactor to the bioreactor. Calculate the specific death rate.

7.28 A 10-L stirred tank is used to cultivate hybridoma cells. Dry air is sparged from the bottom at 1.0 L/min to supply the oxygen. The gas composition in the outlet is 16% oxygen, 3.5% CO_2, and 6.5% water vapor; the balance is nitrogen. The ambient pressure and temperature are 1 atm and 37°C.

The dissolved oxygen in the reactor is controlled at 20% of saturation with ambient air.

(a) Assuming the reactor is at steady state, what is the volumetric oxygen transfer coefficient (K_{La})?

(b) If the specific oxygen uptake rate of hybridoma cells is 5×10^{-11} mmole/cell-hr. What is the cell concentration?

(c) What is the respiratory quotient (R.Q.) (i.e., mole CO_2 produced/mole O_2 consumed)? The mole fraction of CO_2 in the ambient air is negligible. The universal gas constant $R = 0.082F$ (L atm/moleK).

7.29   Prostaglandin E_2 receptors, subtype EP_1 (PGE_2EP_1), have been linked to several physiologic responses, such as fever, inflammation, and mechanical hyperalgesia. Local anesthetics modulate these responses, which may be due to direct interaction of local anesthetics with a PGE_2EP_1 receptor signaling [15].

7.30   Tetrandrine, a bisbenzyltetrahydroisoquinoline alkaloid extracted from the Chinese medicinal herb *Radix stephania tetrandrae*, is known to possess a wide spectrum of pharmacological activities. Situated at the interface between blood and muscular media of the vessel, endothelial cells play a dynamic role in the regulation of vascular tone. Endothelial cells are known to lack voltage-dependent Na^+ or Ca^{2+} channels but to exhibit a Ca^{2+} entry mechanism that depends on the electrochemical driving force and/or Ca^{2+} stores. It has recently been shown that the vascular tone can be modulated by either the change of K^+ concentration in myoendothelial gap junctions or electric spreading of the membrane potential.

References

[1]   Lehninger, A., D. L. Nelson, and M. M. Cox, *Lehninger Principles of Biochemistry,* 5th ed., New York: W.H. Freeman, 2008.

[2]   Sung, Y. H., S. J. Hwang, and G. M. Lee, "Influence of Down-Regulation of Caspase-3 by siRNAs on Sodium-Butyrate-Induced Apoptotic Cell Death of Chinese Hamster Ovary Cells Producing Thrombopoietin," *Metabolic Engineering,* Vol. 7, No. 5-6, September-November 2005, pp. 457–466.

[3]   Shuler, M. L., and F. Kargi, *Bioprocess Engineering: Basic Concepts,* 2nd ed., Upper Saddle River, NJ: Prentice-Hall, 2001.

[4]   Harrison, R. G., et al., *Bioseparations Science and Engineering,* Oxford, U.K.: Oxford University Press, 2002.

[5]   Polge, C., A. U. Smith, and A. S. Parkes, "Revival of Spermatozoa After Vitrification and Dehydration at Low Temperature," *Nature,* Vol. 164, 1949, p. 166.

[6]   Barjavel, M. J., et al., "Differential Expression of the Melatonin Receptor in Human Monocytes," *The Journal of Immunology,* Vol. 160, 1998, pp. 1191–1197.

[7]   Ross, R. M. S., et al., "Binding and Functional Studies with the Growth Hormone Receptor Antagonist, B2036-PEG (Pegvisomant), Reveal Effects of Pegylaton and Evidence That It Binds to a Receptor Dimer," *The Journal of Clinical Endocrinology & Metabolism,* Vol. 86, No. 4, 2001, pp. 1716–1723.

[8]   Changying, Y., et al., "Binding Analysis Between Human PPARgamma-LBD and Ligands," *Eur. J. Biochem.,* Vol. 271, No. 2, January 2004, pp. 386–397.

[9]   Ray, D. W., et al., "Structure/Function of the Human Glucorticoid Receptor: Tyrosine 735 Is Important for Transactivation," *Molecular Endocrinology*, Vol. 13, No. 11, November 1999, pp. 1855–1863.

[10]   Sun, S. F., "Study of Binding of Warfarin to Serum Albumins by High-Performance Liquid Chromotography," *Journal of Chromatography*, Vol. 288, No. 2, April 24, 1984, pp. 377–388.

[11]   Bell, G. I., "Models for the Specific Adhesion of Cells to Cells," *Science*, Vol. 200, 1978, pp. 618–627.

[12]   Palecek, S. P., et al., "Integrin-Ligand Binding Properties Govern Cell Migration Speed Through Cell-Substratum Adhesiveness," *Nature*, Vol. 6, No. 385, 1997, pp. 537–540.

[13]   Powers, M. J., et al., "Cell-Substratum Adhesion Strength as a Determinant of Hepatocyte Aggregate Morphology," *Biotechnology and Bioengineering*, Vol. 53, No. 4, 1997, pp. 415–426.

[14]   Fox, S. R., et al., "Maximizing Interferon-Gamma Production by Chinese Hamster Ovary Cells Through Temperature Shift Optimization: Experimental and Modeling," *Biotechnology and Bioengineering*, Vol. 85, No. 2, 2003, pp. 177–184.

[15]   Honemann, C. W., et al., "The Inhibitory Effect of Bupivacaine on Prostaglandin E_2 (EP_1) Receptor Functioning: Mechanism of Action," *Anesth. Analg.*, Vol. 93, No. 3, September 2001, pp. 628–634.

Selected Bibliography

Karlsson, J. O. M., and M. Toner, "Cryopreservation," Chapter 28, in Lanze, R., R. Langer, and J. Vacanti, (eds.), *Principles of Tissue Engineering*, San Diego, CA: Academic Press, 2001.

Lauffenburger, D. A., and J. J. Linderman, *Receptors*, Oxford, U.K.: Oxford University Press, 1993.

Morgan, J. R., and M. L. Yarmush, *Tissue Engineering Methods and Protocols (Methods in Molecular Medicine)*, Totowa, NJ: Humana Press, 1999.

Palsson, B. O., *Systems Biology: Properties of Reconstructed Networks* Cambridge, U.K.: Cambridge University Press, 2006.

Palsson, B. O., and S. N. Bhatia, *Tissue Engineering*, rev. ed., Upper Saddle River, NJ: Prentice-Hall, 2003.

Vunjak-Novakovic, G., and I. R. Freshney, (eds.), *Culture of Cells for Tissue Engineering (Culture of Specialized Cells)*, New York: Wiley-Liss, 2006.

Walsh, G., *Biopharmaceuticals: Biochemistry and Biotechnology*, New York: Wiley, 1998.

CHAPTER 8

Biomedical Imaging

8.1 Overview

Light is the fundamental source due to which we can see things around us. Photographical techniques use the properties of light and its interaction with objects to capture images of interest. Using the same concept along with optical magnification, light microscopes were constructed, which opened the gates of modern biology and medicine through the discovery of the cell. However, most tissues and organs inside the body do not transmit visible light and cannot be examined for disease or damage. For routine examination of internal parts of the body, developing techniques that allow visualization of tissues or organs nondestructively or with minimum surgical procedures are necessary. Medical imaging techniques are developed to facilitate examination of internal body parts. Most imaging technologies are developed similar to light microscopy but use alternative energy sources such as X-rays, infrared rays, radio waves, and ultrasound based on their interaction with different body parts. Some of the biomedical imaging techniques are shown in Figure 8.1.

Development of less invasive, less expensive, more robust imaging technologies has revolutionized modern medical care. Imaging techniques such as magnetic resonance imaging (MRI), X-ray computed tomography (CT), and positron emission tomography (PET) have become important tools for the early detection and monitoring of diseases and understanding of the basic molecular aspects of living organisms. Imaging capabilities at the cellular level and tracing biomarkers of drug response are also playing an important role in the pharmaceutical industry. It is possible to generate 3D images and movies of various parts of the body using data processing algorithms. Image analysis has also evolved from measurements of hand drawings and still photographs to computational methods that semiautomatically quantify objects, distances, density of cells, and subcellular structures. Further, the new imaging modalities produce information about anatomical structures that is linked to functional data. This chapter introduces basic concepts in biomedical imaging such as the interaction of matter with the electromagnetic spectrum and the basics of medical imaging.

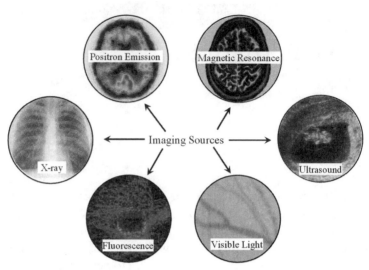

Figure 8.1 Biomedical imaging technologies.

8.2 Properties of Light

8.2.1 Electromagnetic Spectrum

Some properties of light can be explained with the assumption that light is made of tiny particles called photons, and other properties of light can be explained with the assumption that light is some kind of wave. Photons travel at the speed of light (V_{light}) which is 3.00×10^8 m/s in vacuum (the commonly used notation is c—since c is used to denote concentration in this book, a different notation is adapted). Nothing can travel faster than light in vacuum, and V_{light} is the ultimate speed limit in the universe. The visible light is a small part of spectrum of energy sources referred as the electromagnetic (EM) spectrum. The EM spectrum encompasses many types of light or radiation (Figure 8.2). The EM spectrum includes the microwaves commonly used in the kitchen to warm food, and the radio waves that are broadcast from radio stations.

EM waves are characterized by one of three properties: wavelength (λ), which is the distance between two adjacent crests of the wave; frequency (f), which specifies how often the fields return to their original configuration of the wave and measured in hertz; or the energy (E) of the individual photons in the wave (discussed in Section 8.2.2). The wavelength and frequency of the wave are related via the speed of light:

$$V_{light} = f \lambda \tag{8.1}$$

EM waves are grouped into different categories based on their frequency (or on their wavelength). Visible light, for example, ranges from violet to red. Violet light has a wavelength of 400 nm, and a frequency of 7.5×10^{14} Hz. Red light has a wavelength of 700 nm, and a frequency of 4.3×10^{14} Hz. Any EM wave with a frequency (or wavelength) within those two extremes can be seen by the human eye.

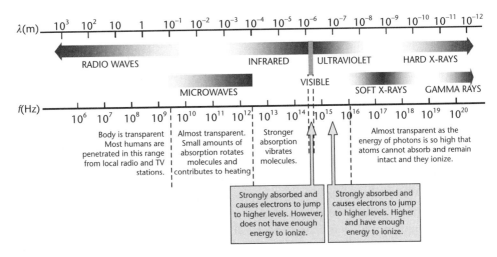

Figure 8.2 Electromagnetic spectrum.

8.2.2 Energy in an EM Wave

No medium is required for EM waves to travel through, unlike sound waves, which cannot travel through vacuum. An EM wave carries energy although it carries no mass. It also has momentum, and can exert pressure (known as radiation pressure). Tails of comets point away from the Sun primarily due to the radiation pressure exerted on the tail by sunlight (and other forms of radiation). Plank proposed that the quantum of energy carried by an EM wave is proportional to the frequency, f.

$$E = hf \tag{8.2}$$

where h is the Plank's constant (6.626×10^{-34} J.s or 4.136×10^{-15} eV s). An electron volt (eV) is defined as the energy that an electron (or a proton) gains when accelerated through a potential difference of 1V. One eV is 1.602×10^{-19} J.

EXAMPLE 8.1

Calculate the energy of a photon of blue light of wavelength 450 nm.
Solution:
From (8.1),

$$f = V_{light} / \lambda$$

From (8.2),

$$E = 4.136 \times 10^{-15} [eV.s] \times \frac{3 \times 10^{8} [m/s]}{450 \times 10^{-9} [m]}$$

$E = 2.76$ eV of energy.
Blue light occurs in quantized packets of 2.76 eV, and the energy always occurs in precisely the same sized energy packets.

Infrared light, microwaves, and radio/television waves are lower energy waves (lower frequency, longer wavelength) relative to visible light. Ultraviolet light, X-rays, and gamma rays possess higher energy than visible light (higher frequency, shorter wavelength). Most gamma rays are somewhat higher in energy than X-rays. It is convenient to describe X-rays in terms of the energy they carry, in units of thousands of electron volts (keV). X-rays have energies ranging from less than 1 keV to greater than 100 keV. Although the distinction between hard and soft X-rays is not well defined, hard X-rays have the highest energy (greater than 10 keV) while the lower energy X-rays are referred to as soft X-rays. The energy and spin of an atom or a nucleus can be changed by the absorption or emission of a photon. *An X-ray photon with sufficient energy can interact with and remove electrons bound to an atom (the process of ionization). This is why X-rays and g-rays are also referred to as ionizing radiation.*

The EM waves are made up of two parts: an electric field and a magnetic field. The two fields are at right angles to each other. EM waves propagate perpendicular to both electric and magnetic fields. If the EM wave propagates in the x-direction, then the electrical field is in the z-direction and the magnetic field will be in the y-direction. Since EM waves have both electric and magnetic fields, the energy in the wave is linked to those fields. The magnetic field is analogous to the electric field (discussed in Chapter 3), and the same field model is used to describe magnetic field. Further, the SI unit for magnetic field strength is Amperes per meter. A constant current produces a constant magnetic field, while a changing current produces a changing magnetic field. Conversely, a magnetic field produces current, as long as the magnetic field is changing, which is termed as induced electromotive force. A steadily changing magnetic field induces a constant voltage, while an oscillating magnetic field can induce an oscillating voltage.

Similar to the energy density (energy per unit volume) in an electric field given by (3.43), the energy density in a magnetic field of strength H is written as

$$w_m = \frac{1}{2\mu_0} H^2 \tag{8.3}$$

where μ_0 is the permeability of free space (1.2566×10^{-6} Weber/Ampere.meter or N/A^2). The number of magnetic lines of force cutting through a plane of a given area at a right angle is known as the magnetic flux density (B) or magnetic induction. The flux density is a measure of the force applied to a particle by the magnetic field. The unit of magnetic flux density is tesla (equivalent to N/A.m or N.s/C.m). Since tesla is a very large unit, the smaller magnetic flux density unit is the gauss (1 tesla = 10,000 gauss), or dyne/A.cm. The Earth's magnetic field is typically a fraction of a gauss. The relation between magnetic flux density and magnetic field strength in vacuum as well as in the air or other nonmagnetic environments is constant ($\beta = \mu_0 H$). Using (3.43), the total energy density associated with an EM wave is written as

$$w = \frac{1}{2}\varepsilon_0 E^2 + \frac{1}{2\mu_0} H^2 \tag{8.4}$$

For an EM wave, the energy associated with the electric field is equal to the energy associated with the magnetic field. That is,

$$\varepsilon_0 E^2 = \frac{H^2}{\mu_0} \text{ or } E = \frac{H}{\sqrt{\varepsilon_0 \mu_0}} \tag{8.5}$$

Hence, the energy density is written in terms of just one or the other. J. C. Maxwell, who showed that electricity and magnetism could be described by four basic equations, also showed a connection between v, μ_0, and ε_0 (the permeability of free space):

$$V_{light} = \frac{1}{\sqrt{\varepsilon_0 \mu_0}} \tag{8.6}$$

Substituting the above relation in (2.5),

$$E = V_{light} H \tag{8.7}$$

EXAMPLE 8.2

A certain plane EM wave has a maximum electric field strength of 30 V/m. Find the maximum magnetic field strength and magnetic flux density.
Solution:
From (8.7), 30 [V/m] = 3.00×10^8 [m/s]*H
Hence $H = 1.0 \times 10^{-7}$ [V.s/m^2] = 1.0×10^{-7} A/m
$B = 1.2566 \times 10^{-13}$ T = 1.2556 nGauss

One way to assess the energy of an EM wave is to evaluate the energy carried by the wave from one place to another. The radiative power ($P_{radiative}$) or radiant flux is a physical quantity expressed in units of watts. If the number of photons per second is n with a wavelength of λ, then

$$P_{radiative} = \frac{nh V_{light}}{\lambda} \tag{8.8}$$

A common measure of the photometric quantity of light is the intensity of the wave or luminous flux, which is represented in lumens. Luminous flux correlates to the visual sensation of light. Intensity (ι) is the power that passes perpendicularly through the unit area seen by an observer. The intensity [units are W/m^2] is written as

$$\iota = \frac{P_{radiative}}{A} = \frac{nbV_{light}}{A\lambda} \tag{8.9}$$

Since it is difficult to count the number of photons, intensity is expressed as energy density.

$$\iota = \frac{V_{light}H^2}{\mu_0} = V_{light}\varepsilon_0 E^2 \tag{8.10}$$

EXAMPLE 8.3

Determine the intensity of the EM wave in Example 8.2.
  Solution:
  $H = 1.0 \times 10^{-7}$ [V.s/m^2] $= 1.0 \times 10^{-7}$ A/m
  $\mu_0 = 1.2566 \times 10^{-6}$ [N/A^2]
  $V_{light} = 3.00 \times 10^8$ [m/s]

From (8.10), $\iota = \dfrac{3.00\times10^8 \ [\text{m/s}]\times\left[1.0\times10^{-7} \ [\text{A/m}]\right]^2}{1.2566\times10^{-6}[\text{N/A}^2]} = 2.387 \left[\text{W/m}^2\right]$

Generally it is useful to use the average power or average intensity of the wave. To find the average values, the root mean square (rms) averages for the electric field E and the magnetic field H are used; the relationship between the peak and rms values is:

$$E_{rms} = \frac{E}{\sqrt{2}} \text{ and } H_{rms} = \frac{H}{\sqrt{2}}$$

8.2.3   Generation of EM Radiation

EM radiation is emitted from all matter with a temperature above absolute zero. Temperature is the measure of the average energy of vibrating atoms and that vibration causes them to give off EM radiation. As the temperature increases, more EM radiation of shorter wavelengths is emitted. However, a nonthermal EM source is used in biomedical applications as the human body is sensitive to temperature. Nonthermal EM waves are generated by moving electrons (or charges) back and forth or oscillating in conductors by modulating the voltage applied to a length of wire. From the basic properties of atoms, we know that the nucleus is at the center, and negatively charged electrons are rotating in orbits. The electrons occupy distinct orbits and thus energy levels along with an intrinsic angular momentum (spin). When electrons move, they create a magnetic field. When electrons oscillate, their electric and magnetic fields change together forming an EM wave, which travels at the speed of light. The oscillation can come from atoms being energized either thermally or nonthermally or from an alternating current. The frequency of the emission is directly related to the traveling velocity of the electrons. The longer the electrons stay in the magnetic field, the more energy they lose. As a result, the electrons make a wider spiral around the magnetic field, and emit EM radiation at a longer wavelength. This can be related to the initial velocity of the electron, or it can be due to the strength of the magnetic field. A stronger magnetic field creates a tighter spiral and therefore greater acceleration. The frequency of the radiation is usually controlled by taking advantage of the natural oscillating frequencies of tuned circuits or quartz crystals. This method is used for generating EM radiation in microwave ovens, TV signals, and garage door openers.

X-rays are generated by accelerating a beam of electrons through a high voltage gradient and then colliding them into a metal plate (Figure 8.3). X-rays are produced when the electrons are suddenly decelerated—these waves are called bremsstrahlung (German word meaning braking) radiation—and when the electrons make transitions between lower atomic energy levels in heavy elements. The X-rays that emerge have a range of energies, however the maximum energy (i.e., the shortest wavelength) is determined entirely by the voltage difference through which the beam of electrons is accelerated (Figure 8.3). For example, a beam of electrons accelerated through a potential difference of 50 keV will produce X-ray photons up to but never exceeding 50 keV. Typically, a heated cathode tube serves as the source of the electrons, which are accelerated by applying a voltage of about

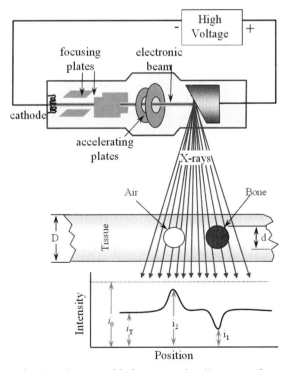

Figure 8.3   Schematic showing the assembly for generating X-rays nonthermally and intensity attenuation through a tissue of various densities.

100 keV between the cathode and the tungsten anode. A major fraction of the energy in the electron beam is converted to heat in the anode; therefore, high-energy X-ray tubes are often water cooled to dissipate the excess heat.

Many elements with an atomic number 90 and above occur in unstable conditions and spontaneously emit radiation to reach a stable state. This process is called radioactivity, during which the nucleus disintegrates. A measure to assess the rate of disintegration is the half-life ($t_{1/2}$), the time required to reach 50% of the initial mass. Emitted radiation can be grouped into three types: alpha particles, beta particles, and gamma rays.

1. An alpha particle is a ^{4}Helium nucleus (two protons and two neutrons), which is produced by the breakdown of the nucleus (or nuclear fission) into two less massive nuclei—one of them the alpha particle.
2. A beta particle is an electron, emerging from a weak decay process in which one of the neutrons inside an atom decays. A proton and an antielectron-type neutrino are also produced.
3. A gamma particle is a photon. It is the same as light except with very high energies. It is distinguished from X-rays by the fact that it comes from the nucleus.

Radiation activity is measured in an international (SI) unit called a becquerel (Bq). The becquerel counts how many particles or photons (in the case of wave radiation) are emitted per second by a source (3.7×10^{10} becquerels = 1 curie). The natural ^{40}K activity in the body of an adult human of a normal weight is 4,000–6,000 Bq. The device used for measurement is the Geiger-Mueller counter, which also detects g-rays. If a Geiger counter is placed over a gram of substance and count 3 clicks per second, the radioactivity of that substance would be 3 becquerel. Another commonly used unit is the roentgen (R), which is a measure of the exposure dose to an ionizing radiation ($1R = 2.58 \times 10^{-4}$ C/kg of dry air at a standard temperature and pressure) but it is not used for particle radiation.

8.3 Interaction of Radiation with Matter

8.3.1 Absorption of EM Waves

A beam of EM wave in vacuum or in an isotropic medium propagates in a fixed direction until it encounters a different medium. When an EM wave hits a different medium, it might get absorbed, bounced off (reflected), or go right through it (transmitted). The outcome depends on the EM energy and the property of the medium. One method is to compare the amount of energy a photon yields when it collides with the material. Some photons yield most of their kinetic energy over short distances; others deplete their energy gradually over longer distances. For each type of photon, the energy dissipated per unit length traversed is calculated, which is referred to as the *linear energy transfer* (LET) and given by

$$\text{LET} = dE/dl\,(\text{J/m or keV/m}) \tag{8.11}$$

where dE is the energy of the incident photon and dl is travel distance. The LET depends upon the chemical composition of the absorbing material and the density of the material. Water is often chosen as the reference absorbing medium to which the LET of various types of radiation are compared. The LET is only the average kinetic energy given up to the absorbing material over the particle's path and it does describe the effects of added energy on the absorbing material. LET does not include the effects of radiative energy loss.

In general, the larger LET value indicates more biologically reactive radiation. Radiation exposure is expressed in several ways to account for different levels of harm caused by different forms of radiation and the different sensitivity of body tissues. An absorbed dose (also referred to as exposure dose) is the amount of radiation energy absorbed into a given mass of tissue. Absorbed dose is measured in "gray" (Gy) units which is equivalent to joules per kilogram; 1 Gy is equivalent to 100 rad (abbreviated form of radiation absorbed dose). This allows unification of different types of radiation (i.e., particles and waves) by measuring what they do to materials. However, not all radiation has the same biological effect even for the same amount of absorbed dose. To determine the effect of radiation exposure on human tissue, a quantity called the the equivalent dose is used. This relates the absorbed dose in a tissue to the effective biological damage of the radiation. The equivalent dose is measured in an SI unit called the sievert (Sv). One Sv is equivalent to 100 rem (abbreviated for roentgen equivalent man). To determine the equivalent dose (Sv), the absorbed dose (Gy) is multiplied by a radiation-weighting factor (W_R), which is unique to the type of radiation. The radiation-weighting factor takes into account that some types of radiation are inherently more hazardous to biological tissue, even if their energy deposition levels are the same. For X-rays and gamma rays, W_R is 1 but the W_R value changes periodically with advances in approximation techniques. The more sieverts absorbed in a unit of time, the more intense the exposure. Alternatively, a Q factor is defined based on *LET*. Although the relation between W_R and Q is controversial, for *LET* less than 10 keV/mm, the quality factor is 10; for *LET* between 10 and 100, Q is given by 0.32L-2.2 and for *LET* greater than 100, Q is $300L^{-0.5}$.

Actual exposure is expressed as an amount over a specific time period, such as 5 mSv per year. This is called the *dosage rate*. The effectiveness of the dose is dependent on the *dosage rate*. The average dose from watching color television is 2 mrem per year. The probability of a harmful effect from radiation exposure depends on what part or parts of the body are exposed. Some organs are more sensitive to radiation than others. One way to account for tissue specific harm caused by radiation is to use a tissue-weighting factor (W_T). When an equivalent dose to an organ is multiplied by W_T for that organ, the result is the effective dose to that organ. The unit of effective dose is the sievert (Sv). If more than one organ is exposed then the effective dose, E, is the sum of the effective doses to all exposed organs.

8.3.2   Scattering of EM Waves

Light scattering is a physical process where the incident EM wave is forced to deviate from a linear trajectory due to localized nonuniformities in the medium. Apart from absorption, light scattering contributes to the visible appearance of most objects. If photons bounced or reflected off the object and then were received by the eye, one could see the object they have bounced off. Black objects absorb the photons so one cannot see black things. Human eyes are capable of distinguishing color only in the visible portion of the EM spectrum (from violet, to indigo, to blue, to green, to yellow, to orange, to red) and cannot perceive ultraviolet or infrared rays. The eye can sense differences in brightness or intensity ranging from black to white and all the gray shades in between. Thus, for an image to be seen by the eye, the image must be presented to the eye in colors of the visible spectrum and/ or varying degrees of light intensity. The front of the eye, including the iris, the curved cornea, and the lens are the mechanisms for admitting light and focusing it on the retina. For an image to be seen clearly, it must spread on the retina at a sufficient visual angle. Unless the light falls on nonadjacent rows of retinal cells (a function of magnification and the spreading of the image), one cannot distinguish closely lying details as being separated (resolved). Further, there must be sufficient contrast between the adjacent details and/or the background to render the magnified, resolved image visible. Because of the limited ability of the eye's lens to change its shape, objects brought very close to the eye cannot have their images brought to focus on the retina. The accepted conventional viewing distance is 10 inches or 25 centimeters. Reflection can also be seen with some sound phenomena. *Ultrasound imaging* is based on sending high-frequency sounds to the body part or organ being studied (see Section 8.5.4).

The different wavelengths of EM waves cause the radiation to react differently with different materials. The wavelengths of X-rays are on the order of or smaller than the size of atoms. As a result, they appear less like waves that can be reflected and more like photons that can effectively pass between the atoms. They penetrate deeply into a material before interacting with an individual atom. When electrons collide with photons, they jump to a higher energy level. All these excited electronic states are unstable, and the electrons will lose their excess energy and fall back to lower energy states. This excess energy can be dissipated in several ways. The most common is increasing the atomic vibrations within the molecule. However electrons in some atoms fall back to their original energy level and reemit the light (Figure 8.4). Some molecules emit light near the visible light, which is called luminescence. The luminscent effect can be referred to as fluorescence or phosphorescence. Fluorescence is luminescence, which has energy transitions that do not involve a change in electron spin; therefore the reemission occurs much faster. Phosphorescence is stored and released gradually. Consequently, fluorescence occurs only during excitation while phosphorescence can continue after excitation.

When the energy of the incident radiation exceeds the ionizing potential of the atoms in the medium, the absorbed energy causes ionizations to occur through particle-particle collisions. From the kinetic theory of the collision of solid bodies (discussed in Chapter 5), scattering can be grouped into elastic and inelastic. Elastic scattering involves negligible change in the radiation energy, whereas inelastic scattering does involve change in the radiation energy. Rayleigh scattering is an example of elastic scattering, and depends on tissue type; this scattering typically accounts for 5–10% of total tissue interaction. In a photoelectric interaction, the entire energy of the photon is transferred to an electron and the photon ceases to

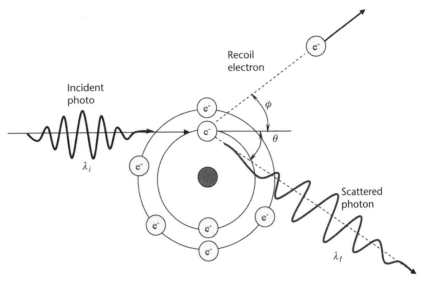

Figure 8.4 Collision of photons with electrons. If the collision energy is higher than ionization energy, then electrons are ejected from the orbit.

exist. The scattering of photons from charged particles is called Compton scattering, named after American physicist Arthur H. Compton, who first observed the scattering of X-rays from electrons in a carbon target and found scattered X-rays with a longer wavelength than those incident upon the target. Compton explained and modeled the scattering by assuming a photon and applying conservation of energy and conservation of momentum principles to the collision between the photon and the electron. In a Compton scattering interaction, only a partial amount of the photon's energy is transferred to an electron, and the photon continues to travel, but at an angle with respect to its initial direction. Compton scattering is of a slightly lower energy level but still enough to travel through tissue. The energy shift depends on the angle of scattering and not on the nature of the scattering medium. The shift of the wavelength increases with scattering angle according to the Compton formula:

$$\lambda_f - \lambda_i = \Delta\lambda = \frac{h}{m_e V_{light}}(1 - \cos\theta) \tag{8.12}$$

where λ_i is the wavelength of incident X-ray photon, λ_f is the wavelength of the scattered X-ray photon, m_e is the mass of an electron at rest, and θ is the scattering angle of the scattered photon.

EXAMPLE 8.4

An EM wave with a wavelength of 5×10^{-12} m hits an electron at rest and the Compton scatter is observed at 45° to the incident angle. Calculate the wavelength of the scattered photon and the energy associated with it.
Solution:

$$\lambda_f - 5 \times 10^{-12} = \frac{6.626 \times 10^{-34} \ [\text{J.s}]}{9.11 \times 10^{-31} [\text{kg}] \times 3 \times 10^{8} [\text{m/s}]} (1 - \cos 45)$$

$$\lambda_f = 5.71 \times 10^{-12} \text{m}$$

$$E = 4.136 \times 10^{-15} [\text{eV.s}] \times \frac{3 \times 10^{8} [\text{m/s}]}{5.712 \times 10^{-12} [\text{m}]} = 2.48 \times 10^{5} \text{eV} = 0.248 \ \text{MeV}$$

Compton scattering occurs in all materials and is dominant with photons of medium energy (i.e., about 0.5 to 3.5 MeV). It is largely independent from the tissue type, increases with high energy levels, and causes poor contrast in high energy biomedical images. Since the scattered X-ray photon has less energy, it has a longer wavelength and is less penetrating than the incident photon. Together with the photoelectric effect and pair production, Compton scattering contributes to the attenuation of energy in matter.

8.3.3 Transmission Imaging

With the exception of nuclear medicine, most imaging techniques require that energy penetrates the body tissues and interacts with those tissues in the form of absorption and scattering. The resulting medical image reflects the interaction of the energy with the tissues. A beam of EM wave is passed through the body structure being examined. The images record how much light passes through the specimen, such that darker regions indicate more tissue and/or darker tissue. The beam passes through less dense types of tissue such as watery secretions (Figure 8.3), blood, and fat, leaving a darkened area on the X-ray film. Attenuation is the rate at which the signal light decreases in intensity, ι, and related by

$$\iota_x = \iota_0 \cdot e^{-\mu_x x} \tag{8.13}$$

where ι_0 is the intensity of incident X-ray, ι_x is the intensity at distance x, and μ_x is the linear attenuation coefficient, which describes the fraction of a beam that is absorbed or scattered per unit of thickness of the absorber. μ_x accounts for the number of atoms in a cubic cm volume of material and the probability of a photon being scattered or absorbed from the nucleus or an electron of one of the atoms. μ_x depends on the photon energy and the chemical composition and physical density of the material. Differences in μ_x values among tissues are responsible for image contrast and identification of different tissues. Another term commonly used is the mass attenuation coefficient obtained by dividing μ_x with density [g/cm^3] of the medium. μ_x/ρ for some tissues are given in Table 8.1, from which one can notice that μ_x/ρ for different tissues are similar at a given incident energy. However, tissue densities are significantly different. Hence, μ_x is approximately proportional to the physical density (kg/m^3), and μ_x tends to increase with an increasing atomic number at the same photon energy.

EXAMPLE 8.5

A tissue containing bone and air of equal thickness (5 cm) are exposed to a 40-keV X-ray. If the attenuation coefficients are 0.665 cm^2/g and 0.095 cm^2/g, what is the expected difference in energy at 5 cm? Assume the density of bone to be 2 g/cm³ and density of air to be 1.1×10^{-3} g/cm³.

Solution: Since the number of photons are the same and assuming the same cross-sectional area, (8.13) can be written in terms of energy.

$$\mu_{air} = 0.095[cm^2/g] * 1.1 \times 10^{-3}[gm/cm^3] = 9.9 \times 10^{-5} cm^{-1}$$

$$\mu_{bone} = 0.665[cm^2/g] * 2[gm/cm^3] = 1.33 \ cm^{-1}$$

$$E_{air} = 40[keV] * e^{-9.9 \times 10^{-5} * 5} = 39.980 \ keV$$

$$E_{air} = 40[keV] * e^{-1.33 * 5} = 0.052 \ keV$$

Clearly, there is a significant difference in the transmitted energy.

X-rays, CT scans, and fluoroscopy images are produced by transmission. For example, the X-ray tube is placed one or more meters from the patient. As the emitted X-rays pass through the patient, some X-rays pass straight through the body (primary radiation), some are scattered by the tissue (secondary radiation) and some are completely absorbed by the tissue (absorbed radiation). Transmitted X-rays are recorded by a detector on the opposite side. Hence, the intensity of photons reaching a detector decreases as μ_x increases for the same absorber thickness and photon energy. Muscle and connective tissues (ligaments, tendons, and cartilage) appear gray. Bones appear white. Scattered light may interfere constructively with the incident light in certain directions, forming beams that have been reflected and/or transmitted. The constructive interference of the transmitted beam occurs at the angle that satisfies Snell's law, while the after reflection occurs for $\theta_{reflected} = \theta_{incident}$.

Fluorescence-based techniques are widely used for studying cellular structure and function, and interactions of biomolecules in the detection and quantitation of nucleic acids and proteins, microarray technology, and fluorescence spectroscopy. *Emission* also occurs when tiny, nuclear particles or magnetic energy are detected by a scanner and analyzed by a computer to produce an image of the body structure or organ being examined. In nuclear medicine, radioactive agents are ingested, inhaled, or injected into the body, and the resulting image reflects

Table 8.1   Attenuation Coefficients for Typical Tissues [1]

$\mu_x/\rho \ [cm^2/g]$	Density [g/cm³]	15 keV	30 keV	50 keV	100 keV	150 keV
Air	0.0013	1.614	0.354	0.208	0.154	0.136
Adipose tissue	0.95	1.083	0.306	0.212	0.169	0.150
Whole blood	1.06	1.744	0.382	0.228	0.169	0.149
Cortical bone	1.92	9.032	1.331	0.424	0.186	0.148
Lead	11.35	111.6	30.32	8.041	5.549	2.014
Lung	1.05	1.721	0.382	0.227	0.169	0.149
Skeletal muscle	1.05	1.693	0.378	0.226	0.169	0.149
Water	1.00	1.673	0.376	0.227	0.171	0.137

metabolic or physiologic interactions between the agent and the tissue. Emission of nuclear particles from nuclear substances introduced into the body is utilized in the evaluation.

8.4 Basics of Imaging

8.4.1 Image Acquisition

Biomedical images are evaluated by a human analyst. For this purpose, the interactions of the energy source should be presented in a form so that the eye can distinguish the intended observation. Acquisition implies recording the interaction of the energy source and the sensing of the environment. Sensing can be accomplished either passively or actively. Sensing is passive when the radiation originates in the surrounding environment, is reflected or absorbed and re-emitted from objects, and then sensed by eyes, or cameras. Sensing is termed active when the radiation is artificially introduced into the environment as an integral part of the vision system (e.g., X-ray, sonar, and radar). In both cases, basic elements in the image acquisition system (Figure 8.5) include detectors, signal processors, display devices such as films and cathode ray tube (CRT) monitors, and storage devices such as hard disks. The first stage of any imaging system is detecting raw image data and transforming it into a form suitable for digital, optical, or biological processing. Detectors help in registering the radiation behind the object that bears the information of interest. The interaction in the detector in most cases is absorption, either by a directly converting semiconductor or by a scintillator followed by a light sensor (e.g., a photodiode). Many types of detectors are used based on the type of energy source used. X-ray imaging as an example is discussed later, but other detectors specific to an imaging technique are described under those medical devices.

The conventional method of acquiring the image is adapted from photography—the screen/film detector—that is, to capture the emitted 3D distributed radiation by a normal photographic film in a 2D projection. For example, when X-rays collide with atoms of phosphorous compounds (scintillators) embedded in a glass plate, the electrons in the compound are raised to higher energy levels causing the phosphorous plate to glow, the dark regions being zones of high X-ray absorption (i.e., this forms a positive image). The phosphor screen converts the X-ray to visible light. Alternately, the glass plate is replaced by a photographic plate. The transmitted X-ray photons expose the photographic film, which is physically separate from

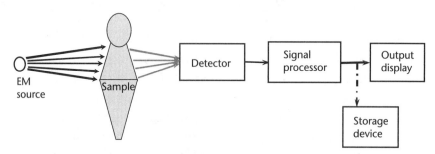

Figure 8.5 Basic components in biomedical image acquisition.

(but adjacent to) a phosphor screen in the same way that light exposes an ordinary photograph. In this case, the regions of high X-ray absorption leave the film unexposed and they appear white on the film; that is, it is a negative image (bright areas correspond to high attenuation). Ribs and the vertebral column appear white while the lungs with little attenuation appear black. In the film, photons are absorbed in the silver halide (e.g., AgBr) crystals, generating very small amounts of free silver.

$$AgBr + hf \rightarrow Ag + Br$$

To increase sensitivity and thus lower radiation dose, the photosensitive film emulsions are made thicker and occasionally coated on both sides of the film, in contrast to normal photographic film. The X-ray image itself is formed by tiny black silver particles, just as in a normal black and white photographic negative. During film processing, any grain with small amounts of free silver are completely converted to metallic, nontransparent silver, while the remaining unreduced silver halide is removed by the fixative.

$$2Ag + 2HCl \rightarrow 2AgCl + H_2$$

After exposing AgCl to metallic Ag dark spots appear (negative image). Ideally, the intensity of the image should be logarithmic with incident X-ray intensity such that it measures the summed attenuation along a line. Improved efficiency is obtained by photographing the light using a fluorescent screen made of luminescent ceramics such as terbium-doped gadolinium oxisulfide (Gd_2O_2S:Tb). The screen is incorporated in a light-tight cassette; the film is loaded into the cassette in a dark room and the cassette is then positioned behind the object or body part to be imaged. The silver content of the film makes X-ray films rather expensive.

Any film has a specific range of optimal sensitivity (exposure range from complete white to complete black). Although equipment is normally assisted by electronic exposure meters, the correct choice of film, exposure time, exposure current, and high voltage is decided by the judgment of the technician. To improve the sensitivity and thus lower radiation exposure to the patient, the film is often brought in contact with a sheet of intensifying screen. The screen contains chemical compounds of the rare Earth elements, which emit visible blue-green light when struck by X-rays or other ionizing radiation. This permits the use of photographic film with thinner emulsions and more normal sensitivity to visible light. While increasing the sensitivity, the use of the intensifying screen blurs the images as the registration of X-ray radiation becomes an indirect process. The patients or the object is not only the source of X-ray absorption but also of X-ray scattering, mainly due to Compton scattering. The scattered radiation carries no direct information about the object and thus only reduces the quality of the image. An antiscatter grid, which suppresses radiation from directions deviating from the direct connecting line between the X-ray source and detector element, is necessary in most cases to suppress scattered radiation and thus improve the quality of the image.

Another subsystem in the acquisition typically includes a signal processor, which amplifies and filters the output. Signals are amplified using *photomultiplier tubes* (PMTs), also called phototubes in the past. PMTs are light detectors with incorporated electron multipliers, which are useful in low intensity applications such as fluorescence spectroscopy and positron emission tomography. PMTs operate with a very high gain as well as reasonable efficiency. PMTs consist of a photocathode and a series of dynodes in a vacuum glass enclosure. Photons that strike the photoemissive cathode emit electrons due to the photoelectric effect. Instead of collecting these few electrons at an anode like in the phototubes, the electrons are accelerated towards a series of additional electrodes called dynodes. These electrodes are maintained at a more positive potential. Hence, additional electrons are generated at each dynode. This cascading effect creates 10^5 to 10^7 electrons for each photon striking the first cathode depending on the number of dynodes and the accelerating voltage. This amplified signal is finally collected at the anode where it can be measured.

8.4.2 Digitizing Images

With the advances in computer science, many medical images are acquired digitally instead of on film. Digitizing images opens many new possibilities of rapid viewing, interactive image availability, postprocessing, and digital transmission. Data can be manipulated to enhance image quality (see Section 8.4.4), and data can be stored in a systematic manner, which speeds up the retrieval process, and can be used as a means of developing expert diagnostic systems. The information for images can also be transmitted to other locations over the internet or telephone lines, and various images, even those from different types of machines, can be compared and fused into composites. However, the correct handling and display of such image information without loss or compression is a critical factor. Further, detectors must be *accurate* and *fast* if real-time processing is required.

Digitization requires a data acquisition system to digitize the signals provided by the detector. Digitization means expressing the signal as various discrete units in two respects: its spatial organization and value. Hence, the image processed and stored in a digital image processor is a 2D array of numbers. Typically, signals are expressed in a matrix with m rows and n columns as a continuous function $f(x,y)$ of two coordinates in the plane. The image quantitation assigns to each continuous sample an integer value. If a digital image is seen with a very high magnifying glass, tiny squares or cells can be seen arranged in an array. These cells are called pixels, an abbreviation for picture element. Most digital image processing devices use quantitation based on what is called k equal intervals. If b bits are used, the number of brightness levels is $k = 2^b$. Eight bits per pixel are commonly used; specialized measuring devices use 12 or more bits per pixel. The finer the sampling (i.e., larger m and n) and quantitation (larger k), the better the approximation of the continuous image function $f(x, y)$. Although the pixel size depends on the objects to be imaged, the number of quantitation levels should be high enough for human perception of fine shading details in the image.

An important component of digital imaging is charge-coupled devices (CCD). CCD is a light-sensitive integrated circuit consisting of a type of metal oxide semiconductor, arranged in a 2D array. Each element in the array stores (due to their capacitance) and displays the data for an image in such a way that each pixel in the image is converted into an electrical charge. Charge is created when photons strike the semiconducting material dislodging electrons. As more photons fall on the device, more electrons are liberated, thus creating a charge that is proportional to the intensity of light. The challenge lies in reading the charges out of the array so they can be digitized. CCD must perform four tasks to generate an image:

1. *Generate charge*: Convert photons into free electrons through the photo-electric effect.
2. *Collect charge*: Gather generated electrons using an array of electrodes, called gates, and group them into pixels.
3. *Transfer charge*: Apply a differential voltage across the gates. Signal electrons move down vertical registers (columns) to horizontal registers. Each line is serially read out by an on-chip amplifier.
4. *Detect charge*: Individual charge packets are converted to an output voltage and then digitally encoded.

To do this, each individual CCD detector (Figure 8.6), or pixel, consists of three transparent polysilicon electrode layers (termed as gates) over a buried channel of doped photosensitive silicon that generates the charge. Gates are separated from a silicon semiconductor substrate by an insulating thin film of silicon dioxide. After electrons are collected within each photodiode of the array, a voltage potential is applied to the polysilicon gates to change the electrostatic potential of the underlying silicon. The silicon substrate positioned directly beneath the gate electrode then becomes a potential well, capable of collecting locally generated electrons created by the incident light. Neighboring gates help to confine electrons within the potential well by forming barriers (i.e., zones of higher potentials surrounding the well). By modulating the voltage applied to polysilicon gates, they can be biased to either form a potential well or a barrier to the integrated charge collected by the photodiode. To read and digitize a particular charge, the voltages of the three gates are cycled in a sequence that causes the charge to migrate down the channel to the next gate, then to the next pixel. The charge ultimately moves down the row until it reaches the end column, where it is read out into a serial register and sent to an analog-to-digital converter. This charge transfer occurs with an efficiency greater than 99.9% per pixel. The sequence of moving the charge from one gate to the next is called coupling.

Processing the information requires specialized computer software and hardware, which are constantly under development and refinement. Further, one has to address some of the problems associated with digitizing images including:

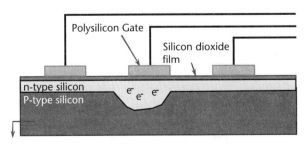

Figure 8.6 Schematic of CCD camera showing different layers.

scattering, imperfect manufacture of CCDs, loss of higher bits during analog-to digital conversion (called clipping or wrap-around), imperfect isolation of detectors from each other (called blooming), effect of quantitation as a sample of intensity from a discrete area of scene, and geometric aberration due to lenses. Computer vision is a discipline that focuses on information extraction from the output of sensors, and on the representation of this information in a computer model. Image registration enables a computer to register (apprehend and allocate) objects as they appear in an internal computer model. A parametric image registration algorithm specifies the parameters of a transformation in a way that physically corresponding points at two consecutive time steps are brought together as close as possible. Fourier transform provides the mathematical and algorithmic foundation for a number of imaging techniques including X-ray CT, PET, SPECT, optical microscopy, and MRI. A more recent development is the wavelet transform that appears to be ideally suited for signal and image processing. Although one class of algorithms operates on previously extracted surface points, other algorithms register the images directly based on the gray-value changes. Most commonly, a cost or error function is defined and an optimization method is chosen that iteratively adjusts the parameters until an optimum is reached. Other approaches extract specific features (e.g., correspondence between points) that serve as a basis for directly calculating the model parameters.

8.4.3   3D Image Reconstruction

In a typical 2D image formed on a film, the information sought could be hidden behind other structures in the object. In these cases, the desired information may be obtained by repositioning the object in the beam several times to see the desired image in detail. To overcome these difficulties, 3D image reconstruction (or rendering) or medical image volume visualization has been developed. In order to access this information efficiently, a model is necessary for the integration of knowledge, which is extracted from the images. In order to have enough data to mathematically reconstruct virtual slices, one needs projections from different angles. The quality of the reconstruction and resolution in an image increases with the number of projections.

The three basic operations performed in 3D medical imaging are data collection, data analysis, and data display. Data collection is similar to image acquisition in the digital form at different angles. For example, in X-ray CT, a series of 2D X-ray images (Figure 8.7) are obtained by rotating an X-ray emitter around the patient, and measuring the intensity of transmitted rays from different angles. From the 2D images at different angles, a projection function of the image is developed by mathematically (using Radon transform) transforming images with lines into a domain of possible line parameters (refer to [2] for more information). To reconstruct the images, the Fourier slice theorem is used to convert the projection function to the 2D Fourier transform of the image. From the 2D Fourier transform of the image, the inverse 2D Fourier transform projection function is developed. Then, projections are run back along the same angles from where the images were collected (hence the name back projection) to obtain a rough approximation to the original. The projections interact constructively in regions that correspond to the

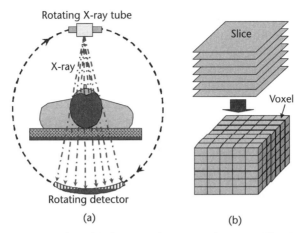

Rotating X-ray tube

X-ray

Rotating detector

(a)

Slice

Voxel

(b)

Figure 8.7   3D image processing. (a) Computed tomography setup. The X-ray source and the detector rotate around the axis, which facilitated the collection of images at different angles. (b) Reconstruction of 3D images using multiple slices.

emitted sources in the original image. A problem in the back projection is blurring that occurs in other parts of the reconstructed image. A data filter (a mathematical function) is used to eliminate blurring. As 2D filtering is computationally intensive, a better approach is to reverse the order of the filtering and the back projection. The multiple transmitted beams are registered and back-projected to generate an "image slice" of high resolution and contrast in digital form. CT systems are usually configured to take many views of the object, often more than 100 to generate reconstructed images of good quality, with excellent density discrimination.

All the digital images are combined in a computer using different algorithms to build a 3D image of any part of the body. A rendering technique is the computer algorithm used to transform serially acquired axial CT image data into 3D images made of voxels (volume pixel), similar to pixels. The volume of the voxels is a product of pixel size and the distance between slices. The brightness of each voxel is dependent on the intensity of signal from the corresponding location in the object. Each voxel has a corresponding pixel pattern in the display panel. A typical MRI 3D image consists of 256×256 or 512×512 voxels. The voxel is usually considerably larger in the third dimension, as the slice thickness is typically 1–5 mm. This means that the in-plane resolution is typically 0.5–2.0 mm. One could then instruct the computer to display 2D slices (each measurement lasts only a fraction of a second and represents an "image slice" of tissue) from any angle.

There are a number of different rendering techniques, but most can be divided into two classes: surface-based (binary) techniques and volume-based (continuous) techniques. Rendering with either technique consists of three steps: volume formation, classification, and image projection. Volume formation involves the acquisition of the image data, the stacking of the resultant data to form a volume, and preprocessing, which varies according to the rendering technique used. Typical preprocessing includes resizing (by interpolation or resampling) of each voxel, image smoothing, and data editing (e.g., deletion of the CT table on which the patient was positioned). The classification step consists of determining the types of tissue (or other classifying quality) represented in each voxel and using that information to assign color and other visual properties to the voxel.

In surface rendering, only part of the 3D dataset is utilized for the reconstruction of an image. The data to reconstruct the image is selected by defining a threshold. Typically, the voxels with a value below the threshold are discarded from the data, and the voxels with a value equal to or above the threshold are selected for the rendering. The main advantage of surface rendering is its speed, since a comparatively small amount of computational power is needed to generate images in a reasonable amount of time. However, the thresholding technique is based on the assumption that each volume element represents only one type of tissue. Hence, it incorrectly classifies voxels that represent mixed tissue interfaces. The thresholding technique is also susceptible to noise introduced during scanning. A small amount of noise can modify attenuation values and create the appearance of soft tissue that actually represents mostly bone.

Unlike surface rendering, volume rendering does not make use of a surface representation but uses a technology originally developed for motion picture computer animation. When a volume rendering algorithm is used, certain properties are assigned to each voxel on the basis of its value. In CT, it is the average of all the attenuation values contained within the corresponding voxel. This number is compared to the attenuation value of water and displayed on a scale of arbitrary units named hounsfield units (HU) named after British electrical engineer Godfrey N. Hounsfield. When properties are assigned a certain value, volume rendering uses the histogram of these values. In a typical CT histogram, the x-axis represents the possible voxel values in the data, and the y-axis, the number of voxels with that specific value. Certain values can be related to specific tissue compositions when the properties of a voxel value are determined. Each of these tissue compositions has specific properties. Partial rendering enables mixing two different tissue compositions (totaling 100%) to establish the properties of a border voxel. Thus, a voxel can partially belong to the surface of interest and can have properties based on the percentages of the properties of the two tissues involved.

8.4.4 Image Quality

Quality of the image produced during the imaging process limits visibility of anatomical details and other small objects in the body. The level of detail in an image is related to the size of objects. The general range of details for each imaging modality is determined by the limitations of the technology, design characteristics, and the selected operating factors. Problems associated with digitizing images could compromise the quality of the image. For example, some of the weaknesses of CCDs that could result in decreased image quality are:

1. *Fading*: Although the coupling process is quite efficient, moving the charges along a row of many pixels adds up to a noticeable loss of charge.
2. *Blooming*: If too many photons strike a CCD element, it gets saturated and some of the charge leaks to adjacent pixels.
3. *Smearing*: If light strikes the sensor while a transfer is taking place, it can cause some data loss and leave streaks behind bright areas of the image.

Medical image quality can better be measured by human performance in visual tasks that are relevant to clinical diagnosis. The standard method of evaluating diagnostic methods is a receiver operating characteristic (ROC) study, which is time consuming and costly because it requires a large number of human observations. Although a detailed discussion of ROC is beyond the scope of this book, image quality is assessed by three basic factors.

8.4.4.1   Spatial Resolution

Spatial resolution of an imaging system is defined as the smallest spacing between two objects that can be imaged clearly. In computer monitors, resolution is the number of pixels contained on a display monitor, expressed in terms of the number of pixels on the horizontal axis and the number on the vertical axis. The sharpness of the image on a display depends on the resolution and the size of the monitor. The same pixel resolution will be sharper on a smaller monitor and lose sharpness on larger monitors if the same number of pixels is spread out over a larger number of inches. A computer display system will have a maximum resolution depending on its physical ability to focus light (in which case the physical dot size matches the pixel size). Spatial resolution is determined by characteristics of each imaging modality and operating factors for that modality. In each case, the distance between separate objects that a device can record is a measure of its spatial resolution. Hence, the intrinsic resolution of the detectors is the critical factor in determining the ultimate resolution of the image. The spatial resolution of a conventional X-ray system with direct film exposure is approximately 0.01 mm while that of a CT scanner is approximately 1 mm.

8.4.4.2   Image Contrast

Image contrast is the difference in film density or image brightness between areas within an image. The medical imaging process sees physical contrast in the body and transfers it into image contrast. The contrast sensitivity of the imaging process determines the lowest physical contrast that is visible. When the contrast sensitivity of an imaging process is low, many things in that image are not distinguishable. Contrast sensitivity is determined by characteristics of each imaging modality and operating factors for the modality. For example, the physical contrast that can be visualized with X-ray imaging is differences in physical density. Although spatial resolution is low in CT, it has low contrast resolution, enabling small changes in the tissue type. The physical contrast that is visible in radionuclide imaging is usually different concentrations of radioactivity. An MRI can visualize many forms of physical contrast. Some are characteristics of tissue and others are characteristics of fluid movement.

In digital images, the differences in gray shades are used to distinguish different tissue types, analyze anatomical relationships, and quantify physiological function. The larger the difference in gray shades between two adjacent tissue types, the easier it is to make these important distinctions. The objective of an imaging system is to maximize the contrast in the image for any particular object of interest, although there may be design compromises accounting for noise and spatial resolution.

8.4.4.3   Image Noise

Image noise is defined as the uncertainty or the imprecision with which a signal is recorded. Empirically noise is quantified with the signal-to-noise ratio (SNR). SNR quantifies the ratio of signal power to noise power. Noise can occur during image capture, transmission, or processing, and may be dependent or independent of image content. For example, an image recorded with a small number of photons generally has a high degree of uncertainty or is very noisy, while an image recorded with a large number of photons is very precise or is less noisy. Increasing contrast may also increase noise intensities. In a film-screen cassette, the film contains individual grains, which are sensitive to the radiation. Therefore, the exposure of the grains in the film produces random variations in film density on a microscopic level, which is a source of noise. The noise source in the imaging process can be increased in a number of ways. In general, noise can be grouped into three categories:

1. *Impulse noise*: Outliers due to faulty electronics.

2. *Gaussian noise*: This noise is often due to electronic or thermal fluctuations. Some detectors generate currents from thermal sources, which introduce random fluctuations into the signal.

3. *Poisson noise*: This noise occurs due to limited photon counts. For example, an X-ray beam emerging from an X-ray tube is statistical in nature (i.e., the number of photons emitted from the source per unit time varies according to a Poisson distribution). Other sources of random fluctuation are introduced by the process of attenuation. Attenuation occurs in the materials present in the path of the radiation beam (patient, X-ray beam filtration, patient table, film holder, detector enclosure), which is also a Poisson noise. The detectors themselves often introduce noise.

After a large number of individual measurements x_i, the mean $(\bar{x}_i)$ and the standard deviations (σ) are calculated using

$$\bar{x}_i = \frac{1}{N}\sum_{i=1}^{N} x_i \text{ and } \sigma^2 = \frac{1}{N-1}\sum_{i=1}^{N}\left(x_i - \bar{x}_i\right)^2 \qquad (8.14)$$

where N is the number of measurements. The square of the standard deviation σ^2 is called the variance, and both the standard deviation and the variance quantify the uncertainty, or noise, in the measurement.

Digital data can be filtered in order to smooth out the statistical noise. There are many different filter functions (lowpass cosine filter, Hanning filter, Butterworth filter, and Weiner filter) available, which have different characteristics. Some will smooth too much so that there are not any sharp edges, and hence will degrade the final image resolution. Others will maintain a high resolution while only smoothing slightly. Regardless of the filter used, the end result is to display a final image that is relatively free from noise.

8.5   Imaging Devices

8.5.1   X-Ray Imaging

The penetrating power of X-rays makes them ideal for medical diagnostics. However, X-ray photons have energies, which can cause ionization and therefore are biologically hazardous if the absorbed dose is not kept below certain minimums. The amount of absorption depends on the tissue composition. Dense bone will absorb more than soft tissues, such as muscle, fat, and blood (see Section 8.3.3). The amount of deflection depends on the density of electrons in the tissues. Tissues with high electron densities cause more X-ray scattering than those of a lower density. Thus, the X-ray image will appear brighter for bone or metal, as less photons reach the X-ray film after encountering bone or metal rather than tissue. As the emitted X-rays (accelerated electrons) pass through the patient, they are deflected by the tissues and recorded by a detector on the opposite side. The positives of imaging using X-rays is a high resolution as small as 0.1 mm and the ease of use. Its limitation are an inability to discriminate tissues of similar densities and rendering only 2D structures.

An X-ray CT does not use film to detect the transmitted rays (see Section 8.4.3). Instead, the photons are collected by an electronic device, which converts the X-ray photons into an electric current. In a CT scanner, the X-rays enter crystal scintillators and are converted to flashes of light. Since the presence of more detectors improves the resolution of the image, multidetector technology has made remarkable progress. A "single slice" CT has a row of these detectors positioned opposite to the X-ray tube and arranged to intercept the fan of X-rays produced by the tube. In some cases, the detector row rotates with the X-ray tube; in others there is a complete ring of stationary detectors. A "multislice" (or multidetector) CT has several rows of small scintillator detectors.

Two basic designs for CT detectors are the ionization chamber, and scintillation with photodiodes. Ion chambers are built on the ability of X-rays to ionize gases such as Xe_2, N_2, and Ar_2. An electric field attracts electron and ions, and measured current is proportional to X-ray intensity. A scintillation detector converts X-rays into light, which is detected by photodiodes. This arrangement is more sensitive than ion chambers. The requirements for X-ray CT scintillators are different than planar imaging, and should have: a low afterglow; high stability (chemical, temperature, and radiation damage); high density (> 6 g/cm^3); an emission wavelength well matched to photodiode readout (500–1,000 nm); and high luminous efficiency (> 15,000 photons/MeV). Commonly used scintillators are made of cesium iodide (CsI) and cadmium tungstate, as well as rare earth oxides and oxysulfides such as cadmium tungstate ($CdWO_4$) and doped gadolinium oxisulfide. Detectors are made of either single crystals or ceramic bars, and their surfaces painted with a metallic paint for light reflection. CT detectors do not count individual photons, but integrate the energy deposited by many photons.

8.5.2 Positron Emission Tomography (PET)

PET relies on the interactions of radioactive substance (also referred as radionu-clides) that decay via positron emission. The positron (e^+) is the antiparticle of the electron (e^-) and behaves nearly the same as electrons. PET was originally devel-oped for research purposes but has found clinical applications. In PET, radionu-clides in tracer amounts (hence they are referred as tracers) are used to image and measure biological processes. The radiation dose is usually less than the dose that individuals receive from their natural environment each year. A patient ingests (or is injected with) the positron-emitting radionuclide (Figure 8.8) and depending on the biodistribution properties of the radionuclide, it is taken up by different organs and/or tissue types. The regional concentration of the labeled compound (in mCi/cc) is imaged as a function of time. For example, oxygen and glucose accumulate in brain areas that are metabolically active.

Commonly used radioisotopes are ^{18}fluorine ($t_{1/2}$ = 109.8 minutes), ^{11}carbon ($t_{1/2}$ = 20.4 minutes), ^{13}nitrogen ($t_{1/2}$ = 9.96 minutes), and ^{15}oxygen ($t_{1/2}$ = 2.07 minutes). These radionuclides have short half-lives (e.g., 2 to 109 minutes) and are produced in cyclotrons or reactors. Radionuclides are selected by first identifying the process to be studied and then synthesizing a molecule with a radioisotope through which the assessment is performed. ^{11}C is a natural choice since it is an iso-tope of an atom that is present in organic molecules, while many of the ^{18}F charac-teristics are similar to those of hydrogen. ^{13}N and ^{15}O have the advantage of being

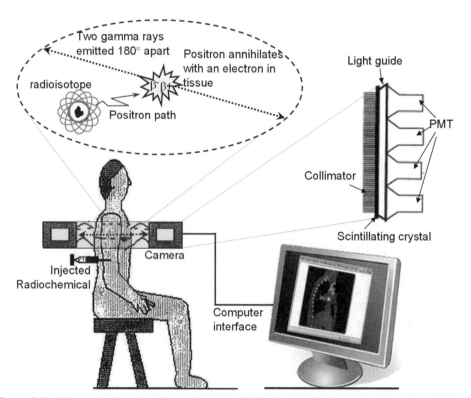

Figure 8.8 Schematic showing the mechanism of PET along with the interactions. In addition, components of the camera are also shown.

the atoms present in all the basic building blocks proteins, fat, and carbohydrates. Further, their $t_{1/2}$ matches reasonably well the biological $t_{1/2}$ of many processes of interest. The radionuclides used in PET can be classified into three categories:

1. Direct imaging where the visualization and intensity of a radiolabelled probe is directly related to its interaction with the target molecule (which can be a protein or RNA or DNA). The expression of cell-surface receptors (described in Chapter 7) is imaged using radiolabelled ligands or receptor antagonists. An example is the imaging of vascular endothelial growth factor (VEGF) using anti-VEGF monoclonal antibodies.

2. Indirect imaging, which involves multiple components. These strategies are adapted from imaging techniques used at the single cell level. An example is the herpes simplex virus type 1 thymidine kinase (HSV1-tk) system where 9-(4-fluoro-3-[hydroxymethyl] butyl) guanine (FHBG) is used as a reporter tracer. FHBG stays inside the cells after phosphorylated by the thymidine kinase in HSV1-tk. Hence, PET imaging identifies sites of thymidine kinase activity.

3. Surrogate imaging, which reflects an event that occurs downstream of a particular process. For example, ^{18}F-2-fluoro-2-deoxy-D-glucose (^{18}F-FDG), a glucose analog utilized to image functional changes in a glucose metabolism.

8.5.2.1   Theory of PET

When the radionucleide emits a positron from the nucleus, the positron (e^+) interacts with an electron (e^-) in the tissue and the masses are annihilated to two high-energy photons that travel in opposite directions,

$$e^+ + e^- \rightarrow \gamma_1 + \gamma_2 \qquad (8.15)$$

The high-energy gamma rays have an average path length of about 10 cm in the tissue before they are scattered or absorbed; thus, the rays have a high probability of escaping the body. The detection of the emitted positron annihilation radiation allows functional and structural imaging. Hence, PET gives images of the distribution of an injected positron-emitting radionuclide. Gamma or annihilation photon detectors record the activity to produce analog signals. An electron-positron annihilation is possible only if the momentum and energy conservation laws are met (discussed in Chapter 4). Photons have momentum just as massive particles do although their momentum is not calculated by mv as usual, which would be zero, but rather by hf/V_{light}.

Consider an electron and a positron with M_1 and M_2 momenta. The net momentum M is the sum of the two momenta, that is,

$$M_1 + M_2 = M = 2mv \qquad (8.16)$$

After the collision, the energy (E) carried by the positron-electron pair is transferred to the photons. The energy carried by positron-electron pair can be calculated using the relation

$$E = M_1 V_{light} + M_2 V_{light} \qquad (8.17)$$

where V_{light} is the light velocity. Net mass is calculated using relation, $m = m_0 \sqrt{1 - v^2 / V_{light}^2}$ where m_0 is the electron rest mass (9.109×10^{-31} kg), and v is the velocity of the pair's center of mass. If the pair's center of mass is fixed ($v = 0$), the two photons fly apart in opposite directions (180° apart) with the same energy:

$$M_1 V_{light} = M_2 V_{light} = m_0 V_{light}^2 = 511 \text{ keV} \qquad (8.18)$$

In the center-of-mass frame of the *positron-electron* pair, the total energy (E_t) of the annihilation gamma rays is

$$E_t = 2E_0 - E_B \qquad (8.19)$$

where E_B is the electron binding energy and E_0 is the rest energy of the electron, which is

$$E_0 = m_0 V_{light}^2 \qquad (8.20)$$

However, when there is a net center of mass energy associated with the annihilating pair, this total energy is not split equally to 511 keV between the two gamma rays. One gamma ray is upshifted while the other is downshifted from the center energy of $m_0 c^2 - E_B/2$ by an amount equivalent to $\Delta E = p_L c/2$, where p_L is the longitudinal component of the electron-positron momentum along the direction of the gamma ray emission. In the case of two-photon annihilation, measuring the deviation angle q of the photons from 180° or the Doppler shift ΔE of the annihilation line (511 keV) makes it possible, at least in principle, to determine the momentum of the electron-positron pair.

The lifetime of a positron is directly related to the electron density in the annihilation site (i.e., the probability it will run into an electron); the higher the electron density, the shorter the lifetime. The inverse of positron lifetime is referred as annihilation rate, λ_D. Theoretically, λ_D is proportional to the effective electron density n_e sampled by the positron, namely,

$$\lambda_D \approx \pi r_e^2 V_{light} n_e$$

where r_e is the electron radius. The process of gamma emission is also characterized with the annihilation rate λ_D. Since λ_D is dependent on the type of isotope used, and each positron emitter has a different E_{max} for positrons and hence a different range, this imposes an isotope-dependent limitation on spatial resolution in PET imaging.

8.5.2.2   PET Detectors

Since the basic event in PET is the simultaneous detection of the two 511-keV γ rays, PET requires instrumentation and image reconstruction methods that differ from those used in other medical imaging modalities. Typically, a gamma camera system is used as the imaging device (Figure 8.8). Conventional gamma cameras have been designed with planar detector geometry since the development of Anger-type position encoding systems. Main components of the PET camera (or detector) are a large-area scintillator, an array PMTs, and a collimator in front of the scintillation crystal to localize radiation. The role of the scintillation crystal is to convert as much energy as possible of the incident gamma rays or photons into visible radiation, which is converted to an electric current by the PMTs. Photons deposit energy within the scintillation crystal by a photoelectric or a Compton scattering (described previously in Section 8.3.2) interaction. A small portion (10%) of the incident photon's energy deposited in the scintillation crystal is converted into visible light photons ~3-eV energy. Commonly used scintillation crystals include bismuth germanate crystals (BGO), lutetium oxy-orthosilicate (LOS, that result in shorter deadtimes and improved countrate responses), GSO, and NaI. Visible light photons are guided towards the photocathodes of an array of PMTs where they are converted into electrons, multiplied, and finally converted into an electrical signal at the anode of each PMT. The amplitudes of the anode signals from each anode are then examined by either an analog or a digital positioning circuitry to estimate the position at which a photon interacts with the crystal.

The purpose of the collimator is to mechanically confine the direction of incident photons reaching the scintillation crystal and thereby provide a means to localize the site of the emitting sources. The collimator is usually made out of a plate of lead or a similar high atomic number substance such as tungsten in which a large array of apertures (circular, triangular, square, or hexagonal shaped) are drilled close to each other with a narrow septal thickness. By physically restricting the region from which gamma radiation can transmit through a given aperture, the collimator attached to the scintillation crystal ensures that the position of events on the scintillation crystal correspond to the position of the 2D projection image of the object. If the apertures of the collimator are all parallel to each other, the collimator is then called a parallel hole collimator. Most clinical examinations are conducted with a parallel hole collimator since it provides the ideal combination of resolution and sensitivity for most regions of the body with no geometrical distortion. The resolution of scintillation imaging is mainly determined by the intrinsic resolution of the camera and the resolution of the collimator, and the net resolution is obtained from a quadrature addition of these two factors. In some applications, the apertures may be angulated to form converging or diverging hole collimators, providing certain advantages in imaging relatively small regions (converging collimator) or relatively large regions (diverging collimator) with appropriate magnification of the images. Alternatively, just one or a limited number of holes may be used to form a pin-hole collimator, which is particularly useful in imaging very small regions or small organs such as the thyroid gland.

Tomographs consist of a large number of detectors arranged in many rings around the subject. A linear combination of energy information from each PMT is used to identify the crystal element where the g interaction occurred. Each detector is connected in coincidence with opposing detectors. The line connecting the two detectors giving a signal is called a line of response (LOR). The events occurring along each possible LOR are histogrammed into sinograms, or projections, which are then converted into radioactivity images using analytical or statistical image reconstruction algorithms. A number of small scintillation crystals are mounted on PMT. Increased spatial resolution requires smaller scintillation crystals. The imaging properties, such as uniformity and spatial resolution, have improved with time as a result of a more efficient camera designs. Commercial tomographs using BGO crystals reached a spatial resolution of approximately 4 mm. An LSO crystal has a five fold higher light output and an eight times faster light decay time. This allows the spatial resolution to be close to its physical limit of 1–2 mm, which is determined by the positron range in tissue and the small noncollinearity of the annihilation photons. The introduction of empirical digital corrections of energy and position of events has also improved the image properties.

In order for PET images to provide quantitative information, which is necessary to obtain biologically meaningful results, proportionality between image count density and radiotracer concentration must be preserved. This requires the tomograph to be calibrated for effects such as dead time and detector nonuniformity. In addition, accurate corrections for effects that alter the perceived source distribution must be performed. These effects include attenuation of the γ rays (which leads to source intensity underestimation), Compton scattering of the detected γ ray, and detection of random coincidences (which contribute to source position misidentification by assigning events to incorrect LORs). When all the corrections are appropriately applied to the data, PET images show the distribution of the radiotracer in units of radioactivity concentration. In addition to providing static images of radioactivity distribution, PET can also provide information on the change of radioactivity distribution as a function of time by performing sequential scans (dynamic scanning). Compartmental modeling (described in Chapter 10) is used to convert a process observed in terms of radioactivity distribution into biologically meaningful variables.

8.5.2.3 Single-Photon Emission Computed Tomography (SPECT)

Conventional planar imaging techniques of PET suffer from artifacts and errors due to superposition of underlying and overlying objects, which interfere with the region of interest. The emission CT approach called SPECT has been developed based on detecting individual photons emitted at random by the radionuclide to be imaged. A major difference between the SPECT and PET is that SPECT uses a radioactive tracer such as technetium-99 (product of the longer lived molybdenum-99 ($t_{1/2} = 2.8$ days) and has a half-life of about 6 hours) that emits a single γ-ray photon instead of a positron-emitting substance. Photons of energy 140 KeV are released, which are easily detected by gamma cameras. The same PET camera technology is used in SPECT where conventional gamma cameras are mounted on a frame that permits 360° rotation of the detector assembly around the patient. Further, the principles of CT (see Section 8.4.3) are applied to provide a 3D quantitative estimate of the distribution of gamma-emitting radionuclides. Projection data are acquired in SPECT at a given view angle, but over multiple planes simultaneously, with a gamma camera. The camera is then rotated around the subject in steps of $\Delta\theta$ covering the full 360° range. Images are reconstructed from these multiple-view projections similar to CT using either a filtered backprojection technique or an iterative reconstruction algorithm. In some situations, for example imaging the heart with a low energy emitter like Tl-201, a 180° scan may be preferred to a 360° rotation.

Despite certain imperfections in some early combined devices, their performance made SPECT a widespread clinical tool within a short period of time. In SPECT, radioactivity distribution of the radionuclide inside the patient is determined instead of the attenuation coefficient distribution from different tissues as obtained from X-rays. Further, the radiation source is within the body instead of outside the patient. Since the source distribution is unknown, an exact solution to the attenuation correction problem is theoretically very difficult to attain. Hence, the technique was varied from CT. Further, the cameras have improved, providing better mechanical stability, more efficient scatter and attenuation corrections, and shorter reconstruction times. However, the amount of radionuclide that can be administered is limited by the allowable dose of radiation to the patient. Furthermore, SPECT with a single rotating gamma camera suffers from low sensitivity compared to PET. If the spatial resolution of PET device is in the order of 5 mm, a SPECT single rotating camera can be made with a spatial resolution in the order of 15 to 20 mm. To increase the detection efficiency while improving the spatial resolution of the imaging system, a three-headed gamma camera has been developed. Nonetheless, several approaches leading to approximate corrections for both iterative and direct reconstruction techniques are also developed.

Emitted gamma radiation interacts with the body by photoelectric absorption and Compton scattering processes, producing a significant attenuation in the primary beam at energies used in SPECT. Since attenuation depends on the properties of the medium interposed between the point of origin of the photons and the object boundary, it is necessary to know the distribution of attenuation coefficients corresponding to the energy of the emitted radiation within the object and the source distribution to accurately compensate for attenuation. In addition to attenuation of photons due to photoelectric absorption, Compton scattering of the emitted photons within the object introduces another error in the acquired data. Due to the finite energy resolution of the detection system, a portion of the scattered photons is indistinguishable from the primary photons and is recorded under the photopeak. Since the scattered photons originate mostly from regions outside the region delineated by the collimator line spread function, counts produced by the scattered photons cause blurring and a reduction of contrast in the image. Although the effects of scattered photons are not as prominent as attenuation losses, corrections for scatter become necessary when higher quantitation accuracy, for example in compartmental modeling, is sought in SPECT. Several methods, depending on the type of image reconstruction algorithm, are proposed to correct for scattering in SPECT.

8.5.3 Magnetic Resonance Imaging (MRI)

MRI is based on the principles of nuclear magnetic resonance (NMR), a spectroscopic technique used to obtain chemical and physical information of molecules. MRI relies on the effect of strong magnetic field on the nucleus of different atoms and a radio signal. Those together, trigger atoms in the body to send out signals of their own [Figure 8.9(a)]. In an atom, both the nucleus and electrons possess intrinsic angular momentum, or spin. Protons present in the nucleus also spin, which is directly related to its nuclear spin. Typically, nuclear spin is represented by the total angular momentum of a nucleus. However, for electrons in atoms, a clear distinction is made between electron spin and electron orbital angular momentum and they are combined to obtain the total angular momentum.

Nuclear spin is related to the angular momentum of the atomic nucleus, which in turn originates from that of the constituent nucleons. Protons and neutrons each have 1/2 spin quantum numbers. When a nucleus contains an even number of protons and neutrons, the individual spins of these particles pair off and cancel out and the nucleus is left with zero spin. However, in a nucleus containing an odd number of protons or neutrons (such as hydrogen), pairing is incomplete and the nucleus has a net spin of 1/2. For example, the six protons in ^{13}carbon provide no net spin [Figure 8.9(b)]. However, the unpaired neutron endows the nucleus as a whole with a spin quantum number of 1/2. All such nuclei undergo NMR. In medical MRI the hydrogen nucleus, consisting of a single proton, is used because of its high NMR sensitivity and natural abundance, that is, very high concentrations of water found in the body. Both protons and electrons are charged particles. Since the charge is spinning, a small magnetic field is created, and in essence such particles are tiny bar magnets that are spinning. Associated with each nuclear spin is a nuclear magnetic moment, which produces magnetic interactions with its environment. The magnetic field, $B(r)_x$, due to a current (I), carrying wire of unit length (Δl is a vector with the direction that of the current) is given by

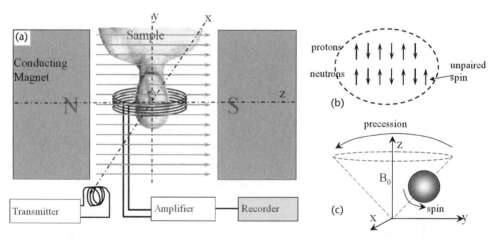

Figure 8.9 Basics of nuclear magnetic resonance analysis. (a) Schematic showing the basic components in magnetic resonance imaging along with the orientation of the axes. (b) Origin of nonzero spin in carbon-13 nucleus. (c) A precessing proton. Due to the torque exerted by the external magnetic field on the spinning magnetic moment, the proton will precess around the direction of the magnetic field.

$$B(r)_x = \frac{\mu_m I (\Delta l \times r)_x}{4\pi r^3} \qquad (8.21)$$

where μ_m is the "magnetic permeability" and $\mu_{m0} = 4\pi \times 10^{-7}$ N/A^2 is the magnetic permeability of the vacuum. Equation (8.21) is referred as the Biot-Savart Law, named after two French physicists. The factor $\Delta l \times r$ is the cross product of the vector Δl with the vector r (radius direction is pointing to the magnetic field point). The effect of the cross product is to make B perpendicular to the direction of the current and radius direction. The direction of B at any point is tangent to a circle centered on and perpendicular to the current element. B is said to be an axial vector field because of its axial (cylindrical) symmetry with respect to its source. The orientation of the B field around the source current (clockwise or counterclockwise) is determined by the direction of the current (the flow of positive ions). The magnitude of the magnetic field is

$$B(r)_x = \frac{\mu_m I \Delta l \sin(\theta)}{4\pi r^2} \qquad (8.22)$$

where θ is the angle between the current element and the position vector of the field point. The magnetic force exerted by B on a length L of test current I is

$$F = ILB \sin\theta \qquad (8.23)$$

where θ is the angle between B and the test current. If B is parallel to I, then F is zero. The direction of F is perpendicular to both B and the test current. This situation is different from that of the electric and gravitational forces, which act along a vector parallel to their respective fields (pointing from one source to another). In further contrast with the case of the electric force, parallel currents attract and antiparallel currents repel. Hence, magnetic fields and forces are inherently 3D. The point in the center of the magnet is called the isocenter of the magnet. The magnetic field at the isocenter is B_0.

8.5.3.1   Frequency Response

Each magnetic moment also experiences a torque due to the field, resulting in precession about the z-axis, and the moment describes a cone in the xy plane [Figure 8.9(c)]. The angular frequency of precession, known as the Larmor frequency (f_R), is proportional to the external magnetic field, B, that is,

$$f_R = \gamma B \qquad (8.24)$$

where γ is the gyromagnetic ratio of the nucleus, a quantity unique to each atom. For hydrogen, γ is 42.57 MHz/T or 2.675×10^4 rad/s-gauss. The nuclei of different elements and even of different isotopes of the same element have very different frequencies (Table 8.2). In biological materials, the magnetic nuclei of ^{1}H, ^{13}C, ^{23}Na,

Table 8.2 Nuclear Magnetic Resonance Properties of Nuclides Important in Biomedical Imaging

Isotope	Spin Quantum Number	Magnetic Moment	Natural Abundance	Relative Sensitivity	f_R (MHz at 1T)
^{1}H	1/2	2.79	99.98	1	42.58
^{2}H	1	0.86	0.02	0.0096	6.54
^{13}C	1/2	0.7	1.11	0.0159	10.71
^{17}O	5/2	−1.89	0.04	0.03	5.77
^{19}F	1/2	2.63	100.00	0.834	40.06
^{23}Na	3/2	2.22	100.00	0.09	11.26
^{31}P	1/2, −1/2	1.13	100.00	0.0664	8.341

^{31}P, and ^{39}K are all abundant. The hydrogen being the abundant material with the largest magnetic moment is the best element for biomedical applications. Particles with nonzero nuclear spins can *split* their energy levels in a magnetic field. The difference in the energy (ΔE) of the two states, which dictates the difference in the populations, depends on the strength of the magnetic field:

$$\Delta E = \frac{h}{2\pi}f = \frac{h}{2\pi}\gamma B_0 \tag{8.25}$$

By Boltzmann's law, named after Austrian physicist Ludwig E. Boltzmann, the proportion of protons in the two states can be calculated as

$$\frac{N_2}{N_1} = e^{-\frac{\Delta E}{kT}} \tag{8.26}$$

where N_1 and N_2 are protons in the lower and higher states, respectively, and k is the Boltzmann's constant ($1.3806505 \times 10^{-23}$ J/K). The populations of the two states differ slightly, with a few more nuclei per million aligned with B_0 than against. This difference in population leads to a bulk magnetic moment.

EXAMPLE 8.6

10 moles of water at 37°C is placed in a magnetic field of 1T. Calculate the proportion of protons in two states of energy.
   Solution: For $1H$, $\gamma = 42.58$ MHz/T
   From (8.25),

$$\Delta E = \frac{6.626 \times 10^{-34}[J.s]}{2\pi} 42.58*10^6[s^{-1}]*1 = 44.89 \times 10^{-28} J$$

From (8.26),

$$\frac{N_2}{N_1} = e^{-\frac{44.89\times10^{-28}}{1.3806505\times10^{-23}*310}} = 1.000001049$$

The nuclear magnetic moment of spinning charged nucleus of an atom is

$$M = \gamma h \frac{n_p e}{4\pi} \frac{m_s}{m} \tag{8.27}$$

where m_s is the quantized spin, n_p is the number of protons, and m is the total mass of the nucleus. The quantum of energy it carries is given by (8.2). If the nucleus is immersed in a magnetic field, its spin (and therefore its magnetic moment) gives it potential energy. Equating

$$\Delta U = -B\Delta M = -B\gamma h \frac{n_p e}{4\pi} \frac{\Delta m_s}{m} = hV_{light} \tag{8.28}$$

If m_s changes from (for instance) $-1/2$ to $1/2$, a photon of frequency f is emitted. By flipping the spin from $1/2$ to $-1/2$ via radiating the nucleus with photons of frequency f, the radiation can be used to image objects. If the sample is exposed to a radio frequency (RF) EM wave whose energy would exactly match the difference between the two energy levels, nuclei in the lower energy level may gain energy by absorbing photons and those in the higher level may be induced to emit photons to lose energy. These processes cause nuclei to flip between the two orientations, modifying the population distribution from Boltzman population. Hence, the magnitude and direction of M is changed as long as the RF wave is present to keep the system.

However, considering the large number of nuclei present in a macroscopic sample, it is convenient to describe the sample in terms of interactions between the bulk magnetization (B_1) and the magnetic field component of the RF wave. The bulk magnetization is forced to tilt away from the alignment with B_0 by the oscillating RF magnetic field (ignoring the electric component of this field). While the bulk magnetization is tipped away from alignment, it will precess around B_0 for the same reason that a spinning top wobbles around the direction of gravity (replace the force of gravity with the force due to B_0) [Figure 8.9(c)].

If the RF frequency, and hence the frequency at which B_1 rotates in the xy plane, matches the precessional frequency of the magnetic moments, f_0. A resonant condition (resonance describes the maximum transfer of energy when the system is tuned to a particular frequency, termed *resonance frequency*) is achieved in which M tips, (or nutates), from the z-axis towards the xy plane. The nutation continues until B_1 is removed, so the amplitude and duration of B_1 can be tailored to nutate M through specific angles. A special case is that in which M is nutated entirely into the transverse plane, leaving no magnetization along the z-axis. An RF wave of sufficient duration and amplitude to achieve this is known, for obvious reasons, as a 90° pulse. Similarly, a 180° pulse inverts M or, if applied after a 90° pulse, rotates M through 180° in the transverse plane. These pulses are the basic building blocks of NMR, and by stringing them together in different combinations, a large of number of experiments can be performed.

A 90° pulse generates net transverse magnetization, which precesses about the direction of B_0. If a conducting coil (i.e., antenna) is positioned appropriately with respect to this precession, an oscillating current will be induced in the coil (given by Lenz's law). This precessing magnetization is detected by means of EM induction (i.e., by placing a tuned antenna close to the sample). Such an antenna is commonly known as an RF coil, and similar coils are used to transmit RF and to detect the emitted NMR signal.

8.5.3.2 Acquiring MRI Images

The NMR experiment described thus far results in the acquisition of an undifferentiated signal, with no means of incorporating spatial information so as to produce an image. A study of a specific tissue can only be achieved by excising the tissue from the body, or if the tissue is superficial, by placing a small RF coil (known as a surface coil) against the body. For acquiring MRI images, the method of spatial encoding introduced by Paul C. Lauterbur (discussed in Chapter 1) is used. Spatial encoding is based on using gradient coils, which generate spatially varying magnetic field so that spins at different locations precess at frequencies unique to their location. The temporary imposition of an additional static magnetic field, which lies parallel to B_0 but varies linearly in strength with position along the x-, y-, or z-axis (or indeed some direction oblique to these axes) [Figure 8.10(b)] results in a linear variation in the Larmor frequency as a function of position:

$$f_R = \gamma(B_0 + zG_x) = f_0 + \gamma xG_x \qquad (8.29)$$

where G_x is the strength of the gradient (in mT/m) applied along the x-axis, and x is the position relative to isocenter. Rearranging (8.24),

$$x = \frac{f_R - f_0}{\gamma G_x} \qquad (8.30)$$

This principle forms the basis for magnetic resonance imaging. This frequency variation can be used in three ways to achieve spatial localization.

EXAMPLE 8.7

A sample contains water at two locations, $x = 0$ cm and $x = 2.0$ cm. A one-dimensional magnetic field gradient of 1 G/cm is applied along the x-axis during the acquisition of an FID. The frequency encoding gradient is 1 G/cm. What frequencies (relative to the isocenter frequency) are contained in the Fourier transformed spectrum?

Solution: The frequency encoding equation is $\Delta f = \gamma G$, where Δf is the frequency offset from isocenter γ is the gyromagnetic ratio, 42.58 MHz/T, and G is the frequency encoding gradient = 0.0001 T/cm

$$x = 0 \text{ cm}, \Delta f = 0$$
$$x = 2.0 \text{ cm}, \Delta f = 8.516 \text{ kHz}$$

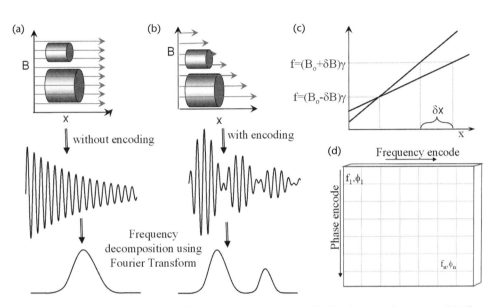

Figure 8.10   Spatial encoding. (a) Effect of constant magnetic field and output frequency. (b) Effect of a gradient magnetic field and the output frequency. (c) Use of the magnetic field gradient for selective excitation of a slice. (d) Slice image formation using both frequency encoding and phase encoding.

Slice Selection

If the gradient is applied along the z-direction [Figure 8.10(c)] simultaneously with the initial RF pulse, and the pulse is modified to contain a narrow band of frequencies (δf) rather than a single frequency, a thin slice (δx) of spins can be selectively excited. Using this approach, a 2D tomographic imaging is obtained. The bandwidth of the RF excitation pulse is obtained by substituting $\Delta f = \Delta \omega / 2\pi$ in (8.32) and then the slice thickness is selected. There are two ways to select different slices: change the position of the zero point of the slice selection gradient with respect to the isocenter, or change the center frequency of the RF to correspond to a resonance frequency at the desired slice. The second option is usually used as it is not easy to change the isocenter of a given gradient coil.

Frequency Encoding

From (8.29), the resonance frequency is a linear function of the magnetic field strength. If a gradient field is applied during image acquisition (i.e., after activating the RF excitation pulse), then the position of a nucleus along that gradient can be recorded by shifting its resonance frequency. This is called frequency encoding. The signal can be decomposed using Fourier transform methods to determine the amount of signal at each frequency and hence at each position, resulting in a projection through the sample. The acquired signal originates from the entire object under study, and spatial information is obtained only by breaking that signal down.

Phase Encoding

If the gradient field is turned on for a short term after activating the RF excitation pulse and turned off before image acquisition, then the effect of the gradient field is no longer time-varying. However, the phase of a nucleus's signal is dependent on its position and the gradient field is a fixed phase accumulation determined by the amplitude and duration of the phase-encoding gradient. This method is called phase encoding. By repeating the experiment a number of times with different gradient amplitudes, it is possible to generate a set of data in the third dimension and have the data Fourier transformed to yield spatial information along the direction perpendicular to the frequency encoding axis [Figure 8.10(d)]. During signal acquisition, the phase of the xy-magnetization vector in different columns will systematically differ. When the x- or y-component of the signal is plotted as a function of the phase-encoding step number n, it varies sinusoidally. As each signal component has experienced a different phase encoding gradient pulse, its exact spatial reconstruction is located by the Fourier transformation analysis. Spatial resolution is directly related to the number of phase-encoding levels (gradients) used.

All three methods are applied to generate multiple 2D slices through the patient. To allow time for this process, acquisition of the NMR signal is delayed for up to a few hundred milliseconds after excitation by using magnetic field gradients and RF pulses to produce an echo at the desired time. Unlike X-ray CT, MRI can image slices in any desired plane by appropriate use of gradients. Alternatively, the technique is modified to produce 3D volumes of data that can be viewed using various software or postprocessed to generate arbitrary slices. Spatial resolution is determined by the number of frequency encoded projections and phase-encoded projections for a given field of view.

As described thus far, an MRI of the body is essentially a map of water distribution. Such an image would be of limited clinical value, since water density varies relatively little between tissues. One of the major strengths of MRI is the ability to manipulate image contrast by tissue relaxation times. Once the RF exposure is discontinued, the high-energy state protons in the material would continue to dump their transition energy (or relax) at the same frequency as was previously absorbed until the Boltzmann equilibrium is reestablished. Nuclei return to their initial population distribution via a variety of relaxation processes, categorized according to whether they cause loss of energy from the spin system or just the exchange of energy between spins. These two phenomena are known as spin–lattice and spin–spin relaxation, and are characterized by the relaxation times τ_1 and τ_2, respectively. τ_1 relaxation results in exponential recovery of z-magnetization, M_z, while τ_2 processes cause exponential decay of the precessing transverse magnetization, M_{xy}:

$$M_z = M(1 - e^{-t/\tau_1})$$

(8.31)

$$M_{xy} = Me^{-t/\tau_2} \tag{8.32}$$

Thus the NMR signal detected by the RF coil rapidly decays due to τ_2 relaxation. Relaxation behavior is strongly dependent on the physicochemical environment of the nucleus and hence on the tissue type in which the nuclei are located. This is an important source of image contrast in biomedical MRI. During the interval between excitation and acquisition of an echo, magnetization in different parts of the sample will be undergoing τ_2 decays at different rates depending on the environment (i.e., the tissue in which it is located). Thus by varying this interval, known as the echo time, τ_E, it is possible to vary the degree of τ_2 weighting (i.e., the extent to which differences in τ_2 affect the appearance of the image). In addition, as noted above, it is necessary to repeat acquisition a number of times to allow phase encoding. During the interval between repetitions, known as the repetition time, τ_R, magnetization undergoes differential τ_1 recovery. By varying this interval, the user can alter the extent of τ_1 weighting. τ_1 and τ_2 weighting result in very different image appearances. τ_1-weighted images provide excellent soft tissue contrast, and material with long τ_1, such as cerebrospinal fluid (CSF) in the ventricles of the brain, appears dark. This type of imaging is often used to depict anatomical structures. On τ_2-weighted images, tissues with increased water content appear bright. This includes CSF, but also pathological processes such as neoplasm, inflammation, ischemia, and degenerative changes. Thus τ_2-weighted images are often used to highlight areas of disease.

The advantage of MRI is balanced by the fact that MRI images have less resolution than those from CT. MRI complements X-ray CT in providing different information. X-ray CT offers details that depend on the density of body structures: the denser an object, the more X-rays it blocks and the whiter its appearance in an X-ray or a CT image. MRI responds to the prevalence of particular types of atoms in the body. Fatty tissues, which have little water, appear bright, while blood vessels or other fluid-filled areas are dark. MRI is particularly useful for seeing details in the brain. Gray matter has more fluid than white matter, making it easy to distinguish between the two. However, the resolution of a conventional clinical MRI is in the order or 1–2 mm and it is not ideal for cellular-level imaging. In 1992, the functional MRI (fMRI) was developed to allow the mapping of the various regions of the human brain. fMRI detects changes in blood flow to particular areas of the brain. It provides both an anatomical and a functional view of the brain. Three types of dynamic brain functions are quantifiable by noninvasive imaging: blood flow changes in response to sensory or mental activity, neurochemical activity, and metabolic activity of energy consumption (e.g., glucose and oxygen consumption). Radiotracer techniques have shown that blood flow in the capillary bed of small regions of the brain increases 2% to 30% when that region is called on to increase nerve activity by some external or internal stimulus. The more recent advent of MRI methods for detecting a signal change associated with this local activity makes it possible to measure mental functioning noninvasively without the use of radiation.

8.5.4 Ultrasound Imaging

The term *ultrasound* refers to any sound wave above the audible frequencies over 20,000 Hz. Ultrasound imaging (also referred as echocardiography or diagnostic medical sonography) depends on the behavior of acoustic waves in various media and uses mechanical energy to penetrate tissues, unlike EM spectrum. An advantage of diagnostic ultrasound besides its safety is the ability to perform live active and/or passive range of motion studies. The ability to move the probe to a particular part of the anatomy and observe and record motion studies such as partial or complete tendon, muscle, and ligament tears should not be understated. MRI and X-rays do not allow for motion studies. Finally, the cost of an ultrasound imaging system is much cheaper than an MRI.

Frequencies used for medical diagnostic ultrasound are near the 10-MHz range. These sound waves bounce off the various types of body tissues and structures. Images are obtained by directing a probe that emits ultrasound. These sound waves penetrate the body and encounter different tissues and tissue planes. Part of the wave is reflected (echoes) back at varying speeds, depending on the density of the tissues. As the sound waves echo from different tissues, the sensitive microphone records tiny changes in the sound's pitch and direction. In conventional use, this also includes deviation of reflected radiation from the angle predicted by the law of reflection. Reflections that undergo scattering are often called diffuse reflections and unscattered reflections are called specular (mirror-like) reflections. Different tissue interfaces cause signature reflective patterns. For example, when the sound wave encounters a dense object such as bone, most of the sound is bounced back to the probe and little is allowed to pass through the tissue. This is seen as bright white on the video screen, and known as *hyperechoic* (very *echogenic*). In contrast, when the sound wave encounters a lean object such as ganglion cyst (a fluid-filled sac), very little sound is bounced back to the probe. This is seen as an oval dark area within tissues and known as *hypoechoic* and anechoic.

Resolution and accuracy of an ultrasound based imaging system are limited by the inherent characteristics of the ultrasound. Depending on frequency and propagation speed of the ultrasound, the wavelength determines the minimum duration of an ultrasound pulse and the thickness of an ultrasound beam. High frequency thus brings high resolution but results in a low sensitivity due to the increase in absorption with the frequency. The depth of tissue penetration is dependent on the frequency of the ultrasound being used. For diagnostic musculoskeletal tissue, a 7.5 MHz is used, which produces an image that is 4.5 cm long and penetrates into the tissues 7 cm deep at maximum. This 7.5-MHz probe known as a high-resolution extremities probe, is used for carpal tunnel syndrome, torn rotator cuff, and many podiatric applications. Probes such as the 5.0 MHz are for deeper structures (for example, examination of hip, knee, and spine), and 3.0 MHz for still deeper structures. Ultrasound is used to examine the arterial/venous system, heart, pancreas, urinary system, ovaries, spinal cord, and more. Ultrasound imaging can be used for identifying cysts in soft tissue because usually they contain fluid. However, many of the longitudinal or transverse scans may initially look the same, unlike that of X-rays. It is essential to become familiar with the different tissue pattern (e.g., a tendon versus bone versus muscle). The way a particular type of tissue responds to an ultrasound wave must be learned. Once the normal structure is recognized, the diagnosis of pathology becomes apparent.

8.5.4.1 Detectors

The instrumentation of ultrasonic imaging consists of the transducer, a beam former, and a signal analysis and display component. When a small handheld probe is pressed against the skin, transducers located in the probe (Figure 8.11) convert the electrical energy to acoustic energy, and vice versa. The conversion of electrical pulses to mechanical vibrations and the conversion of returned mechanical vibrations back into electrical energy is the basis for ultrasonic imaging. The position of the probe and its anatomical location is important. As the probe is moved back and forth over the skin, transducers produce sound waves at a known frequency based on the input electrical signals. Since the ultrasound produced cannot travel through the air and then into the body, the transducer must be in direct contact with the skin, so a transmission jelly is used to insure a complete union. Ultrasound waves are sent into the tissues and reflected signature waves are converted back to electrical signals from which images are reconstructed. The signal processing involved in forming images is based largely on linear models for sound transmission and reflection in tissue.

In order to get the best performance out of an ultrasound imaging system, the transducer must not introduce distortion or have any negative influence on the ultrasound signals. Sensitivity, pulse length, and beam width are the major parameters, which directly depend on the design and performance of the transducer. The transducer is basically a piece of polarized material (i.e., some parts of the molecule are positively charged, while other parts of the molecule are negatively charged) with electrodes attached to two of its opposite faces. When an electric field is applied across the material, the polarized molecules will align themselves with the electric field, resulting in induced dipoles within the molecular or crystal

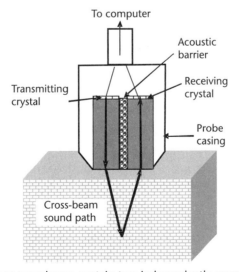

Figure 8.11 Dual element transducers contain two independently operated elements in a single housing. One of the elements transmits and the other receives the ultrasonic signal.

structure of the material. This alignment of molecules will cause the material to change dimensions. This phenomenon is known as electrostriction. In addition, a permanently polarized material such as quartz (SiO_2), barium titanate ($BaTiO_3$), lead zirconate titanate (PZT), or piezo-polymers produce an electric field when the material changes dimensions as a result of an imposed mechanical force. This phenomenon is known as the piezoelectric effect. The active element of most acoustic transducers used is a piezoelectric ceramic, which can be cut in various ways to produce different wave modes. Piezoelectric ceramics have become the dominant material for transducers due to their good piezoelectric properties and their ease of manufacture into a variety of shapes and sizes. The thickness of the transducer is determined by the desired frequency of the transducer by the relation

$$f_x = \frac{nk_x}{l} \tag{8.33}$$

where f_x is the frequency in the x direction, n is the harmonic order, l is the thickness of the crystal in the x-direction, and k_x is known as the "frequency constant" in the x-direction. A thin wafer element vibrates with a wavelength that is twice its thickness. Therefore, piezoelectric crystals are cut to a thickness that is 1/2 the desired radiated wavelength. The higher the frequency of the transducer, the thinner the active element. The primary reason that high frequency contact transducers are not produced is the necessary thin element, which is too fragile to handle. An active area of research is developing novel materials with better performance.

Traditional 2D ultrasound visualization produces planar tomographic slices through the object of interest. However, 3D images are formed by volume visualization methods, significantly improving the spatial relationships. To create 3D images, transducers containing piezoelectric elements arranged in 2D arrays are developed. To aid in construction of complex arrays, simulations are performed to analyze and optimize their beam pattern for a particular application. Array success is based on the *radiation pattern*, which determines how well the ultrasound energy is focused, the lateral resolution, and the contrast, and the *pulse-echo response*, which determines the axial resolution.

However, high frequency arrays are very difficult to build because of fragility.

8.5.4.2   Effects of Ultrasonic Waves on Tissues

Ultrasound interacts with tissues in a few different ways including cavitation, chemical, mechanical, and thermal interactions. Cavitation is unique to ultrasound due to the mechanical nature of the ultrasound wave, which uses a change in a pressure of the medium to propagate. This change in pressure can cause the tissue to collapse when the mechanical wave cycles too far negative. It can also change the chemical composition of a tissue. For example, exposure to ultrasound can cause the depolymerization (or breakdown of long chains of polymers). Ultrasound can also cause oxidation, hydrolysis, and changes in crystallization. Ultrasound has been linked to some increased permeability of tissues, due to the rise in temperature of the tissue as the mechanical waves interact with the tissue at the different interfaces and add energy to the tissue. This added energy translates mainly into an increase in the temperature of the tissue. The increase of temperature, however, is nearly eliminated at lower intensities. Exposure to ultrasound adds more energy to tissue than other types of radiation such as infrared radiation or ultraviolet radiation. Energy added to the tissue must be absorbed (rather than reflected or simply transmitted) in order for the energy to alter the tissue. The added energy can result in a mechanical alteration of the cells, which can result in cells being ripped apart. Unlike the probes used for physical therapy, which are designed for deep heating of tissues, diagnostic ultrasonography probes cause no heating or tissue damage.

8.5.5   Optical Coherence Tomography (OCT)

OCT is a noninvasive imaging technique based on the principle of low-coherence interferometry, where distance information concerning various structures is extracted from time delays of reflected signals. In a way, OCT is analogous to ultrasonic imaging. OCT uses near infrared (700 to 1,200 nm) coherent (single wavelength, parallel plane, in-phase) light instead of sound, which results in higher resolution images (<50 μm) compared to conventional ultrasound. This gain in resolution is unfortunately matched by a reduced depth of tissue penetration (only a few centimeters). It has been applied in:

- Ophthalmology for retinal imaging and the diagnosis and monitoring of macular diseases;
- Developmental biology for repeated in vivo imaging of a developing system to track developmental changes and an assessment of structure and function.

OCT generally involves shining laser light on the object or scene, and measuring the properties of the reflected light as recorded by an optical sensor. The term laser optical tomography is sometimes used, as data are collected by applying a laser source. The OCT system is based on a Michelson interferometer with light split into two arms (Figure 8.12): a sample arm (containing the item of interest) and a reference arm (usually a mirror). The combination of reflected light from the sample arm and reference light from the reference arm gives rise to an interference pattern, but only if light from both arms have travelled the "same" optical distance ("same" meaning a difference of less than a coherence length). By scanning the mirror in the reference arm, a reflectivity profile of the sample is obtained (this is time domain OCT). Areas of the sample that reflect back a large amount of light create greater interference than areas that do not. Any light that is outside the short coherence length will not interfere. Comparison of the reference beam and the reflected light yields:

- The amplitude difference between the reference and reflected beams, which indicates the optical absorption of the surface or the surface reflectance;
- The phase difference, which indicates the additional distance traveled by the reflected beam, from which the distance to the surface can be measured.

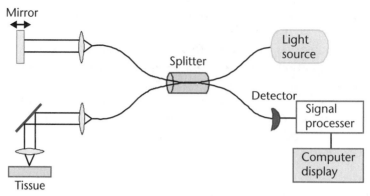

Figure 8.12 Schematic diagram of the optic system of OCT.

An entire image is obtained by scanning the laser light over the entire area and computing the above quantities at every location. The result is two images containing reflectance information and distance (range) information. The reflectivity profile, called an A-scan contains information about the spatial dimensions and location of structures within the object of interest. A cross-sectional tomograph (B-scan) is achieved by laterally combining a series of these axial A-scans. Enface imaging at an acquired depth is possible depending on the imaging mode used. A continuous beam is used, which is modulated in amplitude by a signal of much larger wavelength than the laser radiation.

Instead of lasers, the use of infrared light waves enables improved penetration through highly scattering structures, improved resolution by 10–100 (less than 100 μm) times, and imaging deeper tissues located several centimeters beneath transparent structures and of a few millimeters in highly scattering media or tissue. Light in the near-infrared range penetrates tissue and interacts with it in different ways; the predominant effects are absorption and scattering. Many of the substances of interest, such as hemoglobin and cytochromes, exhibit characteristic absorption spectra that depend on whether the molecule is in its oxidized or reduced state. An early tool of medical optical imaging was the *oximeter* devised in the 1930s to detect the amount of oxygen in blood by measuring the ratio of the light absorbed at two wavelengths. Great improvements to this concept came in the 1970s with the advent of microprocessors and light-emitting diodes that permitted the use of many more wavelengths, thus allowing measurement of the absolute amount of oxygen and elimination of background effects. Assessment of the oxygen content of arterial blood through such methods is a major diagnostic tool for monitoring acutely ill patients. The potential of imaging with light was reinforced with the successful application of optical tools to determine the levels of oxygen in the brain of a cat. Later, this concept was used in monitoring brain and muscle oxygenation in humans, as well as in other applications. Substances such as NAD/nicotinamide adenosine diphosphate (NADH) exhibit fluorescence properties that allow for their detection after excitation by light. As these substances play crucial roles in metabolic processes at the cell level, the ability to discern them through indirect measurements has many medical implications.

8.5.6   Endoscopes

Endoscopes are imaging modalities that are used to attain images of human organs such as the bladder, colon, gastrointestinal track, and bronchus for diagnostic purposes. Similar medical imaging modalities are also used in laparoscopic surgical procedures for viewing the internal parts of the human body. Endoscopes operate under white light illumination to provide color images via a video system that allows for the identification of alterations in normal tissue conditions through changes in color and blood concentrations as well as the observation of structural changes. Because the penetration depth of visible light through tissue is small, the information obtained using optical endoscopes represent only the superficial layers of the tissue or organ under investigation. Development in fiber optics has resulted in a considerable advance in endoscopy. Fiber optics are plates or rods (rigid or flexible) made up of arrays of transparent fibers (e.g., glass) having a high index of refraction with a layer of low index. Each fiber acts as a light pipe, containing the light, which enters within a predetermined aperture angle. Since the individual fibers are quite small (the order of a micron), high-definition images can be transmitted through these fiber-optics elements. Fiber-optics disks are used to couple the phosphor layer with the photocathode in multistage intensifiers.

8.5.7   Fluorescence Imaging

A fluorescent molecule (referred as fluorophores) can be excited to a higher energy level by incident light and as a result it emits light of longer wavelength. The measurement of fluorescence is used for both diagnostic and monitoring purposes. These uses include the clinical in vivo fluorescein angiography to image the retinal vasculature, and guidance of surgical resections. Based on the specific localization of administered fluorescent molecules in tissue or cell structures, the use of fluorescence imaging for in vivo and ex vivo characterization of biological materials is well established. These techniques are routinely used in fluorescence microscopy and immunohistochemisty. In contrast to reflected or transmitted light microscopy, which mainly images tissue structures, fluorescence microscopy can provide information about tissue composition and can indicate the state of individual tissue constituents through analysis of the fluorescence properties. Improved resolution is obtained by deploying either confocal microscopy or laser scanning confocal microscopy. Both variants are capable of submicron resolution, thus allowing subcellular imaging. Imaging can be further enhanced by recording dynamic cellular events within live cells using various contrast agents.

The approaches to characterize tissues by light-induced fluorescence spectroscopy (LIFS) and imaging can be classified according to the type of fluorophore(s) investigated and the principle of the instrumentation used. Typically, the fluorophores are grouped into three types:

1.   Endogenous fluorophores that are responsible for native tissue fluorescence (autofluorescence). This technique can derive intrinsic contrast from autofluorescence emitted from endogenous tissue fluorophores (such as elastin, collagen, tryptophan, nicotinamide adenine dinucleotide, porphyrins, fla-

vins, and the like), which have absorption bands in the near ultraviolet or blue and characteristic spectra in the visible region.

2. Fluorophores synthesized in the tissue after external administration of a precursor molecule, such as protoporphyrin IX induced by 5-aminolevulinic acid (ALA).

3. Fluorophores administered as exogenous drugs, including fluorescein, indocyanin green (ICG), and photosensitizers such as hematoporphyrin derivative (HpD) and tetra(m-hydroxyphenyl)chlorin (mTHPC).

These materials have been used in clinical studies of fluorescence diagnostics. The exogenous fluorophores may also be subdivided according to whether or not a delivery vehicle is used to target them to specific tissue/cell compartments (e.g., coupling to monoclonal antibodies or encapsulation into liposomes). Exogenous fluorescent markers with a high quantum yield can also be used to achieve a stronger extrinsic fluorescence contrast. In this case, information is obtained from the change of the fluorescence properties of the marker during interaction with the different tissue components, or by selective localization of the marker in certain tissue constituents.

There has been increasing interest in fluorescence-based techniques with the discovery of colloidal semiconductor nanocrystals called quantum dots (QDs). QDs are generally composed of atoms from groups II–VI or III–V of the periodic table, and an example is zinc sulfide-capped cadmium selenide (CdSe-ZnS). Their size range of 2–10 nm or 10–50 atoms, often referred to as a size less than the Bohr radius in physics, leads to a quantum confinement effect dictated by the rules of quantum mechanics. This effect endows QDs with unique optical and electronic properties such as: exceptional photochemical stability, high photobleaching threshold, continuous absorption profiles, readily tunable emission properties (from the UV to the IR) allowing simultaneous excitation of several particle sizes at a single wavelength, and size-tunable narrow spectral line widths. For example, in comparison with commonly used organic fluorophore rhodamine, QD luminescent label is 20 times brighter (high luminescence), 100 times more stable against photobleaching, and one-third as wide in spectral line width. In addition, large surface area-to-volume ratio QDs makes them appealing for the design of targeted molecular probes. Such fluorophores allow disease demarcation by fluorescence imaging and, in some cases, can also be exploited for photodynamic therapy. However, side effects of extrinsic fluorophores on the processes studied and toxicity issues make autofluorescence microscopy more desirable.

8.5.7.1 Targeted Imaging Using QDs

To minimize side effects, targeted molecular probes are very useful for diagnostic purposes or as a drug delivery system, and antigen/antibody immunoassays. Specificity in targeting could be obtained by conjugating the surface with small molecules, such as receptor ligands or enzyme substrates (described in Chapter 7), or higher-molecular-weight affinity ligands, such as monoclonal antibodies or recombinant proteins. Targeting specific tissues is possible by tagging specific peptides to QDs and imaging the tissue. Although producing stable QD-biomolecule complexes with clearly defined characteristics has been a challenge while engineering targeting capabilities, advances in molecular engineering will allow developing the targeted QDs of various diseases and will potentially revolutionize fluorescent imaging. If fully exploited, unique QD properties will allow development of fluorescence resonance energy transfer (FRET)-based nanoscale assemblies capable of continuously monitoring target (bio)chemical species in diverse environments.

While using these particles, one has to be careful about the fate of the particles when administered into the body. The main problem encountered by all particles once injected into the bloodstream is adsorption of biological elements, especially circulating plasma proteins. This process is known as opsonisation, and is critical in dictating the fate of the injected particles. Normally opsonisation renders the particles recognizable by the body's major defense system (i.e., recognized by the immune cells as a foreign material) and eliminates the particle. However, if the particles are large in size, then they are typically removed by the liver. Knowing immune surveillance mechanism(s), suitable surface modifications can be attempted to render the particles to remain in the blood circulation for a longer time or be directed to sites of interest. For example, uncoated poly(lactide-co-glycolide) (PLGA) particles are removed rapidly, while particles coated with a block copolymer of polylactide-poly(ethylene glycol) (PLA-PEG) remained in the blood stream for a longer period of time. This is attributed to rendered surface hydrophilicity and sterical stabilization of the particle.

Problems

8.1   Compare the wavelength of a radio wave with a frequency of 1.42 GHz to the wavelength of a visible light wave with a frequency of 6×10^{14} Hz.

8.2   Calculate the frequency and the wavelength of: (a) a 4.5-MeV gamma ray from the alpha decay of radium $_{88}Ra^{226}$; and (b) a 185.7-keV gamma ray from the alpha decay of radon $_{86}Rn^{222}$.

8.3   A certain plane EM wave has a maximum electric field strength of 80 V/m.

(a) Find the maximum magnetic field strength and magnetic flux density.

(b) Determine the intensity of the EM wave.

8.4   Show the energy carried by a photon is 0.511 MeV.

8.5   An ultrasound transducer uses a frequency of $f_i = 5$ MHz. It is held at an angle of $\theta = 60°$ to measure the blood flow in an artery.

(a) If the instrument measures a Doppler shift of 1,290.32 Hz, what is the velocity of the blood (in m/s)? What is the wavelength of the ultrasound signal?

(b) Assume that the compressibility κ of the blood could be increased by a factor of four by replacing it with a different liquid of the same density.

How would this affect the sound propagation inside the vessel? Please compute the speed of sound in the liquid (in m/s).

If the liquid travels at the same speed as the blood, what frequency would be measured as a result of the Doppler shift?

(c) Explain in one or two sentences why it is difficult to scan objects located behind bone using ultrasound.

8.6 Derive an equation for Compton scattering using the conservation of energy and momentum principles.

8.7 We are interested in using a 30-keV energy source. The goal is to separate the view of lungs from the ribcage. The thickness of the bone is 2 cm and lung thickness is 10 cm. What are the differences in the transmitted energies.

8.8 A tissue containing bone and adipose tissue of equal thickness (8 cm) are exposed to a 15-keV X-ray. What is the expected difference in energy at 8 cm?

8.9 We are using a 50-keV X-ray source for the imaging bone. If the emitted X-ray is 38 keV, what is the effective thickness of the bone?

8.10 A primary solar cosmic ray proton travels 1 km in the atmosphere. If its kinetic energy is 2 GeV, what is its LET value in air?

8.11 Calculate the LET rate for the following cases:

(a) A 600-eV electron gives up its energy in water within a distance of 109 nm.

(b) A 800-keV proton gives up its energy in water within a distance of 17.7 μm.

(c) A 5.3-MeV alpha particle gives up its energy in water within a distance of 0.482 mm.

8.12 A 100-MeV fission fragment gives up its energy in water within a distance of 1.11×10^{-5}m. Calculate its LET rate.

8.13 If the ionizing potential of a hydrogen atom is 10.6 eV, how may ionizations will be created by an absorbed alpha particle if its incident kinetic energy of 100 MeV is totally absorbed? 9.4×10^4

A general rule of thumb used by physicists is that about 30 eV is needed for each ionization in liquid water, whereby H_2O is broken into H+ and OH–. How many ionizations are created in water by a single 800-keV proton?

8.14 How many ionizations are created in water by a single 2-GeV proton?

8.15 The electric field associated with a plane EM wave is given by $E_x = 0$, $E_y = 0$, $E_z = E_0 \sin k(x-ct)$ where $E_0 = 2.34 \times 10^{-4}$ V/m and $k = 9.72 \times 10^6$ m^{-1}.The wave is propagating in the +x direction. Write expressions for the components of the magnetic field of the wave. Find the wavelength of the wave.

8.16 The intensity of direct solar radiation not absorbed by the atmosphere on a particular summer day is 130 W/m^2. How close would you have to stand to a 1.0-kW electric heater to feel the same intensity? Assume that the heater radiates uniformly in all directions.

8.17 The maximum electric field at a distance of 11.2m from a point light source is 1.96 V/m. Calculate: (a) the amplitude of the magnetic field; (b) the intensity; and (c) the power output of the source.

8.18 What is the frequency of precession of ^{1}H and ^{31}P at 0.15T? 0.5T? 1.5T? 3.0T?

8.19 Determine the energy difference, ΔE (in eV), of the parallel and antiparallel spin states under the influence of a 3T magnetic field and compare that value with an X-ray photon of 50 keV.

8.20 State five different types of imaging devices. Explain in two sentences their basic principle of operation. State two limitations of each technique.

8.21 It is suggested that mammograms are not sensitive enough for early diagnosis of breast cancer. If this is true, what steps would you consider in improving early detection using mammograms?

References

[1] "Tables of X-Ray Mass Attenuation Coefficients and Mass Energy-Absorption Coefficients," http://physics.nist.gov/PhysRefData/XrayMassCoef/cover.html.

[2] Cho, Z., *Foundations of Medical Imaging*, New York: John Wiley & Sons, 1993.

Selected Bibliography

Cobbold, R. S. C., *Foundations of Biomedical Ultrasound*, New York: Oxford University Press.

Gonzalez, R. C., and R. E. Woods, *Digital Image Processing*, 3rd ed., Upper Saddle River, NJ: Prentice-Hall, 2007.

Najarian, K., and R. Splinter, *Biomedical Signal and Image Processing*, Boca Raton, FL: CRC Press, 2005.

Prince, J. L., and J. M. Links, *Medical Imaging Signals and Systems*, Upper Saddle River, NJ: Prentice-Hall, 2005.

Sonka, M., V. Hlavac, and R. Boyle, *Image Processing, Analysis, and Machine Vision*, 2nd ed., Pacific Grove, CA: PWS Publishing, 1999.

Webb, A., *Introduction to Biomedical Imaging*, New York: IEEE Press, 2003.

CHAPTER 9

Biosensors

9.1 Overview

Sensing and measuring various components are essential in medicine. Apart from a general visual observation, vital signs are continually monitored before, during, and after surgical procedure or severe illness. Heart rate, ECG, blood pressure, respiratory rate and quality, the oxygen saturation of blood, and body temperature are typically tracked. Concepts involved in the development of these tools were discussed in previous chapters. A common biosensing tool is a simple electrode to measure and record biopotentials in the body noninvasively for applications such as ECG. Apart from these measurements, other molecules are tracked to evaluate the healthiness of a person. A glucose biosensor is used by many diabetic patients in the routine monitoring of blood glucose. Other biosensing tools including home pregnancy test units, cholesterol monitors, and oxygen monitors. These analytical devices incorporating a biorecognition system are called biosensors in general. Biosensing tools are essential in a number of other applications. In the defense industry, sensing biowarfare agents such as anthrax spores are essential. Other applications are in environmental pollution monitoring and in process monitoring in the food and pharmaceutical industries (for *E. coli* or Salmonella contamination) where there is a demand for in situ determination of levels of contaminants.

A biosensor consists of a biological sensing element, a transducer, and a data processor that generates a signal (Figure 9.1). The sensing element could be cells, micro-organisms, ions, cell receptors, enzymes, antibodies, nucleic acids, natural products, biologically derived materials or biomimetic materials (e.g., synthetic catalysts, combinatorial ligands, imprinted polymers). Transducers convert chemical/biochemical activity into a measureable signal such as electrical or light intensity signals. A previously established relationship between transducer output and analyte concentration (called a calibration curve) allows for the quantitation of the analyte. The obtained data is processed into a useful format that can be either displayed or manipulated. Miniaturized biosensing tools also offer a number of potential benefits including reduced sample sizes, a reduction in cost, and the possibility of single-use disposable devices. Advances in the development of lab-on-chip devices shrink and potentially simplify laboratory tests like DNA analysis, antigen detection, and high-throughput genotyping. Biological high-throughput assays based on microarray technology, capable of simultaneous analysis of thousands of

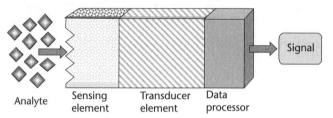

Figure 9.1 Components of a biosensor.

gene and gene products, have revolutionized the discovery of therapeutic agents. In this chapter, the fundamentals of various biosensing elements and their integration with different transducers are discussed. Furthermore, microfabrication technologies used in miniaturization of biosensing tools are discussed along with high-throughput microarray technology.

9.2　Bioelectrodes

The key issue in the development of a biosensor is converting target recognition into a measurable signal. The transducer part of the sensor serves to transfer the signal from the output domain of the recognition system. The transducer part of a sensor is also called a detector, s sensor, or an electrode, but the term transducer is preferred to avoid confusion. Among the various sensing devices developed thus far, the electrochemical method is, in general, superior to optical methods because of its rapid response, simple and easy handling, and low cost.

Bioelectrodes function as an interface between biological structures and electronic systems to measure biopotentials (discussed in Chapter 3) generated in the body due to ionic current flow. Bioelectrodes carry out a transduction function and convert ionic current flow in the body to electronic current in the biosensor. To understand how bioelectrodes work, the basic mechanisms of the transduction process and the effect on the bioelectrode characteristics are discussed next.

9.2.1　Electrode-Electrolyte Interface

When an electrode (material capable of transporting charge such as metals) is placed into an electrolyte (an ionically conducting solution where the charge is carried by the movement of ions), an electrified interface immediately develops. Chemical reactions occur at the surface whereby electrons are transferred between the electrode and the electrolyte. Reaction in which loss of an electron (e^-) occurs is called the oxidation reaction (Figure 9.2), that is,

$$M \leftrightarrow M^{m+} + me^-$$

where M is the metal atom (e.g., silver), M^{m+} represents the cation of the metal, and m is the valence of the cation. If the metal has the same material as cation in the electrolyte, then this material gets oxidized and enters the electrolyte as a cation

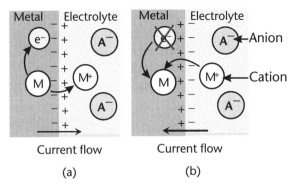

Figure 9.2 Electrode-electrolyte interface: (a) oxidation and (b) reduction.

and electrons remain at the electrode and flow in the external circuit. Reactions are reversible and the reduction (gain of e⁻) reactions also occur. Similarly, an anion in the electrolyte is oxidized at the electrode to form a neutral atom with the release of electrons given to the electrode.

$$A^{n-} \leftrightarrow A + ne^-$$

where A^{n-} represents an anion (e.g., Cl⁻), and n is the valence of anion. Oxidation and reduction reactions are two competing reactions which eventually reach an equilibrium condition whereby the currents due to electron transfer to and from the metal are equal. This equilibrium exchange current density flows across the interface in both directions resulting in a net current of zero. The dominating reaction can be inferred from whether current flows from electrode to electrolyte (oxidation) or current flows from electrolyte to electrode (reduction).

Local concentration for the M^{m+} at the interface changes and electrolyte surrounding the metal is at a different potential from the rest of the solution. Oxidation or reduction reactions at the electrode-electrolyte interface lead to an electrical double layer, similar to that which exists along electrically active biological cell membranes (discussed in Chapter 3). This results in the formation of an electric field between the electrode and the electrolyte. The potential difference established by the electrode and its surrounding electrolyte is called the half-cell potential. Half-cell potential depends on the electrode and the concentration of ions in solution and temperature.

Measuring electrical potential developed by an electrode in an electrolyte solution when the net current flow (or at equilibrium of reduction and oxidation reactions) is zero across the electrode and the electrolyte interface is called potentiametry. The Nernst equation (discussed in Chapter 3) relates potential to concentration of some ion in solution. For the half-cell reaction, the half-cell potential under any conditions is written as

$$\Delta\Phi_{hc} = \Delta\Phi_{hc,0} + \frac{RT}{nF}\ln\frac{\alpha_{Ox}^b}{\alpha_{Red}^r} \text{ or } \Delta\Phi_{hc} = \Delta\Phi_{hc,0} + \frac{RT}{nF}\ln\frac{\alpha_{M^{m+}}}{\alpha_M} \quad (9.1)$$

where $\Delta\Phi_{hc,0}$ is the half-cell potential under standard conditions (25°C and 1 atm) and α represents activity (i.e., ideal thermodynamic concentrations of oxidized and reduced states). In dilute solutions, activity is nearly equal to concentration. However, activity is smaller than concentration in concentrated solutions due to intermolecular effects. Often the reduced state is a metal in which case α_M is a constant equaling to 1. It is more convenient to consider concentrations rather than activities. These parameters are related by the activity coefficient, γ,

$$\alpha_{Ox} = \gamma_{Ox} C_{Ox}$$

The Nernst equation is therefore rewritten as

$$\Delta\Phi_{hc} = \Delta\Phi'_{hc,0} + \frac{RT}{nF}\ln\frac{C_{Ox}}{C_{Red}} \tag{9.2}$$

where $\Delta\Phi'_{hc,0}$ is the formal potential and is related to the standard potential by the equation:

$$\Delta\Phi'_{hc,0} = \Delta\Phi_{hc,0} + \frac{RT}{nF}\ln\frac{\gamma_{Ox}}{\gamma_{Red}}$$

Equation (9.2) relates the concentration of the electrolyte to the electrical potential. One could calculate the unknown concentration of the electrolyte knowing the potential. However, half cell potentials cannot be measured because the connection between the electrolyte and one terminal of the potential measuring device cannot be completed. Hence, the potential difference is measured between the half-cell potential of the electrode and a second electrode called the reference electrode. The reference electrode refers to an electrode that provides a particular reference voltage for measurements recorded from the working electrode. The half-cell potential of the normal hydrogen electrode (NHE) is used as the reference electrode. NHE consisting of a platinum wire dipped into hydrochloric acid (1.18M corresponding to unit activity of H+) solution and acts as the electrode over which hydrogen gas (1 atm pressure at 25°C) is bubbled. The reaction is represented as

$$H_2(g) \leftrightarrow 2H^+(aq) + 2e^-$$

The two electrodes are connected internally by means of an electrically (Figure 9.3) conducting bridge (called a salt bridge) and externally to a voltmeter. When the reference electrode potential is incorporated, the net potential of the cell is

$$\Delta\Phi_{cell} = \Delta\Phi_{hc} - \Delta\Phi_{ref} - \Delta\Phi_{LJ}$$

where $\Delta\Phi_{LJ}$ is the liquid junction potential. One aspect that is often overlooked is the variation of the reference electrode potential with temperature. For example, the half-cell potential of the silver/silver chloride at 25°C is +0.222V versus NHE.

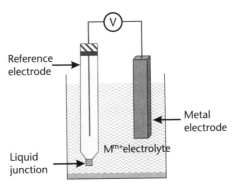

Figure 9.3   Measurement of a half-cell electrode potential.

It typically changes by 0.5–1 mV/°C. Consequently, precise potential measurements require the use of a constant temperature apparatus. In addition, the temperature at which the measurements are made should always be reported. The absence of any temperature control limits the accuracy of the measurements to ~5–10 mV, although this level of precision may be acceptable for some experiments.

EXAMPLE 9.1

A saturated silver-silver chloride electrode contains 0.01M KCl. Calculate the half-cell potential developed at 25°C, assuming a complete dissociation of KCl.

Solution: There is only one variable contributing to the Nernst equation, the concentration (activity) of chloride ion, since Ag and AgCl are in the pure states and thus have activities of 1.

From (9.2),

$$\Delta\Phi = 25.70[\text{mV}]\ln\frac{1}{0.01} = 116 \text{ mV} = 0.116\text{V}$$

9.2.2   Polarization

If there is a flow of current across the interface, then the observed half-cell potential is altered. The difference is due to what is called the polarization of the electrode. The difference between the observed half-cell potential with and without the current flow is known as overpotential. Overpotential is contributed by three mechanisms (Figure 9.4):

1.   Ohmic overpotential, that is, the voltage drops along the path of the current, the current changes the resistance of an electrolyte, and thus, a voltage drop does not follow Ohm's law;
2.   Concentration overpotential, that is, the current changes the distribution of ions at the electrode-electrolyte interface;
3.   Activation overpotential, that is, the current changes the rate of oxidation and reduction. Since the energy barriers for oxidation and reduction

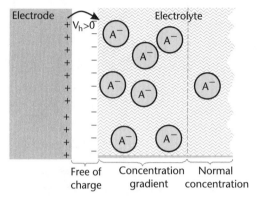

Figure 9.4 Polarization of electrodes.

reactions are different, the net activation energy depends on the direction of current and this difference appears as voltage.

Perfectly polarizable electrodes are those in which no actual charge crosses the electrode-electrolyte interface when a current is applied. The electrode behaves like a capacitor and the current across the interface is a displacement current. An example is noble metals like platinum electrode. Perfectly nonpolarizable electrodes are those in which the current passes freely across the electrode-electrolyte interface, requiring no energy to make the transition. These electrodes see no overpotentials, and the potential does not vary with the current. Nonpolarizable electrodes usually consist of a metal coated with its relatively insoluble salt immersed in a solution containing an ion of the salt. As the ion can be plated or released at each electrode, the current can flow with a very low resistance. Examples include a silver/silver chloride (Ag/AgCl) electrode and a mercury/mercurous chloride (Hg/Hg_2Cl_2) (also referred as calomel electrode).

EXAMPLE 9.2

Using a 10-μM NADH solution on a glassy carbon electrode, the half-cell potential (with SHE) of 700 mV was measured. Calculate the overpotential with the Nernst half-cell potential developed at 25°C, assuming a complete dissociation of NADH.
  Solution: From (9.2), the potential developed by NADH is

$$\Delta\Phi = 25.70[mV]\ln\frac{1M}{10\times10^{-6}M} = 295.882 \text{ mV}$$

The measured potential is 700 mV.
Hence, Overpotential = $700 - 295.882 = 404.112$ mV

9.2.3   Potential Monitoring Electrodes

A commonly used bioelectrode in biomedical applications is the Ag/AgCl electrode (Figure 9.5), which consists of a silver wire coated with silver chloride and immersed in a solution containing chloride ions, typically a KCl solution. AgCl is slightly soluble in water and remains very stable in a liquid that has a large quantity of Cl⁻ such as the biological fluid. The two chemical reactions of Ag/AgCl are:

$$AgCl(s) + e^- \rightleftharpoons Ag(s) + Cl^-(aq)$$
$$Ag(s) \rightleftharpoons Ag^+(aq) + e^-$$

The net reaction is $AgCl(s) + e^- \rightleftharpoons Ag^+(aq) + Cl^-(aq)$.
The Nernst equation for the Ag/AgCl electrode is

$$\Delta\Phi = \Delta\Phi_0 + \frac{RT}{nF} \ln \frac{1}{[Cl^-]} \qquad (9.3)$$

where $\Delta\Phi_0$ is the standard potential (i.e., the potential of the electrode at unit activity under standard conditions). Ag/AgCl is manufactured in a reusable form, but is most often used as a disposable electrode. Bioelectrodes attachable at a skin surface can be grouped into two categories. The first category includes the bioelectrodes prepackaged with the electrolyte contained in the electrode cavity or receptacle. In a prepackaged electrode, storage and leakage are major concerns. Leakage of contents from the receptacle results in an inoperative or defective state. Furthermore, such prefilled electrodes are difficult to apply because the protective seal which covers the electrode opening and retains the fluid within the receptacle cavity must be removed prior to application to the skin surface. After removal of this protective seal, spillage often occurs in attempting to place the electrode at the skin surface. Such spillage impairs the desired adhesive contact of the electrode to skin surface and also voids a portion of the receptacle cavity. The consequent loss of electrolyte fluid tends to disrupt electrical contact with the electrode plate contained therein and otherwise disrupts the preferred uniform potential gradient to be applied.

The second type of bioelectrode is a dry-state electrode comprised of an electrode plate with an upper surface having the means for electrically connecting the

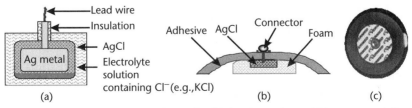

Figure 9.5   Ag-AgCl electrode: (a) wet electrode, (b) dry electrode, and (c) a commercially available electrode. (Courtesy of Vermont Medical, Inc.)

electrode plate to a lead wire and a lower body-contacting surface made of a conductive material for enhancing an electrical connection with the skin. The conductive material comprises a dermally-nonirritating synthetic hydrophilic polymer containing a salt of a carboxylic acid. Dry-state electrodes have numerous advantages in the ease of storage and a greater adaptability for various types of electrode applications.

For monitoring biopotentials generated in the human body, different shapes of bioelectrodes are manufactured based on the requirement. There are metal-plate electrodes, floating electrodes, suction electrodes, flexible electrodes, internal electrodes (for example, detecting fetal electrocardiogram during labor, by means of intracutaneous needles), and microelectrodes. Insulating electrodes consist of a metal or semiconductor with a thin dielectric surface layer, so that the bioelectric signal is capacitively coupled from the skin to the substrate. A metallic suction electrode is often used as a precordial electrode on clinical electrocardiographs. There is no need for strap or adhesive and it can be used frequently. There is a higher source impedance since the contact area is small. More information for each type can be found in [1].

9.3　Biosensing Elements

Bioelectrodes respond directly to the changing activity of electrode ions or changing ion activity in a solution through the formation of complexes. However, they are not very selective. Hence, sensing other biological elements is necessary for various applications. An approach to incorporate selectivity is introducing various elements based on the requirements. Sensing elements in biomedical applications can be broadly grouped into the following categories.

9.3.1　Enzyme-Based Biosensors

Enzymes are used as biochemical recognition elements to selectively catalyze the reaction of a substrate to produce a product. The analyte could be a substrate or a product of the enzyme reaction. The specificity of enzymes with regard to substrate recognition enables sensors incorporating enzymes with a limited range of substrates to achieve much greater selectivity for a target species. Typically, the enzyme participates actively in the transformation of the substrate to a product but remains unchanged at the end of the reaction. Besides amino acids, many enzymes also contain tightly bound nonamino acid components called prosthetic groups such as nicotinamide adenine dinucleotide (NAD), flavin adenine dinucleotide (FAD), heme, quinine, Mg^{2+}, and Ca^{2+}.

The most commonly used enzymes in the design of enzyme biosensors contain oxidoreductases, which catalyze the oxidation or reduction of a substrate. Since redox reactions are easily monitored by bioelectrodes, oxidoreductases are coated onto a bioelectrode and used to monitor the analyte. Enzymes of this type are the oxidases such as glucose oxidase (GOX) and cholesterol oxidase (ChOX) and the pyrroloquinolinequinone (PQQ) dependent dehydrogenases. Examples include blood glucose biosensors, cholesterol monitoring devices, and urea biosensors. Some prosthetic groups enhance enzyme activity and serve as temporary traps of electrons or electron vacancies and remains bound to the protein throughout redox cycle. For example, GOX selectively catalyzes the following two reactions:

$$\text{Glucose} + \text{GOX-FAD} \rightleftharpoons \text{Gluconolactone} + \text{GOX-FADH}_2$$

$$\text{GOX-FADH}_2 + O_2 \rightleftharpoons \text{GOX-FAD} + H_2O_2$$

GOX-FAD represents the oxidized state and GOX-FADH$_2$ represents the reduced state of the flavin active site within glucose oxidase enzyme. Oxidase enzymes (Figure 9.6) act by oxidizing their substrates, accepting electrons in the process and thereby changing to an inactivated reduced state. These enzymes are returned to their active oxidized state by transferring the electrons to oxygen (called a co-substrate), resulting in the production of hydrogen peroxide (H_2O_2). This mechanism is readily utilized in an amperometric biosensor device. Because both O_2 and H_2O_2 are electrochemically active, the progress of the biochemical reaction can be tracked by either reducing O_2 or oxidizing the H_2O_2. Electroenzymatic reactions follow the receptor-ligand interactions (discussed in Chapter 7) and the reaction kinetics is represented by

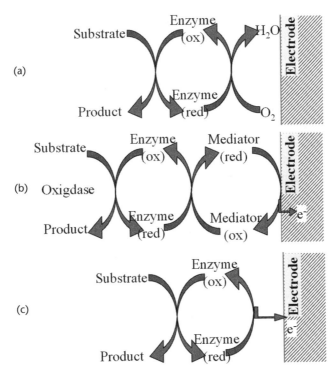

Figure 9.6   Enzyme biosensors: (a) first generation, (b) second generation, and (c) third generation.

$$I = \frac{I_{\max} \cdot S}{K_M + S} \qquad (9.4)$$

where I is the steady state amperometric response of the enzyme electrode at specified substrate concentrations S, k_M is the electrochemical Michaelis-Menten constant. The maximum response $I_{\max} = nFAk_{cat}S$ obtainable from the biosensor is a direct function of the number of electrons transferred n, the Faraday's constant F, the surface area of the electrode A, the turnover rate constant of the enzyme k_{cat}, and the substrate concentration. K_M depends upon the diffusion properties of species in the biosensor system. The $I_{\max}$ and K_M values are evaluated, and the sensitivities are estimated as $I_{\max}/K_M$. Alternative strategies such as mediators (second generation electrodes) and direct transfer biosensors (third generation biosensor) have been developed to improve the performance of enzyme-mediated biosensors. These are discussed in conjunction with amperometric electrodes in Section 9.4.2.

9.3.2 Antibody-Based Biosensors

Antibody-based biosensors are also referred to as immunosensors and depend on antibody-antigen interactions. Antibodies are proteins, called immunoglobulins, produced by the immune system of higher-order animals in response to the entry of foreign materials (viruses, bacteria, and implanted medical devices) into the body. In contrast to enzymes, antibodies do not (usually) catalyze chemical transformations but rather undergo a physical transformation. They bind tightly to the foreign material (the antigen) that provoked the response and mark it for attack by other elements of the immune system. Antibodies are also very specific in recognizing and binding to the foreign substance only and not to materials native to the organism. This specificity is utilized in recognizing various analytes.

Each antibody consists of four polypeptides: two heavy chains and two light chains joined to form a Y-shaped structure [Figure 9.7(a)]. The end regions of the light and heavy chains have different amino acid sequences among different antibodies and this region is referred to as the variable region. However, amino acid sequence in some regions are conserved (i.e., constant in all antibody molecules and this region is referred to as the constant region). The variable region gives the antibody its specificity for antigen binding and the constant region determines the mechanism used to destroy antigens. Treating the antibody with a protease can cleave the variable region, producing fragment antigen binding (*Fab*) that includes the variable ends of an antibody. Antibodies are divided into five major classes, IgM, IgG, Iga, IgD, and IgE, based on their constant region structure and immune function.

Typically, antibodies specifically directed against the substance of interest (i.e., the desired analyte) are immobilized on the transducer of the biosensor. Then the sensor is exposed to the intended medium. If the antigen is present in that medium, it will bind to the antibody to antigen-antibody complex. This will change some physicochemical parameters (usually a mass or an optical parameter) of the environment at the transducer surface of the sensor [Figure 9.7(b)] and that change is subsequently detected. One such immunosensor is developed to detect human

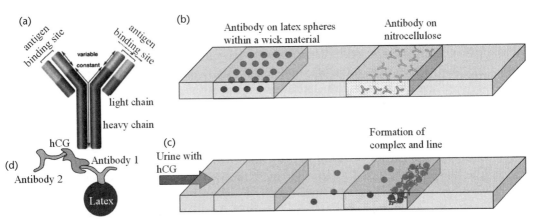

Figure 9.7 Antibody-based biosensor. (a) The structure of an antibody. (b) The basic structure of a ClearBlue Pregnancy test consists of two compartments in which antibodies are dispersed. (c) The introduction of a sample containing the antigen will bind to the antibody on the latex sphere and move due to fluid motion. They encounter the antibody in the second compartment to form a complex that is retained in the test unit to produce a blue line. (d) The formed antigen-antibody complex. The absence of the antigen in the sample will not result in this complex.

chorionic gonadotrophin (hCG) protein (particularly, its free beta-subunit, hCGβ) in urine [Figure 9.7(c)]. hCG is an enzyme, the concentration of which increases geometrically postconception during fetal growth. Hence, monitoring hCGβ provides information concerning the progress of a pregnancy and the health of a fetus. Monoclonal antibodies which bind specifically to hCG have been developed: anti-α-hCG binds to the α domain and anti-β-hCG binds to the β domain on hCG and both are used in home pregnancy test devices. The urine is collected and led through an absorbent sampler protruding from one end. A biosensor is held for few seconds in the urine stream. If hCG is present in the urine, it is carried towards hCG antibodies labeled with a dye. An hCG-hCG antibody/dye complex is formed and moves towards a window in the case where there is another set of hCG antibodies. At the window, these combine to form an hCG antibody/dye-hCG-hCG antibody complex. The accumulation of complexes shows up as a line of the dye at the window, indicating a positive pregnancy test. Commercial pregnancy test kits can detect nano molar (or 10^{-9} M) concentrations of hCG. If the line does not appear, this could be due to the absence of the hormone or insufficient urine on the absorbent tip. To account for insufficient sample size variations, positive controls are built within the device.

9.3.3 Nucleic Acid-Based Biosensors

Nucleic acid biosensors utilize the complementary nature of the nucleic acids for selective recognition capability. They are designed for the detection of DNA or RNA sequences usually associated with certain bacteria, viruses, or given medical conditions. The DNA probes could be used to diagnose genetic susceptibility, diseases, and paternity tests. Nucleic acid biosensors generally immobilize single-strands (called probes) from a DNA double helix (or RNA) onto a surface as the recognition species. The nucleic acid material in a given test sample is then denatured and placed into contact with the binding agent. If the strands in the test sample are complementary to the strands used as binding agent (Figure 9.8), the nucleic acid will attach due to the base-pairing ability. This process is called hybridization. The nucleic acid is then labeled by attaching a fluorescent dye and is brought into contact with the probes by flooding the chip with the sample solution. DNA can be extended, cut, or joined in a very precise manner by a suite of enzymatic tools polymerases, nucleases, and ligases that nature has developed in the course of evolution.

The interaction is monitored by various means, such as a change in mass at the sensor surface or the presence of a fluorescent or radioactive signal. Different strategies are used to transduce this hybridization event into a measurable signal, including the popular and effective method of labeling the target with a fluorescent molecule, exploiting differential binding of a small reporter molecule to double-stranded DNA relative to single-stranded DNA or by using label-free techniques such as surface plasmon resonance or the quartz crystal microbalance. All these approaches indicate that adsorption of target DNA onto the surface has occurred, but in themselves afford little or no information on the nature of the adsorbate, such as whether a DNA duplex has formed at all or whether stable mismatches may be present. This issue is partially addressed by the use of comparative perfect match/mismatch (PM/MM) probe sets for known mismatches or to use modifications such as peptide nucleic acids which display improved discrimination behavior.

9.3.4 Cell-Based Biosensors

Throughout their life cycle within specific tissues and organs, various cell types respond to a variety of chemical and physical signals by different means (i.e., either producing specific proteins or specific free radicals). Some of these events can be re-created using an in vitro cell culture technique (discussed in Chapter 7). Then the cell-specific responses can be utilized to obtain pharmaceutical and chemical safety information. Knowing the desired output, a specific cell type is selected and cultured on a substrate. Sensors containing living cells (mammalian cells, bacteria, and yeast) can be used to monitor physiological or metabolic changes induced by exposure to environmental perturbations such as toxics, pathogens, or other agents

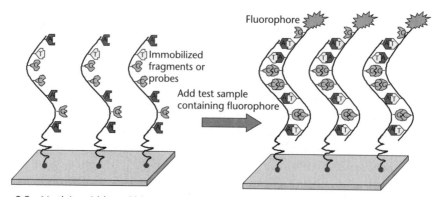

Figure 9.8 Nucleic acid-based biosensors.

that are under development. They are used for the functional characterization and high-throughput drug discovery or detection of pathogens, toxics, and odorants and clinical diagnostics. Unlike other biosensors such as nucleic acid or antibody-based sensors, cell-based biosensors are not specific for certain compounds but are capable of responding to a wide range of biologically active compounds and offer the potential to gather greater information content than biomolecular-based sensors.

To be used as a biosensor, the cellular signal generated in the transducer needs to be determined in a noninvasive manner. Electrically excitable cells such as neurons and cardiomyocytes are particularly useful in this sense, since the activity of cells can be monitored by the extracellular recordings using microelectrodes. For electroanalytical measurements and biosensing, electrode and biosensor devices can be miniaturized down to a cell-size scale using microfabrication technology and can be positioned directly at the vicinity of the cell surface, where cellular signaling substances are captured before they diffuse. In the case of biosensors, highly sensitive in situ monitoring can be performed. Although quantitatively the signal is very small, it is enhanced through either circuit amplifiers or catalytic reactions on the biosensor. Cell-based biosensors may have a longer response time and less specificity to a single analyte of interest due to the presence of other enzymes in the cells.

Many nonexcitable cell types are used for the detection of various classes of materials that do not react with the excitable cells. Transducers could be optical instead of electrical. Immune cells and hepatocytes are good examples of nonexcitable cells used for cell-based biosensors. For example, hepatocytes have been used to assess and predict the effects of toxicants. Many compounds do not result in immediate response or toxicity. In this case, cell motility and adhesion can be also used as a cellular signal. Genetically engineered cells that recognize and report the presence of a specific analyte have also been developed. Cells are produced with a nucleic acid sequence in which genes that code for luciferase or galactosidase enzymes are placed under the control of a promoter that recognizes the analyte of interest. Because the organism's biological recognition system is linked to the reporting system, the presence of the analyte results in the synthesis of inducible enzymes which then catalyze reactions resulting in the production of detectable products.

Many micro-organism-based biosensors are used for assessing toxins, as they are less expensive to construct relative to other cell types. Furthermore, micro-organisms are more tolerant of some assay conditions that would be detrimental to an isolated protein as micro-organisms have mechanisms to regulate their internal environment based on external conditions. Micro-organism-based biosensors are based on using the analyte either as a respiratory substrate or as an inhibitor to the respiration. Biosensors that detect biodegradable organic compounds measured as biological oxygen demand (BOD) are the most widely reported micro-organism-based biosensors using this mechanism. The general limitations of cell-based biosensors are the long assay times, including the initial response and return to baseline.

9.4   Transducing Elements

A biosensing element is closely integrated within a physicochemical transducer or transducing microsystem. The nature of the interaction of the biological element with the analyte of interest impacts the choice of transduction technology. For example, signals from the ionic movement can be transduced using bioelectrodes. However, if there is no ionic movement in analyte-biosensing interactions, then alternative strategies have to be developed. A successful biosensing system will require the development of effective methods of sampling, alongside a biological model of the process. Different transducer technologies are described next.

9.4.1   Electrochemical Biosensors

An approach to exploiting the bioelectrode technology is to incorporate a sensing element that could also play a role in electron transfer. Electrochemical transducers are the most established transduction methods. Although some electrodes approach ideal nonpolarizable behavior at low currents, there is no electrode that behaves in an ideal nonpolarizable condition. Consequently, the interfacial potential of the counterelectrode in the two-electrode system varies as the current is passed through the cell. This problem is overcome by using a three-electrode system (Figure 9.9), in which the functions of the counterelectrode are divided between the reference and working electrode. The reference electrode ensures that the potential between the working and reference electrodes is controlled and the current passes between the working and auxiliary electrodes. The current passing through the reference electrode is further diminished by using a high input-impedance operational amplifier for the reference electrode input. One previously used reference electrode was the saturated calomel electrode (with a large surface area mercury pool). However, since the current passing through the reference electrode in the three-electrode system is many orders of magnitude lower than the current that passes through the two electrode system, the requirements for the reference electrode are less demanding;

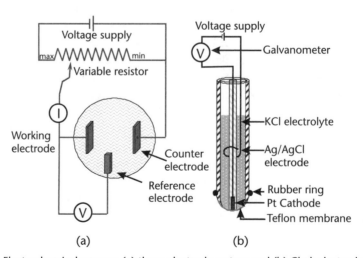

Figure 9.9   Electrochemical sensors: (a) three-electrode system and (b) Clark electrodes.

hence, smaller, more polarizable electrodes are used. The requirements for the counterelectrode of the two-electrode system include a high exchange current (fast electron transfer kinetics), a very large surface area (to lower the current density) and
a high concentration of the species involved in the redox reaction, such that the
concentrations are not significantly changed by the passage of a current.

Two approaches have been developed in biosensor applications as the current
and voltage change while using polarizable electrodes. Electrochemistry techniques
are based on the current (i) measurement as a function of voltage ($\Delta\Phi_{appl}$). The key
electronic component for electrochemical measurement is a potentiostat, which
can be configured for potentiometric (apply fixed current and measure output voltage) or amperometric (apply fixed voltage and measure output current) readout.
A potentiostat is used to apply a constant potential to the working electrode with
respect to a counterelectrode (a reference electrode). A potentiostat is a simple electronic circuit that can be constructed using a battery, two operational amplifiers,
and several resistors.

A problem with potentiometric sensors in sensor applications is the requirement of rapid electrode kinetics for faster response. The standard redox potential
of two coupled half-cells generating a voltage $\Delta\Phi_0$ is related to the free energy
change of the reaction, by the following equation:

$$\Delta G_0 = zFE_0 \tag{9.5}$$

where F is Faraday's constant and z is the number of electrons transferred. Since
ΔG can be related to the K_D value, E_0 is a function of the kinetics of the reaction.
However, many biological reactions are slow interactions.

EXAMPLE 9.3

Using yeast alcohol dehydrogenase, which contains NAD, Mr. Dickerson tested the conversion of alcohol to acetaldehyde. At the end of the reaction, NAD (the oxidized form)
is reduced to NADH. At 25°C, the dissociation constant was determined to be 50 mM.
Calculate the electrode potential.

Solution: From (7.3),

$$\Delta G = RT \ln K_D$$

Equating to (9.4),

$$-zFE_0 = RT \ln K_D$$

Hence,

$$E_0 = \frac{RT}{-zF} \ln K_D$$

$$E_0 = -25.70[\text{mV}] \ln 0.05 = 77 \text{ mV}$$

9.4.1.1 Amperometric Transducers

Voltammetric biosensors are based on measuring a current with a voltage change. The subtechnique in voltammetry is the amperometric biosensors where the current at a fixed potential gives a steady value with time. In the amperometric category, a biosensing element is typically coupled to an amperometric electrode and as the biosensing element reacts with the substrate, a current is produced that is correlated to the analyte concentration. The electrode potential is used to drive an interfacial redox reaction and the current resulting from that reaction is measured. The current flowing is directly proportional to the analyte concentration.

Amperometric transducers are more versatile than potentiometric devices. Amperometric detection is based on measuring the oxidation or reduction of an electroactive compound at the working electrode (sensor). The rate of the analyte reaction is monitored by the variation of the current. The measured signal is a variation in the current, at constant potential, depending on the variation in the reactive species concentration: the relation is linear. In continuous operation there is a risk of surface poisoning; nonsteady-state measurements are preferred to avoid this problem. Sensitivity is greater for an amperometric sensor (LOD of about 10^{-8} M compared with 10^{-6} M for a potentiometric device). Typically, Ag/AgCl or a saturated calomel electrode (SCE) is used as the reference electrode so that reversible oxidation/reduction occurs at a fixed potential at the reference electrode. The applied potential is an electrochemical driving force that causes the oxidation or reduction reaction. With large Cl^- concentration, the Ag/AgCl reaction produces a stable potential. According to the Faraday's law of electrochemistry, the amount of substance consumed or produced at one of the electrodes in an electrolytic cell is directly proportional to the amount of electricity that passes through the cell. The current response, I,

$$I = zF\left(\frac{dn}{dt}\right) \tag{9.6}$$

where F is Faraday's constant, z is the valency of the reagent, and dn/dt is the oxidation or reduction rate [mols]. Equation (9.4) can also be derived from (3.17). The reaction rate depends on both the rate of electron transfer at the electrode surface and analyte mass transport.

EXAMPLE 9.4

A group of researchers are trying to develop an electrochemical biosensor with a requirement of detecting a constant current of 50 mA for 1 hour. If the valency of the analyte is +2, calculate the minimum number of moles of analyte required to be present.

Solution: Rearranging (9.6),

$$dk = \frac{1}{zF}Idt$$

Integrating over the time period,

$$k = \frac{1}{zF} \int_0^1 I \, dt$$

Since a constant current is required, *I* is independent of time and the equation reduces to

$$k = \frac{1}{zF} It \quad k = \frac{1}{+2 * 96,484.5[C/mol]} 0.05[A] * 1[hr] * 3,600[s/hr]$$

$$n = 0.933 \times 10^{-3} \text{ moles or } 0.933 \text{ mmole}$$

With most redox reactions, the rate of electron transfer can be accelerated by increasing the potential at which the electrode is poised. Thus, the quality of reaction rate is less important than in potentiometric bioelectrodes as sluggish electrode reactions are switched on via the applied electrode potential. Typically the rate constant for ET increases by a factor of 10 for every 120-mV increase in electrode potential. As the potential is increased, the reaction reaches the point where the rate is limited by the mass transport of reactant to the electrode. When the reaction at the electrode surface is sufficiently fast, the concentration of analyte at the electrode is zero, and a maximum overall rate of reaction is reached. This overall rate is limited by the rate of mass transfer and can be obtained by combining Fick's law of diffusion:

$$I = nAFD \left(\frac{dC}{dx} \right)_{x=0} \tag{9.7}$$

where *C* is the concentration of the electroactive species, *A* is the electrode area, *D* is the diffusion coefficient, and *x* is the thickness through which diffusion has to occur. Furthermore, zero thickness ($x = 0$) represents the electrode surface. The rate of mass transport to the electrode surface depends on the bulk concentration of analyte, the electrode shape and area, and diffusion conditions. The use of amperometric electrodes is complicated by the regeneration of the electroactive component from the analyte. The concentration of the electroactive component at the surface of the electrode is affected by diffusion of the product through the enzyme layer, the activity of the enzyme, and diffusion of analyte. Potentiometric sensors do not perturb the analyte concentration at the interface since there is no net consumption of the latter. Hence, the mass transfer of analyte is not important. The potentiometric signal can be corrupted by electronic noise. However, amperometric sensors have a significant depletion of the analyte next to the sensor surface so the mass transfer of analyte to the sensor from the bulk solution must be controlled.

9.4.1.2 Clark's Oxygen Electrode

Clark's dissolved oxygen sensor is the most well-known amperometric biosensor of this type. Monitoring oxygen consumption is important during cell culture and microbial development. Among the various tools to determine oxygen, the Clark's (named after Leland C. Clark, Jr.) oxygen sensor is the most widely used and has been applied in clinical analysis, fermentation monitoring, and biosensor development. The Clark electrode consists of a working electrode (cathode) maintained at a negative external potential relative to a reference electrode (anode) and the electrolyte [Figure 9.9(b)]. The cathode is made from noble metals, such as Pt or Au, so that the electrode surface does not participate in the chemical reactions. The current is produced by the chemical reduction of oxygen on the cathode surface, expressed by:

$$O_2 + 2e^- + 2H_2O \leftrightarrow H_2O_2 + 2OH^-$$
$$H_2O_2 + 2e^- \leftrightarrow 2OH^-$$

In many samples, there often exist other electroactive species reducible at the same potential and/or surface-active species adsorbable on the cathode surface that could interfere with the reduction of oxygen. Hence, the electrode compartment is isolated from the reaction chamber by a thin polymeric (Teflon, polypropylene) membrane, which allows oxygen diffusion to reach the cathode. As a result, dissolved oxygen (DO) in the sample has to diffuse through the membrane and the electrolyte between the membrane and the cathode.

Once a negative external potential is applied to the working electrode, the surface will provide electrons to the oxygen molecule. The reduction allows a current to flow. At low potentials, the electrode current is governed by the exchange current density and the overpotential of the reactions above. At higher potentials, the oxygen concentration (activity) and the surface become negligible. The specific value of the polarization potential varies with the material used for the cathode. At the polarization potential the reaction rate of oxygen reduction is fast and there is little or no oxygen accumulation on the cathode surface. Therefore, the reaction rate is limited only by the oxygen diffusion rate from the sample to the cathode surface. Consequently, the limiting current is linearly proportional to the oxygen partial pressure (or activity) in contact with the external surface of the membrane. The current flowing is proportional to the activity of oxygen provided the solution is stirred constantly to minimize the formation of an unstirred layer next to the membrane. Under these conditions, the current becomes independent of the voltage. Clark-type electrodes have undergone significant development and miniaturization since their first development. Using the principle of Clark-type oxygen-sensing electrodes, various biosensors have been developed by immobilizing enzymes, antibodies, or micro-organisms which catalyze the oxidation of biochemical organic compounds.

9.4.1.3 Amperometric Enzyme Electrode

Common strategies that an amperometric enzyme electrode is constructed for monitoring a substrate are as follows:

1. The first-generation enzyme biosensors [Figure 9.6(b)] were the extensions of the oxygen electrode where an oxygen consuming enzyme is immobilized to a platinum electrode and the reduction of oxygen at the electrode produces a current that is inversely proportional to the analyte concentration. Typically, oxidase enzyme is immobilized on to membrane at the surface of a platinum electrode. The consumption of O_2 or the formation of H_2O_2 is subsequently measured at a platinum electrode. The detection limit for H_2O_2-based sensors is generally better than for oxygen sensing systems. A problem with this arrangement is the loss in selectivity between the biorecognition event and the amperometric H_2O_2 detection. Furthermore, it suffers from slow response characteristics, difficulties in miniaturization, and low accuracy and reproducibility. However, a major limitation of the H_2O_2 detection approach is the high oxidizing potential (700 mV versus Ag/AgCl reference electrode) necessary for H_2O_2 oxidation results in substantial interference from the oxidation of other compounds such as ascorbic acid (also called vitamin C), uric acid, and acetaminophen in complex matrices.

2. The second generation is focused on lowering the working potential by means of an artificial electron mediator [Figure 9.6(b)]. The electron acceptor is replaced by the mediator, which shuttles the electrons involved in the redox process from the enzyme toward the electrode or vice versa. Most oxidases are not selective with respect to oxidizing agent, allowing the substitution of a variety of artificial oxidizing agents. An example of glucose oxidase is given in the following reaction:

$$GOX\text{-}FADH_2 + Mediator_{ox} \rightleftharpoons GOX\text{-}FAD + Mediator_{red}$$

These mediators catalyze the oxidation of H_2O_2 at the electrode of biosensors. Reagents such as ferrocyanide, ferrocene derivatives, quinones quinoid-like dyes, organic conducting salts, and viologens have been coimmobilized as mediators between H_2O_2 and electrodes. A kind of ferric enzyme, horseradish peroxidase (HRP), is also used. In these electrodes a reduction current, resulting from either the direct or mediated electron transfer, is measured at low applied potential, thereby circumventing the interference problems encountered during the electrochemical oxidation of H_2O_2.

The selection of mediators with appropriate redox potentials allows a better performance of the working electrode in a potential range where other components in the sample matrix are not oxidized or reduced. Low O_2 solubility in aqueous solutions and the difficulty associated with controlling the O_2 partial pressure were disadvantages of biosensors based on the O_2/H_2O_2 reaction. When a highly soluble artificial mediator is used, the enzyme turnover rate is not limited by the cosubstrate (O_2) concentration. Hence, eliminating the O_2 dependence facilitates the control of the enzymatic reaction and sensor performance. Furthermore, the use of mediators other than O_2 allows an exploitation of other oxidoreductase enzymes such as dehydrogenases and peroxidases. Unlike oxidases, these enzymes do not use O_2 as an electron-accepting cosubstrate. An example is the lactic acid biosensor base on lactate dehydrogenase, which catalyses the following reaction:

$$Lactate + NAD^+ \xrightleftharpoons{\text{Lactate dehydrogenase}} Pyruvate + NADH$$

Another way to measure H_2O_2 at a low oxidation potential is by using carbon with dispersed rhodium ruthenium or iridium particles. Although effective in lowering the operating potential, most mediated electrodes still suffer from some ascorbic acid and uric acid interference. Furthermore, mediators are small molecules, and excessive diffusion out of the film immobilized on the electrode surface results in a mediator loss, which results in a loss of catalytic activity. In addition, there can be competition between the oxidized mediator and oxygen for the oxidation of the active site.

3. The third generation is based on the direct electron transfer between the active site of a redox enzyme and the electrochemical transducer [Figure 9.6(c)]. Hence, the signal transduction eliminates the oxygen consumption at the electrode. One approach involves binding redox-active centers (mediators) and enzymes in a polymeric matrix immobilized on an electrode surface. A series of such enzyme-based systems are developed and are generally referred to as "wired" enzyme electrodes. The enzyme was, in effect, wired by the mediator to an electrode. The wired enzymes were able to transfer redox equivalents from the enzyme's active site through the mediator to an electrode. Wired enzyme electrodes were originally developed by Adam Heller as a solution to prevent the diffusion of the mediators out of the film. The mediators for these systems are osmium bipyridine complexes, which are cationic and hence bind electrostatically to the anionic glucose oxidase.

The wired-enzyme principle resulted in subsequent development of enzyme-immobilizing redox polymers. The coimmobilization of enzyme and mediator is accomplished by the redox mediator labeling of the enzyme followed by enzyme immobilization in a redox polymer or an enzyme and mediator immobilization in a conducting polymer (such as polypyrols). Different configurations of a cholesterol biosensor with a cholesterol oxidase entrapped in a polypyrrole film have also been developed. These polymers effectively transfer electrons from glucose-reduced GOX flavin sites to polymer-bound redox centers. A series of chain redox reactions within and between polymers transfer the equivalents to an electrode surface. This allows the easy exchange of electrons between the osmium centers of the complexes and the active site of the enzyme.

For these possibilities, immobilization technique is important to ensure that the redox center is sufficiently close to the electrode to allow rapid electron transfer. For large redox enzymes, such as glucose oxidase, this is difficult to realize as their active sites are hidden inside the protein structure. For these enzymes it becomes important that they are immobilized on a compatible electrode surface in a way that makes electron transfer from the catalytic center to the electrode feasible without the denaturation of the enzyme. Additionally, enzyme immobilization (discussed in Section 9.5.2) is important in terms of biosensor operational stability and long-term use. Since this factor is, to some degree, a function of the strategy used, the choice of immobilization technique is critical. A large number of reports, as reviewed elsewhere, can be found involving enzymes physically or chemically (covalently) entrapped on the transducer.

Since neither mediator nor enzyme must be added, this design facilitates repeated measurements. Sensor use for multiple analyses minimizes cost pressures on sensor design. It also follows that such a sensor could allow for continuous analyte monitoring. The redox enzyme and wire are immobilized by cross-linking to form three-dimensional redox epoxy hydrogels. A large fraction of enzymes bound in the 3D redox epoxy gel are wired to the electrode. These wires provide a general approach to third-generation biosensors, sensitive to glucose and other components such as sarcosine, L-lactate, D-amino acids, L-glycerophosphate, cellobiose, and choline.

9.4.2 Optical Transducers

Optical transducers form a group that directly display features, making them advantageous over other systems such as electrochemical, mass-sensitive, thermal, acoustic, or other transducers. Generally, these methods use a substrate that turns into a different color or generates a color due to a reaction with the analyte. The developed color is assessed by a number of techniques such as absorption spectrocopy (from the UV to the deep infrared in EM spectrum), conventional fluorescence (and phosphorescence) spectroscopy, bioluminescence, chemiluminescence, internal reflection spectroscopy (evanescent wave technology), and laser light scattering methods.

9.4.2.1 Absorption-Based Techniques

The absorption-based techniques involve the well colorimetric test strips and enzyme-linked immunosorbent assays (ELISAs) routinely used in antigen-antibody interactions. Colorimetric strips are disposable, single-use cellulose pads impregnated with sensing elements and reagents. An example is a glucose biosensor for whole-blood monitoring in diabetes control. The reactions utilized in the production of electrochemical signals are used. However, produced peroxide in the reaction is coupled to a chromogen (for example, o-toluidine or 3,3',5,5'-tetramethylbenzidine) rather than to an electrode, which is converted to a dye in the presence of horseradish peroxidase (HRP).

$$\text{Chromogen (2H)} + \text{H}_2\text{O}_2 \xrightarrow{HRP} \text{dye} + 2\text{H}_2\text{O}$$

Dyed strips are useful when only positive or negative signals are necessary. The strips include glucose oxidase, HRP, and a chromogen. A wide variety of test strips involving various enzymes are commercially available. However, spectrophotometers are used in conjunction with ELISA for the quantification of the concentration of a specific molecule (e.g., a hormone or drug) in a fluid such as serum or urine.

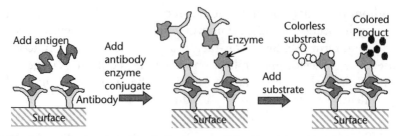

Figure 9.10   Schematic showing steps in the enzyme-linked immunosorbant assay.

The antibody is fixed to a solid surface (Figure 9.10) and then the sample containing the antigen is added. Furthermore, the same antibody coupled to an enzyme such as HRP or β-galactosidase is also prepared. This complex produces a colored product from a colorless substrate. Indoxyls and some of their derivatives are employed as substrates in spectrophotometric assays. However, the spectrophotometric detection may have drawbacks because some measurements take an excessively long time or the sample mixtures are turbid.

An example of absorption-based biosensor is the pulse oximeter which is the most common device used in critical care to measure blood oxygen saturation. Unlike other oxygen sensors, this pulse oximeter is a noninvasive technique and allows continuous monitoring. It is based upon the difference in the infrared light absorption characteristics of oxygenated (HbO_2) and deoxygenated (Hb) hemoglobin. Typically, light emitting diodes (LEDs) are capable of emitting two lights (Figure 9.11) of different wavelengths: infrared (800–1,000-nm wavelength) light and red light (650-nm wavelength), placed on a finger. The lights are partly absorbed by hemoglobin, and the intensity of light passing through is detected by a photodiode placed on the opposite side. The absorptions by oxyhemoglobin and deoxygenated hemoglobin at each wavelength are significantly different. The oxygenated hemoglobin absorbs more infrared light while transmitting a majority of the red light, while the deoxygenated hemoglobin transmits a majority of the infrared light, but absorbs more red light. By calculating the absorption at the two wavelengths, the percentage of oxygenated hemoglobin is calculated. However, apart from absorption from hemoglobin, the light intensity detected by the photodetector also depends on the opacity of the skin, reflection by bones, and tissue scattering. More importantly, the absorption corresponds to the amount of blood in the vascular bed, which varies according to the heartbeat and produces a pulsatile light intensity in the detector, hence, the name pulse oximeter, although direct light sources are used. However, readings of the oximeter are affected by abnormal blood flow conditions such as hypothermia, shock, and vigorous movement. Furthermore, a

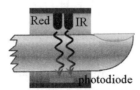

Figure 9.11   Pulse oximeter.

pulse oximeter cannot discriminate between O_2 and CO binding, as the changes in hemoglobin are similar.

9.4.2.2 Fluorescent-Based Techniques

Fluorescent (described in Chapter 8) labels render a biomolecule (or a biological system) fluorescent so to make it amenable to fluorescence spectroscopy. Labeling proteins with fluorophores is an approach in developing a number of immunoassays. The human genome project is based on the fluorescent labeling of nucleic acids, which enables a faster sequencing than other methods. Labels are attached to the species of interest by covalent binding via a reactive group that forms a chemical bond with other groups such as amino, hydroxy, sulfhydryl, or carboxy. Labels are expected to be inert to other chemical species present in the environment, for example, to pH. Labels preferably have long-wave excitation and emission to reduce background luminescence of biological matter, and/or long decay times so that background luminescence decays much faster than the luminescence of the label. As a result, there is a substantial interest in the design of long-wave and long-decay luminescent labels.

A subtechnique within fluorescence is based on the *fluorescence resonance energy transfer* (FRET), which is a process in which the energy from one fluorophore (the energy donor, D) in an excited state is transferred to another chromophore (the energy acceptor, A). The donor and acceptor are selected such that the absorption spectrum of the acceptor is in the same wavelength region as the emission spectrum of the donor. FRET biosensors are used to study molecular events from single cells to whole organisms. They are unique because of their spontaneous fluorescence and targeting specificity to both organelles and tissues. German physicist Theoder Förster, in whose honor some refer to FRET as the Forster resonance energy transfer, developed a quantitative theory based on the assumption that the transfer of energy occurs through dipole-dipole interactions of the donor and the acceptor. The efficiency of the energy transfer (η_{tr}), which can be experimentally measured by the fluorescence intensities or by the lifetime of the fluorophore, is expressed in terms of a Forster distance constant R_0 as

$$\eta_{tr} = \frac{R_0^6}{\left(R_0^6 + \delta^6\right)} \tag{9.8}$$

where δ is the distance between the acceptor and the donor. The advantage of FRET is the dependence (and thus sensitivity) of the energy transfer efficiency on the sixth power of the distance (δ) between the chromophores. Hence, FRET is also used to measure changes in distances in the molecular interactions. R_0 depends on the degree of overlap of the donor emission and the acceptor absorption spectrum and calculated using the relation

$$R_0[\mathring{A}] = 9,772.42 \left(\frac{J\kappa^2 q_d}{n^4}\right)^{1/6} \tag{9.9}$$

where J is the degree of energy overlap between the emission spectrum of the donor and the absorbance spectrum of the acceptor, κ^2 is the relative orientation between the donor and acceptor dipoles (for random orientation, $\kappa^2 = 2/3$ is assumed), q_d is the quantum yield of the donor in the absence of acceptor, and n is the refractive index of the medium.

When δ equals R_0 [Å units], the energy transfer efficiency is 50% of maximum. A labeled macromolecule wherein the donor fluorophore is approximately within $0.5R_0$–$1.5R_0$ from the active site would exhibit a more rapid apparent decay of its fluorescence due to energy transfer; energy transfer is thus a quenching mechanism, and a decrease in fluorescence intensity is observed as well. By placing the donor much closer to the active site, quantitative quenching results, whereas if the donor is further away, the quenching is modest due to the sixth power dependence. An example is quantum dots (QD), described in Chapter 8, conjugated to a maltose-binding protein (MBP) function as the resonance energy-transfer donors. Nonfluorescent dyes bound to cyclodextrin serve as the energy-transfer acceptors. In the absence of maltose, cyclodextrin-dye complexes occupy the protein binding sites. Energy transfer from the QD to the dyes quenches the QD fluorescence (Figure 9.12). When maltose is present, it replaces the cyclodextrin complexes, and the QD fluorescence is recovered. Since fluorescence intensity is proportional to the amount of maltose in the sample, the concentration of maltose is determined.

Chemiluminescence occurs by the oxidation of certain substances, usually with oxygen or hydrogen peroxide, to produce visible light. Bioluminescence is produced by certain biological substances, such as luciferins produced by the firefly. A biosensor involving luminescence uses the firefly *Photinus*-luciferin 4-monooxygenase (ATP-hydrolysing) to detect the presence of bacteria in food or clinical samples. Bacteria are specifically lysed and the ATP released (nearly proportional to the number of bacteria present) react with D-luciferin and oxygen in a reaction that produces yellow light in high quantum yield.

$$ATP - Mg^{2+} + D - luciferin + O_2 \xrightarrow{\;luciferase\;} oxyluciferin +$$
$$AMP + pyrophospate + CO_2 + h\nu(562 \text{ nm})$$

Luminescence spectroscopy utilizes measurement of light intensity, its decay time, polarization, quantum yield, and quenching efficiency, radiative and nonradiative energy transfer, and numerous other combinations. Luminescent fluorochromes can be excited by light (including laser light), electrochemically, by biochemical energy, pressure, or sound. Luminescence is measured in gaseous, liquid,

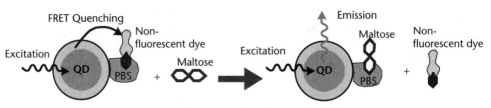

Figure 9.12 Used of FRET in the development of a maltose-biosensor. (*After:* [2].)

solids, tissues, and cells, by directly illuminating the object or by using waveguide optics including optical fibers. Firefly luciferase is a very expensive enzyme, only obtainable from the tails of wild fireflies. Thus, the use of immobilized luciferase greatly reduces the cost of these analyses.

Optical biosensors have been adopted throughout clinical diagnostics and life science research due to their short response time and sensitivity compared with other techniques. Furthermore, using optic fiber technology, some internal parts of the body can be accessed that are inaccessible by other techniques. Antibodies are immobilized at the terminus of a fiber optics probe for use in vitro and in vivo assessment. A single fiber is used to transmit the excitation radiation into the sample and collect the fluorescence emission from the antigen. However, optical biosensors have drawbacks in sample mixtures that are turbid. Some chromophores are also unstable; thus, assay procedures involving nonchromogenic species may be useful. Some optical biosensors require the use of target amplification such that the signal is enhanced to a measurable level. Of these amplification methods, the polymerase chain reaction (PCR) has emerged as a technique for the detection of nucleic acids. However, PCR has several limitations including a time requirement for sample processing and amplification, false positives from sample or reagent contamination, and high instrumentation and labor costs.

9.4.3   Acoustic Transducers

Piezoelectric crystals (discussed in Chapter 8) utilized in ultrasound imaging are also used as transducers. The utility of the piezoelectric crystal as a mass sensor arises from the linear relationship between the change in the mass (or viscosity) at the crystal surface and the change in its oscillating frequency. The vibration of piezoelectric crystals produces an oscillating electric field in which the resonant frequency of the crystal depends on its chemical nature, size, shape, and mass. By placing the crystal in an oscillating circuit, the frequency is measured as a function of the mass. When the change in mass (Δm) is very small compared to the total mass of the crystal, the change in vibrational frequency (f) of the crystal in the circuit relates to m as follows:

$$\Delta f = \frac{bf^2 \Delta m}{A} \tag{9.10}$$

where A is the area of the crystal and b is a constant determined by the crystal material and thickness. Since b and A are constant for a crystal, the oscillating frequency of the crystal changes linearly with the change in mass on the crystal at a particular frequency. Acoustic transducers utilize this unique property in a wide variety of configurations. When antibodies or antigens are immobilizing on to the surface of the crystal, the binding of the antigen or the antibody in the sample changes the crystal mass. The analyte amount is determined by correlating a frequency change to the mass. Immobilized proteins (enzymes and antibodies) for the detection of atmospheric pollutants have proven useful in a number of applications. Formaldehyde in the concentration range of 1–100 parts per billion is assayed in the gas phase using formaldehyde dehydrogenase with no interference from other aldehydes or alcohols.

An advantage of acoustic techniques is the detection, in real time, of binding reactions of chemical compounds (in a gaseous form or dissolved in solution) with the solid surface of the crystal. This feature allows kinetic evaluation of affinity interactions (typically between antibodies and antigens). In addition, the cost of the apparatus is low. Limitations for this transduction method involve format and calibration requirements. Each crystal should be calibrated because its frequency depends on the crystal geometry and the immobilization technique used to coat the surface (generally gold-plated quartz) with the antigen or antibody. The main source of variability is the uniformity of the protein immobilized on the surface.

9.4.4   Other Transducers

The calorimetric approach to signal transduction exploits thermal changes in a solution. Calorimetry is a general detection scheme and is a convenient method for sensing any chemical, biochemical, or physical reaction. Some enzyme-mediated reactions produce heat during the conversion of the substrate to a product. Released heat changes the temperature of the solution, which can be monitored using thermistors. Miniaturized thermistors can also be formed. The sensitivity (10^{-4} M) and range (10^{-4}–10^{-2} M) of thermistor biosensors are low for the majority of applications relative to other transduction methods.

An alternative approach for sensing temperature and/or stress is based on the micromachined cantilever. Microcantilevers are micromechanical devices with dimensions on the order of a 100-μm length with 30–40 μm in width and less than 1 μm in thickness (Figure 9.13). The deflection of the cantilever is measured precisely and exploited for sensing temperature by coating one side of the cantilever with a metal layer to create a bimetallic beam. The different coefficients of thermal expansion associated with each layer cause the cantilever to deflect with temperature change. Considering the low mass of a micromachined cantilever and the sensitivity with which deflection can be measured, the temperature sensitivity is on the order of microdegrees and the energy sensitivity is on the order of 10^{-12} Joules. The small thickness of the cantilever also enables sensitive responses to mechanical stress. Hence, the distinction between thermally and mechanically induced stresses can be deciphered.

Microcantilevers can also be excited into resonance by a number of means including the ambient thermal motion. The resonance frequency of a microcantilever

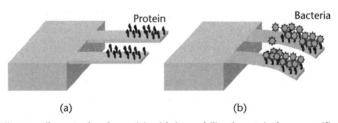

(a)   (b)

Figure 9.13   Microcantilever technology: (a) with immobilized protein for a specific bacterium and (b) bending after adsorption of bacteria to the protein.

varies sensitively with molecular adsorption. For example, the exposure of micro-cantilevers coated with an antibody to the antigen causes a mechanical deflection due to binding of the antigen. The amplitude of deflection is compared to that of a control cantilever. In addition, microcantilevers also undergo bending (Section 5.3.5) due to adsorption of an immobilized biosensing element. If the two surfaces of the cantilever are chemically different (bimaterial, e.g., silicon on gold), a differential molecular adsorption results in a differential surface stress between the top and bottom surfaces of the cantilever. Surface stress produced as molecules adsorb on a cantilever is observed as changes in cantilever deflection. The differential surface stress between cantilever surfaces due to molecular adsorption is augmented by choice of bimaterial elements or by changing the chemical selectivity of one surface over the other.

EXAMPLE 9.5

Using silicon nitride (E = 385 GPa), a microcantilever of a 200-μm length, 20 μm wide and 0.6 μm thick, is developed for use in a bacterial detection. Antibodies specific to those bacteria are immobilized only at the tip. If a deflection of 10 mm is observed, how many bacteria should attach to the cantilever? Assume that each bacterium weighs 53 ng.

Solution: For loading at the tip of the cantilever, using (5.31) $\delta_{max} = \dfrac{FL^3}{3EI}$

Rearranging for F, $F = \dfrac{3\delta_{max}EI}{L^3}$

From Table 5.5, the moment of inertia for a rectangular cross-section is

$$I = \frac{1}{12}bh^3 = \frac{1}{12}0.02[\text{mm}]*\left(0.6\times10^{-3}\right)^3\left[\text{mm}^3\right] = 3.6\times10^{-12}\text{mm}^4$$

Then,

$$F = \frac{3*0.01[\text{mm}]*3.85\times10^8[\text{dyne/mm}^2]*3.6\times10^{-12}[\text{mm}^4]}{0.2^3[\text{mm}]^3} = 51.97\times10^{-4}\ \text{dynes}$$

$F = mg$

Hence,

$$m = \frac{F}{g} = \frac{51.97\times10^{-4}[\text{dynes}]}{981[\text{cm/s}^2]} = 0.053\times10^{-4}\text{g or }5.3\ \mu\text{g}$$

Number of bacteria attached = 5.3/0.053 = 100

The recognition capabilities of single-stranded DNA (or RNA) for its complementary sequence makes DNA an ideal molecule for molecular-scale fabrication of nanostructures such as cubes and molecular grids. There are changes to both the mechanical and electrical properties of nucleic acids. The persistence length of DNA changes from 1 nm for single-stranded DNA to approximately 50 nm for double-stranded DNA depending upon the base sequence and ionic conditions.

Approximately 100 nucleotides of single-stranded DNA are regarded as flexible, while the equivalent length of double-stranded DNA is similar to a rigid rod. Equally important to the construction of nanodevices and biosensors is the fact that electron transfer properties change upon hybridization, with single-stranded DNA possessing considerably greater charge transport properties than double-stranded DNA.

9.5 Manufacturing Technology

9.5.1 Performance Parameters

Manufacturing technology should allow a biosensor of an appropriate size for use in large numbers and at a low cost and a rapid response time, that is, result within the timescale of the process/diagnostic test, self-calibrating (minimal action by user), and a noninvasive mode of detection of indicators. Some of the basic parameters through which biosensors are characterized are described later.

Selectivity is primarily the ability to discriminate between different substrates or analytes. Sensors should only respond to the analyte being quantified. For example, the presence of an enzyme that selectively promotes a reaction introduces selectivity. The biosensors' sensitivity is increased significantly by conversion of the substrate. Many times there may be other components or contaminants that could react with the selective material; this reduces selectivity. In those scenarios, selectivity could be improved with membranes with selective permeability against interfering substances. For example, H_2O_2-based glucose sensor suffers from the presence of ascorbic acid. Adding a membrane that excludes ascorbic acid from the reaction surface is necessary. Antibody-based sensors are generally more specific than enzyme-based ones, but cross-reaction still occurs. Biosensor selectivity is expressed as the ratio of the signal output with the analyte alone to that of the interfering substances.

Sensitivity is the measure of how much the signal changes for a given change in analyte concentration. It is measured by the slope of the response plot of the biosensor output signal against concentration of analyte. For example, sensitivity is described by mA per mmol/L of an analyte for an amperometric biosensor. The bigger the slope, the more sensitive the biosensor. Sensitivity up to the femtomolar limit of detection is achieved in some biosensors. Sensitivity is also improved by improving selectivity.

In some cases, response curves may not be linear. Then sensitivity is quoted over a defined range of analyte concentrations. Furthermore, other effects have to be considered. For example, while using fluorescence-based transducers, photo bleaching, that is, the photochemical destruction of fluorophores, is a critical factor in determining the sensitivity of the biosensor. This is addressed with an appropriate choice of the fluorescent label and conveniently adjusting the power of the laser source. The determination of many biological nitrogen compounds has relied on monitoring the ammonia released from enzymatic reactions by a direct potentiometric monitoring using ammonia gas or ammonium-sensitive electrodes. Alternatively, this ammonia can be transformed into spectrophotometric and fluorometric active derivatives. A common problem encountered in such methods is the presence of endogenous ammonia in the test samples, which means that the measured response corresponds to the total ammonia (i.e., the ammonia produced by the enzymatic reactions and that initially present in the sample). Hence, those sensing systems that monitor the ammonium ions or ammonia are of poor interest in the analysis of real samples containing endogenous ammonia. The choice of other types of electrochemical transducers is preferred in order to avoid these serious drawbacks.

The detection limit is defined as the lowest and the highest analyte concentration that can be detected by the biosensor considering the signal to noise ratio. For a Clark electrode, a lower detection limit is up to 0.5 mg/L at 25°C. Lowering the detection limit is the goal of many manufacturing technologies. The dynamic range is calculated as the measured or extrapolated response of the probe to a zero analyte concentration plus two (or three) standard deviations. Sometimes it is advantageous to have a sensor that can measure a large concentration range. For example, sensors giving a logarithmic signal output could go to multiple orders of magnitude. However, a sensor should give the same signal each time it is put into a fresh analyte solution. A measure of reproducibility is the coefficient of variation, the ratio of the standard deviation to the mean value.

The response time is the time necessary to reach a steady state value. The response time depends on the nature of receptor element. Biological receptors are more sluggish than simple chemical receptors. The response time also depends upon the analyte and product transport rate through different membrane layers. Hence, the thickness and permeability of different layers are essential parameters. For a Clark electrode, 90% of the steady state response is obtained within 20 seconds after adding the analyte into the cell. For single enzymes, 1–3-minute response times are reported. The electrochemical biosensors have a fast response time and a relatively high signal-to-noise ratio. If a sensor is reused, then another parameter to consider is the recovery time (i.e., the time before a sensor can analyze the next sample). It should be reasonable for high throughput screening.

The biosensor stability is designed by the lifetime, which is defined as the storage or operational time necessary for the sensitivity, within the linear concentration range, to decrease by a factor of 10% or 50%. There are three aspects of lifetime: (1) the active lifetime of the biosensor in use, (2) the lifetime of biosensor in storage, and (3) the lifetime of the biocomponent in storage prior to being immobilized. Hence, it is necessary to specify whether the lifetime is a storage (shelf) or operational (use) lifetime and to specify the necessary storage and operational conditions. In addition, the mode of assessment to determine the lifetime should be specified. The active lifetime of a biosensor depends upon the type of biocomponent used and is application-dependent. A sensor lifetime can vary from a few days to a few months. Generally, pure enzymes have the lowest stability, while cell and tissue preparations have longer lifetimes.

9.5.2 Immobilization Strategies

Immobilization is defined as the attachment of a biosensing element to a surface resulting in reduction or loss of mobility. It is critical that the biocomponent be properly immobilized to the transducer. The way in which biosensing elements are immobilized determines the properties of the biosensor. In some cases, immobilization leads to partial or complete loss of bioactivity. In order to fully retain biological activity, biosensing elements should be attached onto the transducer without affecting function. Generally, the choice of a suitable immobilization strategy is determined by the physicochemical and chemical properties of both the surface of the transducer and the biosensing element. For example, the protein activity or function is critically dependent on its 3D structure, which is very sensitive to the local physical and chemical environment. Keeping an immobilized protein molecule in a native state and preserving its function are a major challenge. Many immobilization techniques have been developed, which are mainly based on the three mechanisms of physical, covalent, and bioaffinity immobilization.

9.5.2.1 Physical Entrapment

This technique involves the simple adsorption of the biocomponent on to the electrode surface. Proteins adsorb on surfaces via intermolecular forces, mainly ionic bonds and hydrophobic and polar interactions [Figure 9.14(a)]. Adsorption is classified as physical adsorption (physisorption) and chemical adsorption (chemisorption). Physisorption is usually weak and occurs via the formation of van der Waals bonds or hydrogen bonds between the substrate and the enzyme. Chemisorption is much stronger and involves the formation of covalent bonds. The adsorption capacity of flat surfaces is limited by the geometric size of the immobilized proteins. However, physical adsorption shows random orientation and weak attachment.

An alternative is to physically encapsulate the biosensing element using thin polymer films [Figure 9.14(b)]. A suitable monomer is polymerized in the presence of an enzyme. The resulting polymer can be conducting (such as polypyrrols) or nonconducting (such as polyphenols), depending on the monomer employed. The film thickness is controlled by adjusting the monomer concentration. In addition to immobilizing the biosensing elements, they could act as membranes to improve selectivity and to provide a barrier against electrode fouling. Thin microporous membranes (using materials such as nylon and cellulose nitrate) are also used to encapsulate a biocomponent. Due to very small membrane thickness, the biocomponent and the transducer are very close to each other, which maximize biosensor response. Further bonding of the biocomponent to the transducer surface may be done using a conductive polymer (polypyrrole). The membrane is selected for its

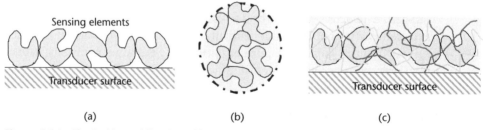

(a)   (b)   (c)

Figure 9.14 Physical immobilization of biosensing elements to the surface: (a) physical adsorption, (b) encapsulation, and (c) entrapment in a matrix.

ability to serve additional functions, such as selective permeability, enhanced electrochemical conductivity, or the mediation of an electron transfer.

Alternatively, cells or pure enzymes are physically constrained (entrapped) to stay inside a 3D matrix [Figure 9.14(c)]. Suitable materials (both natural and synthetic) for entrapment include those that permit uniform cell distribution and have biocompatibility and good transport mechanisms. As hydrogels provide a hydrophilic environment for the biocomponent, they are used as an agent for biosensor entrapment. The naturally occurring polysaccharides such as agar, agarose, alginate, and carrageenan and synthetic polymers such as polyacrylamide, polystyrene, and polyurethane have been explored. Synthetic polymers generally have a smaller pore size, which can lead to less leakage of the biocomponent and hence longer stability; however, synthetic polymers are generally toxic and the immobilization process is accompanied by the generation of heat and the production of free radicals. Natural polymers are generally nontoxic, and the immobilization process is less stressful to the biocomponent. However, natural polymers provide less mechanical strength and stability, and their larger pore size leads to the leakage of the biocomponent.

Biosensors prepared by physically adsorbtion and electron transfer mediator on the electrode surfaces are used in hydrophobic solvents in which the enzyme or other electrode materials are not soluble. However, in some hydrophilic solvents or in solvent water media, the enzyme film on the electrode surface may gradually lose contact with the electrode surface as the solvent medium competes with the electrode surface for electrostatic interaction with the enzyme film. Physical methods of enzyme immobilization have the benefit of applicability to many biosensing elements and may provide little or no perturbation of the native structure and the function of the enzyme. Biosensors based upon direct physical adsorption of a biosensing element onto a surface generally show poor long-term stability due to enzyme leakage.

9.5.2.2 Covalent Bonding

To improve long-term stability, an approach is covalently (which are strong bonds) bonding the sensing element with the transducer. Covalent bonds are particularly useful as reagent-less biosensors because in systems that require electron transfer mediators, this can be incorporated as part of the sensor material. The tailoring of bioelectronic systems requires the assembly of materials on solid conductive supports and the design of the appropriate electronic communication between the biological matrix and the support element. A particular group present in the biocomponent, which is not involved in catalytic action, is attached to the supporting matrix (transducer or membrane) by a covalent bond. In order to retain biological activity, the reaction should be performed under mild conditions. In order to protect the active site, the reaction is carried out in the presence of a substrate. Covalent bonds are mostly formed between side-chain-exposed functional groups of biocomponents with suitably modified supports, resulting in an irreversible binding and producing a high surface coverage. Proteins are covalently bound to the immobilization support through accessible functional groups of exposed amino acids.

Covalent linkage of proteins to conductive, semiconductive, or nonconductive supports often utilizes the availability of functional groups on the surface of the solid support. Metal oxides such as TiO_2 and SnO_2 contain surface hydroxyl groups that are useful in the coupling of organic materials. Noble metals (Au, Pt) are chemically or electrochemically pretreated to generate surface functionalities. Some of the functional groups potentially available in proteins for immobilization and the functionalities required on the surfaces are shown in Figure 9.15.

Chemical binding via side chains of amino acids is often random, since it is based upon residues typically present on the exterior of the protein. As a result, heterogeneity in the modes of protein linkage and orientation with respect to the surface is obtained. The structural alignment of the redox protein relative to the conductive support is essential for many bioelectronic applications, and so the development of methods for specific binding and alignment of proteins on surfaces is important. In some cases, oriented immobilization is obtained if the protein possesses a single reactive amino acid in the structure. For example, cysteine that contains a unique thiol (SH) functional group is an uncommon amino acid residue in proteins. A single cysteine residue on the protein periphery or a genetically engineered cysteine component in the protein allows the linkage of the protein to the surface at a single point. Chemical attachment is also guided in an orderly manner to attain oriented immobilization. Site-specific immobilization requires functionalization of the molecule or altering the surface or both. Thiolated DNA is also utilized for self-assembly onto gold transducers. Covalent linkage to the gold surface via functional alkanethiol-based monolayers is an approach used in developing highly organized DNA monolayers. Proteins are also attached to surfaces in an ordered structure, allowing reproducibility and conformational stability.

9.5.2.3 Affinity

Affinity interactions between an enzyme and its substrate, a receptor protein and its recognition pair, or antigen/antibody pairs are often characterized by high association constants of the resulting complexes. This has enabled the use of specific recognition interactions to construct protein layers on solid supports. Biochemical affinity reactions offer an oriented immobilization of proteins, providing an important advantage over other immobilization techniques. Moreover, not only is

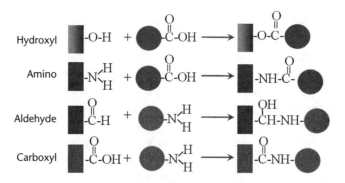

Figure 9.15 Surfaces with functionalities that can couple with biomolecules.

oriented and homogeneous attachment obtained, but it is also possible to detach proteins and make repeated use of the same surface. For example, antigen-antibody [Figure 9.16(a)] interactions are used to organize monolayers and multilayers of GOX on glassy carbon electrodes. Alternatively, glassy carbon electrodes are coated with gelatin and rabbit IgG antibodies are adsorbed onto the gelatin. A conjugate consisting of GOX linked to the antirabbit IgG antibody is added to interact and form a complex with the modified glassy carbon electrode surface. Enzyme multi-layer electrodes are generated by the application of an anti-IgG-GOX conjugate X and an IgG and anti-GOX antibody Y as linking components. Homogeneous mediation is used using ferrocenylmethanol as a freely diffusing mediator. The current increases with an increase in the enzyme loading on the surface.

A popular technology is based on avidin-biotin affinity interactions. Avidin is a tetrameric glycoprotein soluble in aqueous solutions and stable over wide pH and temperature ranges. Biotin (or vitamin H) is a naturally occurring vitamin found in all living cells. Only a small fragment of biotin is required for the interaction with avidin. Avidin can bind up to four molecules of biotin via a strong noncovalent bond ($K_D = 10^{15}$ M^{-1}). Streptavidin is a closely related tetrameric protein, with a similar affinity to biotin. Since biotin is a small molecule, its conjugation to macromolecules does not affect conformation, size, or functionality. Both biotin and avidin/streptavidin are attached to a variety of substrates. A typical biotin/avidin/biotin multilayer [Figure 9.16(b)] is composed by directly immobilizing biotin and avidin, creating a secondary layer for binding biotinylated molecules. This approach is generally preferred due to the higher organization obtained in comparison to that of the direct immobilization of avidin. Use of biotinylated DNA for complex formation with a surface-confined avidin or strepavidin is also an option. A number of high affinity recombinant proteins are also produced by genetic engineering, which are then coupled with affinity tags. Tags are placed at defined positions on proteins, preferably far away from the active site in order to achieve optimal accessibility of the ligands. Poly(His) is the most popular tag used in these situations due to the advantages of small size, compatibility with organic solvents, low immunogenicity, and effective purification under native and denaturing conditions.

9.5.3 Microelectromechanical Systems (MEMS)

Microfabrication technology is borrowed from microelectronic integrated circuits and is utilized to miniaturize devices and develop portable biosensors.

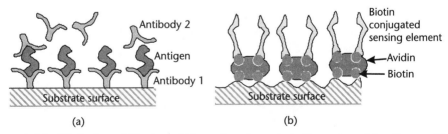

Figure 9.16 Affinity immobilization of biosensing elements to the surface: (a) antigen-antibody based immobilization and (b) avidin-biotin based immobilization.

Microfabrication allows micron level manipulations in the design patterns. The development of capabilities to miniaturize analytical devices and components offers a number of potential benefits, including reduced sample sizes, single-use disposable devices, reduced consumption of reagents, the capability of building integrated systems, the reduction of power consumption, parallel devices, and faster processes that lead to high throughput, the possibility of integrating multifunctionality, and portable devices.

The miniaturization of electronics over the years has resulted in a small, two-piece device used in cochlear implants. One part, consisting of a receiver and stimulator, is implanted under the skin behind the ear. The other part, consisting of a microphone, a sound processor, and a transmitter, is placed externally over the receiver. Both parts are held in place magnetically, requiring no wire connection, reducing the risk of infection and damage to the device. In the case of young children, the sound processor is worn in a hip pack or harness. To create sound, the microphone picks up and amplifies noises that the sound processor filters, giving priority to audible speech. The processor sends electrical signals to the transmitter, which, in turn, sends the processed sound signals to the internal receiver electromagnetically. The receiver and stimulator convert the signals into electric impulses, which are sent to an array of up to 24 electrodes. They, in turn, send the impulses to the hair cells and into the brain via the auditory nerve. The two dozen electrodes must fill in for the 16,000 hair cells normally used for hearing.

9.5.3.1 Photolithography

The microfabrication processes establish specific patterns of various materials that can be deposited on or removed from the planar surfaces in a fashion that allows high reproducibility and control over the dimensional features. The process of defining these patterns on the wafer is known as photolithography. First, the planar surfaces, typically silicon wafers, are cleaned and layered with a light-sensitive material called a photoresist (Figure 9.17). A mask containing the required pattern is generated and transferred on to the wafer by exposing it to light. The transferred

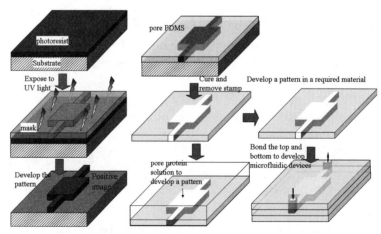

Figure 9.17 Schematic illustration of microfabrication technology. Formed patterns could be used either to pattern surfaces or to develop microfluidic devices.

pattern could be positive (i.e., an exact copy of the mask patterns) or negative, based on the type of photoresist used. Photoresists are classified as positive or negative types.

Positive photoresists become more soluble in chemical solvents upon exposure to radiation, forming a positive image on the wafer. Positive photoresists are made up of a photoactive compound dissolved in a solvent and a low molecular weight resin. Once this photoactive dissolution inhibitor is destroyed by light, however, the resin becomes soluble in the developer. On the other hand, negative photoresists become less soluble (or more durable) in chemical solvents upon exposure to radiation and form negative images of the mask patterns on the wafer. Negative photoresists consist of polymeric organic materials and consist of a chemically inert polyisoprene rubber and a photoactive agent. When exposed to light, the photoactive agent reacts with the rubber, promoting cross-linking between the rubber molecules that make them less soluble in the developer. Positive photoresists are the dominant type used in the industry as negative photoresist processes are more susceptible to a decreased resolution and a distortion of patterns.

9.5.3.2   Soft Lithography

Photolithography is the dominant technology in electronic manufacturing. It has less usefulness in biosensing applications. For example, generating a pattern is expensive as it has to be performed in rooms (referred to as Cleanrooms) that have a negligible amount of contaminants. More importantly, wafer surfaces offer very little flexibility in generating patterns of specific chemical functionalities on surfaces (e.g., for immobilizing proteins). To work with other materials, stamps are generated from elastomeric polymers using a photolithographic patterned master. The use of elastomer such as polydimethylsiloxane (PDMS) has many advantages over silicon or glass substrates. PDMS is cheaper than silicon, it is more flexible, and it bonds more easily to other materials than silicon or glass does. PDMS conforms to the surface of the substrate over a large area and can adjust to surfaces that are nonplanar. PDMS is a homogenous and optical transparent material up to 300 nm. PDMS is waterproof and permeable to gases. The surface properties of PDMS can be changed by exposure of the surface in oxygen plasma that allows cross-linking with other materials. Stamps are created by pouring liquid PDMS (Figure 9.17), prepolymers on the master, degassing to remove trapped air bubbles inside the mold, and curing them at temperatures between 20°C and 80°C for up to 48 hours. These patterned elastomeric templates can be used for generating porous membranes of defined pore characteristics or patterned cells or microfluidic devices. For example, patterns of proteins can be formed onto a variety of different surfaces and polyurethane is then molded against the secondary PDMS master. In this way, multiple copies are made without damaging the original master template.

Such reproducibility and dimensional control allow the large-scale production of micromechanical devices including microcantilevers. Different components of the biosensor can be built layer by layer. For example, a structure comprising a suitable base sensor is established. Additional structures such as a semipermeable solid film or a selectively permeable layer are then established over the resulting base sensor. The semipermeable solid film may further comprise compounds or molecules which serve to sensitize the base sensor to a preselected species (e.g., ammonium ion). Also, such selective layers may function as adhesion promoters by which the preselected ligand receptor is immobilized to the wholly microfabricated ligand/ligand receptor-based biosensor embodiment. DNA electrochemical biosensors are developed using graphite or carbon electrodes. Carbon-based electrodes, however, are generally not adaptable to MEMS technology when small (less than a micrometer) dimensions are needed.

9.5.3.3 Microfluidic Devices

Microfluidics refers to the fluid flow in microchannels (i.e., one of the dimensions of flow is measured in micrometers). They are developed using the elastomeric templates of the required channel patterns by sealing both the ends and creating inlet and outlet points in the top layer (Figure 9.17). Microstructured substrates provide control over the flow of fluid through a geometric shape and the surface chemistry. For example, the passive manipulation of fluids' centrifugal forces can be used to control flow. A number of devices are under development for use in various applications including bioseparations, microdialysis, high-throughput drug screening, DNA analysis, mass spectroscopy ensuring the safety of air, food, and water, and combating terrorism and biowarfare.

There are technical barriers that must be overcome for microfluidic devices to reach their full potential for biosensing applications. A common problem often encountered in microfluidics is the clogging of channels due to particle contamination, which requires the prefiltering of liquid samples. Since microfluidic devices engage fluid in motion, capillary effects, micropumping, and surface tension have to be considered along with the design and fabrication of the micro-sized flow conduits. Surface area is an important factor at the microscale. For example, a 35-mm diameter dish half-filled with 2.5-mL water has a surface area-to-volume ratio of 4.2 cm^2/cm^3, whereas a microchannel that is 50 μm tall, 50 μm wide, and 30 mm long filled with 75-nL water has a surface area-to-volume ratio of 800 cm^2/cm^3. A very large surface area-to-volume ratio makes capillary electrophoresis more efficient in microchannels by removing excess heat more rapidly. Adhesive forces (van der Waals force, electrostatic forces, surface tension) are more dominant than gravity in the microscale. Microfluidic flows have a normally low Reynolds number (N_{Re}) due to the very small length scales. This makes turbulent flow virtually impossible to achieve and the mixing of fluids is generally slow and diffusion controlled. Mixing can be enhanced in the laminar flow regime by subjecting the fluid to a chaotic flow pattern. In a chaotic flow, complex patterns are created that allow the fluid to stretch and fold. Mixing is greatly enhanced by the tendency of fluid particles to become homogeneously dispersed and by a decrease in the length scale for diffusion between unlike components.

There are two common methods by which fluid actuation through microchannels is achieved. In a *pressure-driven flow*, the fluid is pumped through the device via positive displacement pumps, such as syringe pumps. With the assumption of a no-slip boundary condition (the fluid velocity at the walls must be zero), the laminar flow produces a parabolic velocity profile within the channel (Chapter 4). The parabolic velocity profile has significant implications for the distribution of molecules transported within a channel. The pressure-driven flow is relatively inexpensive and quite reproducible to pumping fluids through microdevices. With the increasing efforts at developing micropumps, the pressure-driven flow is also amenable to miniaturization. The resistance to the fluid flow in microchannels is approximated using the Poiseuille law [(4.9)] (i.e., the flow rate within a microchannel is given by $\Delta P = Q*R$ where ΔP is the pressure drop across the channel, Q is the flow rate, and R is the channel resistance). For a rectangular microchannel with a very high aspect ratio, the resistance is approximated by

$$R = \frac{12\mu L}{wh^3} \tag{9.11}$$

EXAMPLE 9.6

A rectangular microfluidic device that is 2 mm long, 0.5 mm wide, and 0.1 mm tall is costructured. If water is flowing through the device at 0.2 mL/min, what is the pressure drop in the device?

Solution:

$$R = \frac{12*0.01[g/cm.s]*0.2[cm]}{0.05[cm]*0.01^3[cm]^3} = 480,000 \ dyne/cm^5.s$$

$$\Delta P = QR = 0.2[cm^3/min][min/60s]*480,000[dyne/cm^5.s]$$

$$= 1,600 \ dyne/cm^2 = 1.2 \ mmHg$$

Another technique for pumping fluids through microchannels is the *electro-osmotic driven flow*. If the walls of a microchannel have an electric charge, an electric double layer of counterions will form at the walls (Figure 9.18). When an electric field is applied across the microchannel, the ions in the double layer move towards the electrodes of opposite polarity. This creates a motion of the fluid near the walls and transfers via viscous forces into convective motion of the bulk fluid. The flow rate is controlled by adjusting the potential at different points in a device. The electro-osmotic flow has a flat flow profile compared to a pressure-driven flow. This results in a reduced broadening of discrete solution plugs as they move through a microfluidic system. The flat velocity profile avoids many diffusion-induced nonuniformities that are observed in a pressure-driven flow. However, sample dispersion

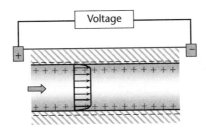

Figure 9.18 Electro-osmotic flow in a microchannel.

in the form of band broadening is a concern for electro-osmotic flow. Another advantage to electro-osmotic flow is the ease of coupling other electronic applications on-chip. However, the electro-osmotic flow often requires very high voltages, making it a difficult technology to miniaturize without off-chip power supplies. Another significant disadvantage of the electro-osmotic flow is sensitivity to both solution chemistry and the chemistry of the channel surface. For example, the protein adsorbtion to the walls substantially changes the surface charge characteristics and changes the fluid velocity. This often leads to undesired changes in the flow during an analysis, which compromise reproducibility and quantitation. When transporting fluids using electro-osmotic flow, the large surface area to volume ratio allows macromolecules to quickly diffuse and adsorb to channel surfaces, reducing the efficiency of pumping. It is important to develop a fundamental understanding of the interplay of surface chemistry, solution chemistry, and fluid mechanics in microfluidic devices in order to realize the full potential of these systems.

9.5.4   Microarray Technology

The development of arrays of surface-immobilized molecules for use as sensors is helpful in increasing the capacity of testing. Molecular selectivity is achieved through a binding reaction with a sensor element often comprising a biological macromolecule (e.g., nucleic acid and protein). In conventional biosensors, sets of related elements (typically DNA fragments, peptides, or drugs) are tested in batch. The microarrays use a precise, spatially ordered arrangement of elements that allow them to be examined side by side. Microarrays offer many advantages over the conventional biosensing tools: the ability to simultaneously analyze a variety of analytes in the same sample, the required sample quantities with a minimal, low consumption of scarce reagents, and a high sample throughput. Microarray technology is utilized to assess the large size of genetic information (genomics) or proteins.

9.5.4.1   Genomics

Genomics is the systematic study of the genetic information of an organism or a cell. Applications of genomics include the identification of complex genetic diseases, drug discovery and toxicology studies, new diagnostic methods of mutations and polymorphisms, the analysis of pathogens, the specific genotype-targeted drugs for food processing, and agriculture product development. The major tools and methods related to genomics are genetic analysis, measurement of gene expression, and determination of gene function. One can analyze differing expression of genes over time, between tissues in healthy and disease states. However, the DNA microarray technology based on the parallel processing is used to monitor the large-scale gene expressions simultaneously. When combined with the metabolic activity, one can understand the changes in varying conditions.

The Human Genome Project predicted 100,000 to 150,000 genes. The actual number of genes was nearly 30,000 to 400,000. Having information on all the genes is useful, but the interaction between the genes should also be understood. Microarray technology allows the analysis of many genes at once and allows the determination of the variations in the gene expressions. DNA microarray technology evolved from the use of solid substrates in performing Southern blot experiments. The basic principle of operation of a DNA-microarray consists of three main steps (Figure 9.19):

1. *Development of the microarray chip.* That is, the immobilization of different DNA sequences onto different positions (probes). DNA probes representing many (hundreds to thousands) genes are positioned by two approaches:

    a. The most common approach consists of an array of contact-printing needles that are first immersed into a multiwell plate holding the several probe solutions and then is positioned onto the chip on an activated substrate such as a glass slide where the needles leave the probe spots by contact printing.

    b. To improve the signal-to-noise ratio, it is crucial to maximize the label density and minimize nonspecific binding with stringency wash conditions. To achieve higher probe densities, more sophisticated techniques are used and in these cases the DNA probes are synthesized directly in situ (i.e., onto the chip). The two most representative techniques are based on photolithographic processes, where protection groups are used to control the growth of the DNA molecules, and removed by means of UV light.

2. *Hybridization.* After carrying out the biological experiment of interest, in patients, animals, or in vitro cell culture, the nucleic acid to be analyzed is isolated. The nucleic acid is then labeled by attaching a fluorescent dye and brought in contact with the probes by flooding the chip with the sample solution. Due to the base-pairing ability, nucleic acids hybridize to the gene.

3. *Readout.* After washing away the microarray to remove all nonbound molecules, the hybridized microarray is scanned to acquire the fluorescent images as emission from the label incorporated into the nucleic acid(s) hybridized to the probe array. Generated colored images are then analyzed through the use of image analysis software. Since the sequence and position of each probe on the array is known, the identity of the bound target nucleic acid after hybridization reactions is determined. Based on fluorescence intensities that are proportional to the number of genes bound in each spot, the expression levels of particular genes are determined.

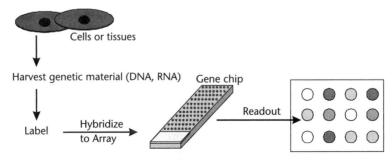

Figure 9.19   A schematic showing the steps in the DNA microarray technology.

Depending on the aim of the study, one can infer the statistical significance of the differential expression, perform various exploratory data analyses, classify samples according to their disease subtypes, and carry out pathway analysis. This allows the elucidation of patterns of gene expression that may be associated with states of health and disease. Because of the scale of information generated, the analysis and management of this data sometimes limit microarray techniques. Bioinformatic techniques (discussed in Section 9.6) enable the meaningful interpretation of microarray data.

9.5.4.2 Proteomics

This term was coined to make an analogy with genomics. Proteomics concerns the entire protein complement in a given cell, tissue, or organism, and assesses protein activities, modification localizations, and the interaction of proteins in complexes. By studying the global patterns of protein content and activity and how they change during development or in response to disease, proteomics research is posed to boost our understanding of systems-level cellular behavior. However, proteomics is much more complicated than genomics. A finding of the Human Genome Project is that there are far fewer genes that code for proteins in the human genome than there are proteins in the human proteome (~22,000 genes versus ~200,000 proteins). The entirety of proteins in existence in an organism throughout its life cycle, or on a smaller scale, and the entirety of proteins found in a particular cell type under a particular type of stimulation are referred to as the proteome of the organism or cell type, respectively. While the genome is a rather constant entity, the proteome is constantly changing through its biochemical interactions with the genome and the environment. One organism will have a radically different protein expression in different parts of its body, in different stages of its life cycle, and in different environmental conditions. Although gene expression profiling helps identifying differences in protein expression between cells, it involves the assumption that mRNA levels correlate well with protein expression levels.

The large increase in protein diversity is thought to be due to alternative splicing and posttranslational modification of proteins. Hence, the behavior of gene products is difficult to predict from a gene sequence alone. Even if a gene is transcribed, its expression may be regulated at the level of translation, and protein products are subject to further control by posttranslational modifications, varying half-lives, and compartmentalization in protein complexes. This limitation of gene expression screening has led to an increasing interest in the systematic analysis of protein expression.

Proteomics bridges the gap between genome sequence and cellular behavior and has its roots in traditional biochemical techniques of protein characterization, particularly 2D gel electrophoresis. Essentially, the proteins in a cell or tissue extract are separated first in one dimension on the basis of charge, and then in a second dimension on the basis of molecular size, resulting in a defined pattern of spots. Spots are isolated from 2D gels and analyzed by mass spectrometry or digested chemically/enzymatically to produce unique protein degradation products that can then be analyzed by mass spectrometry. Using this process, it is possible to derive amino acid sequence information, which can then be used to search protein or genome databases to identify the protein in the original gel spot.

The interest in protein microarray technology also stems to a great extent from the hugely successful DNA microarray technology. The tremendous variability in the nature of proteins and consequently in the requirement of their detection and identification also makes the development of protein chips a particularly challenging task. Typically, protein microarrays are developed using robotic dispensers such as those developed for creating DNA microarrays. The protein samples are made to adhere to the glass slides by various immobilization strategies described above. Soft substrates such as polystryrene, poly(vinylidene fluoride), and nitrocellulose membranes, which have been used to attach proteins in traditional biochemical analyses (e.g., immunoblot and phage display analysis), are often not compatible for protein microarrays. These surfaces often do not allow a suitable high protein density, the spotted material may spread on the surface, and/or they may not allow optimal signal-to-noise ratios. Thus, glass microscope slides or other materials that have been derivatized to attach proteins on their surface at high density are chosen. These slides have low fluorescence background and are compatible with most assays. The detection of bound targets to proteins is considerably more complex than that of DNA microarray detection. Among a variety of detection methods, the preferred method of detection is the fluorescence, because it is generally safe, extremely sensitive, and simple, can have very high resolution and is compatible with standard microarray scanners. Four major barriers in protein microarray development are: (1) minimizing background noise; (2) preserving protein native state and orientation; (3) protein detection and identification; and (4) speed of protein or antibody production and purification.

Fluorescence detection methods are generally the preferred detection method because they are simple, safe, and extremely sensitive and can have very high resolution. They are also compatible with standard DNA microarray scanners. Typically, a chip is either directly probed with a fluorescent molecule (e.g., protein or small molecule) or in two steps by first using a tagged probe (e.g., biotin), which is then detected in a second step using a fluorescently labeled affinity reagent (e.g., streptavidin). One of the challenges that present itself for direct protein detection in matrices such as whole blood, serum, or plasma is the range of pathogen load and protein concentrations within the sample. Protein levels differ by as much as nine orders of magnitude in these conditions, but the detection range is only three to four orders of magnitude (i.e., only a fraction of the proteome is examined by a given experiment). As antibody-binding constants are commonly in the nanomolar range, they are ill equipped to directly measure low abundance targets, which may be present in picomolar range, without sample preconcentration.

When measuring low abundance targets such as virus particles, alternative strategies are required. In this case, diagnostic methods take advantage of either surrogate markers or signal amplification by the immune system and measure the antibodies expressed against the pathogen. Thus, constructing an array of antigens is a convenient method of measuring antibody concentration and relating it back to infection. Another alternative is to use probe molecules with higher binding affinity. An excellent candidate for this approach is aptamers, which are single-stranded oligomers of DNA or RNA that exhibit extremely high affinity for binding molecules such as small organic molecules, drugs, and proteins. Aptamers are generated by an in vitro selection process called SELEX (systematic evolution of ligands by exponential enrichment). In SELEX, a large library of random oligomer sequences of the same length are tested for binding affinity with a given target ligand. The sequences with highest affinity are separated and amplified by PCR and the process is repeated until the library has been refined into a subset of sequences with the highest binding affinity. Thus, aptamers can be generated with a high affinity for either antibodies or antigens and applied to a microarray for testing. Binding affinities of aptamers could be in the picomolar range, much higher than many antibodies.

9.6 Bioinformatics

With the possibility of accumulating large amount of data, new systematic approaches are required to analyze complex interactions between genes and proteins. To understand the output results from DNA microarrays, for example, one has to develop sophisticated statistical tools that can help establish the network of information involved in the regulatory pathways controlling the activity of living organisms. Looking through standard text files (called flat files) and tools is easy and has the advantage that no special tools are required. However, standard text files are less efficient when the data sets get very large (i.e., more than tens and hundreds of gigabits of file sizes). Designing efficient databases that are optimized for efficiency and operating in a high volume so that information can be found quickly based on certain indexes is essential.

Bioinformatics deals with organizing and understanding the biological data using the power of computational tools so that data analysis and the sharing of large sets of data are possible. A fundamental task concerns the creation and maintenance of databases of biological information that are already available. For example, nucleic acid sequences for many genes and amino acid sequences for many proteins are known. The storage and organization of millions of nucleotides is complex. Designing a database and developing an interface whereby researchers can both access existing information and submit new entries is only the beginning. There are common methodological statistical principles such as hidden Markov models and Bayesian statistics for designing and applying information systems that underlie all work in informatics. For further reading on methodical principles and development of database, refer to [3]. The databases are predominantly government-funded such as the National Center for Biotechnology Information (NCBI, www.ncbi.nih.gov) and are accessible to the public with a typical Internet browser. The basics of using different search engines can be accessed at the Web site. An example is to compare a gene or a gene product as given next.

Once a new sequence of DNA is identified, two or more gene sequences can be compared for similarity to get a measure of their relatedness. This can be used to group genes into subsets that might give an indication of the function or activity of the members of these subsets based upon what is known about the proteins encoded by the membership of that subset. A comparison also allows taxonomy to be examined, as well as the drawing of trees of relatedness, and insights can be made into sequence evolution. One could use a public nucleotide sequence database such as the GenBank database, built and distributed by the National Center for Biotechnology Information (NCBI) at the National Institutes of Health (http://www.ncbi.nlm.nih.gov/Genbank/GenbankSearch.html). There are two ways to search the GenBank database: a text-based query can be submitted through the Entrez system, or a sequence query can be submitted through the Basic Local Alignment Search Tool (BLAST) family of programs (see http://www.ncbi.nlm.nih.gov/BLAST/).

BLAST is the most widely used tool for doing sequence searches. One can either use the Web interface directly or BLAST can be downloaded to a personal computer and run it locally. The advantages of the Web interface include ease of use and the fact that the databases are always up to date. The disadvantage of the Web interface is that large number of searches cannot be done easily. Another database in Europe is the European Molecular Biology Laboratory (EMBL) Nucleotide Sequence Database (http://www.ebi.ac.uk/embl/).

Starting with a gene sequence, one can determine the protein sequence using a protein database such as the SWISS-PROT (http://us.expasy.org/sprot/) and the Entrez Protein Database with strong certainty. Protein sequences can also be compared using a number of methods, their relatedness can be measured, and closely related proteins can then be assigned families. One could identify or design ligands that may bind the identified protein and use them for developing a new biosensor or to develop therapies. For further information on biological databases, genome analysis, proteome analysis, and the bioinformatics revolution in medicine, refer to [4].

Problems

9.1 A company has prepared an electrode based on the Quinhydrone solution.

(a) If the company reports a Nernst potential of 59.2 mV at 25°C and a pH of 7.0, what is the concentration of the Quinhydrone solution?

(b) The potential of this cell with respect to an Ag/AgCl electrode (contains a salt bridge concentration of 3.8 M KCl) reference electrode is 202 mV at 25°C. Calculate the potential with respect to a SHE.

9.2 A pH electrode is recording a pH of 6.1. Acid added to the solution and the potential of the pH electrode increases by 177 mV. What is the pH of the new solution?

9.3 The potential of a half-cell with respect to an SCE reference electrode is: −0.793V. Calculate the potential with respect to the cell potential using the Ag/AgCl (0.014V less negative than SCE).

9.4 FAD is a prosthetic group present in glucose oxidase enzyme. It is reduced to FADH and the dissociation constant for the reaction is determined to be 25 μM at 25°C.

(a) What is the potential energy developed at equilibrium?

(b) Is this reaction spontaneous?

(c) If not, what potential will make it spontaneous?

9.5 Calculate the electrode potential by a reaction whose equilibrium rate constant is 10 nM if the valency is +2.

9.6 A group of researchers want to immobilize a 0.5-mm-thick HRP onto an electrode surface that is 1 cm wide and 5 cm long on both sides. They can use 1A of constant current for extended period of time. Determine the time required to develop that thickness. (Hint: Use Faraday's law; you need to know the valency of the HRP and the molecular weight and density.)

9.7 Repeat Problem 9.5 for a material with $E = 5$ GPa and deflection of 20 μm.

9.8 In an application microcantilevers are used with the possibility of absorption of the analyte on the entire cantilver. Derive the equation to correlate a mass of adsorption to the tip deflection. Using the values in Example 9.5, determine the number of bacteria.

9.9 Consider a glucose monitor based on the electrochemical detection of H_2O_2 formed on oxidation of glucose to gluconic acid by the GOX. The GOX preparation used in the sensor is designed using Michaelis-Menten equation M $v(0) = \dfrac{v_{max}[s]}{K_M + [s]}$ and has K_m (substrate dissociation constant) = 8 mg/L and v_{max} is 45 mg/(L-min) at pH 7.4 and 25°C. Note that it takes 2 moles of electrons to electrochemically reduce each mole of H_2O_2 formed by GOX and that an electron flow rate of 1 mole/sec will give a current of 9.649×10^4 amperes. You may assume that the electrochemcial reduction of H_2O_2 is extremely rapid.

(a) Suppose the maximum anticipated glucose concentration in a given blood sample was 90 mg/dL of whole blood and all blood samples are diluted by a factor of 10 in the device. Is this a problem? Why or why not?

(b) What is the maximum current, in amperes, that this monitor can produce? The maximum current occurs when the enzyme produces H_2O_2 as fast as it can (i.e., when $v = v_{max}$).

(c) Suppose that the initial diluted glucose concentration in a given assay was 1.5 g/L. At what time from the start of the assay, in seconds, would the measured current fall to 90% of the initially measured value? [Hint: First calculate what the substrate concentration would be when the current falls by 10%.]

9.10 List the advantages and disadvantages of electrochemical transducers with optical transducers.

9.11 Write the differences between three immobilization techniques.

References

[1] Webster, J. G., *Medical Instrumentation: Application and Design*, New York: John Wiley & Sons, 2000.

[2] Mendintz, I. L., et al., "Self-Assembled Nanoscale Biosensors Based on Quantum Dot FRET Donors," *Nature Materials*, Vol. 2, No. 9, 2003, pp. 630–638.

[3]   Waterman, M. S., *Introduction to Computational Biology: Maps, Sequences and Genomes,* London, U.K.: Chapman & Hall/CRC Press, 1995.

[4]   Buehler, L. K., and H. H. Rashidi, *Bioinformatics Basics: Applications in Biological Science and Medicine,* 2nd ed., Boca Raton, FL: CRC Press, 2005.

Selected Bibliography

Cunningham, A. J., *Introduction to Bioanalytical Sensors*, New York: Wiley-Interscience, 1998.

Eggins, B. R., *Biosensors: An Introduction*, rev. ed., New York: John Wiley & Sons, 1997.

Eggins, B. R., *Chemical Sensors and Biosensors*, New York: John Wiley & Sons, 2002.

Homola, J., and O. S. Wolfbeis, *Surface Plasmon Resonance Based Sensors,* New York: Springer, 2006.

Narayanaswamy, R., and O. S. Wolfbeis, *Optical Sensors: Industrial, Environmental and Diagnostic Applications*, New York: Springer, 2004.

Nguyen, N. -T., and S. T. Wereley, *Fundamentals and Applications of Microfluidics,* 2nd ed., Norwood, MA: Artech House, 2006.

Schena, M., *Microarray Analysis*, New York: Wiley-Liss, 2002.

Spichiger-Keller, U. E., *Chemical Sensors and Biosensors for Medical and Biological Applications*, New York: Wiley-VCH, 1998.

CHAPTER 10
Physiological Modeling

10.1 Overview

Modeling is a mathematical representation of a process, action, or an experimental outcome. Thus far, various concepts have been described in previous chapters that help in developing strategies, techniques, and understanding the function of living organisms. These can be considered as models. Apart from those uses, there are many applications where modeling is utilized in biomedical engineering. Some of the applications of physiological models are shown in Figure 10.1. Models are also utilized in the analysis of experimental data, predicting the overall response of a system to stimuli, deducing underlying principles of system behavior or developing simulators to train surgeons. Physiological models are developed to explain or understand experimental data so that it can be applied to other similar scenarios by simulating those processes, developing therapeutic strategies or designing biomedical devices. The feasibility of using physiology-based models to describe, predict, and extrapolate the extent and magnitude of the occurrence of interactions for various dose levels, scenarios, species, routes, and mixture complexities, arises from the very nature and basis of these models.

Models are used to quantify the relationship between exposure and internal dose of various therapeutic agents, toxins, and traces used in biomedical imaging. They provide a sound theoretical basis for the extrapolation and generalization of results obtained in a reduced number of subjects to the general population. Significant developments in dialysis and heart-lung machine or action potential can be attributed to the success of model development. Knowledge of the fundamental mechanisms of synaptic transmission is, to a large extent, based on model of Hodgkin and Huxley. For developing a tracer for PET, understanding distribution characteristics and the effect of the drug on the body is necessary. Developing models helps in estimating the parameters such as the distribution concentration and volume, or clearance rate of the agent, and compare it to existing therapeutic agents. A compartmental modeling approach to understand the fate of a tracer administered into the body during PET is very useful to measure cerebral blood flow or to understand cerebral glucose metabolism or to obtain quantitative information about the distribution of the target receptor throughout the brain and receptor bindings. This chapter provides an introduction to compartmental modeling and

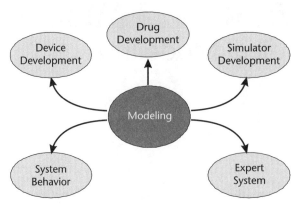

Figure 10.1   Uses of physiological modeling.

data analysis. Two special cases of compartmental models are described in addition to the diffusion-limited processes.

10.2   Compartmental Modeling

Compartmental modeling is widely used in a broad spectrum of biomedical areas to investigate the distribution of materials in living systems, interpreting data from sophisticated imaging techniques to design biomedical devices. A compartment is any specific part of the system being modeled and always refers to an entity, separated from the rest of the system, which allows controlled exchange of mass, energy, and information with the environment. The basic assumption is that within a compartment, the component under consideration is assumed to be uniformly distributed with variable inputs and outputs and the variations with time are described by ordinary differential equations. For example, the amount of urea in the body can be considered as a compartment. Tagged and untagged urea may represent two different compartments in a tracer study. The number of compartments of each model is based on specific knowledge of the physiological and/or biochemical compartments into which the tracer distributes. Choosing a model is dependent on the information desired, the experimental design, the complexity of the system under study, and theoretical considerations.

10.2.1   Chemical Compartmental Model

When a therapeutic agent is administered, modeling the change in its concentration over time as a substance moves from one area to another within the body is important to develop effective doses while minimizing side reactions. Consider a tank (Figure 10.2) that contains Ψ_D volume of water. Into the tank, a drug of mass m is dissolved and the initial concentration is C_0 at time zero. Within the tank, the basic assumptions are that it is well-mixed and homogeneous and all particles have the same probability of leaving the tank. Into the tank, pure water flows in at a volumetric flow rate of Q_0 (typical units is L/min) and leaves the tank at the same flow rate. While leaving the tank, it carries some of the drug. To understand how much

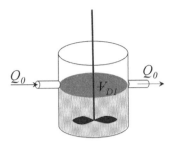

Figure 10.2 Flow of water through a tank filled with drug-containing water.

of a drug is remaining in the tank after certain time, one can use the conservation of mass principle. The general form is

$$\text{Input [I]} - \text{Output [O]} + \text{Generation [G]} = \text{Accumulation [A]} \qquad (10.1)$$

Generation or disappearance is due to a reaction occurring in the system where a product is generated or a reactant is consumed. There is no reaction occurring in the tank. However, concentration of the drug decreases due to that carried by water leaving the tank. Let $C(t)$ be the concentration of the drug at any time, and ΔC is the concentration of the drug carried by the water leaving at time Δt. Then,

$$Q_o * 0 - Q_o C(t)\Delta t = V_D \left[\left(C(t) + \Delta C(t) \right) - C(t) \right]$$

To obtain a continuous function, Δt has to be very small or approaches zero. From the definition of differential calculus, the above equation can be simplified and written as

$$\frac{dC(t)}{dt} = -\frac{Q_o}{V_D} C(t) \qquad (10.2)$$

An assumption is that the volume of water does not change during the time of study. The ratio of Q_0 to V_D is termed as a rate constant and represented by k with units of time^{-1}.

$$\frac{dC(t)}{dt} = -kC(t) \qquad (10.3)$$

Rearranging and integrating with the initial condition (i.e., at time =zero), $C(t)$ = C_0 to obtain

$$\ln C(t) - \ln C_0 = -kt \text{ or } C(t) = C_0 e^{-kt} \qquad (10.4)$$

Thus, knowing the rate constant, the concentration of the drug within the tank at any time can be determined. This forms the basis for a single compartmental model where the compartment is the tank. The half-life is found from the time required to reach 50% of the final value, similar to radioactivity (discussed in Chapter 8). The solution is used to solve for t as

$$\frac{C_0}{2} = C_0 e^{-kt_{1/2}} \text{ or } t_{1/2} = \frac{\ln 2}{k} \tag{10.5}$$

Thus, the definition of the half-life does not depend on the injection route of the drug. The half-life is found from the same formula as that used for a simple injection model. The same analogy can be extended to multiple tanks in series. Consider a system consisting (Figure 10.3) of two tanks containing with V_{D1}, and V_{D2} volume of water. They also have the same drug at concentration $C_{1,0}$ and $C_{2,0}$. Pure water flows at the rate Q_0 (typical units is L/min) enters the first tank and is mixed, then flows at the rate Q_0 into the second tank and is mixed and finally flows out of the third tank at the rate Q_0. The rate constants for the three reactors are defined as $k_1 = \frac{V_{D1}}{Q_o}, k_2 = \frac{V_{D2}}{Q_o}$, and $k_3 \frac{V_{D3}}{Q_o}$. Conservation of mass principle is used to arrive at the differential equation for the system as

$$\frac{dC_1(t)}{dt} = -k_1 C_1(t) \tag{10.6}$$

$$\frac{dC_2(t)}{dt} = k_1 C_1(t) - k_2 C_2(t) \tag{10.7}$$

Equation (10.6) is integrated to obtain $C_{1(t)} = C_{1,0} e^{-k_1 t}$ and then substituted into (10.7) to obtain

$$\frac{dC_2(t)}{dt} = k_1 C_{1,0} e^{-k_1 t} - k_2 C_2(t) \tag{10.8}$$

Using (10.8), the two tanks in series can be considered as one tank with an input of exponentially decaying drug concentration (an example explained later).

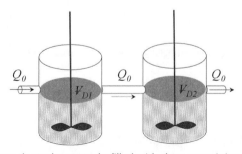

Figure 10.3 Flow of water through two tanks filled with drug-containing water.

A model with a number of these reactors with different interconnections can be similarly developed. Each tank is considered as a compartment (Figure 10.3). The interconnections represent transfer of material (mass/time) which physiologically represents transport from one location to another, chemical transformation, or both. However, it is difficult to have physiological/anatomical correlates for every scenario. For example, if one were to use a single compartmental model for clearance of a component in the kidney, the solution is identical to the single-tank example. However, the physiological phenomenon is glomerular filtration not dilution within the tank. Nevertheless, the amount of a component absorbed per unit time can be calculated within the model; alternatively, the change in chemical concentration in blood or the tissue representing the port of entry may be simulated using appropriate equations.

There are various ways to classify compartmental models. However, typically there can be either mechanistic models or physiological models. In physiologically based compartmental models, compartments are described, as far as possible, with respect to the actual anatomy and physiology of the test animal. Typically, physiological systems are more complex and need more measurements to validate the model. Very often analytical solutions are not possible and need computer programs to solve compartmental systems.

With the mechanistic compartmental models, concentration in tissue versus time is determined by experiment and then described by the equations associated with a particular model. For example, the body is described as consisting of two compartments with appropriate intercompartmental transfer rate constants. The model parameters, including the volumes of the compartments and the transfer rate constants, are estimated by fitting the model to the data hence the term, data-based compartmental models. The model parameters bear no direct correspondence with actual tissue volumes or blood flow rates in the test animals.

10.2.1.1 Apparent Volume of Distribution

In the human body, determining the volume of the compartment is a significant challenge. For example, when a therapeutic agent is administered to a 70-kg patient, one could assume 3L of plasma to be the volume of the compartment. However, not all of the plasma is available for dissolution of the therapeutic agent as much of it could be circulating in other parts of the body. Further, the therapeutic agent could bind to blood components and/or enters the various tissue compartments where it may be dissolved in lipid and water components and be bound to tissue macromolecules. An approach taken is determining the initial concentration (C_0) of the drug at time zero and then estimating the apparent volume of distribution of the therapeutic agent using the dosage (in mass), given by the relation

$$V_D = \frac{Dose[\text{mass units}]}{C_0}$$

As V_D relates the total amount of drug to the plasma concentration, it is also useful in understanding the process of drug distribution in the body. Another terminology used is total clearance (Q_{cl}), which is the volume of plasma completely cleared of a drug per unit time by all routes of the mechanism. Two major routes of clearance mechanisms are excretion (for example, in the kidney) and metabolism (for example, in the liver). If the intrinsic capacity of an organ to clear a drug is high and exceeds plasma flow to that organ, then the clearance equals plasma flow and is altered by changes in plasma flow. *Renal clearance* (or glomerular filtration) is an excretory route for nonvolatile drugs and metabolites. Renal clearance (Q_{RC}) is the volume of plasma completely cleared of the drug per unit time and given by

$$Q_{RC}[\text{volume/time}] = \frac{\text{excretion rate in urine [mass/time]}}{\text{plasma concentration [mass/volume]}} \tag{10.9}$$

A method to determine Q_{RC} of a drug is to measure the drug excreted in the urine between a time interval. In the midpoint of this time interval, concentration of the drug is determined. Mechanism of renal excretion is inferred by comparison of Q_{RC} to that of an indicator of glomerular filtration (creatinine). For example, normal tubular secretion in a 70-kg person is 120 mL/min and less than that indicates net reabsorption in the absence of plasma binding. However, a number of factors such as extent of plasma binding, renal disease (creatinine clearance or its estimate from serum creatinine provides a useful clinical indicator of impaired renal function and is proportional to drug renal clearance), and urinary pH (reabsorption of drug with ionizable group is dependent on urinary pH; raising the pH promotes excretion of acids, impairs excretion of bases) affect drug distribution. Equation (10.9) is also useful to determine values experimentally for other terminal regions such as bile, or exhaled air in which the amount of therapeutic agent is assessed and correlated to the drug concentration in the plasma.

Substituting the new definitions of V_D and Q_{cl} into (10.5), plasma half-life of a therapeutic agent is obtained as

$$t_{1/2} = \frac{V_D \ln 2}{Q_{cl}} \tag{10.10}$$

Hence, the plasma half-life is directly proportional to the apparent volume distribution and inversely proportional to total clearance. For a given volume of distribution, shorter half-lives are expected for higher clearance rates. For practical purposes of designing various therapeutic agents, $t_{1/2}$ is used as an indicator to understand the total clearance and volume of distribution.

EXAMPLE 10.1

A company has developed a contrast agent useful in magnetic resonance imaging. When 500 mg of the agent is administered, blood concentration is determined to be 31.25 mg/L. Within 30 minutes the concentration decreased to half the initial value. Calculate the apparent volume of distribution and the clearance rate of the contrast agent.

$$V_D = \frac{Dose[\text{mass units}]}{C_0} = \frac{500[\text{mg}]}{31.25[\text{mg/L}]} = 16L$$

$t_{1/2}$= 30 minutes
Rearranging (10.10)

$$Q_{cl} = \frac{V_D \ln 2}{t_{1/2}} = \frac{16[L]\ln 2}{30[\text{min}]} = 0.370 \text{ L/min}$$

Consider the case where a constant intravenous infusion is used in a single compartmental model. Then, using the conservation of mass for the system, one can write

$$\frac{dC}{dt} = k_0 - k_1 C \tag{10.11}$$

where $C(t)$ is the drug concentration in the compartment, k_0 is the constant infusion rate, and k_1 is the drug clearance rate in the first compartment. If a steady state concentration is necessary, then the time-dependent term is zero (i.e., $k_0 - k_1 C = 0$). There is difference in the units k_0 [in mg/L.hr] and k_1 [hr^{-1}]. Integrating (10.10) and using the initial condition $C_0(t = 0) = 0$ as there is no drug in the any portion of the body in the beginning,

$$C(t) = \frac{k_0}{k_1}\left[1 - e^{-k_1 t}\right] \tag{10.12}$$

EXAMPLE 10.2

To treat a certain illness, researchers found out the requirements are a constant drug concentration of 5 mg/L in blood. Initial assessment indicated that the drug is distributed in an apparent volume of 10L and that a single-compartment description is sufficient to model the behavior of the drug. The drug is cleared following a first-order (degradation) rate constant of $k_1 = 0.2$ hr^{-1}.
   (a) If the physician decides to give an intravenous infusion at a constant rate, what rate of infusion is necessary rate (in mg/hr) is needed to achieve this steady-state level?
   (b) How long will it take to reach a drug concentration of 2.5 mg/L?
   (c) What is the half-life of the drug?

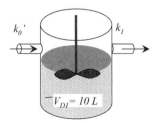

Solution:
   (a) Then, the infusion rate is

$$k_0 = 5(mg/L)*0.2[hr^{-1}] = 1\, mg/hL$$

Since the apparent volume distribution is know to be 10L, infusion rate in mg/hr is

$$1[mg/hr.L]*10[L] = 10\, mg/hr$$

   (b) $C(t) = 2.5$ g/L, $k_0 = 5$ mg/hr.L, $k_1 = 0.2$ hr^{-1} and $t = ?$
   Using (10.10),

$$2.5 = \frac{10}{0.2}\left[1-e^{-0.2t}\right] \text{ or } t = 0.256\, hr$$

   (c) From (10.5),

$$t_{1/2} = \frac{\ln(1/2)}{-k_1} = \frac{0.693}{-k_1} = 0.693/0.2 = 0.256\, hr$$

10.2.2 Single Electrical Compartmental Model

The basic requirement of movement of molecules from one region to another is the concentration difference. The basic requirement for electrical current is the voltage difference. Hence, the concept of compartmental modeling is used to understand the transport of electrical currents in the body. In electrical compartments, an assumption is that each compartment is considered to be isopotential (i.e., at the same potential), similar to concentration uniformity in a single compartment. Each compartment is then equivalent to a simple RC circuit (Section 3.3.3). Using conservation principles, (3.21) is obtained for a single compartment. Rearranging (3.21) in the form

$$\Delta\Phi - \Delta\Phi_{ss} = \left(\Delta\Phi_0 - \Delta\Phi_{ss}\right)e^{-\frac{(t_0-t)}{\tau}} \tag{10.13}$$

shows this is analogous to (10.4) with voltage terms replacing the concentration terms. The time constant is analogous to inverse of rate constant. τ is the time taken for the voltage to decay to $1/e$ or 37% of its initial value in an isopotential cell.

EXAMPLE 10.3

Limbut et al. [1] reported the development of an amperimetric biosensors for measuring endotoxins. The concept is based on the changes in the capacitance of the sensing element after attachment of the endotoxin and one-compartmental model can be assumed. A 50-mV pulse is applied and a voltage of 30 mV was measured after 10 ms. If the resistance is 0.1 MΩ, what is the capacitance of the circuit? If capacitance [nF/cm^{-2}] is related to the concentration (in M) of the endotoxin by 18log Con+1,000 and available area is 2 cm^2, what is the concentration of the endotoxin?

Solution: From (10.13),

$$30 = (50)e^{-\frac{(10)}{\tau}}$$

Then τ = 19.572 ms = $R_m C_m$
Hence, C_m = 1.957 μF or 979 nF/cm^2
Then concentration = Exp(979–1,000)/18 = 0.3M

Synapses can be thought of as electrical connections. The main equation for internal stimulation of the soma or any other compartment where current flow to other processes or neighbored compartments is prevented has always the same form: One part of the stimulating current is used to load the capacity C_m of the cell membrane and the other part passes through the ion channels; that is,

$$I_{stimulus} = C_m \frac{d\Delta\Phi}{dt} + I_{ion} \tag{10.14}$$

where the ion currents I_{ion} are calculated from appropriate membrane models. The rate of change of membrane voltage change, $d\Delta\Phi/dt$, follows as:

$$\frac{d\Delta\Phi}{dt} = \frac{\left(I_{stimulus} - I_{ion}\right)}{C_m}$$

For multiple ions, one can utilize the Hodgkin-Huxley model, which leads to the circuit shown in Figure 3.5(c). Since the time of development of Hodgkin and Huxley model, this model formalism has been applied to a large number of excitable cells. This naturally leads to developing methods to estimate the model parameters. Using the Nernst equation, one can couple the changes in concentration to electrical potential.

10.2.3   Other Single Compartmental Systems

As discussed before, concentration difference is the driving force for movement of molecules and voltage difference is the driving force for electric current. Similarly, temperature difference is the driving force for heat transfer, and pressure difference is the driving force for fluid flow. A fluid moves through a conduit due to the pressure difference (ΔP) that overcomes the attrition after the flow (Q). A battery or other power supply develops a "potential difference" that moves electrons around the circuit to a position of least energy. Thus, a fluid pressure drop (energy per unit volume) corresponds to a potential difference or voltage drop (energy per unit charge). Understanding the analogy between electrical, mechanical, and chemical systems helps in adopting an approach that emphasizes the similarity of modeling principles and techniques for various systems. For example, fluid conservation principles are also applicable to electrical circuits as described in Chapter 4. A water tank stores energy in the form of water raised to a particular height. Similarly, a capacitor stores energy in the form of electrons at a particular voltage. The larger the capacitance, the more energy must be stored to achieve a desired voltage. Defibrillator storage capacitor employs a very thin plastic film dielectric with metallic layer deposited on one side, which acts as the plates for the capacitor. In the cardiovascular system, the pressure drop from the ventricle to the atrium is equal to the pressure at the ventricle.

An analogy between an electrical system and a mechanical system is the resistance of hydraulic systems, similar to the resistance of electrical systems. If velocity is analogous to voltage, force will be analogous to current.

Mobility = Velocity/Force is analogous to Impedance = Voltage/Current

Any lumped damped mass-spring vibration (viscoelastic models described in Chapter 5) system is turned into a lumped resitance-impedence-capacitance system and is solved by a circuit analysis. The advantage of transforming it into a circuit is to use the advanced tools developed for circuit analysis. Thus using these analogies, one can: (1) capture the essential dynamic behavior of the system, (2) employ a unit system so that quantitative statements can be made, and (3) identify pathways and storage compartments of mass, charge, and energy.

10.2.4 Multicompartmental Models

Compartmental models can accommodate mathematical descriptions for different clearance patterns and entrance conditions. Metabolic elimination of a drug by chemical modification could be a constant mass; that is, a linear decrease in drug concentration, proportional to the concentration still remaining in the compartment (referred as first order kinetics) or enzyme-controlled like many biological processes and could obey the Michaelis-Menten type kinetics(described in Chapter 7). Enzymatic activity is generally high in the liver and may convert the therapeutic agent into a more potent drug or a toxic component.

Therapeutic agents could be administered by different routes such as an instantaneous bolus administration through intravenous injection, a constant rate infusion through intravenous infusion, an exponentially decreasing first order oral administration, a diffusion controlled transdermal route, or a complex process like hepatic metabolism-dependent. If the drug is delivered orally, then the drug will dissolve in the digestive cavity before entering the systemic circulation and localizing in the area of therapeutic need. This scenario is identical to Figure 10.3 with two tanks in series: tank 1 is the digestive track and tank 2 is the systemic circulation. Since the effectiveness of the drug is determined by the amount circulating in the systemic circulation, understanding the time-dependent concentration change in the plasma (or in circulating blood) is necessary. In that case, the changes in the digestive track are lumped into one parameter, k_1 and incorporated as an exponential decay function into the second compartment. Thus the first tank is assumed not to exist (shown with dotted line) in Figure 10.4. In other words, (10.8) can be used

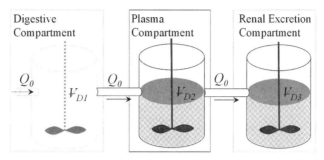

Figure 10.4   Two-compartmental model for the orally delivered drug dynamics in the body.

$$\frac{dC_2}{dt} = C_o e^{-k_1 t} - k_2 C_2 \tag{10.15}$$

where C_0 denotes dosage calculated based on the apparent plasma volume and k_1 represents the drug disintegration rate in the first compartment. Integrating (10.15) with the initial condition of C_2 (time $= 0$) $= 0$,

$$C_2(t) = C_0 \frac{k_1}{k_1 - k_2} \left(e^{-k_2 t} - e^{-k_1 t} \right) \tag{10.16}$$

If the drug is all accumulated in the urine, then an expression can be obtained using

$$\frac{dC_3(t)}{dt} = k_2 C_2 \tag{10.17}$$

Substituting (10.12) into (10.13), and integrating with the initial condition of C_3 (time $= 0$) $= 0$, the following solution is obtained

$$C_3(t) = C_o \left[1 - \frac{k_1}{k_1 - k_2} \left(k_2 e^{-k_2 t} - k_1 e^{-k_1 t} \right) \right] \tag{10.18}$$

EXAMPLE 10.4

A pharmacokinetic problem of importance to many is the body's removal of ingested ethanol. The absorption of ethanol can be described as a first order process with the rate constant $k_a = 1.0$ hr^{-1} for a person of 150 pounds (~68 kg). Ethanol is eliminated by the metabolism in the liver with zero order elimination rate constant of 7 g/hr. Assume $V_D = 40$L and all of the ethanol is absorbed into the blood stream.
  (a) If the person ingested several drinks containing a total dose of 100g ethanol, how long does it take to reach the maximum plasma concentration, and what is its value in g/L?
  (b) When does the concentration fall below a level of 0.6 g/L?
  (c) How does a Breathalyzer assess this?

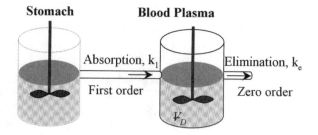

Solution:

(a) Elimination rate constant is given in g/hr and converting to concentration per hour using the apparent volume of distribution

$$k_e = \frac{7[g/hr]}{40[L]} = 0.175 \, g/L.hr$$

$$C_0 = \frac{100[g]}{40[L]} = 2.5 \, g/L$$

Based on (10.15),

$$\frac{dC}{dt} = k_1 C_0 e^{-k_1 t} - k_e \qquad\qquad (E10.1)$$

Integrating with the boundary conditions $C(t = 0) = 0$ and $C(t = t) = C$

$$C(t) = C_0(1 - e^{-k_1 t}) - k_e t \qquad\qquad (E10.2)$$

Maximum is when

$$\frac{dC}{dt} = 0 = \left(k_1 C_0 e^{-k_1 t_{max}} - k_e \right)$$

Rearranging

$$t_{max} = \frac{1}{k_1} \ln\left(\frac{k_a C_0}{k_{e0}} \right) \qquad\qquad (E10.3)$$

Hence,

$$t_{max} = \frac{1}{1[h^{-1}]} \ln\left(\frac{1[h^{-1}]2.5[g/L]}{0.175[g/hr]} \right) = 2.660 \text{ hr}$$

From (E10.2),

$$C_{max} = \left(C_0(1 - e^{-k_1 t_{max}}) - k_e t_{max} \right) = \left(2.5(1 - e^{-1*2.660}) - 0.175 * 2.660 \right) = 1.860 \text{ g/L}$$

(b) Substituting all the values into (E10.2),

$$0.6[g/L] = 2.5(1 - e^{-t}) - 0.175t$$
$$t = 10.8567 \text{ hr}$$

(c) Exhaled vapor is in equilibrium with the body fluids.
Hence, by Henry's law (Chapter 2), $P_B = k_B X_B$, which can convert gas P_B to liquid content.

One compartment model is often used to predict the basic parameters such as clearance rate and apparent volume of distribution. However, the assumption is that there is immediate distribution and equilibrium of the therapeutic agent throughout the body, which may not be possible in many scenarios. If distribution is minimal, the one compartment can be an adequate approximation. If not, the model is altered to either better fit the data or to a more physiological condition. For example, when a positron emitting tracer is injected for a PET scan, understanding the concentration changes with the tissue is important to assess the utility of the tracer. In these scenarios, the human body could be treated as a plasma compartment (also referred as central compartment) and a tissue compartment (also referred as peripheral compartment). Although these compartments do not necessarily have a physiological significance, commonly plasma compartment includes blood and well-perfused organs such as liver and kidney; tissue compartment includes poorly perfused tissues such as muscle, lean tissue, and fat.

For a bolus intravenous injection directly into the plasma (Figure 10.4), conservation of mass principle can be used to arrive at the differential equation for the system as

$$\Psi_{DP} \frac{dC_P(t)}{dt} = k_T \Psi_{DT} C_T - \left(k_P - k_E\right) \Psi_{DP} C_P \tag{10.19}$$

$$\Psi_{DT} \frac{dC_T(t)}{dt} = \left(k_P\right) \Psi_D C_p - k_T \Psi_{DT} C_T \tag{10.20}$$

where Ψ_{DP} is the apparent volume of distribution in the plasma compartment, Ψ_{DT} is the apparent volume of distribution in the tissue compartment, C_p is the concentration of the tracer in the plasma, C_T is the concentration of the tracer in the plasma, k_P, k_T, and k_E are rate constants as shown in Figure 10.5. The initial conditions are $C_T(0) = 0$, and $C_p(0) = D_o/\Psi_p$. Typically, these set of differential equations can be solved using numerical integration techniques. Nevertheless, solving the above two differential equations (using the eigenvalue approach), the solutions are:

$$C_P(t) = \frac{C_{P0}}{A - B}\left[\left(k_T - B\right)e^{-Bt} - \left(k_T - A\right)e^{-At}\right] \tag{10.21}$$

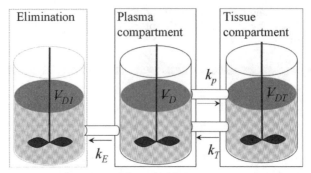

Figure 10.5 Two-compartment system with elimination in the first compartment.

$$C_T(t) = \frac{C_{P0}}{A-B}\left[e^{-Bt} - e^{-At}\right] \tag{10.22}$$

where

$$A = \frac{1}{2}\sqrt{\left(k_p + k_T + k_e\right) + \left[\left(k_p + k_T + k_e\right)^2 - 4k_T k_e\right]}$$

and

$$B = \frac{1}{2}\sqrt{\left(k_p + k_T + k_e\right) - \left[\left(k_p + k_T + k_e\right)^2 - 4k_T k_e\right]}$$

Equation (10.21) is a four-parameter (k_T, A, B) and the lumped $C_{p0} = \dfrac{Dose}{V_D(A-B)}$ model.

Other parameters can be obtained through:

$$k_e = \frac{AB}{k_T} \tag{10.23}$$

$$k_p = (A+B) - (k_T + k_e) \tag{10.24}$$

$$V_D = \frac{Dose}{C_{P0}(A-B)} \tag{10.25}$$

and

$$\mathcal{V}_T = \mathcal{V}_P \frac{k_P}{k_T} \tag{10.26}$$

In a single-compartment model there is a single $\mathcal{V}_D$ term, since the assumption is immediate distribution and equilibrium. Thus, the amount of drug in the body may be at time t, C_{pt} by a similar equation, $\mathcal{V}_D$ is the same whether it is the initial concentration at $t = 0$ or at later time points. For a drug that requires some time for distribution through the body and follows a two-compartment model, this is not the case. There will be different $\mathcal{V}_D$ depending on time. For example, the PET tracer concentration would initially peak in the plasma compartment and later peak in the tissue compartment. If these two compartments (plasma and tissue) are observed continuously over time, then kinetic concentration curves can be obtained. Distribution into the tissue compartment continues until the free concentration in the plasma compartment is equal to the free concentration in the tissue compartment (i.e., there is a net flow of drug out of the plasma). Thus, the concentration decreases rapidly at first and steady-state is not maintained. After steady-state, a concentration gradient is again created but in the opposite direction due to the continual elimination of drug from the plasma compartment. In response to this, drug begins to flow back into the central compartment where it is eliminated. Thus, in the second-phase, the concentration of drug in the tissue compartment is greater than that in the plasma compartment. The concentrations in both compartments decrease proportionally as elimination from the plasma continues.

10.2.4.1   Modeling Based on the Experimental Data

Experimentally, one could monitor the changes in the plasma concentration as a function of time only [Figure 10.6(a)]. With these results, obtaining all the parameters is not possible. For example, (10.9) is not useful in tissues from which drug concentration cannot be assessed repetitively, for example liver, and other methods

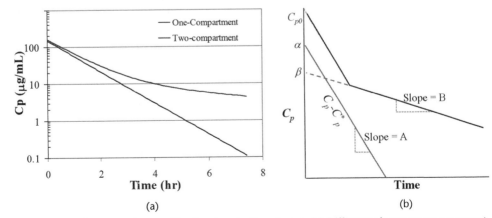

Figure 10.6 Concentration profile of a drug in the plasma. (a) Difference between one-compartment and two-compartment profiles. (b) Estimating parameters in a two-compartment model after intravenous administration.

of assessment are necessary. In these cases, both numerator and denominator terms in (10.9) are integrated with respect to time to obtain total clearance, Q_{Tcle}

$$Q_{Tcle}[\text{volume/time}] = \frac{\int_0^\infty \text{excretion rate in urine [mass/time] } dt}{\int_0^\infty \text{plasma concentration [mass/volume] } dt} = \frac{D}{AUC_\infty} \quad (10.27)$$

where D is the intravenous dose given and AUC_∞ is the total area under the curve in the C_p versus t profile. This may not be true for other routes of delivery. Nevertheless, knowing all the parameters is not relevant in many applications as the main focus is to obtain information on the distribution and elimination of a therapeutic agent to facilitate the formulation of optimum dosing guidelines. Hence, a general form of (10.21) or (10.22) is used

$$C_p(t) = \alpha e^{-At} + \beta e^{-Bt} \quad (10.28)$$

where α and β are lumped constants. Equation (10.28) is used to fit the data and understand few parameters such as apparent volume of distribution, volume of distribution within the tissue, and clearance rate. Depending upon the magnitude of A relative to B, when t becomes large and consequently the e^{-At} in (10.28) becomes negligible [Figure 10.6(b)], α reduces progressively until it reaches zero. Then (10.28) reduces to

$$C_p(t) = \beta e^{-Bt} \quad (10.29)$$

One could assume that the drug concentration between the plasma and tissue compartment have reached a pseudoequilibrium phase. Plot of lnC_p versus *time* will be a straight line and from the y-intercept, the β value can be determined. Further on, the slope of the line is B. Using these values and (10.29), one can calculate various $C_p(t)$ values for different time points. The new $C_p(t)$ is represented as $C^*_p(t)$ to separate from original $C_p(t)$ values. Then one can plot $ln(C_p - C^*_p)$ with various time points. One can determine the α value from the y-intercept of this line and A from the slope of the line. This concept is the basis of "curve striping" (also referred to as feathering or the method of residual), commonly used for the identification of compartmental models. The same equation is also used in many scenarios if a plot of lnC_p versus *time* appears similar to Figure 10.6(b).

10.3   Special Cases of Compartmental Modeling

10.3.1   Modeling Dialysis

Kidney function in patients with partial or complete kidney (renal) failure is insufficient to adequately remove fluid (water) intake (resulting in the retention of fluids that is characterized as edema) and adequately remove excess electrolytes (salts) and waste metabolites generated through ordinary digestive and metabolic processes. As a result, the concentrations of such metabolites will increase to toxic levels unless a device is introduced to help remove them. The major source of waste metabolites is the liver since this is where most of the energy conversion processes in the body take place. The end products of carbohydrate and fat metabolism tend to be water and carbon dioxide, both of which are removed from the body through respiration. The end products of protein metabolism, however, are generally eliminated through the kidneys. Urea is the largest mass of waste metabolite produced in the liver from protein metabolism.

Dialysis is based upon diffusion of a substance such as urea from a region of high concentration (blood) to a region of low concentration (dialysate). Two different models are developed to understand the clearance characteristics: (1) describing the dialysis treatment based on the production and clearance of urea to understand the patient characteristics, and (2) evaluating the performance characteristics of the dialyzers. To understand the urea kinetics within the body, a two-compartment model for generation and elimination of urea in the body (Figure 10.7) is an approach. Urea is generated in the liver and passed into the bloodstream (plasma compartment). The bloodstream is considered a single "compartment" in the body that acts as a reservoir for urea. Urea also distributes intracellularly in some cell types, which is grouped into the tissue compartment. Urea can be removed from the body by the kidney, eliminated with urine. If the kidneys are not functioning, urea is eliminated by dialysis and is lumped as one parameter of urea clearance. Differential equations are written similar to the multicompartmental model described in Section 10.2.4 and production and clearance rates are approximated.

For evaluating the performance of the dialyzers, modeling is also focused on evaluating the clearance rate of urea (or other components such as creatinine, vitamin B_{12}, and β-2-microglobulin). During hemodialysis, blood is perfused from the body and passed through a dialyzer before returning to the body (discussed in Chapter 1). Hollow fiber dialyzers consist of thousands of hollow fibers contained in a shell. Unlike the shell, which is made of impermeable material, the hollow fibers are permeable to the passage of low molecular weight molecules across the fiber. Blood flows through the center (lumen side) of the fibers. Dialysate fluid flows on the exterior of the fibers (shell side). Dialysate is a fluid that contains electrolyte concentrations similar to that found in a healthy person's blood. The difference in

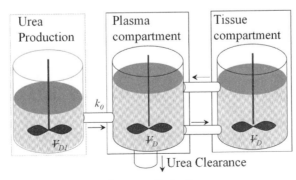

Figure 10.7   Two-compartment system for urea elimination.

waste metabolite concentration in the blood and the dialysate creates a flux of the metabolite from the blood to the dialysate. Electrolytes are not removed because the dialysate and blood have similar electrolyte concentrations. In this model, two compartments are considered (Figure 10.8).

The *blood compartment* is assumed to have a given volume (V_B). To predict the changes in the blood (or plasma) urea concentration (C_{BU}) with time during dialysis, the plasma urea change is written, similar to compartments before as

$$\frac{d(V_B C_{BU})}{dt} = Q_{B,out} C_{B,out} - Q_{B,in} C_B$$

where $Q_{B,in}$ and $Q_{B,out}$ are the volumetric inlet and outlet flow rates of blood through the dialyzer, respectively, and $C_{BU,\ out}$ is the "blood" urea concentration exiting the dialyzer (and entering the CST). For the blood compartment, the inlet urea concentration to the dialyzer is the same as the outlet concentration of the CST ($C_{BU,in} = C_{BU}$). Unlike other two-compartmental models, a certain volume of water is removed in the urine. Hence volume of blood is a function of time and cannot be treated as a constant. Using the chain rule,

$$V_B \frac{dC_{BU}}{dt} + C_{BU} \frac{dV_B}{dt} = Q_{B,out} C_{BU,out} - Q_{B,in} C_{BU} \tag{10.30}$$

For the changing blood volume, material balance with constant density is represented by:

$$\frac{dV_B}{dt} = Q_{B,out} - Q_{B,in} \tag{10.31}$$

The substitution of (10.30) into (10.31) and simplification gives:

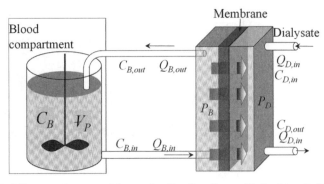

Figure 10.8 Modeling countercurrent (opposite direction flow with the blood) dialysis unit.

$$\frac{dC_{BU}}{dt} = \frac{Q_{B,out}}{V_B}\left(C_{\cdots\cdots} - C\right) \tag{10.32}$$

Both $C_{BU,out}$ and $Q_{B,out}$ are interrelated and (10.31) and (10.32) must be solved simultaneously. Further, to solve (10.31) and (10.32), it is important to know how $C_{BU,out}$ and $Q_{BU,out}$ are related to dialyzer inlet conditions.

10.3.1.1   Dialyzer Compartment

The rate of transfer is governed by the concentration difference across the membrane, the molecular size, and the permeability characteristics of the membrane. The dialyzer consists of two channels for blood and dialysate phases, which are separated by a microporous membrane through which the solute diffuses perpendicularly to its exposed surface area. For a single pass through the dialyzer, the solute transferred from the blood across the membrane over a differential membrane transport area ΔA can be written as

$$Q_B C_{B,i}\big|_A - Q_B C_{B,i}\big|_{A+\Delta A} = K_i \Delta A\left(C_{B,i} - C_{D,i}\right)$$

where $C_{B,i}$ is the concentration of solute i in the blood, $C_{D,i}$ is the concentration of solute i in the dialysate, and K_i is the mass transfer coefficient (units of length per time) describing diffusive transport across the membrane. Rearranging,

$$-\frac{\left[Q_B C_{B,i}\big|_{A+\Delta A} - Q_B C_{B,i}\big|_A\right]}{\Delta A} = K_i\left(C_{B,i} - C_{D,i}\right)$$

To obtain a continuous function, assume ΔA is very small or approaches zero. From the definition of differential calculus, one can rewrite the above equation as

$$-\frac{d\left(Q_B C_{B,i}\right)}{dA} = K_i\left(C_{B,i} - C_{D,i}\right) \text{ or}$$

$$-Q_B \frac{dC_{B,i}}{dA} - C_{B,i}\frac{dQ_B}{dA} = K_i\left(C_{B,i} - C_{D,i}\right) \tag{10.33}$$

Similarly, the solute gained (hence, the opposite sign of losing in blood) by the dialysate can be written as

$$Q_D \frac{dC_{D,i}}{dA} + C_{D,i}\frac{dQ_D}{dA} = K_i\left(C_{B,i} - C_{D,i}\right) \tag{10.34}$$

The blood and dialysate flow are written using Starling's hypothesis and (2.25) as

$$\frac{-dQ_B}{dA} = \frac{-dQ_D}{dA} = L_p \left(\Delta P - RT \sum_i \sigma_i \left(C_{B,i} - C_{D,i} \right) \right) \qquad (10.35)$$

where L_p is the hydraulic conductance and σ_i represents the Staverman reflection coefficient for solute i. For a single solute of urea, (10.35) reduces to

$$-\frac{dQ_B}{dA} = -\frac{dQ_D}{dA} = L_p \left[\Delta P - RT\sigma_C \left(C_{BU} - C_{DU} \right) \right] = K' \qquad (10.36)$$

Substituting (10.36) into (10.33) and rearranging

$$Q_B \frac{dC_{BU}}{dA} = \left[K' - K_U \right] C_{BU} + K_U C_{DU} \qquad (10.37)$$

Similarly,

$$Q_D \frac{dC_{DU}}{dA} = K_U \left(C_{BU} - C_{DU} \right) - K' C_{DU} \qquad (10.38)$$

Based on the relative flow direction of each of the two fluids, blood and dialysate, two scenarios are possible: flow may be cocurrent (same direction) or countercurrent (opposite direction). For analysis involving cocurrent dialysate flow, (10.36) and (10.38) have opposite signs. One could solve (10.32) and (10.36) through (10.38) using numerical integration technique to obtain the loss of urea and water as a function of time. The simplest solution for predicting urea loss as a function of time is when the assumptions of constant flow rates ($K' \approx 0$ and/or K' $\ll K_C$) are valid, such that $Q_{B,out} = Q_{B,in}$. Integrating (10.33) and (10.34) results in

$$C_{BU,out} = C_{BU,in}(1 - E) + C_{DU,in}E \qquad (10.39)$$

where E is the extraction ratio given by

$$E = \frac{1 - \exp\left[-N_T \left(1 + z \right) \right]}{\left(1 + z \right)} \quad \text{cocurrent flow} \qquad (10.40)$$

$$E = \frac{\exp\left[N_T \left(1 - z \right) \right] - 1}{\exp\left[N_T \left(1 - z \right) \right] - z} \quad \text{countercurrent flow} \qquad (10.41)$$

where $N_T = \dfrac{K_U A}{v_{B,in}}$ is the number of transfer units and $z = \dfrac{Q_{B,in}}{Q_{D,in}}$. Parameters E and N_T provide the performance parameters for the dialyzer. Value of E varies from 0 to 1 and depends on the blood flow rate, and dialysate flow rate on a day-to day-operation. It is used to monitor the patient along with clearance rate or adjust the flow rate of the dialysate. The extraction ratio will decrease if the blood flow rate is increased. At any given blood flow rate, a dialysate flow rate increase will increase the extraction ratio, and thereby also increase the clearance.

EXAMPLE 10.5

A clinical engineer has two dialyzer units. One is a small, with $K_U A$ value of 400 mL/min and the other with a $K_0 A$ value of 1,000 mL/min. Two patients (v_b= 200 mL/min) are connected with these dialyzers in countercurrent mode with a dialysis rate of 500 mL/min.
- (a) Which one has higher efficiency of extraction?
- (b) If the blood urea level of a patient is 100 mg/dL and dialysate inlet urea content is zero, what is the urea concentration of blood exiting the dialyzer in a single pass?
- (c) The blood volume of that patient is 4L. If the second dialyzer is used, how much time will the patient save sitting in a dialysis center for 90% clearance of urea.

Solution:

(a) $z = \dfrac{200[\text{mL/min}]}{500[\text{mL/min}]} = 0.4; N_{T1} = \dfrac{400[\text{mL/min}]}{200[\text{mL/min}]} = 2; N_{T2} = \dfrac{1,000[\text{mL/min}]}{200[\text{mL/min}]} = 5$

For countercurrent flow, extraction ratio for

The first dialyzer, $E_1 = \dfrac{\exp\left[2(1-0.4)\right]-1}{\exp\left[2(1-0.4)\right]-0.4} = 0.794$

The second dialyzer, $E_2 = \dfrac{\exp\left[5(1-0.4)\right]-1}{\exp\left[5(1-0.4)\right]-0.4} = 0.970$

Since E_2 is significantly higher than E_1, it has a better extraction efficiency.

(b) From (10.39), $C_{BU,out} = C_{BU,in}(1-E)+C_{DU,in}E$

$C_{D,in}= 0$ and $C_{b,in}= 100$ mg/dL.
Hence in the first dialyzer, $C_{BU,out} = C_{BU,in}(1-0.794) = 20.6$ mg/dL

In the second dialyzer, $C_{B,out} = 100(1-0.970) = 3$ mg/dL

(c) V_B= 4L. $C_{BU,out}$ depends on $C_{BU,in}$, which will change upon reentering the blood. Hence $C_{BU,out}$ should be expressed as a function of $C_{BU,in}$ which is assumed to be C_{BU}.
From above, for the first dialyzer, $C_{BU,out} = C_{BU,in}(1-0.794) = 0.206C_{BU}$
From (10.32) with the assumption that $Q_{B,in}= Q_{B,out}$,

$$\dfrac{dC_{BU}}{dt} = \dfrac{200[\text{mL/min}]}{4,000[\text{mL}]}(0.206C_{BU} - C_{BU}) \rightarrow \dfrac{dC_{BU}}{dt} = -0.0397[\text{min}^{-1}]C_{BU}$$

Integrating and using the limits of $C_{BU}(t = 0) = 100$ mg/dL, $C_{BU}(t = t)$ 10 mg/dL.
$t = 58$ minutes for the first dialyzer.
For the second dialyzer, $C_{BU,out} = 0.03C_{BU}$ and $t = 47.476$ minutes
Time saved $= 58 - 47.476 = 11.524$ minutes.

10.3.2 Cable Theory

As discussed in Chapter 3, the excitable membranes of certain cells have ion channels whose gating is typically voltage-dependent. The dynamic interactions between these ionic channels play a dominant role in the function of the cell. To understand the transmission of signals and changes in membrane potentials, a model describing the electrical properties of the tissue is required. Compartmental models of neurons furnish a means for simulating both the steady-state and the transient electrical activity of a neuron. In the compartmental approach (Figure 10.9), the neuron's volume is partitioned into a finite number of contiguous regions. There could be separate compartments for the cell body, the axon hillock, and several compartments for the axon and each of the dendrites. The membrane is assumed to be a cylinder of uniform radius along its length, although this assumption is not required if resistivities are used while developing the model. A cable model, similar to electrical cables, is used to describe the transmission of signals. A cable is an extended structure consisting of a conducting core bounded by a thin insulating layer or membrane. A neuron is thought of as a cable with particular electrical characteristics. It is often visualized as cylindrical, but the form of the cross section is not

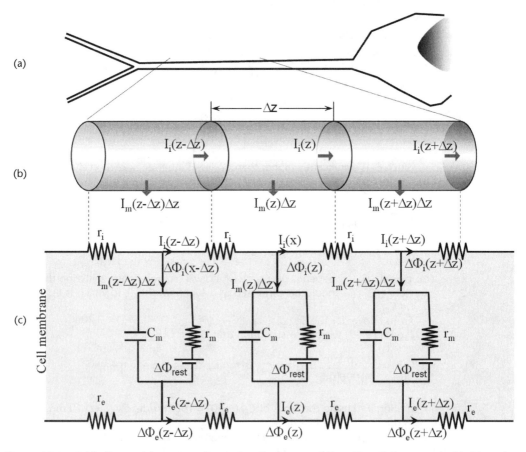

Figure 10.9 Cable theory: (a) portion of an axon of a neuron, (b) portion of the axon devided into three compartments, and (c) electrical equivalent circuit for each compartment.

important if the resistivity of the internal medium is low compared to the specific resistance of the bounding membrane. A small blood vessel can also be treated as a cable. Resistivity of the internal medium is related to the resistance, R_i, by

$$R_i = \frac{\text{Resistivity} \times L}{\pi a^2} \qquad (10.42)$$

where L is the length of the cable and a is the radius of the cable. For simplification of the derivation r_i is defined as resistances per unit length of cylinder, that is,

$$r_i = \frac{R_i}{L} = \frac{\text{Resistivity}}{\pi a^2} \qquad (10.43)$$

Similarly r_e is the external resistance per unit length of the cylinder. Consider a case where a membrane is divided into three compartments of equal length Δz [Figure 10.9(b)]. Each compartment is assumed to be an isopotential patch of membrane. The membrane of each subcylinder is represented by a parallel combination of membrane capacitance $c_m \Delta z$ and a circuit for the ionic conductances in the membrane. The total current through a membrane patch is $I_m(z)\Delta z$ and current flowing in the direction of increasing x is assigned a positive sign. The membrane current varies with distance x down the cylinder. If I_m and c_m are membrane current and capacitance per unit length of the cylinder, then multiplying Δz gives the total current and capacitance in a subcylinder. r_m is the membrane resistance for the length of the cylinder with the units of ohm.length.

The membrane potential inside the cell is represented as $\Delta\Phi_i(z)$ and outside the cell as $\Delta\Phi_e(z)$. Assuming that the potentials also vary with distance down the cylinder, they also become a functions of distance x. The membrane potential is $\Delta\Phi_i(z) - \Delta\Phi_e(z)$. As the potentials vary along the length of the cylinder, there will be currents $I_i(z)$ and $I_e(z)$ flowing between the nodes; $I_i(z)$ is the total current flowing down the interior of the cylinder and $I_e(z)$ is the total current flowing parallel to the cylinder in the extracellular space. In the brain there are many cylinders from different neurons packed together, so there are many extracellular currents. $I_e(z)$ is only the portion of the extracellular current associated with the cylinder under study. The internal current $I_i(z)$ flows through resistance $r_i\Delta z$, which is the resistance of the solutions inside the cylinder between the center of one subcylinder and the center of the next. $r_e\Delta z$ is similarly defined as the resistance in the extracellular space between the center of two subcylinders, that is, as the resistance to the flow of current $I_e(z)$. Ohm's law for current flow in the intracellular and gives: $\Delta\Phi_i(z) - \Delta\Phi_i(z + \Delta z) = I_i(z)r_i\Delta z$.

Rearranging and taking the limit as Δx goes to 0,

$$\frac{\partial \Delta\Phi_i}{\partial z} = -I_i(z)r_i \qquad (10.44)$$

This equation is written as a partial derivative as voltage and current change with time as well as distance. Similarly for the extracellular spaces,

$$\frac{\partial \Delta \Phi_e}{\partial z} = -I_e(z) r_e \qquad (10.45)$$

The conservation of current at the intracellular nodes gives

$$I_i(z - \Delta z) - I_i(z) = I_m(z) \Delta z$$

With the limit as $\Delta x \to 0$,

$$\Delta x \to 0, \quad \frac{\partial I_i}{\partial z} = -I_m(z) \qquad (10.46a)$$

Similarly,

$$\frac{\partial I_e}{\partial z} = I_m(z) \qquad (10.46b)$$

Total axial current is the sum of internal and external currents. Substituting (10.44) and (10.45),

$$-I_T = \frac{1}{r_i} \frac{\partial \Delta \Phi_i}{\partial z} + \frac{1}{r_e} \frac{\partial \Delta \Phi_e}{\partial z}$$

Defining the membrane potential relative to the resting potential as $\Delta \Phi = \Delta \Phi_i - \Delta \Phi_e$

$$-I_T = \frac{r_i + r_e}{r_i - r_e} \frac{\partial \Delta \Phi_i}{\partial z} - \frac{1}{r_e} \frac{\partial \Delta \Phi}{\partial z}$$

Rearranging

$$\frac{1}{r_i} \frac{\partial \Delta \Phi_i}{\partial x} = \frac{r_e}{r_i + r_e} \frac{\partial \Delta \Phi}{\partial x} - \frac{r_e}{r_i + r_e} I_T$$

Substituting (10.46),

$$-I_i = \frac{r_e}{r_i + r_e} \frac{\partial \Delta \Phi}{\partial z} - \frac{r_e}{r_i + r_e} I_T$$

Taking d/dx both sides,

$$-\frac{\partial I_i}{\partial z} = \frac{\partial}{\partial z}\left(\frac{r_e}{r_i + r_e}\frac{\partial \Delta \Phi}{\partial z}\right) - \frac{\partial}{\partial z}\left(\frac{r_e}{r_i + r_e}I_T\right)$$

The second term on the right side is zero due to the conservation of current along the axial direction. Substituting (10.46a),

$$I_m(z) = \frac{\partial}{\partial z}\left(\frac{r_e}{r_i + r_e}\frac{\partial \Delta \Phi}{\partial z}\right) \qquad (10.47)$$

I_m is written as the sum of the ionic current $I_{ion}(x, \Delta\Phi, t)$ through the r_m and the current through the membrane capacitance as

$$I_m(z)\Delta x = I_{ion}(z, \Delta\Phi, t)\Delta z + Cm\Delta z\frac{\partial \Delta \Phi}{\partial t} \qquad (10.48)$$

The ionic current is in general a complex and nonlinear function of membrane potential modeled, for example, by Hodgkin-Huxley type equations. As for I_m, I_{ion} is ionic current per unit length of membrane cylinder. Also, $I_{ion} = I_m - I_e = \frac{\Delta\Phi}{r_m} - I_e$.

Substituting (10.47) into (10.48),

$$\frac{\Delta\Phi}{r_m} - I_e + C_m\frac{\partial \Delta \Phi}{\partial t} = \frac{\partial}{\partial z}\left(\frac{r_e}{r_i + r_e}\frac{\partial \Delta \Phi}{\partial z}\right)$$

Typically the external current is neglected. Rearranging and substituting $\tau_m = r_m c_m$ and $\lambda = \sqrt{\frac{r_m}{r_c + r_i}}$,

$$\tau_m\frac{\partial \Delta \Phi}{\partial t} = \lambda^2\frac{\partial^2 \Delta \Phi}{\partial^2 z} - \Delta\Phi \qquad (10.49)$$

Equation (10.49) is referred to as the cable equation and often external current is used. τ_m is the time constant defined in a single-compartmental model. λ is known as the length constant (or space constant) of the cable and depends on the resistance of cable membrane, the resistivity of the internal medium, and the geometry of the cable cross section. Hence, differences in physical properties (e.g., diameter and membrane properties) and differences in potential occur between compartments rather than within them. These are not always measurable individually. However, the space constant λ is a useful parameter commonly used to quantify the extent to which voltage changes spread; λ is the length of a cable with the same diameter (as the compartment) that has a membrane resistance equal to the cytoplasmic resistance. Knowing the time constant and the length constant, the propagation speed of the impulse is calculated. Consider two synaptic inputs represented by two current steps. The voltage change from the second current step will add to the first current step, if delivered within a short time interval after the end of the first current. If λ is small, then the voltage decay following the first step is quick, and the two inputs do not add. If λ is large, the voltage changes are more likely to add and the voltage reaches threshold for an action potential. Another terminology used is the electrotonic length, which is the ratio of physical length to the space constant.

In steady-state condition (10.49) reduces to

$$0 = \lambda^2 \frac{\partial^2 \Delta\Phi}{\partial^2 x} - \Delta\Phi \text{ or } \lambda^2 \frac{\partial^2 \Delta\Phi}{\partial^2 x} = \Delta\Phi \qquad (10.50)$$

Integration of (10.50) twice will result in

$$\Delta\Phi = Ae^{-\frac{x}{\lambda}} + Be^{-\frac{x}{\lambda}} \qquad (10.51)$$

where A and B are constants.

EXAMPLE 10.6

Under steady-state conditions, the intracellular resistance (r_i) of a nerve cell in the axial direction is 8 MΩ/cm. The resistance of the cell membrane in the radial direction is 2×10^4 Ωcm (r_m) and the capacitance is 12 nF/cm. Calculate the characteristic length and time constant.

Solution: Since the extracellular axial resistance is frequently negligible relative to intracellular axial resistance

$$\lambda = \sqrt{\frac{r_m}{r_i}} \quad \lambda = \sqrt{\frac{2 \times 10^4}{8 \times 10^6}} = 0.05 \text{ cm} = 500 \ \mu\text{m}$$

Time constant of the membrane is

$$\tau_m = r_m c_m = 2 \times 10^4 \ [\Omega\text{cm}] \times 12 \times 10^{-9} \ \text{F/cm} = 240 \ \mu s$$

Appropriate boundary conditions needs to be specified for computing the membrane potential. The boundary conditions specify what happens to the membrane potential when the neuronal cable branches or terminates. The point at which a cable branches out and the point where multiple cable segments join is called a node. At a branching node, the potential must be continuous; that is, the functions $V(x, t)$ defined along each segment must yield the same result when evaluated at x value corresponding to the node. For obeying conservation of charge principle, the sum of the longitudinal currents entering (or leaving) a node along all of its branches must be zero. The potential at each point is determined by the distribution of both active and passive membrane conductances as well as the intrinsic resistance of the intracellular material. With the appropriate differential equations for each compartment, behavior of each compartment can be modeled along with its interactions with neighboring compartments. Compartments are connected to one another via a longitudinal resistivity according to the topology of a tree describing the detailed morphology of the original neuron.

In general the excitation process along neural structures is more complicated than under space clamp conditions. Current influx across the cell membrane in one region influences the neighbored sections and causes effects like spike propagation. Besides modeling the natural signaling, the analysis of compartment models helps to explain the influences of applied electric or magnetic fields on representative target neurons. Typically, such a model neuron consists of functional subunits with different electrical membrane characteristics: dendrite, cell body, initial segment, myelinated nerve fiber, and nonmyelinated terminal.

10.4   Modeling Diffusion-Limited Processes

Compartmental models assume uniform distribution of the driving force within the compartment. However, this assumption is not valid in many circumstances such as diffusion limited processes along with chemical reactions. Distribution of a component could depend on diffusion characteristics (Chapter 2) and the fluid flow characteristics (called convective mass transfer). For example, diffusion plays an important role during transdermal (across the skin) drug delivery. Therapeutic agent diffuses and is simultaneously removed if successfully delivered to the blood vessel. The spatial dependence to the molecule distribution dominates the process. In these instances, the effects on the dynamics of the chemical system are determined by the relative timescales of the diffusion process and the chemical kinetics involved.

Reaction-diffusion models describing the concentration's pattern formation of chemical substances have achieved important results in theoretical studies of morphogenesis. One can generate a great variety of spatial and temporal patterns, simple gradients, spatially cyclic patterns, and temporally oscillating patterns. The spatial pattern generated by the system provides information on the functionality of the tissue or cell or developing embryo. Prior to developing reaction-diffusion models, one can test whether both reaction and diffusion have to be considered by calculating the second Damköhler number, N_{Da} (also referred as the Thiele modulus in catalytic reactions). This second Damköhler number is defined as

$$N_{Da} = \frac{\text{Reaction rate}}{\text{Diffusion rate}} \qquad (10.52)$$

N_{Da} is associated with characteristic diffusion and reaction times therefore scaling is necessary.

For $N_{Da} \gg 1$, the specific reaction rate is much greater than the diffusion rate distribution and is said to be diffusion limited. In other words, diffusion is slowest diffusion characteristics dominate, and the reaction is assumed to be instantaneously in equilibrium.

For $N_{Da} \ll 1$, the reaction rate is much slower than the diffusion rate distribution and is said to be reaction limited. In other words, diffusion occurs much faster than the reaction; thus, diffusion reaches equilibrium well before the reaction is at equilibrium.

By estimating the N_{Da}, one can get an intuitive idea about what process dominates a chemical distribution. In a problem where the reaction rate is proportional to concentration in the bulk of the solution, N_{Da} is equivalent to $kL^2/\alpha D_{AB}$, where k is the rate constant, D_{AB} is the diffusivity, and α is a parameter representing the geometry of the problem. α is 1 in Cartesian coordinates [2].

The generalized Navier-Stokes equation for diffusion-reaction models is written in vector notation as:

$$\frac{\partial C_A}{\partial t} + \nabla \cdot \vec{N}_A - R_A = 0$$

where R_A is the rate of the chemical reaction and N_A is the mass transfer flux. The expanded terms in different coordinates are given here.

Cartesian coordinates: $\dfrac{\partial C_A}{\partial t} + \left(\dfrac{\partial N_{A,x}}{\partial x} + \dfrac{\partial N_{A,y}}{\partial y} + \dfrac{\partial N_{A,z}}{\partial z} \right) - R_A = 0$

Cylindrical coordinates: $\dfrac{\partial C_A}{\partial t} + \left[\dfrac{1}{r}\dfrac{\partial}{\partial r}\left(r N_{A,r} \right) + \dfrac{1}{r}\dfrac{\partial N_{A,\theta}}{\partial \theta} + \dfrac{\partial N_{A,z}}{\partial z} \right] - R_A = 0$

Spherical coordinates: $\dfrac{\partial C_A}{\partial t} + \left[\dfrac{1}{r^2}\dfrac{\partial}{\partial r}\left(r^2 N_{A,r} \right) + \dfrac{1}{r\sin\theta}\dfrac{\partial}{\partial \theta}\left(N_{A,\theta}\sin\theta \right) + \dfrac{1}{r\sin\theta}\dfrac{\partial N_{A,\varphi}}{\partial \varphi} \right] - R_A = 0$

In order to solve the set of nonlinear partial differential equations, initial and boundary conditions for the domain are necessary. Often, properties of diffusion are derived from free diffusion (Chapter 2), and proposed as generic diffusion properties while the actual boundary conditions shape the concentration profiles. Knowing these shapes is important when interpreting experimental results or when selecting curves (i.e., solutions of model problems for fitting experimental data to estimate (apparent) diffusion parameters). Choosing the appropriate boundary conditions is also important. Free diffusion is of use to model interactions in infinite space or in a bulk volume on appropriate timescales (i.e., situations where diffusion is used as a simple passive transport or dissipative mechanism). However, when diffusion is used as a transport mechanism to link cascading processes that are spatially separated or when diffusion is used in conjunction with processes that are triggered at certain concentration thresholds, a detailed description of the geometry of the system is very important. Then the complex geometry imposes the majority of the boundary conditions. These systems of partial differential equations are solved using a finite element method. For demonstration of concepts, three simple cases generally encountered in biomedical applications are discussed next.

10.4.1   Case 1: Reaction-Diffusion in Cartesian Coordinates

Consider the situation (Figure 10.10) where there are two baths with different concentrations of a chemical C. One bath is the chemical source (unlimited supply), maintained at a concentration C_0. The other bath initially contains no chemical and has a width L. At the far edge of the second bath ($x = L$), there is an impermeable boundary. In addition at this boundary the chemical of interest is degraded according to a first-order chemical reaction (surface reaction).

The conservation equation in the bulk of the second bath $C(x)$ reduces to

$$\frac{\partial C}{\partial t} = D_{AB} \frac{\partial^2 C}{\partial x^2} + R_A$$

At steady-state and for a first-order degradation reaction of the reactant,

$$0 = D_{AB} \frac{\partial^2 C}{\partial x^2} - kC \tag{10.53}$$

where k is the rate constant. Equation (10.53) is analogous to (10.50), and one could write the analytical solution similar to (10.51)

$$C = Ae^{-x\sqrt{\frac{D_{AB}}{k}}} + Be^{-x\sqrt{\frac{D_{AB}}{k}}} \tag{10.54}$$

where A and B are integration constants. One boundary condition is $C(0) = C_0$ at $x = 0$ and the other boundary condition is the flux equation at $x = L$, given by:

$$D_{AB} \frac{\partial C(L)}{\partial x} = -kC(L) \tag{10.55}$$

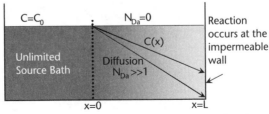

Figure 10.10   Chemical species diffusing into an "empty" bath with a degradation reaction occurring at $x = L$.

as the boundary condition at $x = L$ is impermeable. Substituting these boundary conditions into (10.54), A and B can be determined.

10.4.2 Case 2: The Krogh Tissue Cylinder

Consider the case of drug eluting stent, where drug is radially diffusing in a tube in one-dimension (Figure 10.11). Since the drug has to be dispersed along the thickness of the blood vessel, one has to determine the concentration profile of the drug across the thickness of the artery. Consider the length of the stent to be L and radial thickness Δr at r radius in the flow system. Performing a material balance on the differential element from r to $r + \Delta r$ yields

$$\frac{\partial Ci}{\partial t} 2\pi r L \Delta r = \left(N_{i_r} r - N_{i_{r+\Delta r}} (r + \Delta r) \right) 2\pi L + R_i 2\pi r L \Delta r \tag{10.56}$$

Dividing both sides of the equation by the volume element $2\pi L r \Delta r$, taking the limit as $\Delta r \to 0$, and using the definition of the derivative results in

$$\frac{\partial C_i}{\partial t} = \frac{1}{r} \frac{\partial r N_{i_r}}{\partial r} + R_i$$

Assuming that the solution is dilute and using Fick's first law, flux can be represented as

$$N_{i_r} = D_{AB} \frac{\partial C_i}{\partial t}$$

Substituting the above expression and assuming D_{AB} to be constant with respect to radius

$$\frac{\partial C_i}{\partial t} = \frac{1}{r} \frac{\partial}{\partial r} \left(D_{AB} \frac{r \partial C_i}{\partial t} \right) + R_i \text{ or } \frac{\partial C_i}{\partial t} = \frac{D_{AB}}{r} \frac{\partial}{\partial r} \left(r \frac{\partial C_i}{\partial t} \right) + R_i \tag{10.57}$$

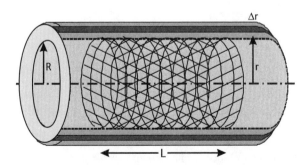

Figure 10.11   One-dimensional radial diffusion.

Let the measured drug concentration in the coronary arteries is C_0 in the outer part of the artery (known as the adventia). The inner radius of the artery is R_b and the outer radius of the vessel is R_0. Neglect chemical reactions, the steady state form of (10.57) is

$$\frac{1}{r}\frac{d}{dr}\left(r\frac{dC_i}{dr}\right) = 0$$

Integration the above equation twice yields, applying the boundary conditions leads to the following solutions for the concentration

$$C = C_b - \frac{(\ln R_b/r)(C_b - C_o)}{\ln(R_b/R_o)}$$

The flux in the artery is

$$N_i = -D_{AB}\frac{dCi}{dr} = -\frac{D_{AB}(C_b - C_o)}{r\ln(R_b/R_o)} \tag{10.58}$$

Notice that the flux is dependent upon location, decreasing in absolute value as r increases. The reason for this change with position is that the cross-sectional area through which transport occurs increases with increasing distance. Since the same amount of solute must transport through each cross-section, increasing cross-sectional area leads to a reduced flux.

For thin vessels ($R_b/R_o \approx 1$), curvature can be neglected and the flux approaches that for transport across a membrane. The logarithmic expression can be written as $\ln(1 + \varepsilon) \approx \varepsilon - 0.5\varepsilon^2 + \ldots$. Then R_b/R_o can be written as $1 + R_b/R_o - 1$ and can be expanded as

$$\ln(R_b/R_o) \approx R_b/R_o - 1 - 0.5(R_b/R_o - 1)^2 + \ldots$$

Substituting this expansion into (10.58), after neglecting higher power terms and approximating r as R_o

$$N_i \approx -\frac{D_{AB}(C_b - C_o)}{R_o - R_b} \tag{10.59}$$

10.4.2.1   The Krogh Tissue Cylinder

The molecules required for tissues are carried in the blood vessels to the capillaries where they diffuse through the capillary wall to the tissue space. The Krogh tissue cylinder model (Figure 10.12) is a simplified model of the tissue surrounding the

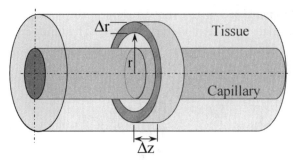

Figure 10.12   The Krogh tissue cylinder showing the capillary and the tissue space.

capillary. It assumes a cylindrical layer of tissue surrounding each capillary with the solute transferred only from that capillary. The capillary is assumed to be cylindrical and of constant radius, R. As the solute move moves along the capillary, its concentration C decreases because of solute transport through the capillary wall. By considering the tissue cylinder surrounding each blood vessel to have a uniform transport behavior, Krogh proposed that the diffusion of oxygen away from the blood vessel into the annular tissue was accompanied by a zero-order reaction, that is $R_{O2} = -m$, where m is a constant. At steady state, (10.57) reduces to

$$d\left(r\frac{dC_A}{dr}\right) = \frac{m}{D_{AB}}r.dr \qquad (10.60)$$

Assuming that the concentration of oxygen in the blood stream is constant, a boundary condition is

$$r = R, \frac{dC_A}{dr} = 0$$

Another boundary condition is $C = C_t$ at $r = R_t$ where C is the concentration of O_2 in the tissue, and R_t is a given position in the tissue. Integrating (10.60) twice and substituting boundary conditions, and rearranging,

$$C_A = C_t + \frac{m}{D_{AB}}\frac{\left(r^2 - R_t^2\right)}{4} + \frac{m}{D_{AB}}\frac{R^2}{2}\ln\left(\frac{R_t}{r}\right) \qquad (10.61)$$

Equation (10.61) is used to understand the concentration change within the tissue and has served as the basis of understanding oxygen supply in living tissue. Due to its simplicity and agreement with some observations, it is extensively used and successfully extended to other applications such as hollow-fiber bioreactors used for the production of monoclonal antibodies, bioartificial liver, to predict and optimize the oxygen exchange performance of hollow-fiber membrane oxygenators, drug diffusion, water transport, and ice formation in tissues.

10.4.3   Case 3: One-Dimensional Radial Diffusion in Spherical Coordinates

Consider the case of unsteady one-dimensional radial diffusion through a sphere of radius R (Figure 10.13). Assume the concentration of the diffusing chemical (could be oxygen, nutrient, enzyme, or a drug encapsulated in a polymeric capsule) is independent of the angle of orientation and a function of radius only, that is, $C = C(r)$. An important factor affecting concentration is the absorption or metabolism of the component that can be considered as a reaction R_i (units are mol/m^3.s). It is conceivable that the reaction will decrease the concentration of the component and affect the diffusion. A material balance on a thin spherical element with an inner surface area of $4\pi r^2$ and an outer surface area of $4\pi(r + \Delta r)^2$ yields

$$4\pi r^2 \Delta r \frac{\partial C_i}{\partial t} = \left(N_{ir} r^2 - N_{ir+\Delta r}(r + \Delta r)^2\right) 4\pi + 4\pi r^2 \Delta r R_i \qquad (10.62)$$

Note the $4\pi r^2 \Delta r$ is the volume of the shell. Divide both sides of the equation by the volume element $4\pi r^2 \Delta r$, taking the limit as $\Delta r \to 0$, and using the definition of the derivative results in

$$\frac{\partial C_i}{\partial t} = \frac{1}{r^2} \frac{\partial N_{ir} r^2}{\partial r} + R_i \qquad (10.63)$$

Substituting Fick's law for diffusion $(N_{ir} = D_{AB} = \frac{\partial C_i}{\partial t})$ assuming D_{AB} to be constant with respect to the radius gives

$$\frac{\partial C_i}{\partial t} = \frac{D_{AB}}{r^2} \frac{\partial r}{\partial r}\left(r^2 \frac{\partial C_i}{\partial t}\right) + R_i \qquad (10.64)$$

Equation (10.64) can be solved for various applications such as change in concentration of a drug entrapped in a spherical particle or oxygen profile within the aggregate of cells with appropriate boundary conditions.

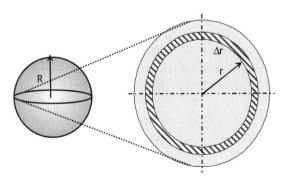

Figure 10.13   One-dimensional radial diffusion in spherical coordinates.

EXAMPLE 10.7

An erodable spherical polymer of radius R (0.5 mm) contains a drug nerve growth factor to treat brain disorder. Experiments were performed in vitro to characterize the release of the drug assuming that the rate of diffusion is fast relative to the rate of dissolution of the polymer. At the surface of the polymer ($r = R$) the drug concentration is C_0 (700 ng/mL). Far from the surface ($r \to \infty$), the drug concentration drops to zero. Determine the steady-state release rate from the polymer.

Solution:

At steady state with no chemical reactions, (10.3) reduces to

$$\frac{1}{r^2}\frac{d}{dr}\left(r^2\frac{dC_i}{dr}\right) = 0$$

Integrating twice results in the expression $C_i = B - A/r$
From the boundary condition at $r \to \infty$, $B = 0$
From the boundary condition at $r = R$, we get $A = -RC_0$
The resulting concentration profile and flux are

$$C_i = \frac{C_0 R}{r} \text{ and } N_{ir} = D_{AB}\frac{C_0 R}{r^2}$$

As the distance from the surface increases, the flux decreases, due to the increase in the cross-sectional area through which transport occurs. The steady state release rate (mol/s) at the polymer surface is the product of the flux times the surface area.

$$\text{Release rate} = N_{ir}\big|_{r=R}\, 4\pi R^2 = 4\pi D C_0 R$$

10.4.4 Complex Model Systems

The comprehensiveness and complexity of models are increasing due to the necessity of developing more realistic and integrative models for physiological subsystems of the human body. The goal is to base the development of diagnostic and therapeutic procedures on mathematical models. The questions being asked are much more demanding and the level of interdependence of the component models has increased dramatically in order to answer complex physiological questions. In many cases several time scales and levels of functionality are considered. Frequently varieties of models require interfacing of different models from many different sources with different local design rules. For example, in order to express the convective component of the mass transport equation, it is necessary to know the velocity profiles. The fluid dynamics can be described by the equation of continuity or mass balance and the equation of motion, which describes the momentum balance (described in Chapter 4). Analytical solutions to these sets of differential equations are difficult to obtain and there may be multiple solutions. One widely used approach is to solve them using numerical integration techniques. The development of decision support systems for clinical treatments also needs extensive modeling. The large number of parameters of the underlying complex models and their large dimension pose challenging problems for model validation. Furthermore, model validation is challenging in clinical applications, because of the difficulty to obtain reliable data by noninvasive or slightly invasive techniques.

A general modeling approach involves the following steps:

- Define the physical phenomena and identify all the variables involved. Choices have to be made between a very accurate but complex model, and a simple but not-so-accurate model.

- Make reasonable assumptions and approximations. The results obtained from an analysis are at best as accurate as the assumptions made in simplifying the problem.

- Apply relevant physical laws that will result in mathematical equations. Equations can be differential (ordinary or partial) or algebraic.

- If equations are differential in form, consider appropriate boundary and initial conditions to eliminate integration constants. If analytical solution is not possible, consider using numerical integration techniques with the help of Mathsolvers.

- Interpret the obtained results. At a minimum, the model should reflect the essential features of the physical problem it represents. Therefore, the solution obtained should not be applied to situations for which the original assumptions do not hold. Thus any solution to a problem should be interpreted within the context of its formulation.

Problems

10.1 Midazolam is a short-acting water-soluble drug that could be used for management of refractory status epilepticus. However, prolonged usage could cause significant side effects. It is constant infused at 0.1 mg/hr.kg of the patient. Two patients at 71 kg and 42 kg received the therapy. The rate constant is determined to be 0.0131 hr^{-1} and 0.0344 hr^{-1} using a single compartmental model. Determine the half-life of the drug for each patient.

10.2 Ultravist (iopromide) is a nonionic, water-soluble X-ray contrast agent for intravascular administration, marketed by the Bayer Corporation. One of the packages provides 623.40 mg at a concentration of 300 mgI/mL. In healthy young male volunteers receiving the Ultravist Injection intravenously, the pharmacokinetics are first order. The compound is predominantly distributed in the extracellular space as suggested by a steady-state volume of distribution of 16L. Iopromide's plasma protein binding is negligible. The mean total and renal clearances are 107 mL/min and 104 mL/min, respectively. After an initial fast distribution phase with a half-life of 0.24 hour, a main elimination phase with a half-life of 2 hours and a slower terminal elimination phase with a half-life of 6.2 hours can be observed. Calculate the concentration of the drug after 3 hours.

10.3   You decided to prescribe a new drug DRUG-4U to a patient Ms. Model, who weighs 50 kg and has a normal renal function. The average kinetic parameters for the drug are a 0.6-L/kg volume of distribution and a 60.6-mL/min clearance time. The therapeutic efficiency occurs at 2.38 μg/mL and side effects begin to occur at 5.0 μg/mL. You decide to administer a single does of 100 mg by IV injection.

   (a) Assuming rapid distribution in the V_D, is your dosage in the effective range or toxic range?

   (b) What is the half-life of the drug? How long before 75% of the dose is eliminated?

   (c) Analysis of the urine collected from Ms. Model after 16 hours of dosing contained 37.5 mg of the drug. To what extent is the renal function of Ms. Model of importance to the total clearance of the drug?

   (d) A week later, you decided to administer DRUG-4U by constant IV infusion to achieve 2.38 μg/mL? What infusion rate would you prescribe?

   (e) At steady-state, what would be the C_{max}?

10.4   Ms. Model has taken 500-mg NOT-U-ME drug. What is the concentration of drug at 0, 2, and 4 hours? The known parameters are 30L apparent volume of distribution and 0.2-hr^{-1} elimination rate constant.

10.5   What IV bolus dose is required to achieve a plasma concentration of 2.4 μg/mL at 6 hours after administration? The elimination rate constant is 0.17 hr^{-1} and the apparent volume of distribution is 25L.

10.6   Data shown in the following table was collected after an IV bolus dose of 500 mg tracer. Calculate the elimination rate constant and the apparent volume of distribution.

Time (h)	1	2	3	4	6	8	10
C (mg/mL)	72	51	33	29	14	9	4

10.7   Gentamicin is an aminoglycoside used to treat a wide variety of infections. However, due to its toxicity, its use must be restricted to the therapy of life-threatening infections. Protein binding is low for gentamicin (depending on the test performed, binding is reported to be 0–30%) with a volume of distribution of 0.25 L/kg body weight, and the elimination rate constant for adults is 0.231 hr^{-1}. Gentamicin follows a one-compartment body model with first-order elimination.

   (a) Calculate the initial IV dose if the desired concentration of drug in the plasma is 6 mg/L.

   (b) To maintain a constant blood level, an infusion needs to be started. What should the infusion rate be to maintain the desired 6-mg/L plasma concentrations?

(c) What is the half-life of the drug?

(d) What is the plasma concentration of the drug after 5 hours?

10.8 An artificial liver is a device that utilizes cultured liver cells to replace the metabolic function of the normal liver. One way to evaluate the performance of the device is to measure its metabolic activity under steady state operating conditions. Consider an artificial liver in a flow system where blood is introduced to the device from an artery and returned to the body through a vein. Blood enters and leaves the device at a flow rate of 250 mL/min. No other fluid flows enter or leave the artificial liver; however, reactions take place. In particular, blood glucose concentrations are measured at some particular time and are found to be 2.00 mg/mL at the arterial inlet and 1.80 mg/mL at the venous outlet. What is the glucose metabolic activity, measured in units of mg/min, of the device?

10.9 In the laboratory, a group of students investigated the modeling of an artificial kidney using a scheme shown in the following table. The thick arrow represents transport across the dialysis membrane that separates the blood fluid from the dialysate fluid. The urea concentration (C_U) is also known the literature as blood urea nitrogen (BUN). The concentrations are typically expressed as mg%, which is equivalent to mg/100 mL. Let the body fluid volume be 50L. The relationship between the outlet (C_{U_0}) and inlet (C_{U_1}) urea concentrations are given by

$$C_{U_0} = C_{U_1} \exp(-V_{Cl}/F)$$

where V_{Cl} is the clearance parameter and F is the inlet blood flow rate. Assume the clearance and flow rates are constant during dialysis, but the inlet concentration changes with dialysis time.

	Flow Rate (mL/min)	BUN (mg)
Arterial (entering) blood	200	190
Venous (exiting) blood	195	175

(a) What is the clearance for the device as shown?

(b) How long does it take for the patient's BUN to reach 50 mg%? How much water is removed in this time?

10.10 Twenty mg of a PET tracer was administered intravenously to a 70-kg patient. The plasma concentrations of the tracer are given in the following table. The minimum plasma concentration required to cause significant enhancement of gastric emptying is 100 ng/mL.

$t(h)$	1	2	4	6	8	10
C_P (ng/mL)	180	136	80	43	24	14

(a) Plot the tracer concentration versus time and draw a compartmental scheme showing the number of compartments involved.

(b) Write the equation describing the disposition kinetics of the tracer.

(c) Calculate the biological half-life of the tracer elimination, the overall elimination rate constant, the apparent volume distribution, the coefficient of distribution, and the duration of action in the tissue.

(d) Comment on the extent of tracer distribution in the body.

10.11 After an intravenous injection of a single 100-mg dose of phenobarbital in a 70-kg patient, the following plasma concentrations were obtained:

t (h)	2	4	6	8	10	12	24	48	72	96
C (mg/L)	1.63	1.30	1.17	1.04	0.94	0.86	0.68	0.54	0.44	0.36

(a) Plot the phenobarbital plasma concentrations versus time.

(b) Determine the number of compartments involved in the disposition of this drug. Draw a compartmental scheme describing the disposition of this drug.

(c) Calculate the biological half-life and the volume of distribution.

10.12 Okusanya et al. [3] reported the compartmental pharmacokinetic analysis of amprenavir (APV) protease inhibitor using compartmental model. Many potent antiretroviral therapy regimens utilize human immunodeficiency virus type 1 protease inhibitors as a backbone of HIV therapy. Retrieve [3] and simulate the model. Discuss the pitfalls in the model.

10.13 Peritoneal dialysis works on the same principle as hemodialysis, but the exchange of waste products between the blood and dialysate fluid takes place inside the body rather than in a machine. The abdomen contains a peritoneal cavity, which is lined by a thin membrane called the peritoneum. The peritoneum is a natural membrane that serves the same purpose in dialysis as the artificial membrane in a dialysis machine. To accomplish this, the peritoneal cavity in the abdomen is filled with dialysis fluid that enters the body through a permanently implanted tube or catheter. Excess water and wastes pass though the peritoneum into the dialysis fluid. After a while, this fluid is drained from the body through the catheter and discarded. Consider the dialysate fluid and the body to be two well-mixed compartments of volumes V_D and V_B, respectively. The flux of species i across the membrane is $N_i = K_i (C_{bi} - C_{Di})$ where C_{Bi} and C_{Di} refer to solute concentrations in the blood and dialysate fluid for species i, respectively, and K_i is an overall mass transfer coefficient for species i.

Suppose that V_B is 50L, V_D is 2L, the initial solute concentrations in the dialysate are zero, and the peritoneal membrane area A is 2 m^2. Find the concentrations in the dialysate fluid as a function of time for the solutes listed in the following table with their mass transfer coefficients. Plot your results. Assume that the blood solute concentrations remain constant and that water ultrafiltration across the membrane is negligible.

Solute	MW	Blood Concentration (mg/dL)	K_i (cm/min)
Urea	60	200	0.001
Creatinine	113	6	0.0005
β_2-globulin	12,000	0.8	0.00004

10.14 The concentration of a urea leaving the dialyzer C_{UM} is related to the concentration of urea in the body (C_{UB}), and the reservoir by the equation $\ln\left(\dfrac{C_{UM}-C_{UR}}{C_{UB}-C_{UR}}\right)=-0.874$. At the start of the dialysis process, the urea concentration in the blood within the body C_{UB} is 1.0 g/L. The volume of blood V_B is 6.0L, the blood flow rate Q_b is 5.94 L/hr, and C_{UR} is constant at 0.05 g/L. Determine the time required for the urea concentration in the body (C_{UB}) to drop to 0.2 g/L.

10.15 Consider the filtration of the urea from blood using a countercurrent flow, a dialysate exchanger with an area of 210 cm². The mass transfer coefficient K for the exchanger is 0.2 cm/min. If the dialysate flow rate is 400 mL/min, the blood flow rate is 200 mL/min and the urea concentration in the blood at the start of dialysis is 150 mg/cm³; then:

(a) After a short transient during which the dialyzer reaches steady state, what are the concentration of the urea in the dialysate outlet and in the blood outlet?

(b) At the start of dialysis, what is the clearance?

(c) How long will it take to bring the urea concentration in the blood to 10% of its initial value if the body fluid volume is 10L?

10.16 Assume that a drug overdose is taken, is rapidly absorbed into the body, and distributes into the extracellular fluids (13L) only. If the drug is neither secreted nor reabsorbed by the kidneys, how long will it take for 90% of the drug to be eliminated in the urine? Model the extracellular fluid pool as a well-mixed compartment, neglect drug metabolism (not a good assumption for most drugs), use a GFR of 125 ml/min, and assume a blood flow of 1,200 ml/min to the kidneys. This example will show you why *forced dieresis* (stimulation of high rates of urine formation by the use of certain agents) is a common technique for treating drug overdose.

10.17 Derive the governing equation for the alveolar partial pressure in the control volume as a function of time. Assume that it has a solubility of β_b in the blood and an apparent solubility β_a in the air. (Note that β_a is a convention that merely allows one to talk about the relative solubility of a gas between air and blood.) Further assume that the inhaled concentration and volumetric flow rate are constant with respect to time and that $C_v(t)=0$. This is a flow-through model (i.e., the ventilation does not change the volume of the alveolus). Assume that the inhaled tracer is not soluble in the tissue volume.

(a) Once you have found the differential equation describing this system, solve it for $P_A(t)$.

(b) What basic parameters are determined in this model and do they have any physical meaning?

(c) What limitations are there in this model?

(d) What can you say about the effects of the volume of inhalation volumetric flow rate and blood volumetric flow rate on the model?

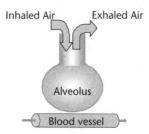

10.18 A radioactive tracer experiment is performed in order to measure the flow rate of blood in a blood vessel (stream 3). The concentration of the tracer in the inlet and outlet blood streams is determined via autoradiography. A compartmental schematic of the experiment is shown in the table. Four different experiments were performed. For this set of experiments, determine the average blood flow rate, in mL/min, with 95% confidence limits for blood out.

Experiment	Tracer in Flow Rate (mL/min)	Tracer in Tracer Concentration (ppm)	Blood In, Tracer Concentration (ppm)	Blood Out, Tracer Concentration (ppm)
1	1.73	4,007	345	732
2	1.67	3,999	472	847
3	1.69	4,013	507	875
4	1.60	4,002	523	900

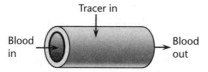

10.19 Show that the units of τ_m and λ are time^{-1} and length.

10.20 For an infinitely long cylindrical element, show that $\Delta\Phi = \Delta\Phi_0 e^{-\frac{x}{\lambda}}$ at steady-state.

10.21 The membrane resistance of a long cylindrical cell (diameter 80 μm) is 120 Ω cm^2. The resistance of the cell axoplasm is 1 Ω cm and the extracellular resistance is 3 Ω/cm. The time constant of the cell axoplasm is 1 ms. What is the characteristic length of the cell and what is the capacitance of the cell membrane per unit area?

10.22 The key to improving the clinical outcomes of gene therapy is optimizing the parameters involved in gene delivery, intracellular degradation and expression. For this purpose, Banks at al. [4] utilized a multiple compartmental model (one for the cytoplasm, one for the nucleus, and one for the extracellular volume) to understand the degradation kinetics and distribution of the gene in a cell. Read [4] and check their model development strategy. Derive the kinetic parameter and check with [4].

10.23 A drug D equivalent to 48.15 mg was given by mouth, and blood concentrations of the drug D and also of its only metabolite M were measured. Also the cumulative amounts of D and M in the urine were measured. The obtained values are given in the following table. We wish to devise a model for the uptake, metabolic conversion, and excretion of this drug, and curve-fit to adjust the model to fit the observed data. You could use a three-compartmental model shown in the figure. Develop kinetic parameters for the model and discuss the implications of that data.

Time (hr)	Blood-D (mg/L)	Blood-M (mg/L)	Urine-D (mg)	Urine (mg)
0.82	0.1746	0.822	—	—
1	—	—	1.87	7.23
1.2	0.166	0.144	—	—
1.4	0.1264	0.152	—	—
2	0.1092	0.860	3.23	15.53
2.4	0.0904	0.649	—	—
2.9	0.0828	0.602	—	—
3	—	—	4.02	—
3.38	0.0704	0.382	—	—
3.92	0.0591	0.403	—	—
4	—	—	4.59	25.88
4.42	0.0511	0.304	—	—
5.18	0.0355	0.252	—	—
6	—	—	5.77	32.42
6.35	0.0148	0.143	—	—
8	—	—	6.3	34.89
8.3	0.0081	0.0636	—	—
10	—	—	6.51	36.16
10.28	0.0047	0.0333	—	—
12	—	—	6.65	37.06
12.4	0.0026	0.0210	—	
24	—	—	6.92	38.7
24.57	0.0009	0.0065	—	—
48	—	—	7.3	40.29
72	—	—	7.38	40.77

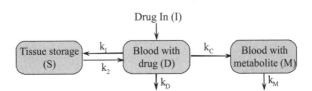

10.24 Nicotrol, a transdermal drug patch, is designed to slowly deliver nicotine to the body through the skin to a blood vessel. The drug patch consists of a sealed reservoir containing the drug encapsulated within a porous polymer matrix. A diffusion barrier attached to the bottom surface of the patch sets the surface concentration of the drug in the body tissue at 2 mol/m^3, which is below the solubility limit. The mean distance from the drug patch to the nearest blood vessel is 5 mm. To be effective, nicotine concentration must be at least 0.2 mol/m^3 at the top edge of the blood vessel. Determine the time (in hours) it will take for nicotine to begin to be effective. The effective molecular diffusion coefficient of the drug through the body tissue is $1*10^{-6}$ cm^2/s.

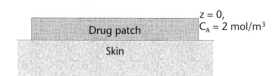

10.25 A drug patch is designed to slowly deliver a drug through the body tissue to an infected zone of tissue beneath the skin. The drug patch consists of a sealed reservoir containing the drug encapsulated within a porous polymer matrix. The patch is implanted just below the skin. It is designed to release 2.0×10^{-11} mol/s of the drug uniformly over a surface area of 1 cm^2. The mean distance from the drug patch to the infected area of tissue is 5 mm. To be effective, the drug concentration must be at least 0.2 mol/m^3 at the top edge of the infected zone. Determine the time (in hours) it will take for the drug to begin to be effective for treatment. The effective molecular diffusion coefficient of the drug through the body tissue is 1×10^{-6} cm^2/s.

10.26 Consider the transfer of oxygen transfer the alveoli to the blood vessel. Since the alveoli expand significantly, the curvature can be ignored and modeled as a plane wall of thickness L. The inhalation process may be assumed to maintain a constant molar concentration $C_{A,0}$ of oxygen (species A) in the tissue at its inner surface ($x = 0$), and assimilation of oxygen by the blood may be assumed to maintain a constant molar concentration $C_{A,L}$ of oxygen (species A) in the tissue at its outer surface ($x = L$). There is constant oxygen consumption in the tissue due to metabolic processes, and the reaction is zero order. Derive an expression for the oxygen concentration profile in the tissue and for the rate of assimilation of oxygen by the blood per unit tissue surface area [5].

10.27 By considering the tissue cylinder surrounding each blood vessel, August Krogh [6] proposed that the diffusion of oxygen away from the blood vessel into the annular tissue was accompanied by a zero-order reaction, that is, $R_{O2} = -m$, where m is a constant. This reaction describes the metabolic consumption of oxygen to produce carbon dioxide. The following boundary conditions were suggested:

$$\frac{dC}{dr} = 0 \, @ \, r = R$$
$$C = C_t \, @ \, r = R_t$$

where C is the concentration of O_2 in the tissue, R_t is a given position in the tissue, and R is the blood vessel/tissue interface. The first boundary condition indicates that the concentration of oxygen in the blood stream is constant. Determine the flux of oxygen that enters the tissue cylinder at R_t. Assume steady-state and that the concentration of O_2 in the tissue cylinder is very small.

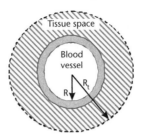

10.28 Consider a capillary-cylinder tissue model to describe transport of glucose (A) taking into account the thickness of the capillary wall and the axial variation of solute concentration inside the capillary. The concentration of A decreases along the capillary owing to the solute diffusion from the capillary into the surrounding tissue.

(a) Assume that radial variations in solute concentration inside the capillary are small, and use an average concentration inside the capillary $\overline{C}_A(z)$. The average velocity in the capillary is V, and the capillary wall thickness is R_0-R_1. The flux (N) across the capillary wall is given by $N = K_0(\overline{C}_A - C_A|_{R_0})$ where $\overline{C}_A$ is the concentration in the capillary, $C_A|_{R_0}$ is the solute concentration in the tissue at the outer capillary wall and K_0 is a mass transfer coefficient. Do a solute balance on a thin slice of capillary of thickness Δz. Take the limit $\Delta z \to 0$ to find a differential equation for $\overline{C}_A(z)$.

(b) Solve the diffusion equation for C_A, the solute concentration in the tissue cylinder. Use as a boundary condition that the concentration at $R = R_0$ is $\overline{C}_A|_{R_0}$, which depends on z. What values do you need to calculate the concentration of glucose in the tissue?

(c) To show how $\overline{C}_A$ varies with z, do a macroscopic mass balance over a length z of the capillary as shown next. The rate at which A enters the capillary at $z = 0$ minus the rate that A consumption per unit time per unit volume of tissue is constant. Write an algebraic equation for the mass balance. Assume $\overline{C}_A$ is $\overline{C}_A^0$ at the capillary entrance.

(d) Take the derivative with respect to z of the result from (C) and substitute it into the result from (b) to find the solute concentration in the capillary.

(e) Find the solute concentration in the tissue.

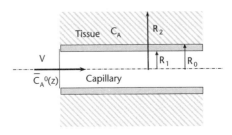

10.29 A fermentation broth consists of an aqueous solution of nutrients and cells. As the cells grow, they cluster into spherical pellets of radius R. On average, the cell density inside a pellet is 0.02g of cell mass per cubic centimeter of pellet volume. The pellets can be considered homogeneous. The dissolved oxygen concentration in the broth is 5×10^{-6} g/cm^3. The cells use oxygen at a rate of 1.2×10^{-3} mol of oxygen per hour per gram of cell mass, via a zero-order reaction (i.e., the reaction rate does not depend on oxygen concentration in the pellet). Assume that the diffusion coefficient of oxygen within the pellet is 1.8×10^{-5} cm^2/s and that the broth external to the pellet is well mixed.

(a) Derive an expression for the concentration of oxygen in the spherical cell cluster.

(b) How large can R become before oxygen concentration becomes zero at the center of the pellet?

10.30 It is shown that liver cells perform better in aggregate form than individual cells when used in extracorporeal assist devices. However, they are highly metabolically active (i.e., need a significant oxygen supply). Since the primary mode of oxygen transport is diffusion, the spherical aggregate of a 400-μm radius were formed and oxygen was measured as shown in the following table. The position "0" corresponds to the center of the cell aggregate. Assume that the temperature of the medium is 37°C.

Distance (mm)	0.10	0.20	0.30	0.35	0.38	0.39	0.40	0.41	0.42	0.45	0.50	0.60	0.70
PO$_2$ (mmHg)	12	18	28	40	52	56	60	64	74	80	99	121	140

(a) Use the mass transport equations to find the general solution for the pO$_2$ profile inside the aggregate assuming that oxygen is consumed at a constant rate m.

(b) Assuming that pO$_2$ is zero at the aggregate center, and assuming that the diffusion constant DO$_2$ inside the aggregate is 10^{-3} mm^2/s, use the value of pO$_2$ at the edge of the aggregate to estimate the rate of oxygen consumption m inside the cell aggregate. How does the value you obtain compare with the value we used in our class example?

(c) Assuming that DO$_2$ is the same for the medium and the cell aggregate, find the complete pO$_2$ profile inside and outside the aggregate.

10.31 You are trying to build a vascular graft of internal diameter 3 mm and of different thickness. Consider the fundamental transport issue of oxygen diffusion and metabolism in the cylindrical region where you would like to grow the cells. The outer boundary of the tissue region whose O$_2$ concentration is governed by the supply from the inside of the graft. To understand cell death due to O$_2$ limitation, the Damkohler number (Da), the ratio of reaction rate to diffusion rate, is used. Assume that the O$_2$ metabolism in the tissue can be represented by a depletion reaction rate of the first order in the O$_2$ concentration, that the concentration at R_1 is constant, and there is no net flux of O$_2$ out of the tissue region R_2. The following values are known: the diffusivity (D) of oxygen is 2×10^{-5} cm^2/sec and $k = 15$/s.

(a) Derive the expression for Damkohler number.

(b) Plot the changes in concentration of oxygen (C/C_0) as a function of the percentage of maximum thickness ($R_2 - R_1$) for a series of values of $R_2 = $ 4 mm, 5 mm, 8 mm, and 13 mm. For plotting graphs, you can use either the Taylor series or any mathematical software such as Mathematica or MathCAD [2].

(c) If the derived Damkohler number expression is $Da = k(R_2 - R_1)^2/D$, then what are the Damkohler number values for each of the above cases?

(d) Explain the relationship between the oxygen concentration plots and the Damkohler number values.

(e) What design modifications would you consider using to improve the oxygen availability in cases where oxygen is the limiting factor?

10.32 Consider a long hollow fiber bioreactor where an enzyme is immobilized within a bundle of n hollow fibers. Each hollow fiber has an inner diameter ID and an outer diameter OD. The wall of the hollow fiber is permeable to the substrate and the product, but not the enzyme. The substrate at concentration S_i is fed at a flow rate F. The feed passes through the shell side of the hollow fiber reactor and exits at the other end with a bulk substrate concentration of S_o. The inside of the hollow fiber is stagnant.

(a) Given the diffusivity of the substrate through the hollow fiber wall D_w and that through the medium D_m and given the intrinsic reaction rate expression $v(s)$, set up the material balance equation(s) that needs to be solved to find the substrate concentration profile within the fiber wall and within the inner space of the hollow fiber.

(b) Provide applicable boundary conditions.

(c) Describe briefly how you would solve the above ODE to arrive at a solution $s(r)$, where r is the radius from the center of the hollow fiber.

(d) Assume that you have solved the equation from the last part, and you now have the function $s(r)$ for $r = 0$ to $r = R_o$, where R_o is the outer radius of the fiber. Find the observed rate of reaction per unit length of the fiber per fiber.

(e) Provide the definition of the effectiveness factor, and find the effectiveness factor as a function the substrate concentration at $r = R_o$ (i.e., at the shell side).

(f) List the equation(s) that must be solved to find the bulk substrate concentration at the exit, S_o.

References

[1]   Limbut, W., et al., "Capacitive Biosensor for Detection of Endotorin," *Analytical and Bioanalytical Chemistry*, Vol. 389, No. 2, 2007, pp. 517–525.

[2]   Deen, W. M., *Analysis of Transport Phenomena*, Oxford, U.K.: Oxford University Press, 1998.

[3]   Okusanya, O., et al., "Compartmental Pharmacokinetic Analysis of Oral Amprenavir with Secondary Peaks," *Antimicrobial Agents and Chemotherapy*, Vol. 51, No. 5, May 2007, pp. 1822–1826, http://www.ncbi.nlm.nih.gov/pmc/articles/PMC1855557/.

[4]   Banks, G. A., at al., "A Model for the Analysis of Nonviral Gene Therapy," *Gene Therapy*, Vol. 10, 2003, pp. 1766–1775.

[5]   Incropera, F. P., *Fundamentals of Heat and Mass Transfer*, New York: John Wiley & Sons, 2002.

[6]   Krogh, A., "The Rate of Diffusion of Gases Through Animal Tissues, with Some Remarks on the Coefficient of Invasion," *Journal of Physiology*, Vol. 52, No. 6, May 1919, pp. 391–408.

Selected Bibliography

Ross, E. C., and C. A. Simmons, *Introductory Biomechanics: From Cells to Organisms*, Cambridge, U.K.: Cambridge University Press, 2007.

Schoenwald, R. D., *Pharmacokinetic Principles of Dosing Adjustments: Understanding the Basics*, Boca Raton, FL: CRC Press, 2000.

Khoo, M. C. K., *Physiological Control Systems: Analysis, Simulation, and Estimation*, New York: Wiley-IEEE Press, 1999.

Calvey, T. N., and N. E. Williams, *Principles and Practice of Pharmacology for Anaesthetists*, 5th ed., New York: Wiley-Blackwell, 2008.

He, B., (ed.), *Modeling & Imaging of Bioelectrical Activity: Principles and Applications (Bioelectric Engineering)*, New York: Springer, 2004.

Fall, C., et al., (eds.), *Computational Cell Biology*, 3rd ed., New York: Springer, 2005.

Hoppensteadt, F. C., and C. S. Peskin, *Modeling and Simulation in Medicine and the Life Sciences*, 2nd ed., New York: Springer; 2004.

Guelcher, S. A., and J. O. Hollinger, *An Introduction to Biomaterials*, Boca Raton, FL: CRC Press, 2005.

CHAPTER 11

Ethical, Legal, and Societal Aspects

11.1 Overview

The intention of biomedical engineers is to develop tools and methodologies that help in the diagnosis, treatment, or monitoring of human disease or injury. In previous chapters, various design considerations were discussed. To maintain homeostasis for which one has to measure, investigate, replace, or support human physiological functions, requires consideration of various human factors. For example, to evaluate a developed prosthetic device, one has to test the efficacy of the device in a human subject. While performing these tests, a biomedical engineer has to consider the safety of the test patient, history (both the health and the cultural) of the test patient, and the potential harm to the test patient if the device fails. However, if the device is not tested before, many times it is difficult to predict the success in the clinical settings. During these interactions, it is paramount to perform the studies legally and ethically so that patient is not abused for the benefit of developing the device. Historical evidence shows that many times monetary benefits invoke the abuse of other human beings while developing and utilizing a device. This has warranted development of many regulatory policies and procedures to be followed during research, and clinical operations that help in safe implementation. These issues can be grouped into five categories as shown in Figure 11.1.

A number of agencies and policy boards have been developed at the local, national, and international levels. Understanding these procedures and policies is critical to the successful development of a biomedical device. For example, if a researcher is interested in testing the developed device on human subjects or animal models, then understanding the procedure to conduct research within the bioethical framework is necessary. The Federal Drug Administration is responsible for the regulation of drugs and medical devices within the United States, and does not regulate the practice of medicine. Thus, many procedures have to be approved by the internal oversight committees within a clinical setting. While developing devices for the international community, different countries have different cultural influences and different ethical reasoning. World population is also very diverse including altered anthrophometry composition and socioeconomic conditions. With the globalization of manufacturing, one has to understand and comply with the minimum standards. Understanding the minimum safety requirements for a different country during the development of a biomedical device is also important. In

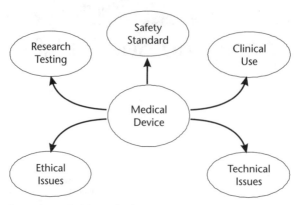

Figure 11.1   Compliance issues in biomedical practice.

this chapter, the basics of biomedical ethics and basic internal review boards are described. Further, some of the safety standards to be adhered to are also discussed.

11.2   Fundamentals of Bioethics

11.2.1   Classical Ethical Theories

Ethics as a philosophical discipline is the study of morality as distinguished from the scientific study of morality. Ethics as applied to medical practice dates back to the ancient civilization by the symbolic adherence to the Hippocratic oath. With the advancement of biomedical technology, ethics in medicine or "bioethics" has become an autonomous discipline. The topic of ethics is usually subdivided into normative ethics and metaethics, although the precise relationship of these two branches is a matter of some dispute. Normative ethics attempts to establish what justifiable moral right is and what morally wrong is with regard to human action. General normative ethics deals with the "right" moral theory and applied normative ethics deals with particular moral problems such as biomedical ethics, business ethics, and legal ethics. Bioethics could be considered as a form of general normative ethics. Metaethics is concerned with tasks such as analyzing the nature of moral judgments and specifying appropriate methods for the justification of particular moral judgments and theoretical systems (e.g., duty, rights, and moral reasoning).

Classical ethical theories are broadly classified into two categories: teleological and deontological. Teleologists (*telos* means goal, end, or purpose in Greek) base their moral theory on the tenet that there are some things such as pleasure, happiness, ideals (including freedom, knowledge, justice, and beauty) and preferences which do not have instrumental measurements, but have value in themselves. This is called intrinsic value. An example of the teleological system is utilitarianism, developed by British philosopher Jeremy Bentham. According to utilitarians, the purpose of morality is to guide people's actions in such a way as to produce a better world. According to Bentham's hedonistic theory, pleasure has a positive intrinsic value and pain has a negative intrinsic value. Others refined the utilitarian principle to a systematic way of calculating goods. Many utilitarian thinkers embrace more elaborate and nonhedonistic theories of intrinsic value. Utilitarianism is classified into two categories:

1. *Act utilitarianism* looks at the consequences of each individual act and calculates utility each time the act is performed.
2. *Rule utilitarianism* looks at the consequences of having everyone follow a particular rule and calculates the overall utility of accepting or rejecting the rule.

Act-utilitarian reasoning would be used in a biomedical context, for example, when considering whether to allocate dialysis services to a derelict, thus prolonging his miserable existence or to allocate these services to a 35-year-old mother, whose life is of benefit to her family and to society, and who has better prospects for a good quality of life. Act-utilitarianism ignores special relationships, such as parent and child, physician and patient. For a physician to damage the interests of an individual patient in an effort to maximize utility seems wrong. Critics of hedonism believe that some pleasures do not have intrinsic value (e.g., malicious pleasures) or things other than pleasure have intrinsic value (e.g., knowledge and justice).

Deontologists (*deon* means duty in Greek) consider that doing one's duties is self-evident for correctness and morality (i.e., they rely not on the consequences of an action but on whether an action is right or wrong for it's own sake). German philosopher Immanuel Kant believed that statements about the moral law were a priori and could be reached through logic alone, independent of experience. Kant explored rational steps that direct him to his conclusions. Kant distinguishes two types of imperatives: hypothetical and categorical. A hypothetical imperative is entirely dependent on the consequences. Do A to ensure that B and, in turn, C happen. According to Kant, any fully rational agent would follow categorical imperatives not based on any consequences but performed for the sake of duty only; they are ends in themselves not simply a means to an end. For example, a physician may never justifiably lie to a patient and experimental subjects must never be used without their voluntary informed consent.

All single-principle ethical theories such as act-utilitarianism and Kant's categorical imperative do not appreciate the complexity of bioethical decision making or the tensions that arise from different ethical obligations pulling people in different directions. For example, what is extraordinary care and who decides it? Who should decide when such care should be withdrawn? Should it be the father whose action caused the irreversible coma, the mother, the hospital, or the society that is forced to pay for the care? Should a baby with a defective heart be given the heart of a pig or none at all? There are mixed theories in regards to the moral value of an act, which can depend on both style and outcome. Bioethical committees comprised of individuals from different backgrounds are usually used in decision making. Hospitals generally have an ethics committee to handle decision making in individual cases, and general policies for the hospital. Universities and institutions (see Section 11.3.2) have committees that generate and insure compliance with ethical standards governing the use of animal and human subjects in experimentation. Every doctor, nurse, or bioengineer must have some training in ethics, ranging from a module of a course to one or two courses.

11.2.2   Difference Between Ethics and Law

Both ethics and laws give rules of conduct to follow, with the fundamental goal of protecting society. Laws stem from legislative statutes, administrative agency rules, or court decisions. Laws often vary in different locales as laws evolve to reflect societal needs and are enforceable only in those jurisdictions where they prevail. Although ethical principles do not change because of geography (at least not within one culture), interpretation of the principles may evolve as societies change. This same evolution occurs within the law. Laws establish the rules that define rights and obligations of a person in the society. Laws penalize people who violate these rules. Ethics incorporates the broad values and beliefs of correct conduct. Good ethics often make good laws, whereas good laws may not make good ethics. Although societal values are incorporated into both the law and ethical principles and decisions, ethical principles are basic to society. Most laws are derived from other laws, although based loosely on societal principles. Significant overlap exists between legal and ethical decision making. Both ethical analysis and the law use case-based reasoning in an attempt to achieve consistency. Justice is a moral standard that applies to all human conduct. Therefore, the laws enforced by the government may have a strong moral standard.

However, in some areas the law and bioethics differ markedly. The law operates under formal adversarial process rules, such as those in the courtroom, which allow little room for deviation. On the other hand, ethics consultations are flexible enough to conform to the needs of each institution and circumstance. Further, ethical consultations are designed to assist all parties involved in the process rather than being adversarial. The law also has some unalterable directives that require specific actions. Bioethics is designed to weigh every specific situation on its own merits although it is based on principles. The key difference between bioethics and the law is that bioethics relies significantly on the individual person's values. Also, even without the intervention of trained bioethicists, medical personnel can and should make ethically decisions. The law does not consider individual values and generally requires lawyers for interpretation.

An example for demonstrating the difference between bioethics and law is the case of Dr. Kevorkian. In Michigan, Dr. Kevorkian was charged for violating the assisted suicide law and delivering a controlled substance without a license in the death of a patient. The patient was suffering from a terminal disease and his family requested Dr. Kevorkian for help to end the pain and suffering of the patient. The patient signed a notice stating the intension to die and chose direct injection. However, Dr. Kevorkian was convicted of the murder and the family spoke against the sentence. According to family members, it was ethical to end the patient's life. However, according to the judicial system, Dr. Kevorkian violated the law.

11.2.3   Influence of Religion and Culture

Religious beliefs shape outlooks on every aspect of life including diet, lifestyle, and moral beliefs. The tradition of medicine has also been strongly influenced by religion. Religion can dictate desired care and the medical decisions made by patients and is important for organ donation. In homogenous societies (with one religion in majority), religion has long been the arbiter of ethical norms. Religion could offer the three essentials for a good health system, namely "care, competence, and compassion." Modern bioethics uses decision-making methods, arguments, and ideals that originated from religion. In addition, clinicians' personal spirituality may allow them to relate better to patients and families in crisis. In multicultural societies, with no single religion holding sway over the entire population, a patient value-based approach to ethical issues is necessary. Although various religions may appear dissimilar, most have a form of the Golden Rule or a basic tenet that holds: do not kill, cause pain, disable, deprive of freedom, deprive of pleasure, deceive, or cheat; keep your promise, obey the law, and do your duty. Moral rules govern actions that are immoral to do without an adequate moral reason and can justifiably be enforced and their violators be punished. Although none of these rules are absolute, they all require one to not cause evil. Somewhat paradoxically, however, they may neither require preventing evil nor doing good.

Problems surface when trying to apply religion-based rules to specific bioethical situations. For example, although "do not kill" is generally accepted, the interpretations of the activities that constitute killing vary with the world's religions, as they do among various philosophers. The morality of embryonic stem cell research and human cloning is another issue. Nearly every religious group has condemned human cloning for reproductive purposes, but the response of religious communities to stem cell research is mixed. The Catholic Church and many Protestants call for a ban on all embryonic stem cell research, saying it is an assault on innocent human life. On the contrary, many other Christian churches and Jewish groups favor embryonic stem cell research, pointing to potential cures for medical conditions such as Parkinson's disease. Therefore several generally accepted secular principles have emerged, such as autonomy, beneficence, nonmaleficence, and fairness, which have guided ethical thinking.

Cultural backgrounds and religion need to be considered while implementing biomedical technology in different parts of the community. For example, the spread of AIDS is rampant throughout the world. However, AIDS is considered as a punishment by God in some parts of the world. The best way to persuade people of the efficiency of a particular program and gain their support for it is to base the program on the people's own culture, traditions, and religious beliefs. Preachers, welfare workers, temples, mosques, and churches have an important role to play in this process. Were it to incorporate the call to good health in its message, a religious order would be perfectly capable of achieving wonders in the area of health and well-being.

11.3   Research Involving Human Participants

11.3.1   The Declaration of Helsinki

Research and experimentation refer to a set of scientific activities whose primary purpose is to develop or contribute to general knowledge about the chemical, physiological, or psychological processes involved in human functioning. Under the Code of Federal Regulation (CFR) of the United States, research is defined by the Office of Human Research Protections [1] as: "[A] systematic investigation, including research development, testing and evaluation, designed to develop or contribute to generalizable knowledge." After World War II, atrocities committed by the Nazi researchers against prisoners in concentration camps drew international attention. A proper set of standards for judging the physicians and scientists who had conducted experiments on the prisoners was drawn up by the presiding international military tribunal. The tribunal formulated a 10-point statement in 1948, called the Nuremberg Code, delimiting permissible medical experimentation on human subjects. According to this statement, humane experimentation is justified only if its results benefit society and it is carried out in accord with basic principles that "satisfy moral, ethical, and legal concepts." The basic ethics of the Nuremberg Code continue to serve as a cornerstone for modern regulations regarding the use of human participants in experimentation.

In the 1950s another disaster occurred in Europe with the usage of thalidomide as a sedative. Thalidomide caused significant toxicity to the nervous system and this incidence led to the adoption of the "Kefauver Amendment" in 1962 to the Food, Drug, and Cosmetic Act of the United States, requiring drug manufacturers to prove the safety and effectiveness of their products and physicians to obtain informed consent from potential subjects before administering investigational medications. The World Medical Association in 1964 responded with the Declaration of Helsinki, built on the Nuremberg code to develop good clinical practices. It must be stressed that the standards as drafted are only a guide to physicians all over the world. Physicians are not relieved from criminal, civil, and ethical responsibilities under the law of their own countries. There have been a number of amendments to the original declaration, which can be obtained directly from the World Medical Association [2].

11.3.2 Belmont Report

In the United States, the National Research Act was passed in 1974, which helped with the creation of the National Commission for the Protection of Human Subjects of Biomedical and Behavioral Research (NCPHSBBR). To codify the requirement that human participants in research must be protected, the NCPHSBBR published guidelines for review of research involving human subject research in 1979 known as *The Belmont Report: Ethical Principles for the Protection of Human Subjects of Research* [3]. *The Belmont Report* considers three "principles" or concepts to be integral to the investigator-subject relationship: respect for persons, beneficence, and justice. Some points from *The Belmont Report* are given here:

1. *Respect for Persons* has at least two ethical considerations: the individual human research participant needs to be treated as an autonomous being, a person who makes decisions or deliberates for himself or herself about personal goals and then acts upon them; those persons who are not able to make and carry out decisions for themselves, such as children or sick people or those who have a mental disorder, must be protected from coercion by others and from activities that harm them. How much to protect them is related to the risk of harm and the likelihood of benefit to them. In

research, respect for people demands that participants enter into a research program voluntarily and with good information about the research goals.

2. *Beneficence* is doing good to the individual. In *The Belmont Report*, beneficence is understood in a sense to do no harm and to "maximize possible benefits and minimize possible harms" to the individual research participant. "Do no harm" is a Hippocratic principle of medical ethics though its extension into research implies that "one should not injure one person regardless of the benefits that might come to others." Sometimes investigators cannot know that something is harmful until he or she tries it and in the process of trying, or experimentation, people may be harmed. The Hippocratic oath also requires that physicians help patients "according to their best judgment," but again learning what help may mean exposing a person to risk. The principle of beneficence obligates both society and the individual investigator. Society has to consider the long-term benefits and risks that result from the development of novel devices or procedures that are the outcome of research. Investigators and their institutions have to plan to maximize benefits and minimize risks.

3. *Justice* refers to the benefits and harms to individual subjects of research in *The Belmont Report*. In nineteenth and early twentieth century hospitals, the burdens of experimentation fell upon the poor charity patients while the rewards of the improved medical care went primarily to the rich private patients. The Nazi researchers' experimentation on concentration camp prisoners provides another example of injustice. The benefits and burdens of research should be justly distributed. The selection of research participants needs to be constantly monitored to determine whether some pools of participants are being systematically selected from simply because they are easily available, vulnerable, or easy to manipulate, rather than chosen for reasons directly related to the research problem being studied.

Regulation and guidelines concerning the use of human research participants in the United States and increasingly so in other countries, are based on *The Belmont Report*. However, several philosophers articulated their thoughts on bioethical methods at around the same time as *The Belmont Report*. Tom Beauchamp and James Childress published their textbook, *Principles of Biomedical Ethics*, which was built upon the "four principles approach," also known as principlism [4]. According to principlism, medical ethical issues should be decided by an appeal to the four conditional principles; which principle prevails in conflict cases "depends on the context."

11.3.3   Institutional Review Board (IRB)

To enforce the protection of the rights and welfare of human subjects in research, many government agencies have adapted the concept of reviewing the research at the level of institutions where the work is carried out. An IRB is an independent ethics committee within an institution (or entity) specially constituted to review and monitor biomedical research involving human subjects conducted by faculty, students, and staff at a facility belonging to that institution. IRB is constituted of members whose interests are in protecting the rights and welfare of human subjects in research. In other words, IRB is consists of members who are not influenced by any political, institutional, professional or market in reviewing and decision-making of individual research.

While working with animals as participants of research, institutional review and approval is also necessary. The Institutional Animal Care and Use Committee (IACUC), part of the IRB in many places, is responsible for providing oversight and evaluation of the institution's animal care and use program and facilities by ensuring compliance related to the proper care, use, and humane treatment of animals used in research, testing, and education. Further, if biological materials such as DNA and proteins that may have the potential to cause adverse effects are utilized, additional approvals are necessary. Many institutions implement the review process through Institutional Biosafety Committees (IBC). The modes of operation of the IACUC and IBC are similar to IRB and locally set up guidelines to ensure compliance. These committees are responsible for creating a culture that assures the humane care and use of animals while facilitating their use in research and teaching. Guiding this process are the federal and state laws, university policies, and ethical principles, particularly that of *The Belmont Report*.

Investigators prepare and submit research projects and related materials (e.g., informed consent documents and investigator brochures) for review to the IRB, according to the guidelines set by the IRB. Investigators cannot commence research involving human subjects until the IRB has approved the study. Researchers must explain the significance of their actions in the wider scientific and human contexts as well as think of their actions in terms of future experimental design. Researchers take the lead in ensuring that the progress of research is both ethical and free from political intervention as only they can do so. Demonstration of a well-conducted process not only protects the investigators from exposure to liability, but increases the patient's autonomy in decisions concerning health and encourages compliance with treatment.

The IRB meets regularly to review submitted research projects. However, special meetings may be called in case of emergency conditions such as timeliness of reviews, patient safety concerns, and compliance issues. At the meeting, each research project is reviewed to determine the ethical conduct of research involving human subjects. The IRB review is to assure, both in advance and by periodic review, that appropriate steps are taken to protect the rights and welfare of humans participating as subjects in the research. IRBs may approve the research project without concerns, send clarification questions prior to approval, or reject the research project due to significant concerns regarding ethical standards. If there are concerns and questions, the investigator addresses those concerns and questions in writing and submits a revised project for IRB approval. Investigators interested in performing research are primarily responsible for assuring that research protocols are planned, conducted in an ethical manner, and consistent with standards established by the IRB. If the IRB approves the research project, then the investigator is notified in writing of the approval decision.

11.3.4   Informed Consent

Common to all the codes drawn up by IRBs is the principle that experimentation cannot be conducted on human subjects without their informed consent. Researchers have an ethical and legal duty to obtain a patient's informed consent before ordering testing and treatment. The consent document is a written summary of the information that should be provided to the participant. Many investigators use it as a guide for the verbal explanation of the study. The entire informed consent process involves:

1. Giving a participant adequate information about the study;
2. Providing adequate opportunity for the participant to consider all options, and responding to the participant's questions;
3. Ensuring that the participant has comprehended this information;
4. Obtaining the participant's voluntary agreement to participate;
5. Continuing to provide information as the participant or situation requires.

11.3.4.1   Information

The informed consent process must allow human participants, as much as they are able, to be given the opportunity to choose what will or will not happen to them. Most codes of research establish specific items for disclosure intended to assure that subjects are given sufficient information. The minimum components of an informed consent form include:

1. A description of the intended research which should include: the purposes of the research, the expected duration of participation, the procedures to be followed, and the identification of any procedures that are still in the experimental stage.
2. A description of any adverse effects (risks) or discomforts to the subject.
3. A description of anticipated benefits to the subject or to others, which may reasonably be expected from the research.
4. Disclosure of alternative procedures or treatments (where therapy is involved) that might be advantageous to the subject.
5. For research involving more than minimal risk, a description to whether any compensation and any medical treatments are available if injury occurs and, if so, what they consist of, or where further information may be obtained.
6. Contact information incase the subject wants to ask questions about the research and his or her rights. Also, contact information in the event of a research related injury to the subject is necessary. The consent document should provide the name of a specific contact person, and information such as telephone numbers and e-mail addresses. It is important for the subject to know what events should necessitate contacting the contact individual. The person named for questions about research subjects' rights should not be the primary investigator since this may tend to inhibit subjects from reporting concerns and discovering possible problems. If a student is con-

ducting the research, the faculty member who is mentoring the investigator should be included, possibly as the contact person to answer questions.

7. A statement that participation is voluntary and refusal to participate or discontinuation of participation involves no penalty or loss of benefits to which the subject is entitled.

8. A description of anticipated circumstances under which the subject's participation may be terminated by the investigator without regard to the subject's consent.

The procedural requirements of informed consent vary as a function of the risks of the tests or treatments. Treatment may be provided in an emergency situation without consent if the treatment given represents the standard of emergency care. Consent of human subjects for participation in research requires that they fully understand their role and risks, not be coerced, and be allowed to withdraw at any time without penalty. Vulnerable research subjects require additional protection.

11.3.4.2 Comprehension

The manner and context in which information is conveyed is important. These technical documents should be written in language understandable to participants.

11.3.4.3 Voluntariness

An agreement to participate in research constitutes a valid consent only if voluntarily given (i.e., the participant must understand the information and volunteer rather than be coerced into participation). Typically, a statement is included in the consent form that suggests that the participant has made a decision whether or not to participate after reading the document and that the participation is voluntary, and refusal to participate or discontinuation of participation will involve no penalty or loss of benefits to which the subject is otherwise entitled. The signature of the participant is necessary. Researchers are not just if they only select disadvantage persons for risky research or only provide beneficial research to groups they favor. Special classes of injustice arise when participants are drawn from vulnerable populations. The three vulnerable research groups used in research are:

1. *Children, mentally disabled patients, or those who are very ill:* Can they really give consent, and does parental consent apply? With these groups, often permission is sought from surrogate decision makers who would be in a position to understand the incompetent participant's situation and act on their behalf. This third person should be able to follow the research and be able to withdraw the participant if it appears to be in the best interest for the individual. Parents are normally given control assuming that they have child's best interest, and they know the child best. Older children and adolescents are asked to provide their consent for treatment in addition to their parents' permission.

2. *Prisoners:* Does their situation make consent sufficiently free and informed?

3. *Paid research subjects:* Does payment take advantage of the economically deprived?

11.3.4.4   Assessment of Risks and Benefits

Assessing risks and benefits means the researcher needs to assemble all data that explains why the research will obtain the benefits that are sought by the research project. The review committee of the researcher's sponsoring institution, upon review of the collected data, can decide whether the risks to the subjects are justified. The prospective participant can determine whether or not to participate. The term "risk" refers to the possibility that harm might occur. Risks could be psychological, physical, legal, social or economic hardship. The term "benefit" in the research context refers to something positive as related to health or welfare. Risk and benefits affect not only individual participants, but also their families and society at large. Importantly, in past regulations about human subjects, the risk to participants has been outweighed by the sum of both the anticipated benefit to participants, and the anticipated benefit to society in the form of new knowledge to be gained by the research.

11.4   Standards

11.4.1   Standards and Guidelines

Hospitals and clinics are constantly exposed to new medical equipment and diagnostic devices. However, the staff in these places seldom gets time to read operating manuals or attend training in the new tools. Although it is up to the manufactures to develop user-friendly devices, clinical failures and adverse reactions expose patients to potential dangers. To ensure the safety of the patient, acceptable standards for medical products and devices in a community are established. Further, new medical products and devices must go through compliance testing and device approval before they can be marketed. For this purpose, standards are created and published by national or international standards organizations or by regulatory authorities within a nation. Some of the national standards organizations are the American National Standards Institute (ANSI, www.ansi.org); the American Society for Testing and Materials (ASTM, www.astm.org); the Association for the Advancement of Medical Instrumentation (AAMI, www.aami.org); the Association Française de Normalisation (AFNOR, www.afnor.org); the British Standards Institute (BSI, www.bsi-global.com); the Comité Européen de Normalisation Electrotechnique (CENELEC, www.cenelec.eu); the Deutsches Institut fuer Normung (DIN, www.din.de); and the Japanese Industrial Standards Committee (JISC, www.jisc.go.jp). Standards are updated frequently as more knowledge is accumulated in using novel devices.

The globalization in the manufacturing of human biomedical products presents common challenges and opportunities for a variety of product regulators to explore cooperative arrangements at the international level. Three bodies are responsible for the planning, development, and adoption of international standards:

1. The International Electromedical Commission (IEC, www.iec.ch) is responsible for electrotechnical sector;
2. The International Telecommunication Union (ITU, www.itu.int/) is responsible for the telecommunications technologies (discussion is beyond the scope of this book);
3. The International Organization for Standardization (ISO, www.iso.org/) is responsible for all other sectors.

In the international standards, medical devices are defined as any instrument, apparatus, appliance, material, or other article (including the software necessary for its proper operation) used for human beings for any of the following purposes:

-   Diagnosis, prevention, monitoring, treatment or alleviation of disease, injury or handicap, or compensation for injury or handicap;
-   Investigation, replacement, or modification of the anatomy or of a physiological process;
-   Control of conception.

Verification and validation should be part of a medical manufacturer's quality management system. The assumption is that a design firm will have a quality management system so that requirements called for in the IEC standard will be included in the design, verification, and validation of the medical device. Although it is not a requirement to follow international standards, there are a number of reasons to comply with the standard:

1. If the company is in compliance with different applicable international standard during the development of a new product, it is presumed to be the best guarantee of a flawless evaluation report.
2. There is no need to reinvent the wheel when standards are designed especially for medical device companies.
3. The certification of compliance automatically improves the perception of the company and its products.

11.4.2 International Electromedical Commission (IEC)

The IEC consists of more than 50 countries. Within the IEC, responsibility for medical electrical equipment standards is within the purview of Technical Committee 62. The IEC has developed a series of safety standards and regulations specifically for electromedical equipment. Standards can enable electrical and electronic equipment to work together no matter where it is designed, manufactured, assembled, or used. The IEC defines medical electrical equipment in IEC 60601-1 clause 2.2.15 as [5]: "equipment, provided with not more than one connection to a particular supply main and intended to diagnose, treat, or monitor the patient under medical supervision and which makes physical or electrical contact with the patient and/or transfers energy to or from the patient and/or detects such energy transfer to or from the patient." Examples of electromedical products include ventilators, pulse oximeters, battery-operated thermometers, MRI imaging systems, endoscopic cameras, and infusion pumps. Accessories used with such equipment also fall under this standard.

The IEC 60601 family of safety standards has a four-level structure:

1. *General Standard.* IEC 60601-1 entitled "Medical Electrical Equipment-Part 1: General Requirements for Safety" addresses the general requirements for electromedical products. It also suggests ways to document designs. The IEC 60601-1 Standard is accepted in nearly all markets for supporting regulatory registrations and approvals either with national deviations (e.g., JIS T 0601-1 in Japan) or in its original form (e.g., in Brazil). Common national deviations include the requirements of the electrical code of the particular country, another national standard that may apply to the product type or its components, and different national component requirements (e.g., modified marking requirements). ANSI/UL 60601-1 is the U.S. national standard for safety testing electrical medical devices. CAN/CSA C22.2 no. 601.1 is the equivalent standard in Canada. The European Economic Council (EEC) publishes the Medical Devices Directives, which declare EN 60601-1 (identical to IEC 60601-1) a harmonized standard.

2. *Collateral Standards.* Standards numbered IEC 60601-1-x contain horizontal issues that deal with many different types of medical devices. Examples of collateral standards are

   a. IEC 60601-1-2, Collateral Standard for Electromagnetic Compatibility;
   b. IEC 60601-1-6, Collateral Standard: Usability;
   c. IEC 60601-1-8, Collateral Standard: Alarms Systems—General Requirements, Tests, and Guidance for Alarm Systems in Medical Electrical Equipment and Medical Electrical Systems.

2. *Particular Standards.* Standards numbered IEC 60601-2-x provide requirements for a specific type of medical device. An example of particular standards is IEC 60601-2-33, entitled "Particular requirement for the safety of magnetic resonance equipment for medical diagnostic," to assure safety in the MRI environment. Particular standards can amend, modify, and/or supersede part of the requirements specified in IEC 60601-1.

3. *Performance Standards.* Standards numbered IEC 60601-3-X lay out performance requirements for specific types of devices. IEC 60601-3-1, for example, contains "essential requirements for the performance of transcutaneous oxygen and carbon dioxide partial pressure monitoring equipment."

11.4.3   International Organization for Standardization (ISO)

The ISO is a worldwide federation of national standards bodies from nearly 140 countries that promotes the development of standardization and related activities to facilitate the international exchange of goods and services and to develop intellectual, scientific, technological, and economic cooperation. ISO members are primarily the national standards bodies, and organizations representing social and economic interests at the international level. ISO is supported by a central secretariat based in Geneva, Switzerland. The ISO's technical committees, working groups, and ad hoc study groups represent the viewpoints of manufacturers, vendors and users, engineering professions, testing laboratories, public services, governments, consumer groups, and research organizations in each member country. The ISO acts for harmonized regulatory processes to assure the safety, quality, and performance of medical devices. The IEC cooperates with the ISO, based on the expertise each organization embodies. They have a common set of rules for the development of standards working on different projects. The working procedures for both organizations are identical, and joint projects and standards often combine expertise from both organizations.

ISO standards can be broadly grouped into three classes: a horizontal standard (also known as basic safety standards), semihorizontal standards (also known as group safety standards), and a vertical standard, which is specific to a device.

11.4.3.1 Horizontal Standards

Horizontal standards indicate fundamental concepts, principles, and requirements with regard to general safety aspects applicable to a wide range of products and/or processes. Standards concerning the safety of medical devices are prepared by the ISO technical committee 194, which meets annually to review progress made. ISO 10993 is recognized as the minimum requirements for biocompatibility testing of all medical devices (e.g., standards concerning risk assessment and control of medical devices). Biological evaluation of medical devices includes many parts; some of them are listed in Table 11.1.

After approval at the international level, regulatory authorities in participating nations adopt the ISO standards as written or with minor modifications. For example, ISO 10993 is adopted as the EN 30993 standard in Europe, and the ANSI/AAMI 10993 in the United States with some exceptions. ISO 10993-1 was adopted by the FDA with modifications for intrauterine and some other types of devices. ISO 10993-1 is recognized by most national regulatory bodies, but does not super-

Table 11.1 Parts of ISO 10993

1. Evaluation and testing

2. Animal welfare requirements

3. Tests for genotoxicity, carcinogenicity, and reproductive toxicity

4. Selection of tests for interactions with blood

5. Tests for in vitro cytotoxicity

6. Tests for local effects after implantation

7. Ethylene oxide sterilization residuals

8. Degradation of materials related to biological testing

9. Tests for irritation and sensitization

10. Tests for systemic toxicity

11. Sample preparation and reference materials

sede the guidelines published by the individual nations or the testing requirements for a specific medical device.

Other examples for horizontal standards include ISO 13485: quality management systems: requirements for regulatory purposes and ISO 14971: The International Risk Management Standard for Medical Devices. ISO 13485 is a modified version of ISO 9001 and is a voluntary tool available to meet the quality system requirement. ISO 13485 was adopted in Europe with additional requirements especially tailored for the medical device industry. Quality system requirements have to be satisfied as outlined in the directives (Annex 2 and Annex 5 specifically) but the process adapted is left to the individual. According to ISO 14971, risk analysis must take into consideration all available data for the device; that is, chemical data (solubility, corrosion), physical–mechanical data (hardness, shrinkage), and available information on its effect on living tissues from historic data. Furthermore, from the intended use of the device, the possible harm must be identified.

11.4.3.2 Semihorizontal Standards

Semihorizontal standards apply to families of similar products and/or processes making reference as far as possible to horizontal standards (e.g., standards concerning sterile or electrically powered medical devices). Medical devices comprise a group of nearly one half of a million different items. Hence, specific requirements concerning the biocompatibility of special groups of medical devices are necessary and handled in semihorizontal standards. Examples are:

- *ISO 7405:* Dentistry—preclinical evaluation of biocompatibility of medical devices used in dentistry—test methods for dental materials;
- *ISO 18153:* In vitro diagnostic medical devices—measurement of quantities in biological samples—metrological traceability of values for catalytic concentration of enzymes assigned calibrators and control materials.

11.4.3.3 Vertical Standards

Vertical standards indicate necessary safety aspects of specific products and/or processes, making reference, as far as possible, to basic safety standards and group safety standards (e.g., standards for infusion pumps or for anesthetic machines). Some examples are:

- *ISO 7197:* Neurosurgical implants—sterile, single-use hydrocephalus shunts, and components;
- *ISO 1135-3:* Transfusion equipment for medical use—part 3: blood-taking set;
- *ISO 3826-4:* Plastics collapsible containers for human blood and blood components.

Both ISO 1135-3 and ISO 3826-4 list requirements for cell culture cytotoxicity, short-term intramuscular implantation, hemolysis in vitro, delayed contact sensitization, intracutaneous irritation, pyrogenicity, and sterility.

In the case of innovative devices, the frequent absence of a specific standard creates the risk of an incomplete evaluation and the unsuitable use of a medical device, and therefore the risk of serious or repetitive incidents that will hinder the development of a valuable technique. Standardization also means selection of tests and adoption of the formal, sometimes rather time consuming, procedure of setting up a standard. The consequences are that methods for very specific scientific problems are not included in a standard and that the most recent developments in testing methods may not be reflected in such a standard. Standards for complex or invasive devices are particularly in need of development. For some of these devices, there is no consensus among existing standards on what are appropriate tests. For example, there is disagreement on the testing of hemodialyzers. The French and German standards require hemolysis testing on an eluate from the hemodialyzer, whereas FDA guidelines require the following tests: cytotoxicity in vitro, hemolysis, complement activation, cell adhesion, protein adsorption, whole-blood clotting time for thrombogenicity, pyrogenicity, genotoxicity, acute systemic toxicity, intracutaneous injection, implantation, guinea pig maximization for delayed sensitization, subchronic toxicity, thrombogenicity by examining platelet and fibrinogen turnover, thrombus formation, and the resulting emboli. The FDA requirements do not delineate the biological system and exposure protocol that are necessary for interpretation of the measurements. For example, a whole-blood clotting time assay may not be meaningful because heparin anticoagulants are used during hemodialysis procedures.

11.5 Regulatory Agencies

11.5.1 The Food and Drug Administration

Standards establish basic requirements for safe development and implementation of medical devices, minimizing the risk caused to the patients when used. The new device must be proven not only safe but also effective in curing or ameliorating a specific disease. Standardized tests should be used while developing medical devices. The advantages of standards are mainly enhanced reproducibility, resulting in a comparison of the data obtained from different laboratories. Standards are used in some countries as regulatory requirements rather than being voluntary guidelines. Regulators play an important role in ensuring safety by specifying standard requirements that medical devices and therapeutics must meet before they can be placed on the market. In the United States, the FDA (www.fda.gov) is responsible for the regulation of drugs and medical devices. In the United Kingdom, the Medicines and Healthcare Products Regulatory Agency (www.mhra.gov.uk/) is involved in ensuring the acceptable safety of medicines and medical devices. The Therapeutic Goods Administration (www.tga.gov.au/) is essentially Australia's counterpart to FDA.

The FDA is responsible for ensuring that human and veterinary drugs, biological products, medical devices, and electronic products that emit radiation are safe and effective; food is safe, wholesome, and sanitary; and cosmetics are safe. However, the FDA does not regulate the practice of medicine and leaves areas such as assisted reproduction unregulated. In addition, the FDA's statute allows it to regulate only on the basis of safety and efficacy, and not on normative grounds. The National Institutes of Health (NIH) does have a broader authority to regulate on the basis of ethical concerns, but its jurisdiction extends only to federally funded research.

The FDA began as the Division of Chemistry and then (after July 1901) the Bureau of Chemistry. The modern era of the FDA dates to 1906 with the passage of the Federal Food and Drugs Act. FDA defines [under 21 CFR 56.102(c)] clinical research as [6]: "Any experiment that involves a test article and one or more human subjects [or specimens] the results of which are intended to be submitted to FDA." The FDA consists of eight centers/offices,

1. Center for Biologics Evaluation and Research (CBER)
2. Center for Devices and Radiological Health (CDRH)—oversees the regulation of electrical medical devices and houses the Office of Device Evaluation (ODE)
3. Center for Drug Evaluation and Research (CDER)
4. Center for Food Safety and Applied Nutrition (CFSAN)
5. Center for Veterinary Medicine (CVM)
6. National Center for Toxicological Research (NCTR)
7. Office of the Commissioner (OC)
8. Office of Regulatory Affairs (ORA)

The FDA also has affiliated organizations including the Joint Institute for Food Safety and the Applied Nutrition National Center for Food Safety and Technology.

11.5.2   Device Classification

For medical device manufacturers, compliance with the regulatory agencies is a critical requirement. Recognizing the economic impact of drawing uniform guidelines for every medical device, regulatory agencies have classified regulatory requirements based on the severity of risk to the patient. The FDA has classified medical devices into three classes: class I, class II, and class III. Similar classifications can be found in other countries. Accessories used with a medical device to support the use of the device are considered the same classification as the medical device. Based on the classification, the FDA determines the needed level of regulatory control to legally market the device. As the classification level increases, the risk to the patient and FDA regulatory control increases.

11.5.2.1   Class I Medical Devices

Class I medical devices are typically those which have a reasonable assurance of safety and effectiveness, are not life supporting or life sustaining, and do not present a reasonable source of injury through normal usage. They can be approved through the 510(k) route to be marketed as general controls. Examples of class I devices are ankle contractures, tongue depressors, bedpans, bandages, examination gloves, grip rest hand splints, hand-held surgical instruments, wheelchairs, and endoscopes.

11.5.2.2 Class II Medical Devices

Class II medical devices are typically those where general controls are not sufficient to assure safety and effectiveness, and existing methods/standards/guidance documents are available to provide assurances of safety and effectiveness. In addition to compliance with general controls, class II devices are required to comply with special controls including special labeling requirements, mandatory performance standards (both internationally and in the United States), postmarket surveillance, and FDA medical device specific guidance. Examples of class II devices include physiologic monitors, X-ray machines, gas analyzers, blood pressure cuffs, oxygen masks, infusion pumps, electrically powered wheelchairs, surgical needles and suture material, acupuncture needles, and surgical drapes.

11.5.2.3 Class III Medical Devices

Class III Medical Devices are those that cannot assure safety and effectiveness through the controls of class I or II. Class III medical devices are those that support or sustain human life, are of substantial importance in preventing impairment of human health, or which present a potential, unreasonable risk of illness or injury (21 CFR 814) [7]. The Federal Food, Drug, and Cosmetic Act (FD&C Act) section 515 requires device manufacturers to submit a PMA application to the FDA if they intend to introduce a device into commercial distribution for the first time, or to introduce or reintroduce a device that will be significantly changed or modified to the extent that its safety or effectiveness could be affected. Under PMA regulations, the FDA requires the submission and review of valid scientific evidence to determine whether a reasonable assurance exists that the device is safe and effective, and has clinical utility. A scientific and regulatory review process is required to allow marketing of a class III medical device unless an essentially equivalent device exists. Class III medical devices need preclinical testing (conducting in vitro experiments and in vivo animal experiments, obtaining preliminary efficacy, and determining toxicity and pharmacokinetic information) and clinical trials (expensive, time consuming, and difficult to have concurrent controls for devices). Examples of class III medical devices are replacement heart valves, silicone gel-filled breast implants, pacemakers, defibrillators, vascular grafts, defibrillators, and drug-eluting stents.

PMA is the most stringent type of device marketing applications required by the FDA. The PMA application approval is a private license granted to the applicant for marketing a particular medical device. Recognizing the fact that to device development companies, time and money are critical factors, the FDA has developed other types of applications which include:

- *Premarket notification* [Section 510(k) of the FD&C Act and popularly referred to as 510(k)] for devices similar to existing products already sold in the United States.

- An *investigational device exemption* (IDE) application (refer 21 CFR 812 for more details) required for many companies to continue forward with a clinical trial, particularly for devices that pose a significant risk to patients participating in a clinical trial. IDE approval allows a company that would otherwise be subject to premarket approval to ship the device for the purpose of conducting a clinical trial.

- A *humanitarian device exemption* (HDE) application if the device qualifies under this category, which is determined by the absence of a comparable device available to treat or diagnose the disease or condition. Institutional review board (IRB) approval is necessary prior to usage and for the FDA-approved indication.

- *Product Development Protocol (PDP)* for class III devices with significant literature data and the existence of recognized protocols.

11.5.3   Compliance Requirements

For the medical device industry, documenting new product development activities from concept, design, and manufacturing to sales and distribution in a controlled manner is detailed in the quality system regulation (QSR) 21 of the Code of Federal Regulations (CFR) 820. The FDA also publishes guidance documents intended to state the current thinking on a topic to manufacturers and FDA reviewers. Guidance documents are written by FDA staff and specify what the FDA believes will assist in the determination of safety and effectiveness. However, guidance documents do not create or confer any rights for or on any person and do not operate to bind the FDA or the public. Alternative approaches can be used as long as applicable statutes and regulations are satisfied. Recognizing the important role of information technology, the FDA issued regulations regarding the use of electronic records and electronic signatures (21 CFR Part 11) [8]. 21 CFR part 11 provides basic principles for the use of electronic records and electronic signatures in complying with other FDA regulations, specifically where record keeping, documentation, or authorization is specified. In the context of 21 CFR part 11, FDA regulations such as those concerning manufacturing practices, laboratory practices, adverse event reporting, and product tracking are referred to by the agency as predicate rules. Whenever a company chooses to comply with predicate rules by means of electronic records or electronic signatures, 21 CFR part 11 must be followed. Manufacturers must be certain that the computer systems they use to replace paperwork are at least as effective in complying with FDA regulations.

A quality system is meant to establish the ways in which the company will comply with applicable regulations. The documents are developed specifically for the ways in which manufacturers actually do business. In practice, these documents consist of a quality manual, standard operating procedures (SOPs), forms, audit records of suppliers, and training files, and describe the procedures to follow in order to comply with the regulations. The design process is a good illustration of an operation that requires quality system documents. Different SOPs may be required for varying operations. If someone is making an engineering drawing or creating a prototype, an SOP for design controls is required. Design control records must be kept for a certain period of time, and an SOP for quality records would govern this aspect. If there are raw materials used in the manufacture of medical devices, which have been tested for certain applications and have been found to have acceptable properties, they should be disclosed. This could speed up the validation process, prevent unnecessary animal tests from being conducted, and save time and money.

The FDA, and in some cases parallel organizations in certain states, routinely audit a firm's quality system. These audits determine whether the quality system is consistent with the regulations and whether it describes actual operations. The quality system is the yardstick against which regulatory authorities measure compliance. Although the primary definition of the content of a quality system is found in 21 CFR 820, other FDA regulations that must be taken into account include those governing labeling, medical device reporting, and reports of corrections and removals (or recalls). If the FDA finds a manufacturer to be in violation of FDA rules, the consequences can be serious: warning letters, mandatory product recalls, inability to ship products, and criminal penalties for individual managers may occur.

Problems

11.1   Recently, the use of embryonic stem cells has been controversial. Write five bulleted points from the perspective of a person who ethically thinks it is: (a) right to use them and (b) wrong to use them in biomedicine.

11.2   The face of a French citizen was mauled by a dog, which led to a few portions being completely lost. Physicians had to reconstruct the entire face using the face of another deceased person. Since the face has many meanings to different people, it raised a number of ethical issues. Summarize the risks and benefits of facial transplant research.

11.3   In your organization, one of the basic steps of bioethical training is to take a multiple choice test through the Internet. Identify if your company has such a test. If you can take the test, provide a printout of your score on various modules.

11.4   You are interested in demonstrating some biomedical applications to high school students visiting your campus. You want to know the effectiveness of the demonstration for which you have to survey the students (the majority of them minors) before and after the demonstration. Write the steps involved in taking this action.

11.5   A bioengineer has developed a novel porous structure. She is interested in testing it in mice. However, she is not sure of the process involved. Please explore the IACUC in your institution and write the steps involved in obtaining an approved protocol.

11.6   Determine how many nontechnical members are present in your institution's IRB. How many of them do not work in the institution?

11.7   Your company is interested in selling their new pacemakers in Italy. Write the initial approach you would pursue in terms of whom you contact or what standards you would follow in obtaining the approval from the Italian government.

11.8   How many sections are in ISO 10993? List the number and corresponding title.

11.9   A new bioengineering company is interested in marketing a tissue-engineered skin product. They hired you to help them in determining the standards necessary for obtaining FDA approval. Provide three minimum standards they need to begin working on the FDA approval process.

11.10 You have developed a new prosthetic device for the hand. List the steps you would take to market this product in the United States, Europe, and Japan.

References

[1]   Office of Human Research Protections, http://www.hhs.gov/ohrp/.

[2]   World Medical Association, http://www.wma.net/e/policy/.

[3]   National Commission for the Protection of Human Subjects of Biomedical and Behavioral Research (NCPHSBBR), *The Belmont Report: Ethical Principles for the Protection of Human Subjects of Research*, 1979.

[4]   Beauchamp, T. L., and J. F. Childress, *Principles of Biomedical Ethics: Reflections on a Work in Progress,* 5th ed., New York: Oxford University Press, 2001.

[5]   IEC 60601-1 subclause 2.2.1, 3rd. ed., 2005.

[6]   Federal Food and Drugs Act, Institutional Review Boards, Code of Federal Regulation Title 21 56.102, Definitions, Revised April 1, 2009.

[7]   Code of Federal Regulation, Title 21, Part 814, Volume 8, Premarket Approval of Medical Devices, Revised April 1, 2009.

[8]   Code of Federal Regulation, Title 21, Part 11, Volume 1, Electronic Records: Electronic Signature, Revised April 1, 2009.

Selected Bibliography

Bankert, E. A., and R. J. Amdur, (eds.), *Institutional Review Board: Management and Function,* 2nd ed., Boston, MA: Jones & Bartlett, 2005.

Bulger, R. E., E. M. Bobby, and H. V. Fineberg, (eds.), "Committee on the Social and Ethical Impacts of Developments in Biomedicine, Institute of Medicine," *Society's Choices: Social and Ethical Decision Making in Biomedicine,* Washington, D.C.: National Academies Press, 1995.

Engelhardt, Jr., H. T., *The Foundations of Bioethics,* 2nd ed., Oxford, U.K.: Oxford University Press, 1995.

FDA, http://www.fda.gov/cdrh/devadvice/.

Hall, R.T., *An Introduction to Healthcare Organizational Ethics,* Oxford, U.K.: Oxford University Press, 2000.

Hamer, D. H., *The God Gene: How Faith Is Hardwired into Our Genes,* New York: Doubleday, 2004.

Kanholm, J., *ISO 13485 and FDA QSR Internal Audit Checklist,* Milwaukee, WI: AQA Press Publisher, 2005.

King, P. H., and R.C. Fries, *Design of Biomedical Devices and Systems,* 2nd ed., Boca Raton, FL: CRC Press, 2008.

Medical Device Link, http://www.devicelink.com/mddi/.

Trautman, K. A., *The FDA and Worldwide Quality System Requirements Guidebook for Medical Devices*, Milwaukee, WI: ASQ Quality Press Publisher, 1996.

Veatch, R. M., *A Theory of Medical Ethics,* New York: Basic Books, 1981.

About the Author

Sundararajan V. Madihally is an associate professor in the School of Chemical Engineering at Oklahoma State University. He received a B.E. in chemical engineering from Bangalore University and worked as a consultant to pharmaceutical companies and as a process engineer to a design consulting company. He received a Ph.D. from Wayne State University in chemical engineering on research related to the development of bioreactors for umbilical cord stem cells. Dr. Madihally held a research fellow position at Massachusetts General Hospital, Harvard Medical School, and Shriners Hospital for Children and worked on burn trauma-induced systemic alterations and therapeutic development. Throughout his career, he had the opportunity to introduce biomedical engineering to a broad spectrum of students ranging from k-12 students to postgraduate students with significant medical experience. As a visiting assistant professor in 2001, he cotaught a new course in the Department of Biomedical engineering at Rutgers University. Dr. Madihally developed a new biomedical engineering course at Oklahoma State University suitable for engineering students. His honors include best teacher award from the chemical engineering students, young teacher award from the college of engineering, and outstanding advisor award from the Oklahoma State University President, and outstanding poster award from the American Society of Engineering Education. He also served as the program chair for the American Society of Engineering Education chemical engineering division annual conference. He has published numerous articles and proceedings related to his research areas as well as novel educational paradigms in biomedical engineering.

Index